FLIGHT

100 YEARS OF AVIATION

FLIGHT

100 YEARS OF AVIATION

R. G. GRANT

DUXFORD
Imperial War Museum

LONDON, NEW YORK, MUNICH,
MELBOURNE, and DELHI

Project Editor David Summers
Project Art Editors Tony Foo, Kirsten Cashman
Editors Nigel Ritchie, David Tombesi-Walton
Designers Ian Midson, David Ball, Jörn Kröger,
Jamie Hanson, Becky Painter, Hugh Schermuly
Design Assistant Paul Drislane
Picture Researcher Louise Thomas
DTP Designer Rajen Shah
Production Controller Elizabeth Cherry

Managing Art Editor Philip Ormerod
Category Publisher Jonathan Metcalf

Special Photography Gary Ombler

Editor-in-Chief Dominick A. Pisano
Project Co-ordinator Ellen Nanney
Picture Researcher Charles O. Hyman

NASM Archives Division Melissa A. N. Keiser, Allan
Janus, B. E. Weitbrecht, Dana Bell, Dan Hagedorn, Brian
D. Nicklas, Paul Silbermann, Larry Wilson, Mark Kahn
NASM Aeronautics Division Dominick A. Pisano,
F. Robert van der Linden, Peter L. Jakab, Thomas J. Dietz,
Dik A. Daso, Dorothy Cochrane, Roger Connor, Russell Lee
NASM Space History Division Allan A. Needell,
Cathleen Lewis, Michael J. Neufeld, Valerie Neal

DUXFORD
Imperial War Museum
Consultant Steve Woolford
Contributors Steve Cross, David Lee

First published in Great Britain in 2002
This updated edition published 2007,
by Dorling Kindersley Limited
80 Strand, London WC2R 0RL

A Penguin Company

ISBN: 978-1-4053-1768-9

Reproduced by GRB, Italy
Printed and bound in China by L-Rex Printing co Ltd

see our complete catalogue at
www.dk.com

FOREWORD

One hundred years ago the world was a very different place. The United States was emerging as a world economic power, but had yet to realize its full potential. Europe was at peace, tenuous as it was, while trouble in Russia was a portent of war and revolution to come. Much of the rest of the world remained as either economic or political colonies of the dominant powers.

Millions of Europe and Asia's poor continued to travel days and weeks by steamship to seek a new life in the

United States. Once there, steam-driven trains were the primary means of long-distance overland transport for immigrant and citizen alike. The vast rail network linked most communities and enabled those who could afford it to ride across the country in less than a week. Nevertheless, once settled, most Americans stayed at home and few ever travelled more than 25 miles from their place of birth. And so it was across the rest of the developed world.

Transportation was taking the first tenuous steps that would soon change the world forever. With the invention of the internal combustion engine, in the late nineteenth century, new possibilities of motive force became available. By 1903 the automobile was set to challenge the horse. Transportation would soon change even

more dramatically because of a new invention – the aeroplane. Within the century that followed, humankind took to the air, led by the pioneering example of Wilbur and Orville Wright. First in frail craft, but soon in sturdy and reliable machines, aviators shattered long-standing barriers of time and distance. By mid-century air travel was common, and by the late 1950s it had replaced the train and steamship as the preferred mode of transport. By the last quarter of the twentieth century, with large, efficient jet-powered aircraft, air travel was commonplace and affordable to all. Flying has become second nature to hundreds of millions of people and is so deeply intertwined into the fabric of society that it is impossible to imagine a world without it.

CAPTIVATING SPECTACLE
Straining to see over the perimeter fence, spectators gaze in wonder at the aircraft on display at the Hendon airfield in 1911. By capturing the public's imagination, early aviators inspired an enthusiasm for flight that would outlast the century.

The aeroplane also rapidly developed as a weapon of war. Used widely during World War I, where the techniques of air power were initially developed, military aircraft became an integral part of warfare by World War II. The advent of jet power, and sophisticated electronics perfected during and after the Cold War, has now turned the aircraft into a feared weapon over the twenty-first century battlefield.

Today, 100 years after the Wright brothers first took to the air in the first powered, controlled, heavier-than-air machine, the political, social, and economic challenges are different, yet, in many respects, remarkably similar. Today, aviation and spaceflight are critical tools for the improvement of the human condition and powerful instruments of positive change.

This book is the story of that most remarkable achievement of the twentieth century – flight. Using superlative historical images and extraordinary new photography to illustrate an excellent text, this book is a fitting tribute to the courage and efforts of the pioneering individuals and organizations that inspired the first 100 years of aviation.

J. R. DAILEY
DIRECTOR

NATIONAL AIR AND SPACE MUSEUM
SMITHSONIAN INSTITUTION

AGE OF THE PIONEERS

1

Having spent centuries watching birds with admiration and envy, when humans eventually took to the air – in the late 18th century – it was, in fact, in a balloon rather than on the wing. Nevertheless, the desire to soar through the air like a bird remained. In the course of the 19th century, scientists and inventors worked on the basic principles of flight, experimenting with gliders and ungainly steam-powered flying machines and models. But it took the persistent efforts of the Wright brothers, in experiments between 1899 and 1905, to finally achieve practical powered aeroplane flight. The period up to 1914 brought spectacular progress. The public was enthralled by long-distance flying races and displays of aerobatics, while new speed and altitude records were posted yearly although at the cost of the lives of many early aviators

TENTATIVE TAKE-OFF
In sand dunes on the shores of Lake Michigan, in the summer of 1896, assistants of American flight pioneer Octave Chanute experiment with a triplane glider. As a result of these experiments, Chanute produced the most influential glider of the pre-Wright era.

THE PREHISTORY OF FLIGHT

THE PATH TO POWERED FLIGHT WAS OPENED UP BY DREAMERS AND ODDBALL INVENTORS WHO BRAVED BOTH PUBLIC RIDICULE AND PHYSICAL INJURY

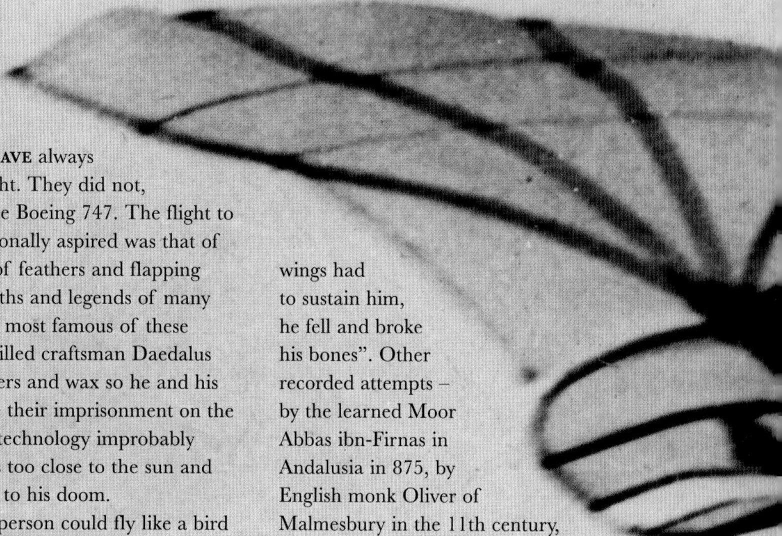

"The desire to fly is an idea handed down to us by our ancestors who... looked enviously on the birds soaring freely through space... on the infinite highway of the air..."

WILBUR WRIGHT

HUMAN BEINGS HAVE always dreamed of flight. They did not, however, dream of the Boeing 747. The flight to which humans traditionally aspired was that of the birds, a business of feathers and flapping wings. To this the myths and legends of many cultures testify. In the most famous of these ancient stories, the skilled craftsman Daedalus makes wings of feathers and wax so he and his son Icarus can escape their imprisonment on the island of Crete. The technology improbably works, but Icarus flies too close to the sun and melts the wax, falling to his doom.

The illusion that a person could fly like a bird or a bat cost some brave and foolish men their lives or limbs. The historical record is scattered with "tower-jumpers" who launched themselves into the air supported only by blind faith and poorly improvised wings. In 1178, for example, in Constantinople, a follower of Islam chose the moment of a visit to the Christian Byzantine Emperor by a Muslim sultan to demonstrate his powers of flight, jumping off a high building in a copious white robe stiffened with willow sticks. In the words of a later flight experimenter, Octave Chanute, "the weight of his body having more power to draw him downward than his artificial wings had to sustain him, he fell and broke his bones". Other recorded attempts – by the learned Moor Abbas ibn-Firnas in Andalusia in 875, by English monk Oliver of Malmesbury in the 11th century, by Giovanni Battista Danti in Perugia, Italy, in 1499 – all had the same result for the same reason.

"Instruments to fly"

Myth and folklore were also rich in tales of airworthy vehicles that might carry the weight of a human, from various "chariots of the gods" to witches' broomsticks. The idea of a "flying machine" was picked up by English philosopher-monk Roger Bacon in the 13th century – a man regarded as one of the founders of the modern scientific tradition. Bacon declared himself certain that humans could build "instruments to fly", involving a mechanism that would flap wings. Such "ornithopters" also obsessed the imagination of Italian Renaissance genius Leonardo da Vinci. "There is in man [the ability] to sustain himself by the flapping of wings", Leonardo wrote. He was wrong. In the many sketches of flying machines found in Leonardo's notebooks, the only truly promising idea is a screw-like propeller that he hoped would spiral into the air – a remote

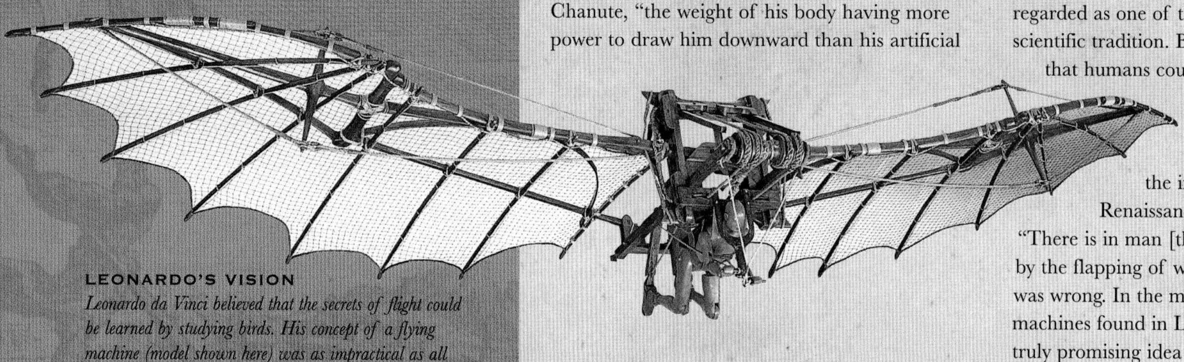

LEONARDO'S VISION
Leonardo da Vinci believed that the secrets of flight could be learned by studying birds. His concept of a flying machine (model shown here) was as impractical as all other devices for muscle-powered, flapping-wing flight.

BAT MACHINE
French engineer Clément Ader took his inspiration for the Avion III flying machine from a bat. Like a bat's, the wings could be folded away for easy storage. However, this machine, built for the French Ministry of War, failed to become airborne when twice tested in 1897.

BIRDMAN PIONEER
German engineer Otto Lilienthal made over 2,000 flights in the 1890s using hang-gliders of his own design. Although he was the author of an influential work on bird flight, Lilienthal believed that "a proper insight into the practice of flying" could only be achieved "by actual flying experiments".

foreshadowing of a helicopter. If no one could see how to make a machine that would fly, they could possibly see why you would want to – especially in militaristic Europe, which was divided into states that were more or less permanently at war with one another. In 1670, proposing yet another impractical design – this time for an airship lifted by spheres from which the air had been pumped to create a vacuum – Italian Jesuit Father Francesco de Lana pointed out that such a vehicle could be used to land troops to capture a city in a surprise attack, or to destroy houses and fortresses by dropping "fireballs and bombs".

Lighter-than-air flight

Although de Lana's vacuum-lifted airship was a non-starter, it did point the way to the first successful human flight. De Lana's aim was "to make a machine lighter than the air itself". This could not be done with vacuum spheres, but it could with a balloon filled with hot air or a light gas such as hydrogen. As usual in the history of invention, the solution to a problem became apparent to several inventors at once. When Joseph and Etienne Montgolfier, paper manufacturers from the French town of Annonay, brought a hot-air balloon to Paris in 1783, they faced competition from gentleman-scientist Jacques Charles, who was ready to demonstrate a hydrogen-filled balloon.

FIRST BALLOON FLIERS
In June 1783 the Montgolfier brothers conducted the first public display of a hot-air balloon, and the following November François Pilâtre de Rozier and the Marquis d'Arlandes made the first manned ascent (above). Early French aeronauts achieved some spectacular flights. In February 1784 Jean-Pierre Blanchard soared to over 3,800m (12,500ft) in a hydrogen balloon.

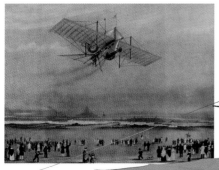

FLIGHT OF FANTASY
In 1843 William Henson formed the "Aerial Steam Transit Company". Despite the circulation of optimistic images such as this one, the Aerial Steam Carriage never flew.

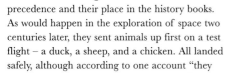

Tailplane

Wing brace

Pusher propeller

Rudder

But it was the Montgolfiers who established precedence and their place in the history books. As would happen in the exploration of space two centuries later, they sent animals up first on a test flight – a duck, a sheep, and a chicken. All landed safely, although according to one account "they were, to say the least, much astonished". The first free manned flight followed on 21 November, when young physician François Pilâtre de Rozier and the Marquis d'Arlandes, an army officer, drifted over Paris covering 8km (5 miles) in about 25 minutes. Ten days later Jacques Charles and a companion flew 40km (25 miles) in a hydrogen balloon.

Ballooning captured the public imagination much as flying machines would in the early 20th century. Crowds flocked to demonstration flights and the fliers became national heroes. French aeronaut Jean-Pierre Blanchard, with American expatriate John Jeffries on board, flew across the Channel from England to France in January 1785, 124 years before Louis Blériot (see page 41).

Scientific progress

One of the many individuals fascinated and inspired by reports of the early balloon flights was a young boy growing up on his father's estate in Yorkshire, England. He was George Cayley, who was to make the first serious practical and theoretical progress towards heavier-than-air flight. Cayley could easily be marked down as an eccentric – a member of the landed gentry using his privileged leisure to pursue a fanciful hobby. But he, in fact, worked within a maturing scientific tradition, which enabled him to precisely define the challenge of heavier-than-air flight: "The whole problem

is confined within these limits," he wrote, "to make a surface support a given weight by the application of power to the resistance of air."

Cayley addressed himself to these problems of lift and drag through careful observation of bird flight, systematic experimentation, and mathematical calculations. He used an ingenious device known as a "whirling arm" – essentially a precursor of the wind tunnel – to test the lift created by different aerofoils, or wings, at various angles and speeds.

IMAGE OF THE FUTURE
Sir George Cayley engraved this image of a flying machine on a silver disk in 1799. Cayley's design was the first to resemble the configuration of a modern aeroplane.

SIR GEORGE CAYLEY
Basing his theories on experimentation and observation, Cayley (1773–1857) pioneered the conquest of flight with his works on aeronautics.

As early as 1799 Cayley engraved on a silver disk an image of a flying machine that marked a crucial step forwards in design from Leonardo-style ornithopters: the wing had ceased to be the means of propulsion, becoming instead purely a device to generate lift. Through the next decade he built both model and full-size gliders. His full-size glider had a wing attached to the front end of a pole, and at the rear of the pole a vertical rudder and horizontal tailplane. "When any person ran forward in it with his full speed," Cayley wrote, "taking advantage of a gentle breeze in front, it… would frequently lift him up and convey him several yards together".

In 1809–10 Cayley made the results of his work public in a three-part paper, "On Aerial Navigation". His calculations of lift and drag, and his comments on how an aircraft could be stabilized and controlled, constituted a solid basis for potential progress towards heavier-than-air flight. Unfortunately, they were largely ignored. As Cayley himself admitted, flight remained "a subject rather ludicrous in the public's estimation".

The awakening of a more sustained interest in heavier-than-air flight did not come for another 30 years. It was provoked by the success of the steam engine applied to transport systems. By the 1840s, railway construction was booming as steam

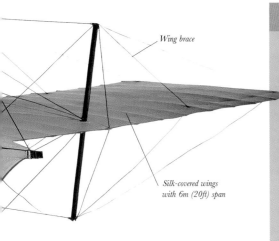

Wing brace

Silk-covered wings
with 6m (20ft) span

Launching wheels

AERIAL CARRIAGE
This reconstruction of William Samuel Henson's proposed Aerial Steam Carriage displays more elements of a modern-day flying machine than any machine before it. Although a full-size version was never built, its cambered wings and separate tail with rudder and elevator were later widely adopted. Its two pusher propellers would have been driven by a light steam engine of up to 30hp.

trains transformed journey times by land. At sea, steam ships were a growing threat to the dominance of sail. Jumping on the bandwagon, in 1843 ambitious English inventor William Samuel Henson patented an Aerial Steam Carriage "for conveying letters, goods, and passengers from place to place".

Steam-driven airline
Basing his ideas on Cayley's published research, Henson imagined a monoplane with a cambered wing for extra lift, a rudder and tailpane for control, and two six-bladed pusher propellers. This contraption was to be powered by a 30hp steam engine in the fuselage. Henson's grandiose plans for an Aerial Steam Transit Company momentarily attracted the interest of investors, the proposal for passenger flights spanning the globe rendered credible by fanciful illustrations of the Steam Carriage soaring over exotic locations. But doubt and ridicule soon followed. Although Henson built a small model of his aircraft, he could not find anyone ready to put up the cash for a full-size version and rapidly abandoned aerial experimentation for good.

LIGHTER-THAN-AIR FLIGHT

THE BALLOON FLIGHTS OF 1783 began a tradition of lighter-than-air flight that ran parallel with – and for a long time ahead of – experiments with heavier-than-air flight. The drawbacks of balloons were obvious. A huge balloon was needed to carry even a small weight, and then it was only marginally controllable and at the mercy of the winds. Yet serious practical uses were found for balloons in the 19th century: they were employed as observation platforms during the American Civil War, and during the Franco-Prussian War they were used to carry messages.

The first controlled powered balloon – a dirigible or airship – was demonstrated by Frenchman Henri Giffard in 1852. Mounting a steam-driven propeller under a cigar-shaped bag filled with coal gas, he flew 27km (17 miles) at around 10kph (6mph). His example inspired other enthusiasts, although they were hindered by the lack of alternatives to the steam engine. In the 1880s electric motors came into vogue, and the La France, built by Charles Renard and A.C. Krebs, managed controlled flights at speeds of around 20kph (12mph).

The advent of the internal combustion engine brought a further leap forwards. In 1898 Alberto Santos-Dumont, the son of a Brazilian coffee-plantation owner, embarked on a series of highly successful experiments in the skies of Paris, France, where he lived. He became a well-known and popular figure, responding to mishaps, such as crashing on the roof of a hotel, with admirable panache. He built 14 airships in all before transferring his enthusiasm to heavier-than-air flight. Meanwhile the Germans entered the airship field when Count von Zeppelin flew his first airship LZ 1 in 1900 (see pages 56–57).

BALLOON MEN
The first manned flight was made in a Montgolfier balloon (left, as shown on a cigarette card) on 21 November 1783, but balloon flight had a limited potential. It could not, for example, be used to create a viable transport system.

TRIP ROUND THE TOWER
In 1901 Alberto Santos-Dumont flew his dirigible from the Parisian suburb of Saint-Cloud round the Eiffel Tower and back in under 30 minutes to win a 100,000-franc prize.

BOY FROM BRAZIL
Paris-based Brazilian Alberto Santos-Dumont stands in his balloon basket ready for an ascent. Although an eccentric dilettante, he proved an outstanding pioneer of airship and aeroplane flight.

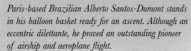

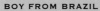

Yet interest in flight had been stimulated afresh – not least in the heart of Sir George Cayley, who now embarked on a new round of experiments that culminated in the world's first manned heavier-than-air flight in a glider in 1853. The "pilot" was Cayley's coachman. He was reluctantly persuaded to climb into the boat-like fuselage of a glider, which then rolled down one side of a valley, lifted into the air, and briefly flew before coming down uncomfortably. The coachman is said to have immediately given his notice, on the grounds that he had been "hired to drive, not to fly". Cayley's success, achieved in the privacy of his estate, had less public impact than Henson's failure. The story of the coachman's flight only came to light long after Cayley's death in 1857.

High society

The growing respectability of flight research was exemplified by the foundation of the Aeronautical Society of Great Britain in 1866, a dignified association of scientists and engineers who staged the world's first exhibition of flying machines at London's Crystal Palace two years later. None of them flew. However, there were some notable efforts to advance understanding of aerodynamics.

Francis Wenham, a distinguished marine engineer and a founder-member of the Aeronautical Society, built the first wind tunnel and produced improved data on the lift provided by different wing shapes. In the 1870s, brilliant young French engineer Alphonse Pénaud made significant progress in wing and tail design through experiments with model aircraft powered by a twisted rubber band. And Louis-Pierre Mouillard's book *The Empire of the Air*, published in 1881, was an inspirational work based on the author's observations of bird flight.

Practical efforts to progress in heavier-than-air manned flight in the late 19th century divided into two main approaches. One focused on power: was it possible to find an engine powerful enough, or rather with a favourable enough power-to-weight ratio, to lift a machine and a man into the air? The other focused on unpowered flight as a

> ### "Give us a motor and we will very soon give you a successful flying machine."
>
> HIRAM MAXIM, 1892

FRENCH CANDIDATE
In the 1890s French electrical engineer Clément Ader built two steam-powered flying machines with wings modelled on a bat. Although his first, the Éole (below) managed to hop for 50m (165ft) the Avion III, seen above, failed to get off the ground.

means of understanding the secret of flight as exhibited by birds. Success would only be achieved when the traditions of powered and unpowered flight came together in the Wright brothers.

Engine power

Early experimenters in powered flight were unfortunate in that their only feasible power plant was a steam engine. The first of the steam-powered experimenters to make a serious attempt to get off the ground was a French naval officer, Félix du Temple de la Croix. In the 1850s, with his brother Louis, he designed and flew a model aeroplane powered first by clockwork and then by a miniature steam engine. He then patented a design for a full-size monoplane with a lightweight steam engine and the surprising refinement of a retractable undercarriage. His man-carrying aeroplane was finally built and ready to test in 1874. With a French sailor on board, it ran down a sloping ramp, briefly lifted into the air, and immediately came back down to earth.

A decade later, du Temple's "hop" was matched by a Russian experimenter, Aleksander Mozhaiskii. In 1884, at Krasnoe Selo outside St Petersburg, Mozhaiskii tested a two-engined monoplane with a mechanic at the helm. Spouting smoke from its ship-like funnel it momentarily lifted, then crashed to the ground.

The first claim to have actually cracked the problem of powered flight came from French electrical engineer Clément Ader. After testing his bat-winged, steam-powered Éole in 1890, he claimed: "I have resolved the problem after much work, fatigue, and money". What he had achieved, as far as can be ascertained, was to skim the ground at a

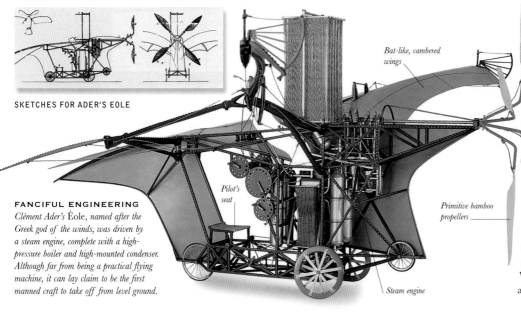

SKETCHES FOR ADER'S EOLE

Bat-like, cambered wings

Pilot's seat

FANCIFUL ENGINEERING
Clément Ader's Éole, named after the Greek god of the winds, was driven by a steam engine, complete with a high-pressure boiler and high-mounted condenser. Although far from being a practical flying machine, it can lay claim to be the first manned craft to take off from level ground.

Primitive bamboo propellers

Steam engine

height of around 20cm (8in) for a distance of 50m (165ft). This could not be called controlled, sustained flight, but it was a start. To fund further experiments, Ader turned to the French Ministry of War, which was keen to explore any "secret weapon" that might give France an edge over its neighbour, Germany. Armed with the first military budget for aeroplane development, Ader built a twin-engined aircraft, the *Avion III*. But when tested in front of military observers in October 1897, it failed to get off the ground. Funding was cut off and Ader's experiments came to an end.

The most prominent advocate of the power-centred approach to flight was Sir Hiram Maxim, the American-born inventor of the Maxim machine gun. He wrote that it was "neither necessary nor practical to imitate the bird… Give us a motor and we will very soon give you a successful flying machine." In the 1890s Maxim devoted a large part of his considerable fortune to developing a huge biplane on his estate in Kent. Maxim did not intend this giant to, in any proper sense, fly. It sat on an ingenious test track consisting of two sets of rails. The aeroplane was to run on the lower rails while gathering

speed for take-off; the upper rails were to prevent it rising into free flight, which would inevitably result in a crash. In July 1894, Maxim's machine rose from its rails after accelerating to 67kph (42mph), only to foul the upper rails, sustaining heavy damage. His experiments progressed no further.

Maxim, Ader, and other power-centred flight experimenters had given little or no thought to how they would actually fly their machines should

LARGE-SCALE BIPLANE
This test rig, built by the American-born Hiram Maxim in the 1890s, was a truly huge aircraft. With a wingspan of 33m (107ft) and weighing more than 3.6 tonnes (3.5 tons), it took two 180hp steam engines, each driving a propeller 5.5m (18ft) in diameter, to drive it along its restraining rails.

SOCIETY VISIT
Hiram Maxim (reclining, centre) poses with members of the British Aeronautical Society alongside his massive steam-driven biplane at Baldwyns Park in Kent, England.

OTTO LILIENTHAL

BORN IN POMERANIA, Otto Lilienthal (1848–1896) was fascinated from an early age by the flight of birds. Although he trained as an engineer and ran a factory building steam engines, he remained convinced that ornithology held the key to human flight – a belief reflected in the title of his seminal publication of 1889, *Birdflight as the Basis of Aviation.* Although he became famous for his experiments with what would now be called hang-gliders, flying his first one in 1891, he never abandoned the idea of flapping wings as a means of propulsion. In his systematic work on wing shapes, Lilienthal showed a genuinely scientific temperament, but he also possessed a streak of showmanship that helped publicize the pursuit of human flight. He was also a man of great physical courage who had a huge impact on the development of flight. Lilienthal died on 10 August 1896 following a crash in his glider.

PIONEERING BIRDMAN
Despite his uncompromising methods and "birdman" reputation, Lilienthal's intelligent and systematic approach to flying made him a powerful inspiration to other serious researchers.

They were light and flimsy structures, made by stretching cotton material over willow and bamboo ribs. But, unlike Maxim's ponderous machine, they actually flew.

In all, Lilienthal carried out more than 2,000 flights, the longest covering a distance of 350m (1,150ft). A key lesson he learned from these experiences was that the air could be a treacherous medium to move through. Since Lilienthal's gliders had no control system, he was obliged to throw his body around to maintain balance and stability amid the shifting air currents.

A visiting American journalist described Lilienthal's energetic mode of flight: "He went over my head at a terrific pace, at an elevation of about 50 feet [15m]… The apparatus tipped sideways as if a sudden gust had got under the left wing… then with a powerful throw of his legs he brought the machine once more on an even keel, and sailed away across the fields at the bottom."

The ultimate sacrifice

Repeated flights in such unstable machines involved an astonishing level of risk. Lilienthal did devise a shock-absorber to protect him if he crashed, but only used it fitfully. On 9 August 1896, caught in a sudden gust of wind, his glider stalled and crashed. He died of his injuries the following day. By then Lilienthal was a famous man. Photographs of him in flight had inspired a great deal of public interest, and his writings had been translated into several languages. His most successful glider, the No. 11 standard monoplane, had been sold to a number of enthusiasts, making it the first aircraft to be produced in quantity.

they take to the air. Contemporary experimenters in unpowered flight, by contrast, hoped to make progress through building up experience of flying. Their acknowledged leader was the German "flying man" Otto Lilienthal.

The flying man

In some respects Lilienthal was a direct descendant of the medieval "birdmen" and "tower-jumpers". Lilienthal's flights – pacing down a hill into the wind, encumbered by his wide bird-like wings, and lifting into a glide that carried him high above the ground – were an impressive spectacle but not consonant with the notion of progress as understood in the late 19th century. It is easy to see why Hiram Maxim, with his powerful state-of-the-art steam engines and expensively constructed test track, dismissed Lilienthal as a "flying squirrel". Yet the apparently eccentric Lilienthal was far more scientific and practical in his exploration of flight than either Maxim or Ader. From a scrupulous study of bird flight and bird anatomy, he concluded that a curved, or cambered, wing was essential to produce lift. He

proceeded to carry out experiments with specially constructed test equipment to see which precise wing shape, or aerofoil, would give maximum lift.

Even more striking than Lilienthal's systematic study of aerodynamics was his commitment to practical experiment through flying himself. He began by trying out ornithopters, but these inevitably futile wing-flappings were soon succeeded by a more fruitful exploration of the potential of fixed-wing gliders. Between 1891 and 1896, Lilienthal designed and built 16 different gliders, mostly monoplanes but some biplanes.

IMPRESSIVE SPECTACLE
Otto Lilienthal's glider experiments attracted substantial crowds and won him a reputation as "the flying man". Most of the tests were carried out at Lichterfelde, outside Berlin, where Lilienthal built an artificial conical hill that allowed him to take off from any side, responding to the direction of the wind.

Inspired by his example, experimenters including Britain's Percy Pilcher and Americans Octave Chanute and the Wright brothers continued to explore glider flight in the last years of the century.

Yet there were serious limitations to Lilienthal's work. Although his flight experiments had demonstrated how essential a control system would be to any flying machine, he had failed to progress beyond control by the shifting of the pilot's body weight. Moreover, his gliders were only suitable for sport. If flight was ever to have practical use, it would have to involve powered machines. But when Lilienthal turned his attention to powered flight, he reverted to the hopeless notion of using an engine to power flapping wings.

CHANUTE'S GLIDER EXPERIMENTS

FRENCH-BORN CHICAGOAN Octave Chanute was a wealthy middle-aged civil engineer when the "flying bug" bit him in the 1880s. A tireless communicator, he corresponded with all the aviation pioneers of his day and built up an impressive knowledge of the subject that informed his influential book *Progress in Flying Machines*, published in 1894.

Inevitably Chanute was drawn to carry out flight experiments of his own. Working with younger collaborators, notably New York aviation enthusiast Augustus Herring, Chanute developed a variety of hang-gliders and, in the summer of 1896, embarked with his team to try them out among the windblown sandhills of Lake Michigan's southern shore. The site was only 50km (30 miles) from Chicago, so the activities of these oddballs soon attracted the attention of curious visitors, including reporters and news photographers.

The experimenters encountered difficulties aplenty. The winds were high and unpredictable – as soon as Chanute's group arrived in the dunes, their tent was ripped to shreds in a storm. The Lilienthal-type glider with which they began proved almost unflyable and was soon abandoned – a decision vindicated by Lilienthal's death.

Desiring a safer, more stable machine that would not require acrobatic movements of the body to fly, Chanute then tested a glider with no fewer than 12 movable wings. His principle was that "the wings should move, not the man". The multiplane glider was not, however, popular with the young collaborators who actually had to fly it.

Genuine success came late in the summer when a biplane glider designed jointly by Chanute and Herring achieved flights of up to 110m (360ft). The two wings were braced by a Pratt truss, a system that was to make an important contribution to flying-machine design. The wings were fixed and a cruciform tail assembly increased stability. The machine proved so reliable and easy to fly that visitors were invited to joy-ride down the dunes.

> "All agreed that the sensation of coasting on the air was delightful."
>
> **OCTAVE CHANUTE**
> ON THE REACTION OF VISITORS WHO WERE ALLOWED TO TRY HIS GLIDERS

DUNE FLIGHT
The Chanute-Herring biplane glides down a dune of the shores of Lake Michigan in 1896. Chanute, who was in his sixties at the time, did not attempt to fly himself.

LAUNCHING THE GREAT AERODROME

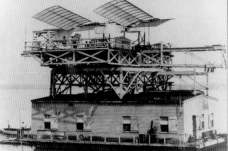

DASHED DREAMS
The first attempt to fly the Great Aerodrome was made in October 1903. Catapulted on a trolley along a track atop a houseboat, the flying machine headed downwards the moment it reached the end of the track and plunged into the Potomac. The "pilot" was rescued and the wreck of the Aerodrome towed away for repair.

ON 7 OCTOBER 1903 engineer Charles Manly climbed aboard a massive flying machine perched on top of a houseboat on a stretch of the Potomac River 65km (40 miles) south of Washington, D.C. The machine was Samuel Pierpont Langley's Great Aerodrome, the product of a costly four-year project to achieve manned heavier-than-air flight. An audience of journalists and military and scientific observers had been invited to witness its first take-off.

Manly was a mechanical wizard who had devised the aeroplane's innovative 52hp lightweight gasoline engine.

He was expected to function as more of a passenger than a pilot, since Langley's machine was supposed to be so stable it would fly itself. With the engine running sweetly, Manly raised his arm and a labourer wielded an axe, cutting a retaining cable. The catapult trolley shot forwards, accelerating the Aerodrome to the end of its launch track. Langley described the fiasco that followed: "Just as the machine left the track, those who were watching… noticed that the machine was jerked violently down at the front… and under the full power of its engine was pulled into the water, carrying with it its engineer." Manly struggled clear of the plane to be rescued.

Langley was now under enormous pressure to deliver on his promise of manned flight. By the time the Aerodrome had been repaired it was winter. On the afternoon of 8 December, conditions were far from favourable but, as Langley wrote, "the funds for continuing the work were exhausted,

UNDER CONSTRUCTION
Workmen assemble Langley's Great Aerodrome on its launch track on top of a houseboat, ready for the first flight. An elaborate catapult mechanism had been created to propel the flying machine along the track and into the air.

rendering it impossible to wait until spring for more suitable weather for making a test…" This time the attempt took place off Anacostia, at the edge of Washington, where crowds of onlookers lined the riverside. The brave Manly once more shot along the launch track and for a moment the huge machine lifted into the air, before the whole tail section sadly crumpled and broke away. Plunged into the icy water, Manly was trapped under the wreckage, but pulled himself free.

Stung by the ridicule heaped upon him, Langley blamed the failure on a faulty launch mechanism, insisting his Aerodrome could have flown.

As the 19th century drew to a close, the attaching of an engine to some form of glider had suddenly become more feasible through the development of the internal combustion engine, which had the potential to generate more power per weight than any steam engine. Lilienthal's disciple Percy Pilcher was the first would-be aviator to develop a gasoline aero-engine. He intended to use the 4hp power plant to drive a propeller attached to one of his gliders. But in September 1899, while the powered machine was still being assembled, Pilcher was killed when his glider fell apart during a demonstration flight.

The deaths of Lilienthal and Pilcher were a major setback for those who believed in gliding as the route to powered manned flight. The road seemed open for followers of the power-focused tradition to triumph, in the person of distinguished American scientist Samuel Pierpont Langley. Dismissive of Lilienthal and his followers, Langley believed that the application of sufficient power to an aerodynamically stable machine would solve the problem of flight. In 1896, he felt he had proved his point by flying steam-powered model aircraft off the roof of a houseboat on the Potomac River. The models, which he called Aerodromes from the Greek for "air runners", had a wingspan of around 4.25m (14ft). One flew for 1 minute 30 seconds and another for 1 minute 45 seconds.

Had Langley ended his work there, his contribution to aviation history would have been a resounding success. But the temptation to pursue manned flight proved too strong. In 1898 the United States went to war with Spain. The US War Department offered Langley generous funding to produce a flying machine, regarded as a potential weapon of war. With a budget of $50,000, plus the resources of the Smithsonian to call on, Langley fondly expected to achieve manned flight by the end of 1899. Yet delay followed delay. Langley settled on a gasoline engine to power his aeroplane, but it took years to develop one with the power-to-weight ratio he required. Building a full-scale version of his Aerodrome models also proved taxing for the Smithsonian

SAMUEL PIERPONT LANGLEY

SAMUEL PIERPONT LANGLEY (1834–1906) rose to prominence as an astrophysicist working at the Allegheny Observatory in Pennsylvania. Recognized as one of America's leading scientists, he was appointed to the prestigious position of Secretary of the Smithsonian Institution in Washington, D.C., in 1887.

Langley began investigating the practicality of flight in the 1880s and continued his experiments at the Smithsonian, exploiting its resources. He progressed from building small models powered by rubber bands to larger steam-powered "Aerodrome" models. In stark contrast to the Wright brothers, Langley developed no hands-on experience of either building flying machines or piloting them. His manned aeroplane, the Great Aerodrome, was the product of money and bureaucratic organization applied to the problem of flight. Its very public failure in 1903 was a crushing blow to the vanity of a proud man.

MAN OF LETTERS
A respected public figure with an impressive reputation as a scientist, Langley was humiliated by the Aerodrome fiasco.

workshops. The project ended up way over budget and four years behind schedule. And the huge flying machine that resulted simply did not work. Aerodynamically and structurally unsound, with no adequate control system, it twice plunged straight from launch into the Potomac, taking Langley's reputation with it (see left).

The failure of this government-funded project conducted by America's leading scientist caused many people to conclude that heavier-than-air flight would never be a reality. Ironically, a mere nine days after Langley's last failed attempt with the Great Aerodrome in December 1903, two bicycle makers from Dayton, Ohio, proved the sceptics wrong.

BRITISH EXPERIMENTER
British pioneer Percy Pilcher demonstrates his Bat glider, which he developed under the influence of Lilienthal. At the time of his death in 1899, he was experimenting with putting an engine on a glider – the path to powered flight that the Wrights would follow.

FIGHT TO BE FIRST

BY 1900, THE DREAM OF ACHIEVING POWERED FLIGHT WAS CLOSE TO BECOMING REALITY. THE QUESTION BEING ASKED WAS – WHO WOULD GET THERE FIRST?

> "For some years I have been afflicted with the belief that flight is possible to man. My disease has increased in severity and I feel that it will cost me an increased amount of money if not my life."
>
> WILBUR WRIGHT, 1900

LATE IN THE AFTERNOON of Saturday, 8 August 1908, on a race track at Hunaudières outside Le Mans in western France, Wilbur Wright unhurriedly settled himself at the controls of a flying machine. Dressed in a grey business suit, a high starched collar, and a golf cap, this phlegmatic man was preparing for his first flight in Europe. He hoped to establish in the eyes of the world that he and his brother Orville had been the first to achieve heavier-than-air powered flight.

Watching from the stands was a handful of flight enthusiasts, most of whom had come in the hope and expectation that the American would fail. They all knew that the Wrights claimed to have flown as long ago as 1903, but there was widespread scepticism about this alleged conquest of the air. France had its own claimants to the title of "first to fly" – including Alberto Santos-Dumont, who had briefly lifted off the ground in a heavier-than-air machine in 1906, and Henri Farman, who had flown a full kilometre (⅗-mile) circuit earlier in 1908.

A RESERVED MAN
Although Wilbur Wright was a man of austere and reserved temperament, his private letters reveal a caustic sense of humour, as well as piercing intelligence.

Since Wilbur Wright's arrival in France, the press had been running articles deriding his claims to primacy; he was, they said, "not a flier, but a liar".

Wilbur Wright cannot have been certain that he was about to prove the sceptics wrong. He had never operated this particular machine, the Type A. He seemed to take forever over his preparations, ignoring the crowd's mounting impatience, until he finally announced: "Gentlemen, now I'm going to fly." Wright's assistants set the two propellers whirling, weights dropped from the catapult derrick, and the flying machine sped along its launch rail and lifted into the air. Travelling at a height of about 10m (30ft), Wright approached the end of the racetrack and put his machine into a graceful banked turn to come back over the heads of the spectators. After completing one more circuit of the track, he brought the machine gently down on its skids. There was uproar. Clapping and cheering, the spectators ran forward to mob the pilot. Lasting just 1 minute 45 seconds, the flight had exceeded any display of flying the French had ever seen.

Since that day at Le Mans, it has been generally – though not always universally – accepted that the Wright brothers were indeed the inventors of the first heavier-than-air machine capable of sustained, controlled, powered flight.

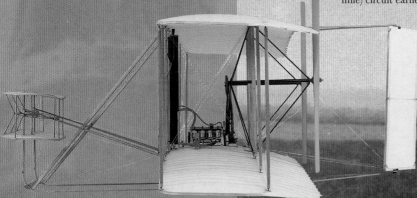

THE WRIGHT BROTHERS' 1903 FLYER
On 17 December 1903, Orville Wright made the world's first powered heavier-than-air flight at Kill Devil Hills, Kitty Hawk, North Carolina. The Flyer flew for 12 seconds over a distance of 37m (120ft). On the fourth and final flight, with Wilbur at the controls, it flew for 59 seconds over a distance of 260m (852ft).

WRIGHT SILENCES CRITICS
Wilbur Wright showed off the Type A Flyer in a series of demonstration flights in Europe in 1908. The flights received popular acclaim and persuaded most European air enthusiasts that the Wrights had indeed been the first to fly. One French journalist wrote of the "masterly assurance and incomparable elegance" of Wilbur's flying displays.

But it still seems one of history's more mischievous twists that a goal that had eluded distinguished scientists and engineers, as well as rich enthusiasts, should have been attained by two brothers who ran a bicycle shop in Dayton, Ohio.

Men of their time

Although they lived far from centres of fashion and power, Orville and Wilbur Wright had grown up very much in touch with contemporary currents of thought and innovation. Their formative years were a time when new inventions proliferated – the telephone, automobiles, electric light, wireless telegraphy, and the cinema. Inventors such as Thomas Edison and Alexander Graham Bell were the heroes of the age. It would have been surprising had the Wrights not taken some interest in the widely publicized flight experiments of the 1890s. Interest seems to have blossomed into committed research through the perception that flight was, as Wilbur wrote, "almost the only great problem that has not been… carried to a point where further progress is very difficult".

From the outset, the Wrights displayed the systematic approach that was to characterize their entire endeavour. Their first need was to absorb existing knowledge. In May 1899, Wilbur wrote a letter to the Smithsonian Institution, asking for any papers it might have on flight and a reading list of books on the subject. "I wish to avail myself of all that is already known", Wilbur wrote, "and then if possible add my mite to help on the future worker who will attain final success." The letter received a prompt and helpful response.

When the Wrights had acquainted themselves with the works of, among others, Cayley, Pénaud, Chanute, Lilienthal, and Langley, they identified an area that seemed to have been neglected: control. Men such as Langley had imagined a flying machine to be rather like a car – an essentially stable machine to be switched on and then steered. However, the Wrights instinctively felt that a flying machine was more like a bicycle, and would need to be flown with constant adjustments of balance. From the start they posed the problem not simply of how to build a flying machine, but also how to fly it.

> "… I have some pet theories as to the proper construction of a flying machine…"
>
> **WILBUR WRIGHT**
> LETTER TO THE SMITHSONIAN INSTITUTION, 1899

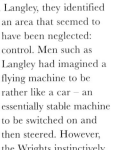

LETTER TO THE SMITHSONIAN
In 1899 Wilbur Wright wrote to the Smithsonian, expressing his interest in aviation and asking for a reading list: "I wish to avail myself of all that is already known and then if possible add my mite to help on the future worker who will attain final success."

First experiments

Their first breakthrough came from a more traditional direction. Watching soaring buzzards, Wilbur was struck by the movement of the feathers on their wingtips, which kept the birds' lateral balance. The brothers puzzled for a long time over achieving a similar effect on an aircraft wing, until Wilbur had a sudden moment of inspiration. Absent-mindedly twisting the ends of a narrow cardboard box in opposite directions, he saw that the same could be done with a wing. "Wing-warping" had been devised.

By 1900, the Wrights were ready to begin experiments with a glider, which they built in their bicycle workshop. Glider experiments required wind, and having contacted the US Weather Bureau, they established that Kitty Hawk, a small beach settlement on the coast of North Carolina, would provide a suitable location. In

TESTING THE GLIDER
Intended as a man-carrying glider, this kite-glider (their second) had a wingspan of 5.2m (17ft) and a forward elevator. However, tests at Kill Devil Hills, near Kitty Hawk, in September 1900, soon revealed that the efficiency of the aerofoil was insufficient to support a man's weight, unless the wind was very strong.

WILBUR AND ORVILLE WRIGHT

WILBUR (1867–1912) AND ORVILLE (1871–1948) were the third and fifth sons of Milton Wright, a bishop in the evangelical Church of the United Brethren in Christ, and his wife Susan. Wilbur would have attended university but for a freak sports accident at the age of 18 that undermined his health for several years. In the event, neither brother had a college education and both stayed at home, running several businesses before moving into the bicycle trade in 1892. Originally setting up just to rent bicycles, they soon expanded into building their own, which they sold for a remarkably low price ($18 compared to the $160 that Orville paid for his first bicycle in 1892). Inspired by the death of Otto Lilienthal in 1896, the Wright brothers started to finance their

> "From the time we were little children, my brother Orville and myself lived together, played together, worked together, and… thought together."
>
> WILBUR WRIGHT

aeronautical experiments from 1899 onwards with the profits from the bicycle business. They calculated that it cost them $1,000 to crack the problem of powered flight. Their experiences dealing with something as inherently unstable as a bicycle, and the insights it gave them into combining lightness with strength to achieve balance and control, gave them a novel approach to the problem of creating a workable heavier-than-air flying machine.

Inventive and self-reliant, the brothers not only had the practical skills to make their own tools and engines, they were also voracious readers with the intellect to work out complex theoretical problems. Their ingenuity and persistence as methodical experimenters was matched by their physical bravery as test pilots. Yet they were cautious individuals – Wilbur, in particular, flew only when absolutely necessary for experimental or demonstration purposes.

Neither brother had any taste for luxurious living. They did not smoke or drink, and they were rarely seen with any women other than their sister. When Wilbur carried out his demonstration flights

in France in 1908, he was cheered by crowds and courted by princes and businessmen, but he cooked his own meals and slept in a hangar with his flying machine.

The brothers combined supreme self-confidence with a deep mistrust of everyone outside their family circle. They were stubborn, hard-headed businessmen, relentless in the legal pursuit of those they thought had wronged them. Essentially private people, they coped very well with the immense fame they earned. Wilbur tragically died of typhoid fever in 1912; Orville lived on into the jet age, dying in 1948.

WRIGHT CYCLE SHOP
The bicycle became fashionable in the 1890s, and the Wright brothers combined their entrepreneurial and engineering skills when they opened up their own "Wright Cycle Co." in Dayton, Ohio, in 1892.

THE TOAST OF EUROPE
Wilbur (left) and Orville are pictured during their 1909 European tour, when they were fêted by kings, politicians and generals in France, Italy, and England.

THE COUPE DE MICHELIN TROPHY
Wilbur Wright won this trophy for the world-record-breaking distance and duration flight he made in December 1908, in France. He flew for more than two hours and covered over 100km (62 miles).

IN THE WORKSHOP
Orville (right) is shown with a worker at the Wright Cycle Company workshop in 1897. The availability of raw materials and machinery in their well-equipped shop helped the Wright brothers with their investigations into the flying problem.

September 1900, the Wright brothers pitched camp at Kitty Hawk and assembled their first glider.

Testing the gliders

In many ways, this biplane resembled the machine in which they would eventually achieve powered flight. The pilot lay face down in a gap in the lower wing, a position that minimized drag. Sticking out in front of him was a movable elevator with which, using a hand lever, he controlled horizontal pitch. The wing-warping mechanism was operated by the pilot's feet. When the weather was right for a glide, the contraption was dragged to the top of a high dune. Either Wilbur or Orville climbed on board, while the other brother and a local assistant – usually the helpful Dan or Bill Tate – each held a wingtip. When the pilot was ready, they ran the glider downhill into the wind until it lifted off into skimming flight. To their immense satisfaction, the Wrights found that the controls operated well, achieving balanced flight and smooth landings.

> "We could not understand that there was anything about a bird that could not be built on a larger scale."
>
> ORVILLE WRIGHT

Back in Dayton, through the winter of 1900–01, they worked on a new glider that would be the largest anyone had yet flown. The most unsatisfactory feature of the 1900 glider had been the lift, which had fallen short of expectations. The Wrights had worked out the wing size and shape – crucially, the degree of camber – in line with Lilienthal's published calculations of lift and load. Now they had a second try, based on the same figures, almost doubling the surface area of the wings and using a deeper camber.

The Wrights returned to the North Carolina dunes for a second, more prolonged stay in the summer of 1901, this time setting up at Kill Devil Hills some miles from Kitty Hawk. They had by then become established members of the scattered fraternity of flight enthusiasts and were in regular correspondence with its doyen, the veteran Octave Chanute, who came to witness their new round of experiments. These did not go smoothly. It took many attempts before the glider would fly at all, and when it did get off the ground, the nose proved liable to pitch dangerously upwards or downwards. Substantial changes to the shape of the wing restored control, but the risks of these experiments were becoming very apparent. On one occasion Wilbur suffered a stall reminiscent of the one that killed Lilienthal. Fortunately the front elevator proved an excellent safety device, producing a cushioned fall instead of a fatal crash.

One of the Wrights' objectives in the 1901 flights was to achieve controlled banked turns using the wing-warping mechanism, but these experiments led to a side-slipping crash in which Wilbur was injured. The brothers struggled to understand why their turns would not work, eventually deciding that the wing-warping was creating drag effects that upset the machine's aerodynamics. Wilbur later wrote: "When we left Kitty Hawk at the end of 1901... we considered our

Wilbur (left) and Orville (at the controls) launch their No. 3 glider, on which they tested and mastered their control systems, in 1902. A movable rear rudder worked with wing-warping to give the craft lateral control, allowing smooth banked turns.

experiments a failure".

But the brothers did not give up. Instead, deciding that "the calculations on which all flying machines had been based were unreliable", they set out to produce their own data for the lift created by various wing shapes moving at different speeds and angles. The results of these experiments with a home-made wind tunnel were then applied to a completely new glider, ready for the 1902 flying season.

AN IMPORTANT VISITOR
The picture above shows the Wrights' camp at Kill Devil Hills, North Carolina, during the visit of Octave Chanute (seated second from left) in August 1901. A pioneer of glider design, Chanute took much interest in the Wrights' glider experiments (right).

THE WRIGHTS' RIGOROUS EXPERIMENTS

IN THE WINTER OF 1901–02, the Wright brothers carried out a remarkable set of experiments that overturned accepted wisdom on wing design. Their aim was to test a wide range of potential wing shapes in order to establish their aerodynamic characteristics.

First, they conducted fairly rough-and-ready experiments on a bicycle. As one of the brothers cycled along to create an airflow, he then adjusted the angle of the aerofoil until it balanced the wind pressure on the flat plate (see far right).

Although the bicycle experiments showed that previous published figures (by Lilienthal and others) were wrong, they lacked precision. The key experimental challenge was to create a perfectly controlled airflow and an exact record of the resulting performance. For this the Wrights built a wind tunnel, in which they tested aerofoils for two months under controlled conditions. This gave them a highly accurate series of figures that they were then able to apply to wing design.

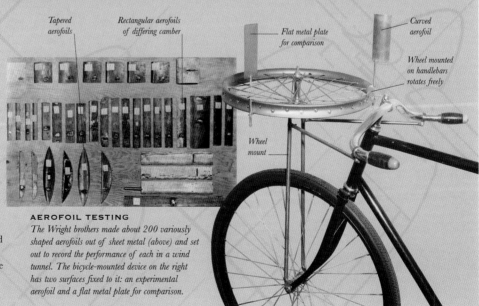

Tapered aerofoils

Rectangular aerofoils of differing camber

Flat metal plate for comparison

Curved aerofoil

Wheel mounted on handlebars rotates freely

Wheel mount

AEROFOIL TESTING
The Wright brothers made about 200 variously shaped aerofoils out of sheet metal (above) and set out to record the performance of each in a wind tunnel. The bicycle-mounted device on the right has two surfaces fixed to it: an experimental aerofoil and a flat metal plate for comparison.

THE WORLD'S FIRST POWERED FLIGHT

WAITING EXPECTANTLY
The Wright's Flyer sits outside the makeshift hangar at their Kill Devil Hills camp (left). Below, the brothers' ground crew and audience are pictured next to the Flyer, which has been transferred to its wooden rail just before Wilbur's flight of 14 December 1903.

ON THE MORNING of 17 December 1903, the wind at Kill Devil Hills was gusting at up to 48kph (30mph). This would help the Wright Flyer get off the ground, but was sure to create problems controlling the untested machine. Although Wilbur Wright's first brief hop into the air with the Flyer three days earlier had ended swiftly and ingloriously, the Wrights were certain their powered machine could fly.

The brothers had agreed to alternate at the controls, so it was Orville's turn first on this occasion. At 10.00am a flag was run up to signal to the helpful personnel of the nearby Kill Devil Hills lifesaving station, who had agreed to act as witnesses and helpers. The brothers then set about laying a wooden launch track alongside their campsite. The Flyer was too heavy to be launched like a glider – by two men holding the wings – and wheels would have sunk into the soft sand. So the machine was to be launched from a

> "It was only a flight of 12 seconds, and it was an uncertain, wavy, creeping sort of flight… but it was a real flight at last and not a glide."
>
> **ORVILLE WRIGHT**
> ON THE FIRST POWERED HEAVIER-THAN-AIR FLIGHT

trolley running along a wooden rail. Wilbur and Orville stood by the machine while the engine warmed up. Then, as one of the lifesavers, John Daniels, later recounted, "they shook hands, and we couldn't help notice how they held on to each other's hand… like two folks parting who weren't sure they'd ever see each other again". A camera had been positioned to capture the scene and Daniels was entrusted with operating the shutter. Orville mounted the machine, lying face down. Then, amid the racket of the engine and excited shouts from the onlookers, the machine was released from its restraining rope and set off along the track. As it lifted into the air, Daniels took the picture below.

Like his brother three days earlier, Orville found it hard to control the Flyer, and after 12 seconds in the air, he came down with a bump. Whether this or two subsequent attempts constituted true powered flights was rendered irrelevant by the fourth attempt. Orville, this time an onlooker, described what happened in his diary: "At just 12 o'clock Will started on the fourth and last trip. The machine started off with its ups and downs as it had before, but by the time he had gone over three or four hundred feet he had it under much better control, and was travelling

AN HISTORIC IMAGE
On 17 December 1903, the Wright Flyer lifted off the sands in the first-ever manned flight. Orville was at the controls and Wilbur, caught mid-stride, watches in amazement. This photograph, taken by John Daniels, became one of the most reproduced images of the 20th century.

on a fairly even course". Wilbur flew for 59 seconds, travelling a distance of 260m (852ft), before the Flyer pitched down to a bone-jarring landing, breaking the elevator support. The Flyer has spent less than a minute in the air, but it was certainly enough to constitute sustained, controlled, powered flight!

Shortly after this momentous event, the Flyer was caught by a gust of wind as it was being carried back to the camp and flipped over, taking with it Daniels, who was lucky to end the day with no more than cuts and bruises. The flying machine was a wreck, but the brothers did not let the incident spoil their delight in their achievement. After lunch the brothers walked over to the Kitty Hawk weather station to telegraph home the news of their success and imminent return.

CAUGHT IN TIME
This hand-held stopwatch was used to time the four flights made by the Wrights on 17 December 1903. Below is the understated telegram that Orville sent their father asking him to inform the press of their successful flights.

HARD LANDING
At the end of the fourth flight, which lasted for almost a minute, Wilbur landed hard and broke the Flyer's elevator support. It was then transported back to the Wrights' campsite, where they intended to repair the damage.

The new data led the Wrights to design a wing that was longer and slimmer, with a flatter camber. For the first time they added a tail – two fixed vertical fins that they hoped would prevent the machine going out of control in banked turns.

The 1902 No. 3 glider proved that the Wrights' calculations were correct. Aerodynamically, it was the most efficient machine yet built, but at first it was even more tricky to control than their previous model. After a few dangerous spills, the Wrights took stock and came up with a solution. The culprit was the fixed tail fins. They needed to be movable so they could be turned to counterbalance drag. With typical ingenuity, the brothers created a control system that linked the rudder to the wing-warping mechanism. By the summer's end they were making controlled glides of up to 200m (600ft), staying airborne for up to 26 seconds.

Now the Wrights were ready to embark on the momentous step to powered flight. For this they needed an engine and a propeller. When automobile companies proved incapable of supplying a suitable engine, the Wrights had one made by their assistant Charlie Taylor, who delivered a remarkable gasoline engine weighing 82kg (180lb) and delivering 12hp. In contrast, the problem of propeller design proved astonishingly complex, forcing the brothers to tackle intricate questions of theoretical physics and mathematics.

The Wrights returned to Kill Devil Hills in late September 1903, well aware that, at that very moment, Samuel Pierpont Langley was preparing for the first flight of his Great Aerodrome. When news came through that his first attempt had failed, Wilbur wrote: "It seems to be our turn to throw now, and I wonder what our luck will be".

For a time, luck seemed to be against them. In stationary tests, the engine proved temperamental and eventually damaged the propeller shafts. They were sent for repair, but when tests resumed at the end of November, one of the repaired shafts was found to be cracked. Orville returned to Dayton to make completely new steel shafts.

PRESS MISREPRESENTATION
Typical of the stories circulating in the days following the Kitty Hawk flights was this headline from the Norfolk Virginian-Pilot. *Such wild inaccuracies provoked the brothers into issuing a press statement containing exact details of their achievement.*

On 8 December Langley's second attempt to fly his Aerodrome failed, and the way was open for the Wrights. In their first attempt, on 14 December, Wilbur could not control the machine and it came down heavily almost immediately after take-off. However, on Thursday, 17 December, the goal of so many dreamers was finally attained (see pages 26–27).

Although the event did not go unreported – the Wrights themselves issued a press statement – the public response was muted. The Langley Aerodrome fiasco had created a climate of scepticism, and most newspaper editors were inclined to dismiss claims of heavier-than-air flight out of hand. The attitude of the Wright brothers themselves did nothing to allay scepticism. In their January 1904 statement they concluded: "We do not feel ready at present to give out any pictures or detailed descriptions of the machine." The Wrights had not originally pursued a policy of deliberate secrecy, but once they achieved powered flight, they were determined to stop anyone else stealing their invention before they could profit from it.

After 17 December 1903, the Wrights still faced a daunting technical challenge. Transferring

Take-off rail

Derrick

ASSISTED TAKE-OFF
From 1904, the Wright brothers used the device shown above to assist take-off. A weight, attached to the front of the aeroplane by a rope, was raised to the top of a derrick. When the aeroplane's engine had started and the pilot was ready, the weight was released and its fall jerked the machine along the rail.

their operations back to Dayton, they worked on building and testing improved models of their flying machine. The 1904 Flyer II had trouble getting off the ground under the very different weather conditions of Ohio. With the help of a catapult-assisted take-off system, Flyer II proved capable of staying in the air for more than 5 minutes. However, between June and October 1905, in the much-improved Flyer III the Wrights made flights of up to 38 minutes' duration, covering more than 30km (20 miles) at a time. If anyone wanted to question whether the Wright

1903 Wright Flyer

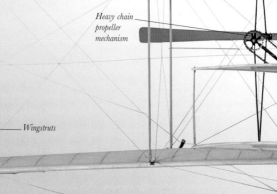

POWERED FLIGHT
The Wright brothers' first powered flying machine was constructed from spruce, ash, muslin, and piano wire and was launched from a wooden monorail. Restored in 1984 and 1985, the Wright Flyer was returned to its place in the National Air and Space Museum's Milestones of Flight *gallery in Washington, D.C.*

Heavy chain propeller mechanism

Wingstruts

EACH OF THE FLIGHTS made by the Wright brothers at Kitty Hawk in December 1903 was marked by instability, since the nose (and therefore the entire aircraft) would slowly bounce up and down. Sharp contact with the ground on the last flight broke the front elevator, ending that season's flying.

Between 1903 and 1908 the Wrights developed their original Flyer into a more robust and powerful machine, without making any fundamental changes to its configuration or control systems. All the Wright flying machines were controlled in pitch by the front elevator, in yaw by twin vertical rudders, and in roll by the twisting of the wing tips, known as wing-warping.

Flying like cyclists, the Wrights kept the aeroplane balanced with continuous small adjustments of the controls and leaning the machine into turns. This required considerable experience – a "feel" for flying that had to be learned. The trickiest feature was the front elevator, which tended to be overly sensitive. Any slight error of judgment could cause the aircraft to climb or dive alarmingly.

UNDER CONSTRUCTION
Orville Wright attaches wing-warping wires to a wing in the primitive hangar at Kill Devil Hills. Wing-warping was devised by Wilbur to control the Flyer in roll. Twisting the wings to lift one side or the other allowed it not only to fly level, but also to make banked turns, rather like a bicycle cornering.

> "I believe the new machine of the Wrights to be the most promising attempt at flight that has yet been made."
>
> OCTAVE CHANUTE, 23 NOVEMBER 1903

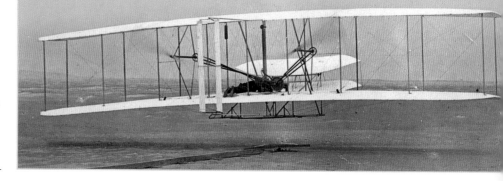

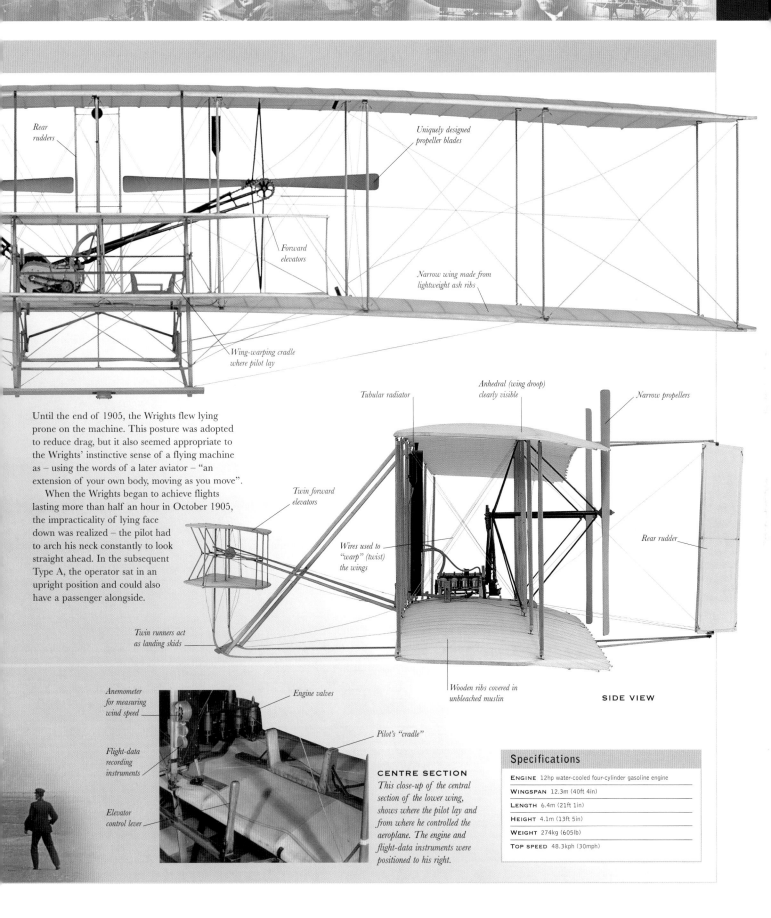

Rear
rudders

Uniquely designed
propeller blades

Forward
elevators

Narrow wing made from
lightweight ash ribs

Wing-warping cradle
where pilot lay

Until the end of 1905, the Wrights flew lying
prone on the machine. This posture was adopted
to reduce drag, but it also seemed appropriate to
the Wrights' instinctive sense of a flying machine
as – using the words of a later aviator – "an
extension of your own body, moving as you move".

When the Wrights began to achieve flights
lasting more than half an hour in October 1905,
the impracticality of lying face
down was realized – the pilot had
to arch his neck constantly to look
straight ahead. In the subsequent
Type A, the operator sat in an
upright position and could also
have a passenger alongside.

Tubular radiator

Anhedral (wing droop)
clearly visible

Narrow propellers

Twin forward
elevators

Wires used to
"warp" (twist)
the wings

Rear rudder

Twin runners act
as landing skids

Wooden ribs covered in
unbleached muslin

SIDE VIEW

Anemometer
for measuring
wind speed

Engine valves

Flight-data
recording
instruments

Pilot's "cradle"

Elevator
control lever

CENTRE SECTION
This close-up of the central
section of the lower wing,
shows where the pilot lay and
from where he controlled the
aeroplane. The engine and
flight-data instruments were
positioned to his right.

Specifications

ENGINE	12hp water-cooled four-cylinder gasoline engine
WINGSPAN	12.3m (40ft 4in)
LENGTH	6.4m (21ft 1in)
HEIGHT	4.1m (13ft 5in)
WEIGHT	274kg (605lb)
TOP SPEED	48.3kph (30mph)

Brazilian pioneer Alberto Santos-Dumont's tiny 19 Demoiselle monoplane, built in 1907, had a wingspan of just 6m (18ft) and was perhaps the first microlight. It was designed as an aerial "runabout" and easily separated into two parts (the tail, and the wings and propeller), to allow for easy transport.

brothers' flights in 1903 deserved to be called "the first", there could be no doubt whatsoever that by the end of 1905 they were the only people in the world with a practical flying machine. At this time, the brothers took the extraordinary decision to cease all further flying experiments, devoting much of their effort to a search for lucrative business contracts. The obvious potential customer for the new flying machine was the army.

The brothers suggested in a letter to their congressman, Robert Nevin, in January 1905, that the machine could be used for "scouting and carrying messages in time of war". But when Nevin raised the matter with the US War Department, the official response was dismissive. Faced with rejection at home, the Wrights approached the British and French military

establishments. A French delegation visited Dayton to negotiate with the Wrights in spring 1906 but no agreement was reached. The crux of the problem was that the Wrights would not demonstrate their flying machine until someone had signed a contract to buy it, but potential buyers were reluctant to commit without seeing the machine in action.

The Wrights' decision to stop flying was extremely risky. Details of most aspects of their work were known to aviation enthusiasts. Other experimenters had a serious chance of catching up with or overtaking them. In 1907 the inventor Alexander Graham Bell set up the Aerial Experiment Association in Hammondsport, New York, bringing together a talented team – including motorbike manufacturer Glenn H.

GLENN H. CURTISS

ONE OF THE FOUNDING FATHERS of American aviation, Glenn H. Curtiss (1878–1930) was born in Hammondsport, New York. Like the Wright brothers, Curtiss started in the bicycle business, before moving on to building and racing motorbikes. His skill in producing lightweight motorbike engines attracted the attention of inventor Alexander Graham Bell, who in 1907 invited Curtiss to join his Aerial Experiment Association – where he played a leading part in designing a series of aircraft controlled by ailerons, rather than the wing-warping used by the Wrights. On Independence Day 1908, Curtiss made the first public flight in the United States in *June Bug*.

A fearless pilot, Curtiss was often found at early aviation meetings, specializing in speed events. He eventually set up his own aircraft manufacturing company, pioneering seaplane and flying boat designs. By 1914, he was the leading aircraft manufacturer in the US.

RACING MAN
Before turning to flying in 1907, Glenn Curtiss was a successful racing motorcyclist. After winning many prizes for his flying skills, Curtiss went on to organize his own flying displays (right). His new career was hampered by a bitterly contested patent dispute with the Wright brothers over their wing-warping technology.

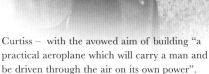

Curtiss – with the avowed aim of building "a practical aeroplane which will carry a man and be driven through the air on its own power".

Main competition

The most potent challenge to the Wright brothers came from France. In a tradition dating back to the Montgolfier brothers, the French considered themselves the natural leaders in world aviation. Reports of the Wright brothers' achievements greatly perturbed the French enthusiasts centred around the prestigious Aéro-Club de France. Some reacted by disparaging what the Wrights had done; all felt that it was their patriotic duty to prove that the French could do better. Fortunately for the Americans, France's would-be aeronauts had more

FIRST EUROPEAN FLIGHT
Renowned for his inventiveness and courage, France's first aviation hero was the Brazilian Alberto Santos-Dumont. He made the first powered heavier-than-air flight (little more than a hop) in Europe three years after the Wright brothers. Standing upright in his unwieldy 14bis box-kite aircraft, Santos-Dumont flew 220m (722ft) on 12 November 1906, in Bagatelle, Paris.

enthusiasm than method. Despite the existence of clear published accounts of the Wrights' wing-warping system, no French aviator understood the need for control in roll. Yet, stimulated by the offer of large cash prizes for variously defined "first flights" from rich enthusiasts, the French began to create successful flying machines.

The first man to claim some of this prize money was the popular Brazilian-Parisian, Alberto Santos-Dumont, already famous for his airship exploits. In 1906, Santos-Dumont built the 14bis, an ungainly, impractical biplane, with its fuselage and front elevator sticking out in front of the pilot, who stood upright in a wicker balloon basket. Its design owed much to the box-kite developed by Australian Lawrence Hargrave in the 1890s – an influence that was present in many early European flying machines. Santos-Dumont's public demonstrations during the autumn of 1906 caused a sensation. Progressing from tiny hops in September to a longer hop of about 50m (70ft) in

October, he ended with a triumphant 220m (722ft) flight on 12 November. Although negligible compared with the Wright brothers' flights of the previous year, Santos-Dumont's efforts were greeted in Europe as a major breakthrough. Le Figaro trumpeted: "What a triumph! … The air is truly conquered. Santos has flown. Everybody will fly."

French engineering

French aviators had at their disposal the excellent aero-engine, the Antoinette, developed by Léon Levavasseur, and the world's first factory dedicated to aircraft

manufacture, set up by the Voisin brothers in 1906. During 1907, both Louis Blériot and Robert Esnault-Pelterie achieved short flights in tractor (powered from the front) monoplanes, a configuration that would soon play a crucial role in the evolution of flight. But the outstanding French performances of 1908 were achieved in modified Voisin biplanes. These basically

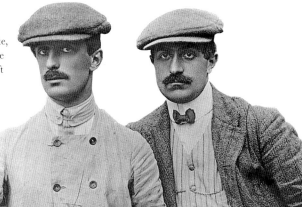

A FAMILY BUSINESS
Brothers Gabriel (left) and Charles Voisin established one of the world's first aeroplane factories, in the Parisian suburb of Billancourt, in 1906. By 1918, it had produced over 10,000 aircraft.

HENRI FARMAN

HENRI FARMAN (1874–1958), the son of a British journalist, was brought up in France. Although he sometimes wrote his first name as "Henry", he never spoke English. Farman originally sought to satisfy his adventurous and unconventional temperament in the Bohemian lifestyle of a Parisian art student, but soon found headier excitement in the pursuit of speed. In the 1890s he took up the new sport of bicycle racing, and from there progressed to automobile racing. In 1907 he transferred his sporting prowess and mechanical know-how to the new craze for heavier-than-air flying machines. His successes as a pilot soon made him one of the most famous men in France and reasonably wealthy. He used his money to set up an aircraft factory, enjoying immediate success with a box-kite biplane. His brother Maurice became a partner in the enterprise and, in 1912, the Farmans became France's largest aircraft manufacturers, producing over 12,000 military aircraft during WWI. The company was taken over by the state in 1936.

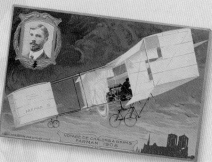

CHECKING THE CONTROLS
Henri Farman, at the controls of a Voisin-Farman biplane, prepares to take two passengers for a ride (left). The postcard above commemorates Farman's historic first "town-to-town" flight from Bouy to Reims on 30 October 1908.

resembled the Wright flying machine – pusher biplanes with a forward elevator – but they had a box-kite tail structure and lacked any form of lateral control. During 1907, Parisian sculptor Léon Delagrange and sportsman Henri Farman each turned up at the Voisin factory, coming away with his own individually modified version of the biplane. Both men quickly taught themselves to fly, logging up a series of increasingly impressive flights.

As a competitive sportsman, Henri Farman's chief target after learning to fly was to win the 50,000-franc Deutsch-Archdeacon prize for the first person to fly a 1-km (3/5-mile) circuit. On 13 January 1908, at Issy-les-Moulineaux outside Paris, a committee of the Aéro-Club de France gathered to witness Farman's attempt. At a signal from the pilot, two assistants holding the aircraft by its wingtips let go and the aeroplane raced forwards, lifting into the air. Using the rudder alone, Farman made a wide, flat turn around a pylon placed 0.5km (1/3 mile) from the start, returning safely to his starting point. The feat was hailed throughout Europe as an historic first, even though the Wrights had achieved the same feat – with smooth, banked turns – in 1904.

In June, Delagrange stayed aloft for more than 18 minutes, and the following month, Curtiss won the Scientific American trophy in his *June Bug*, for the first flight of over a mile. Then on 30 October

CHEERED TO THE FLAG
Henri Farman made aviation history when he won the Grand Prix d'Aviation by completing the first 1-km (⅔-mile) circular flight in Europe on 13 January 1908. Wilbur Wright had already completed a similar circuit – with banked turns – on 20 September 1904.

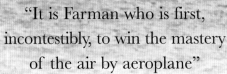

"It is Farman who is first, incontestibly, to win the mastery of the air by aeroplane"

ERNEST ARCHDEACON
FRENCH AVIATION ENTHUSIAST, 13 JANUARY 1908

1908, after further modifications, including the addition of four large ailerons to his wings, Farman made the first cross-country flight (between two points, rather than circuits around a field), covering the 27km (17 miles) between Bouys and Reims, in 20 minutes.

Success at last

In the winter of 1907–08, pushed into action by the increasingly successful flights of other experimenters, the Wright brothers finally agreed deals to market their machines. In the United States the agreement was with the Army Signal Corps, and in France with a business syndicate. Each would buy Wright machines if they successfully fulfilled stringent performance criteria in public trials. While Orville stayed behind to prepare for the US military trials, Wilbur set off for France, having shipped an unassembled Flyer

ahead of him. However, it turned out that the machine had been severely damaged in customs, and Wilbur had to spend weeks making repairs. Had the aircraft not flown on 8 August 1908, Wilbur would have faced utter public humiliation. Instead, his triumphant demonstration at the Hunaudières racetrack brought instant celebrity (see pages 20–21). Over the following months he flew repeatedly, attracting huge crowds. Gradually he extended his time in the air, culminating with an extraordinary flight of 2 hours 20 minutes on the final day of the year. He also set a new altitude record of 110m (360ft), and carried over

ROYAL INTEREST
England's King Edward VII (right), with Orville and Wilbur Wright (wearing his trademark flat-cap), watches a flight in France in March 1909. Wilbur wrote: "Princes & millionaires are as thick as fleas."

60 passengers, demonstrating that flight had become both practical and safe.

Orville's experience in the military trials at Fort Myer, Virginia, was less reassuring. His flights were a resounding success until 17 September, when his Flyer crashed with a military observer on board, killing the passenger and injuring Orville (see page 34). Despite this setback, the army had seen enough to confirm their interest. In France, hostility towards the Wrights largely disappeared, for Wilbur's perfect control over his machine surpassed anything Europe had previously seen. Taken as sufficient proof of the Wrights' claims to earlier flights, aviation journalist François Peyrey expressed the opinion of the overwhelming majority when he wrote: "The Wright brothers are the first men who have succeeded in imitating the birds. To deny it would be childish." French aviators now rushed to incorporate the key Wright characteristic, control in roll, into their machines.

The Wrights were now among the most famous men in the world. In 1909 they were immensely busy, demonstrating flight to the rich and powerful, dealing with the business offers that flooded in from all directions, and training pilots – a necessary part of sales contracts since Wilbur and Orville were the only people who knew how to fly their machines. Early in the year Wilbur was joined in Europe by Orville and their sister Katherine. When they returned to the United States in late spring, they were belatedly fêted at home. Dayton celebrated its local heroes with fireworks and a

ORVILLE WRIGHT'S ARMY TRIALS

WHILE WILBUR WAS DEMONSTRATING in Europe,
Orville prepared for the all-important US
Army tests, arriving at Fort Myer, Virginia, on
20 August 1908. During September, Orville
and his Military Flyer set nine new world
records, including two for altitude and one for
endurance (flying for just under an hour).
However, disaster struck on September 17,
when Orville took an official army passenger,
Lieutenant Thomas Selfridge, for a ride. On his
fourth circuit, he heard a tapping noise,
followed by two loud bangs, announcing the
loss of a faulty propeller. Orville lost control of
the aeroplane and it smashed into the ground,
crumpling into a twisted wreck. Selfridge died
a few hours later from a fractured
skull. Orville was lucky to
escape with serious injuries,
including a fractured
thigh, broken ribs, and
serious scalp wounds.

FIRST CASUALTY
*When Thomas Selfridge became
the first person to die (right) in
a powered aircraft, safety straps
had not yet been thought of. Pilots
and passengers simply grabbed
one of the wing struts to keep
themselves on board.*

parade; the president received them at the White
House; senators adjourned the Senate so that they
could see them fly when they came to Fort Myer
to complete the army trials; and an estimated one
million people turned out to watch Wilbur make a
spectacular flight along the Hudson River. Yet
amid this whirlwind of celebration and publicity,
the Wrights were already showing signs of pulling
back from the aviation circus. They refused to
compete for the large cash prizes on offer for the
first man to fly from London to Manchester, or
across the Channel, showing a dislike of such
publicity stunts. And in August 1909, the Wrights
were the only significant fliers not to attend the
Reims air meeting – a defining moment for the
future of aviation (see pages 44–45). Instead they
tried to put their aircraft-manufacturing business
on to a more solid footing.

In 1910 the Wright aircraft company was set
up by a consortium that included some of the
country's wealthiest businessmen, and a factory
was opened in Dayton. In Europe, Wright flying
machines were made by a number of companies
under licence. But despite successful new models
such as the Baby Wright racer, the Wrights
soon ceased to be market leaders, for instead
of concentrating their efforts on the development

WOWING AMERICA
On his return from a successful tour of Europe, Wilbur Wright made a demonstration flight, on 4 October 1909, along the Hudson river from Governor's Island in New York Harbor to President Grant's tomb and back, witnessed by over a million spectators.

of their aeroplanes, they spent much of their energy on legal action against their competitors in Europe and the United States for infringement of patents. Their most bitterly fought case, against Glenn Curtiss, dragged on until 1914. Curtiss retaliated by taking part in an attempt, backed by the Smithsonian Institution, to prove that Samuel Pierpont Langley should be credited with creating the first viable flying machine. The endless litigation undermined Wilbur's morale and health and he died suddenly of typhoid fever in 1912. Orville continued aeronautical research, but his relationship with the Wright company ended in 1915. He stayed in Dayton, living long enough to see the Smithsonian finally, in 1943, accept the Wrights' claim to have been the first to fly.

AT THE WHITE HOUSE
This photograph shows US president William Howard Taft, flanked by Wilbur, Orville, and Katherine Wright, standing on the White House terrace after an honours ceremony in 1909. Wilbur and Orville received the Congressional gold medal (right) the same year.

SLOW WORK
The Wright brothers' increasing fame led to a steady string of orders for flying machines from their factory in Dayton, Ohio. This picture shows a wing under construction – the wooden frame is being covered with cloth – a labour-intensive and painstaking business.

THE FIRST AIRCRAFT

UNTIL 1907, ONLY ONE PAIR OF experimenters had achieved anything worthy of being described as a powered heavier-than-air flight, rather than a powered hop. Between 1903 and 1905 (when they temporarily suspended their experiments), the Wright brothers made 124 flights, the longest lasting over 38 minutes. In comparison, Clément Ader's *Eole*, one of a number of steam-powered winged vehicles built in the nineteenth century, only remained airborne for a few seconds, and Langley's *Aerodrome A* did not, at least in its manned version, fly at all. After Santos-Dumont's impressive, but still limited, flight demonstrations in the 14bis in late 1906, the Europeans at last began to make machines capable of developing into true fliers, with the likes of Esnault-Pelterie's R.E.P series, Blériot's monoplanes, and the Voisin brothers' biplanes. In 1908 these experimenters began to rival the Wrights' early achievements, as did Curtiss in the the US, but this was five years after the true pioneers.

SCARING THE HORSES
Wilbur Wright and Paul Tissandier go aloft in the Wright Flyer at Pau, southern France, in 1909.

Ader (Clément) *L'Eole*

Built by distinguished French electrical engineer and inventor Clément Ader, who had previously worked on the telephone and stereophonic sound, the bat-like, steam-powered *Eole* was the first aeroplane to take off under its own power. However the "flight", on 9 October 1890, in the grounds of a chateau at Armainvilliers in Seine-et-Marne, only covered some 50 metres (164 feet) and was never recognized as either sustained or controlled. Nevertheless, this success led in 1892, to the first government commission to build a, subsequently unsuccessful, aeroplane, the Avion III.

ENGINE	18–20hp steam-powered engine	
WINGSPAN	14m (46ft)	LENGTH 6.5m (21ft 4in)
TOP SPEED	Unknown	CREW 1
PASSENGERS	None	

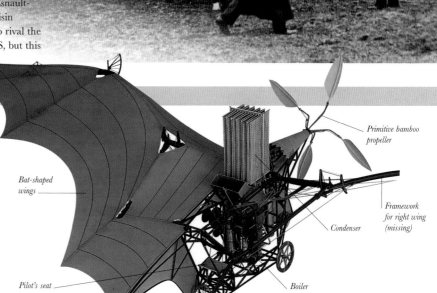

Primitive bamboo propeller

Bat-shaped wings

Framework for right wing (missing)

Condenser

Pilot's seat

Boiler

British Army Aeroplane No.1

ENGINE	50hp Antoinette V8 water-cooled	
WINGSPAN	c.15.8m (c.52ft)	LENGTH c.12m (c.40ft)
TOP SPEED	Unknown	CREW 1
PASSENGERS	None	

On 16 October 1908, British Army Aeroplane No.1 made the first officially recognized powered flight in Great Britain. Its designer was the 62-year-old American showman Samuel F. Cody, who had progressed from building man-lifting kites. Flown at the Balloon Factory, Farnborough, the large biplane flew 424m (1,390ft) in 27 seconds before crash-landing.

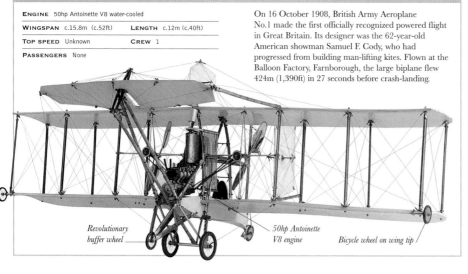

Revolutionary buffer wheel

50hp Antoinette V8 engine

Bicycle wheel on wing tip

Ellehammer Biplane

On 12 September 1906, an odd semi-biplane powered by an excellent air-cooled engine, both built by Danish engineer Jacob C.H. Ellehammer, made a tethered circular 42m (138ft) "flight" around a post on the island of Lindholm. It was incorrectly claimed as the first European flight, since it lacked any means of positive flight control.

ENGINE	Ellehammer 18hp 3-cylinder air-cooled radial	
WINGSPAN	12m (39ft 4in)	LENGTH Unknown
TOP SPEED	Unknown	CREW 1
PASSENGERS	None	

Langley *Aerodrome A*

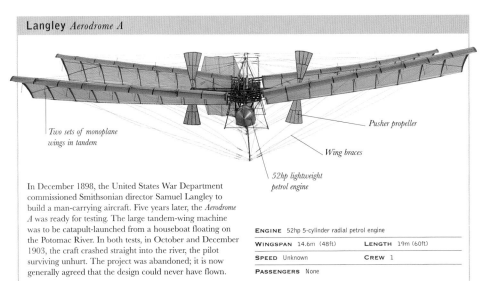

Two sets of monoplane wings in tandem

Pusher propeller

Wing braces

52hp lightweight petrol engine

In December 1898, the United States War Department commissioned Smithsonian director Samuel Langley to build a man-carrying aircraft. Five years later, the *Aerodrome A* was ready for testing. The large tandem-wing machine was to be catapult-launched from a houseboat floating on the Potomac River. In both tests, in October and December 1903, the craft crashed straight into the river, the pilot surviving unhurt. The project was abandoned; it is now generally agreed that the design could never have flown.

ENGINE	52hp 5-cylinder radial petrol engine		
WINGSPAN	14.6m (48ft)	LENGTH	19m (60ft)
SPEED	Unknown	CREW	1
PASSENGERS	None		

R.E.P 1 (1907)

In 1904, French engineer Robert Esnault-Pelterie built a Wright-style biplane glider that was controlled by the first use of ailerons. When he turned to powered craft, his R.E.P 1 – a bird-like monoplane with a tapered wing – used wing-warping, although it did boast an innovative flight control stick. The excellent engine was his own design. The aeroplane made a number of short flights at Buc during November and December 1907. Later designs were more successful.

ENGINE	30–35hp R.E.P 7-cylinder air-cooled		
WINGSPAN	9.6m (31ft 5in)	LENGTH	6.9m (22ft 3in)
TOP SPEED	Unknown	CREW	1
PASSENGERS	None		

Santos-Dumont 14bis

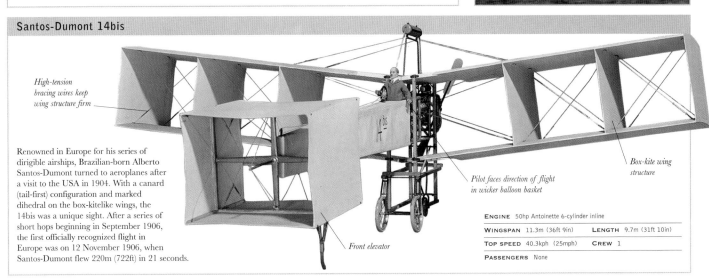

High-tension bracing wires keep wing structure firm

Box-kite wing structure

Renowned in Europe for his series of dirigible airships, Brazilian-born Alberto Santos-Dumont turned to aeroplanes after a visit to the USA in 1904. With a canard (tail-first) configuration and marked dihedral on the box-kitelike wings, the 14bis was a unique sight. After a series of short hops beginning in September 1906, the first officially recognized flight in Europe was on 12 November 1906, when Santos-Dumont flew 220m (722ft) in 21 seconds.

Pilot faces direction of flight in wicker balloon basket

Front elevator

ENGINE	50hp Antoinette 6-cylinder inline		
WINGSPAN	11.3m (36ft 9in)	LENGTH	9.7m (31ft 10in)
TOP SPEED	40.3kph (25mph)	CREW	1
PASSENGERS	None		

Dorand 1910 Biplane

Captain-Engineer Jean Dorand began his study of heavier-than-air flight in 1894. In 1910, he built the biplan-laboratoire using the Maurice Farman plan of forward elevators, staggered wings, tractor propeller, and biplane tail. Dorand used this aircraft as his test model; it included equipment for measuring speed, pitch, and roll, as well as a camera.

ENGINE	60hp air-cooled Renault		
WINGSPAN	12m (39ft 4in)	LENGTH	Unknown
TOP SPEED	Unknown	CREW	2
PASSENGERS	None		

Wright 1905 Flyer III

The Wright Flyer III was the world's first practical powered aeroplane. The Wrights began testing in June 1905 and their final flight on 5 October covered over 39km (24 miles). In 1908, the Flyer III was modified for European demonstrations with an extra second seat.

ENGINE	20hp Wright 4-cylinder water-cooled inline		
WINGSPAN	12.3m (40ft 6in)	LENGTH	8.5m (28ft)
TOP SPEED	c.56kph (c.35mph)	CREW	1
PASSENGERS	1 (from 1908)		

Wright 1902 Glider (No.3)

With their third glider, the Wrights finally achieved their aim of controlled flight. Nearly 1,000 flights were made at Kill Devil Hills by the end of October 1902, with a top distance of 190m (622ft) and duration of 26 seconds achieved during these tests.

ENGINE	None		
WINGSPAN	9.8m (32ft 1in)	LENGTH	4.9m (16ft 1in)
TOP SPEED	Unknown	CREW	1
PASSENGERS	None		

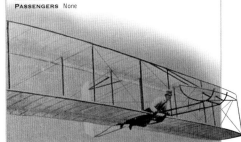

> "Until now I have never really lived! … It is in the air that one feels the glory of being a man and of conquering the elements. There is an exquisite smoothness of motion and the joy of gliding through space."
>
> GABRIELE D'ANNUNZIO
> ITALIAN POET AND NOVELIST, 1909

FLYING TAKES OFF

FROM 1909 TO 1914, AVIATION DEVELOPED FROM AN OBJECT OF CURIOSITY INTO A MODERN CRAZE THAT GRIPPED THE POPULAR IMAGINATION

THE BEGINNING OF this new phase in the conquest of the air was symbolized by two events in the summer of 1909: Blériot's flight across the English Channel and the Reims aviation meeting.

In July 1909, the attention of the world's mass media was focused on the cliffs of Sangatte, outside Calais, on the north coast of France. From there, on a rare clear day, you can see across to the white cliffs of Dover on the other side of the Channel. This narrow strip of water, which had been of such historic importance in separating the insular British from continental Europe, was about to be crossed by a flying machine.

This was an event that had been created for the consumption of the newspaper-reading public. Lord Northcliffe, owner of the *Daily Mail*, was a firm believer in the need for newspapers to create news, rather than just report it. He was also an enthusiastic believer in the future of aviation. In the wake of the sensation caused by Wilbur Wright's flights in Europe in 1908, Northcliffe had put up a prize of £500 (later raised to £1,000) for the first man to fly across the Channel. The challenge captured the public imagination and other newspapers from across

MAN OF THE WORLD
Hubert Latham was a wealthy playboy who raced automobiles and speedboats in France before turning his attention to aeroplanes in September 1909. A month later, he won the altitude prize at the Grande Semaine de L'Aviation de la Champagne, flying an Antoinette IV.

HEADLESS FLYING MACHINE
A favourite of stunt fliers like Lincoln Beachey, this 1910 Curtiss Model D headless pusher had several remarkable features. Easy to assemble for shipment or repairs, it had bamboo tail-struts, which were light and splinter-resistant. Fast for its day, it won numerous prizes at flying meetings.

Europe and North America were obliged to send their journalists and photographers to cover the story.

Three aviators had ambitions to make the first cross-Channel flight: Hubert Latham, the Comte de Lambert, and Louis Blériot. The Comte de Lambert had been taught to fly by Wilbur Wright and was the owner of two Wright biplanes. Latham and Blériot were equipped with new monoplanes, of a very different configuration

A SPORTING RACE
Stunt pilot Lincoln Beachey battles it out with racing driver Barney Oldfield over a racetrack in Davenport, Ohio, in 1914. A former balloon pilot, Beachey – who was renowned for his repertoire of loops, spirals, and dives – is shown flying a Curtiss Beachey Special. Events such as these helped establish in the public's consciousness the versatility and practicality of the aeroplane.

BLÉRIOT'S CHALLENGER
Hubert Latham's Antoinette IV is pictured being pushed across a field at Sangatte, France, just before Latham's first attempt to cross the English Channel on 19 July 1909.

from the Wright aircraft. They had a single propeller at the front and an elevator at the back, forming part of the tail with the rudder. Latham's mount was the supremely elegant Antoinette IV. Designed by Léon Levavasseur, it perfectly fulfilled the French desire for a machine that would be an object of beauty as well as a functioning piece of engineering. The Blériot XI was an altogether plainer machine, and also had a far less powerful engine than the Antoinette (see pages 42–43).

Crossing the Channel

The favourite to win the prize was Latham. Of Anglo-French descent, Latham was a wealthy playboy and adventurer who once gave his occupation as "man of the world". Sated with the excitements of big-game hunting, speedboat racing, and long-distance balloon flights, he had turned his attention to heavier-than-air flight in April 1909. Easy and confident, Latham was soon breaking speed and flight-duration records. The first to put his name forward for the cross-Channel prize, he arrived at Calais with Levavasseur early in July.

The French navy agreed to provide a warship to accompany the flier across the sea. But it was the kind of summer with which residents of the Channel coasts are only too familiar: for days on end, wind and rain, cloud and mist kept Latham grounded. It was not until 19 July that a morning dawned with some hope of a break in the weather. Seizing his opportunity, Latham took off at 6.42am, watched by a substantial crowd of journalists and admirers. One enraptured reporter described his "mechanical bird… diving into the light fog that blurred one's view of the uncertain horizon", like a "new Icarus". Too much like Icarus, it turned out. Latham had just flown over the top of the escorting French warship's funnels when his engine spluttered, coughed, and finally stopped altogether. He glided down to a smooth landing on the sea – the first ever made in an aeroplane. The French seamen sent to rescue the aviator found him sitting nonchalantly on his floating machine enjoying a cigarette.

Blériot's aircraft arrived at Calais on board the same train that brought Latham a replacement Antoinette from Paris. Blériot himself (see left) set up at Les Baraques, a farm not far from Sangatte. He was on crutches because of a severely burned foot (it had been doused with hot oil from his engine during a recent flight). His aircraft was also looking battered, but once it was unpacked, his mechanics set about assembling and repairing it. Meanwhile, the Comte de Lambert, at nearby Wissant, crashed while testing one of his Wright biplanes and was effectively out of the race. Once again the weather closed in. But in the early hours of the morning of Sunday, 25 July, Blériot's team sensed a drop in the wind. Before dawn Blériot made a trial flight around Calais and found his machine in perfect working

LOUIS BLÉRIOT

Louis Blériot (1872–1936) was an engineer and skilled businessman who made his fortune in the automobile accessories business before turning his attention to flight. At the age of 28, he began his lifelong dedication to aviation by designing a flapping machine. By 1906 he was one of the most prominent French aeronautical experimenters, but his early aircraft designs were ungainly failures.

Gallantly supported by his adoring wife, who ran the business in his absence and cared for him after his frequent crashes, he stubbornly persisted. Although almost bankrupted by

CHANNEL HOPPER
Louis Blériot was born in Cambrai, France, and achieved worldwide acclaim after becoming the first person to fly across the English Channel in a Blériot XI monoplane on 25 July 1909.

A HOMAGE
The avant-garde painters who were revolutionizing art in the early 20th century admired aircraft as a symbol of modernity. Robert Delaunay, who painted this Hommage à Blériot (1914), *had written to congratulate the aviator on his cross-Channel flight in 1909.*

the expense of his experiments, he achieved a breakthrough in 1907 with the Blériot VII, a monoplane that flew 500m (1,650ft) and set the basic configuration for his Channel-hopping Blériot XI with its Anzani 25hp engine.

After the Reims meeting, Blériot stopped flying in deference to his wife's reasonable fears for his life – he was a remarkably accident-prone pilot even by the standards of the time. Blériot's future was assured by the huge demand for Blériot XI monoplanes that his spectacular flight generated. The Blériot Company would later produce the SPAD fighter flown by the Allies during World War I, and it was still a thriving business at the time of Blériot's death in 1936.

FIRST ACROSS THE CHANNEL

IN THE LAST WEEK OF JULY 1909, two competitors, Hubert Latham and Louis Blériot were waiting on the cliffs near Calais hoping to win fame, glory, and the *Daily Mail's* £1,000 prize for being the first to cross the English Channel by aeroplane. On Saturday, 24 July, the weather had been very rough, making flights look unlikely all weekend. The following morning, Blériot – who was suffering from a burned foot – woke very early and decided to drive to his hangar at Les Baraques. Then came a stroke of luck as the weather suddenly cleared. Blériot tested his engine, informed the French naval escort of his imminent departure, and waited for dawn.

Blériot was well prepared against the cold: "I was dressed in a khaki jacket lined with wool for warmth over my tweed clothes… A close-fitting cap was fastened over my head and ears." However,

> "A break in the coast appeared to my right, just before Dover Castle. I was madly happy… I rushed for it. I was above ground!"
>
> LOUIS BLÉRIOT

SAFE LANDING
Blériot posing with his wife after crash-landing near Dover castle. He crossed the Channel in just over 36 minutes, travelling at an average speed of 64kph (40mph).

his navigational preparations were far less meticulous. Visibility was poor, yet he had no compass, watch, or map.

At 4.35am, Blériot rose into the air and flew off into the mist. The weather had been so gusty that Latham's friends assumed Blériot was just making a test flight, and when they saw the little monoplane disappear out to sea they realized it was too late to catch him. Blériot soon overtook the French destroyer, but after ten minutes found himself in a disturbing void: "I turn my head to see whether I am proceeding in the right direction. I am amazed. There is nothing to be seen – neither the destroyer, nor France, nor England. I am alone; I can see nothing at all." After a further ten minutes struggling to keep his machine level, Blériot suddenly saw the English coast appear. He realized that he had drifted off his intended course and immediately changed direction. However, this brought him up against the wind

DRAMATIC CROSSING
This image, taken from a cigarette card commemorating Blériot's famous crossing, conjures the drama of Blériot and his trusty steed battling through adverse conditions to victory.

and he battled his way towards the cliffs. Charles Fontaine, a reporter for the Paris newspaper *Le Matin*, was waiting for him just outside Dover Castle, and as Blériot emerged through the mist, Fontaine waved a French tricolour flag to guide him down. Caught by gusting winds as he crossed the cliffs, Blériot had some hairy moments before, as he put it, "I stop my motor, and instantly my machine falls straight upon the ground from a height of 20 metres [65ft]." It was not an elegant arrival – he had broken the propeller and smashed his undercarriage – but it was enough to enter the history books.

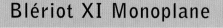

Blériot XI Monoplane

THE BLÉRIOT XI MONOPLANE in which Louis Blériot made his historic crossing of the English Channel on 25 July 1909, established the main monoplane design features for generations to come, and paved the way for the Blériot company's considerable commercial success.

It had a front-mounted modified motorcycle engine, tricycle undercarriage, front-mounted wings, and rear-mounted tailplane, elevators, and rudder. Its fuselage was a simple wire-braced, wooden-box girder, enclosed with fabric at the front to give the pilot some protection. Strips of rubber bungee were stretched down the front undercarriage legs to absorb landing shocks, while the tailwheel had a stiff spiral spring to perform the same function. Facilitated by wires that ran above and below the fuselage, lateral control was provided by warping the wing's trailing edges, directional control was provided by the moving rudder, and longitudinal control was provided by the moving elevators on the tips of the tailplane. Like all early aeroplanes, the Blériot XI had thin wings. They were braced against flying and landing loads by wires attached to the landing gear and the cabane above the cockpit.

> "No pilot of today, no matter how great, could repeat this exploit [the cross-Channel flight] in such an aircraft and with such an engine."
>
> **CHARLES DOLLFUSS**
> AVIATION HISTORIAN, WRITING IN 1932

DESIGN DISADVANTAGE
One of the main disadvantages of all monoplanes, including the Blériot model above, was the weakness of the single wing, which needed strong wire bracing to withstand the loads placed upon it.

After the cross-Channel flight, rich sportsmen queued up to buy Blériot monoplanes. So did armies when they began to investigate the military use of aircraft in 1910–11. Many improvements were made to later versions of the monoplane, including the early replacement of the underpowered Anzani engine with a 50hp Gnome rotary engine. More than 130 Blériot XIs were built, but by 1914 they were beginning to look old-fashioned compared with a new generation of fast, sturdy, manoeuvrable biplanes. Blériot monoplanes flew reconnaisance missions for the French and British armies in the early part of World War I, but they were soon relegated to use as trainers.

Deeply arched wing

Cabane used to brace wings against flying and landing loads

Flexible wooden airframe

Rubberized fabric wing covering

Innovative bungee-sprung undercarriage

Wing-warping control wires

ICONIC AIRCRAFT
The leading role in the design of the Blériot XI was played by Raymond Saulnier, a young engineer employed by Blériot in 1908. The cross-Channel flight made the Blériot monoplane a fashionable aircraft for rich sportsmen to fly and a cultural icon, the image of which was reproduced in art and advertising.

OPEN COCKPIT

Before the development of safety glass in the late 1920s, pilots sat in the open, exposed to howling winds, freezing cold, and damp. Blériot's cockpit had a rudder bar at the pilot's feet, while a control column (patented by Blériot) between the knees operated the wing-warping and elevators.

Specifications

ENGINE	25hp Anzani three-cylinder air-cooled semi-radial
WINGSPAN	7.8m (25ft 6in)
LENGTH	8m (26ft 3in)
HEIGHT	2.7m (8ft 10in)
WEIGHT	300kg (661lb)
TOP SPEED	58kph (36mph)

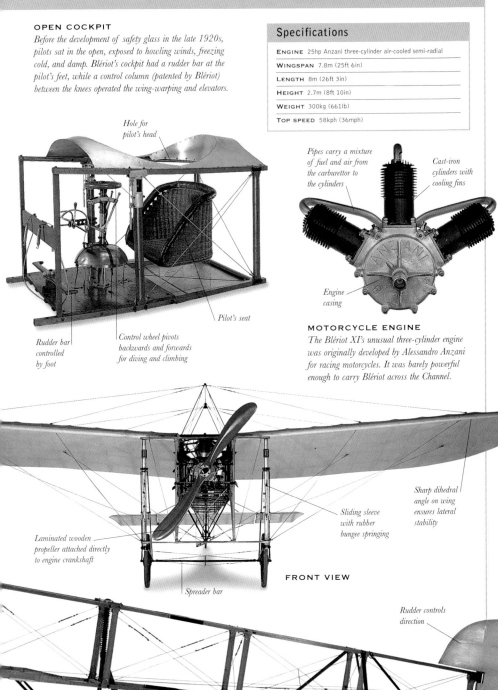

Hole for pilot's head

Pipes carry a mixture of fuel and air from the carburettor to the cylinders

Cast-iron cylinders with cooling fins

Engine casing

Pilot's seat

Rudder bar controlled by foot

Control wheel pivots backwards and forwards for diving and climbing

MOTORCYCLE ENGINE

The Blériot XI's unusual three-cylinder engine was originally developed by Alessandro Anzani for racing motorcycles. It was barely powerful enough to carry Blériot across the Channel.

Sharp dihedral angle on wing ensures lateral stability

Sliding sleeve with rubber bungee springing

Laminated wooden propeller attached directly to engine crankshaft

Spreader bar

FRONT VIEW

Rudder controls direction

Supporting pylon for tailwheel

Elevator control wire

Tailwheel

Bracing wires

Tailplane elevator

---

order. The French warship was alerted and the aviator impatiently awaited sunrise, when his epic journey began. "At half past four we could see all around. Daylight had come. My thoughts were only upon the flight and my determination to accomplish it this morning. Four thirty-five. Tout est prêt! In an instant I am in the air, my engine making 1,200 revolutions – almost its highest speed. As soon as I am over the cliff I reduce my speed. There is now no need to force my engine. I begin my flight, steady and sure, towards the coast of England…" The rest is history.

Following his successful crossing, Blériot was quite unprepared for the sensation his flight would cause. After a celebratory lunch with his wife and friends, he returned across the Channel by boat, patently expecting to continue with his life. The newspapers had other ideas. Blériot was obliged to return immediately to England and attend a glittering dinner set up by the *Daily Mail* at the Savoy Hotel in London. He was then taken back to Paris in the grip of the newspaper *Le Matin*, which suspended his monoplane outside its Paris offices to be gawped at by passers-by. In both capitals, Blériot was mobbed by delirious crowds.

In truth his flight had not been a striking technological achievement – British aviation historian Charles Gibbs-Smith described it as, "a splendid feat of daring, aided by good luck, performed in an unsuitable machine". Much of the response to it was what would now be called "media hype". Not surprisingly, the *Daily Mail* felt the flight marked "the dawn of a new age for man". *Le Matin* risked ridicule by describing this bourgeois Frenchman as reminiscent of the "robust defenders of ancient Gaul", because of "his direct and honest look, and above all his long and powerful drooping whiskers". Yet symbolically, two crucial points had been made. Britain, the world's greatest naval power, had been forced to recognize that its navy may no longer be able to defend it against all future forms of attack from abroad. And France had regained the lead in world aviation that it felt it should rightfully possess, a lead it would hold for some years to come.

The Reims meeting

Although no other event of 1909 could match the Channel crossing for the scale of publicity it attracted, the air meeting officially known as the Grande Semaine d'Aviation de la Champagne, which followed in August, was even more important in establishing the credibility of heavier-than-air

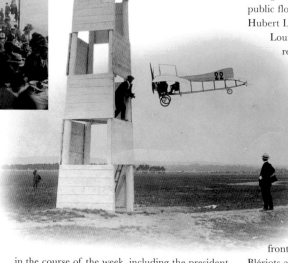

SPECTATOR SPORTS
This picture (left) shows the crowded grandstand (with a buffet in the foreground) during the Grande Semaine d'Aviation de la Champagne, held at Reims, 22–29 August 1909. Below, a Blériot XII is shown flying past a pylon during a race.

French dominance

They saw 23 aviators fly in nine different types of aircraft, representing the entire world of aviation at that time – save for the marked absence of the Wright brothers. It was a sign of the predominance of the French in aviation that only two of the pilots were from abroad: Glenn Curtiss from the United States and Englishman George Cockburn. The French stars whom the public flocked to see included Louis Blériot, Hubert Latham, Henri Farman, and newcomer Louis Paulhan, a mechanic who had recently won an aircraft in a newspaper competition and taught himself to fly.

The aircraft were of two general configurations. One group consisted of pusher biplanes with the elevator at the front and the propeller or propellers behind the pilot. These included the Wright and Voisin biplanes and the Henri Farman III, which was to prove an immensely popular model over the following years. In the other main group were the tractor monoplanes, with the propeller at the front and the elevator to the rear – mostly Blériots and Antoinettes. The one oddity was a tractor biplane designed by innovative Frenchman Louis Breguet. Although it did not perform well at Reims, this was the precursor of many future classic aeroplanes.

flight as a practical new technology. From 22–29 August, on the desolate plain of Bétheny outside the city of Reims, in the Champagne region of eastern France, the world's leading aviators competed for lavish prizes mostly financed by the producers of the region's most famous product. A large investment of capital and the support of high-ranking individuals made this a prestige event. A railway branch line was built to bring spectators from Reims. Stands, with accompanying bars and restaurants, were built to accommodate well-off spectators, while cheap tickets admitted the masses to surrounding open ground. In all, almost 200,000 spectators attended

in the course of the week, including the president of France Armand Fallières, leading British politician David Lloyd George, former US president Teddy Roosevelt, and high-ranking military officers from around the world.

As long as weather conditions were reasonable, these contraptions of wire, wood, and fabric were all viable flying machines. They were capable of flying at around 65kph (40mph) for an hour or more, and were satisfactorily controllable in simple manoeuvres, performing turns and circuits. Some used Wright-style wing-warping for their control in roll, while others employed some

AN INDESCRIBABLE THRILL

THE REIMS MEETING OF AUGUST 1909 provided an extended aerial display of the machinery that was now conquering the skies. The full excitement of the Reims event was experienced by the few individuals who were taken up for a ride as passengers.

Gertrude Bacon, an Englishwoman with a reputation as an adventurous balloonist, went up in a Farman biplane and left a remarkable account of this uncomfortable and exhilarating experience. She first faced a scramble on to the lower plane of the wing, on which were mounted "the engine, its screw-propeller behind; and in front of it, right on the edge of the plane… the little basket seat of the pilot". Since "passenger flight had not been contemplated or arranged for", the only place for Bacon to sit was on the

ARTISTIC ADVERTISING
This coloured lithograph poster by Ernest Montaut is an evocative advertisement for the world's first aviation meeting at Reims in 1909.

wing behind the pilot. When he scrambled in after her the pilot "was very close in front, wedging me tightly between himself and the extremely hot radiator". However, all discomfort was soon forgotten when "the mechanic swung the propeller, the engine… started with the first turn, and we were off across the track. The ground was very rough and hard, and as we tore along… I expected to be jerked and jolted. But the motion was wonderfully smooth – and then – suddenly there [was] a new indescribable quality – a lift – a lightness – a life!"

type of aileron – that is, moving a special control surface that was part of the wing, rather than twisting the whole wing. Only the Voisin biplane still flew completely flat.

All the flying machines suffered from unreliable engines, which were to be a bugbear of aviators for years to come. Even here, however, progress was afoot. The Henri Farman III sported a Gnome rotary engine, designed and manufactured by the Séguin brothers. Whereas in a radial or in-line engine the cylinders would be fixed around a rotating crankshaft, in the rotary engine the cylinders span around with the propeller. The Gnome created some fearsome gyroscopic effects, making the aeroplane relatively difficult to control, as well as giving off nauseous castor-oil fumes and alarming spurts of flaming oil. But it generated a lot of power for its weight, and was to play a leading part in aviation until well into World War I.

The Reims aviation week was almost ruined by the weather which, as during the Channel-crossing bids, was singularly inclement for the summer season. Torrential rain and strong winds kept the aircraft grounded for most of the first day, in front of a damp and restive crowd. But late on the weather cleared and Wright biplanes piloted by the Comte de Lambert, Eugène Lefebvre, and Paul Tissandier were catapulted into the air, amazing the crowd as they changed altitude in easy swoops, and banked around

> "Flying machines are no longer toys and dreams, they are an established fact."
>
> **DAVID LLOYD GEORGE**
> BRITISH POLITICIAN, AT REIMS, 1909

the pylons that marked out the course. Over the following days it became apparent that the long periods of inaction enforced by doubtful weather conditions, and the uncertainty as to whether a favourite pilot would emerge from his hangar, would only increase the expectation of the spectators and the fascination of the spectacle.

Thrills and spills

There was no lack of thrills and spills. Few of the pilots at Reims had yet accumulated much experience in the air, and some were virtual novices. Some crashes were a result of ill-judged manoeuvres – too steep a climb, too tight a turn – while many others were a consequence of air turbulence, especially later in the week when the weather heated up. Glenn Curtiss admitted to a surprise that must have been felt by other pilots: "I had not then become accustomed to the feeling an aviator gets when the machine takes a sudden drop."

True to form, the accident-prone Blériot experienced the most serious incident, when his Type XII monoplane burst into flames because of a ruptured fuel line. Yet no one was seriously injured and the general impression was that flying machines had proved their worthiness. About 120 flights were made in the week, almost three-quarters of them covering more than 5km (3 miles).

The high points of the meeting for spectators were the victories in the three key competitions,

A RISKY BUSINESS
There was no lack of action and drama at Reims. At times the ground was littered with the wreckage of aeroplanes – the result of poorly judged manoeuvres and difficult flying conditions.

for distance, speed, and altitude. The distance contest, for the Grand Prix de la Champagne, was easily won by Henri Farman, who flew a record 180km (112 miles) before running out of fuel. The sight of his aircraft trundling round and round the circuit at low altitude for over three hours might not sound very exciting, but the response of the patriotic French crowd was delirious, to the point of threatening public order.

The prize for the highest speed over a distance of 30km (19 miles) was put up by the owner of the *Paris Herald* newspaper, Gordon Bennett. After other contestants fell out – the English pilot Cockburn ran into a haystack – the speed contest turned into a head-to-head between Blériot and Curtiss in his Reims Racer. Curtiss went first and later described his somewhat rough ride: "The sun was hot and the air rough, but I had resolved to keep the throttle wide open. I cut the corner as close as I dared and banked the machine high on the turns… In front of the tribunes the machine flew steadily, but when I got around on the back stretch… I found remarkable air conditions. The machine pitched considerably, and when I passed

above the 'graveyard', where so many machines had gone down and were smashed during the previous days, the air seemed literally to drop from under me." He survived to complete the distance with an average speed of 75kph (47mph). To the great disappointment of the patriotic crowd, Blériot took six seconds longer and the Stars and Stripes was run up over the grandstand.

In the end, perhaps the spectacle that most struck onlookers was the altitude contest. It was won by Latham in his elegant Antoinette, which rose to the unprecedented height of 155m (508ft). Most pilots still flew close to the ground; seeing an aeroplane shrink to little more than a dot as it climbed into the blue must have created an awesome impression.

Commercialization

The Blériot cross-Channel flight and the Reims meeting set the agenda for the immediate future

THE LONDON-TO-MANCHESTER AIR RACE

IN APRIL 1910, English pilot Claude Grahame-White and Frenchman Louis Paulhan, both flying Farman biplanes, competed for the £10,000 prize put up by Lord Northcliffe for the first flight from London to Manchester – a distance of 296km (185 miles).

Spotting a break in the bad weather on the evening of 27 April, Paulhan got away first with the Englishman in hot pursuit. The French pilot followed a specially hired train along the railtrack between the two cities; Grahame-White also followed the rail line, with his supporters keeping up in a fleet of automobiles. Popular excitement was intense, and crowds, eager for the latest news, gathered outside newspaper offices in London and Paris. A huge map was set up in the Place de l'Opéra in central Paris, with model aircraft showing the race's progress. When night fell,

A COLOURFUL FIGURE
Claude Grahame-White was one of the most colourful characters in the early history of aviation. Automobiles were his first love, but in 1909 he attended Reims as a spectator and immediately fell in love with flying. He asked Blériot to build him an aeroplane and taught himself to fly.

both pilots landed alongside the railway – no one had ever flown cross-country in darkness. Seeing his only chance of overtaking the Frenchman, Grahame-White decided to take off again in the moonlight, using the automobile headlamps to light up the field. Even this bold move was not enough to bring victory. Up at the crack of dawn and braving gusting winds, Paulhan reached Manchester by 5.32am, ahead of the Englishman, who had been forced to land his wind-battered Farman short of the goal. Paulhan had spent 4 hours 18 minutes in the air.

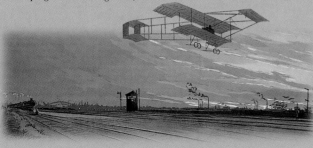

FLIGHT PATH
Louis Paulhan's Farman biplane is shown flying over railway tracks during the London–Manchester aeroplane race. Both he and Grahame-White kept their bearings by following the railway line.

of flight. While military interest remained tentative, there was little official financial support for aviation, but the lure of fame and fortune, in the form of large cash prizes and appearance money, stimulated the drive towards improved performance and fresh achievements. The money mostly came from newspaper magnates – whose papers were engaged in bitter circulation wars with rivals – putting up cash for individual feats or long-distance races, and from ticket sales for the air meetings that proliferated in the wake of the commercial success of the Reims event. Even while the Reims meeting was going on, entrepreneurs were thrusting forwards to sign up the pilots to appear at other hastily arranged air tournaments across Europe. These events were to give hundreds of thousands of people their very first sight of aeroplanes in flight.

The first follow-up meet, at Brescia in northern Italy, was held in September 1909. Curtiss, well on his way to becoming a wealthy man, again won the speed prize. In 1910, aviation meetings proliferated throughout Europe, with some 30 tournaments taking place in cities as diverse and widely scattered as St Petersburg, Barcelona, Florence, Nice, Munich, Bournemouth,

CAL RODGERS AND THE VIN FIZ

IN SEPTEMBER 1911, former college football star, Calbraith Perry Rodgers set out to win the $50,000 prize, put up by the flamboyant publisher William Randolph Hearst, for the first coast-to-coast flight across the United States in under 30 days. Rodgers raised finance from the Armour Company of Chicago, which used the pilot's Wright EX as an advertising board for its carbonated drink Vin Fiz.

Amid a blaze of publicity, Rodgers took off from Sheepshead Bay, Brooklyn, on 17 September, bound for Long Beach, California. He crashed the following day, and from then on his progress across the United States – tracked by a special train carrying, among others, his wife, his mother, and the Wright brothers' mechanic Charles Taylor – was a catalogue of mishaps and error. By the time he reached Chicago, he had already encountered so many delays that he had no hope left of winning the prize money. But Rodgers was not a quitter. He told a reporter: "I am bound for Los Angeles and the Pacific Ocean… and if canvas, steel, and wire, together with a little brawn, tendon, and brain, stick with me, I mean to get there."

COAST TO COAST
This map shows the route of the Vin Fiz during its trouble-plagued transcontinental flight. Rodgers made 69 stops along the way, finishing 84 days after he began, having spent more than 82 hours in the air.

It took Rodgers 49 days to reach California, by which time he had survived 18 crashes. Then, still 14km (9 miles) short of Long Beach, he crashed again, breaking both legs and a collar bone. Hospitalized, he declared his determination to "finish that flight" – and he did, eventually reaching the Pacific 84 days after leaving New York. Only two wingstruts and a rudder remained from his original aeroplane. Rodgers died the following year, when a seagull became jammed in his aircraft's rudder during an exhibition flight at Long Beach in April 1912. Unable to control his plane, Rodgers crashed into the ocean.

CALAMITY CAL
Cal Rodgers poses, grinning, cigar in mouth, sitting in his Wright EX Vin Fiz biplane with a top speed of 88kph (55mph). An established automobile and speedboat racer, he was an accident-prone flier (below), crashing 19 times during the trip and badly injuring himself.

AIR SPECTACLE AT HENDON
The popular obsession with aviation peaked in 1911 with a large number of record-breaking flights. The English public's enthusiasm was catered for at the regular flying meetings held at Hendon, near London, between 1911 and 1939.

and Dublin. Although with less intensity, the craze also gripped the United States, with important meetings at Los Angeles, Boston, and Belmont Park on Long Island, New York.

The Belmont Park meet, held in October 1910, did most to promote the cause of aviation in America. It was attended by prominent American businessmen, political leaders, and military top brass, and its main contests – salted with controversy – excited widespread popular interest. The race for the Gordon Bennett trophy, first contested at Reims and now an annual event, brought spills that filled the newspaper headlines. The field of international contestants was required to fly 20 times around a 5-km (3-mile) circuit. Only two made the distance, as a series of often spectacular crashes left three pilots in hospital. Englishman Claude Grahame-White, flying a Blériot monoplane with a 100hp Gnome rotary engine, was a worthy victor at an average

ALL WRAPPED UP

Jules Védrines, wearing a leather flying suit and large scarf, is surrounded by a small crowd of officials and reporters after winning the second leg of the 1911 Paris–Madrid race at San Sebastian, Spain. Védrines was one of the top racing pilots of the period and went on to fly for the French army in World War I.

speed of 98kph (61mph). The excitement of the Gordon Bennett race was eclipsed, however, by a one-off race from Belmont Park round the Statue of Liberty and back. Given the state of aviation safety, it beggars belief that such an event, flown mostly over a densely populated urban area, could have been permitted. By good fortune it passed off without mishap, but it ended in dispute and recrimination. Grahame-White felt that he had won the race since American pilot John B. Moisant, who bettered his time, had taken off later than the hour laid down in the rules. Moisant, who had been delayed through crashing his own machine and having to buy a substitute mount from another pilot, claimed and was awarded the victory, becoming an American hero. The dispute sputtered on for years.

High flier

In 1910, the young Peruvian Jorge Chávez (who was born and lived in Paris) leapt to the forefront of European aviation with a series of record-breaking, high-altitude flights, eventually reaching 2,479m (8,127ft). In September, he boldly took up the challenge of flying across the Alps from Switzerland to Italy, for which feat a large cash prize was on offer from the Aero-Club of Milan. Flying a Blériot XI, Chavez took off from the Swiss mountain town of Brig to cross the Simplon Pass, which rises at its summit to 2,013m (6,600ft). A cavalcade of cars carrying mechanics, doctors, and Alpine guides tracked him on the winding mountain road as he rose towards a beacon that had been lit at the top of

LET'S GO!

Jules Védrines raises his hand from the cockpit of his Deperdussin racer, signalling to his ground crew to release the aircraft for take-off. The photograph was taken in Calais, France, during the Circuit-of-Europe, which took place between 18 June and 7 July 1911.

the pass. Astonishingly Chávez made it across the mountains in his frail machine and headed down to land at the Italian town of Domodossola, 41 minutes after take-off. But as he glided down to the landing place, his monoplane suddenly plunged into the ground, crushing the pilot in the wreckage. Chávez was gravely injured and died four days later in hospital. His heroism made him a legend in Europe – poetry was dedicated to his "sublime death" – but his crash has never been adequately explained. One theory is that he had become so numbed by cold flying over the Alps in an open cockpit, that he could not operate the controls and let the aircraft stall. Alternatively, the Blériot's flimsy airframe may have collapsed under the cumulative strains of the high-altitude flight.

The cost of flying

The tolerance of injury and loss of life in the early days of flying was often, to a modern view, remarkable. In 1911, a series of long-distance races was arranged in Europe, proving the occasion for some horrendous accidents. The competitors in the Paris–Madrid air race, held in May, took off from Issy-les-Moulineaux outside Paris, in front of a crowd estimated at 300,000. Conditions were chaotic and take-offs were repeatedly impeded by groups of people wandering in front of the flying machines. Among those far too close to the action was a clutch of French politicians and officials, including prime minister Ernest Monice and minister for war Maurice Bertaux. Unsurprisingly, one of the aircraft got into difficulties, suffering a sudden loss of power. The pilot, Emile Train, swerving to avoid a troop of cavalry, plunged straight into the crowd of dignitaries. The minister for war was killed outright and the prime minister was seriously injured, along with some 50 other spectators. Despite this incident, the race was restarted the following day and proceeded as planned. After a welter of crash-landings and mechanical failures, only one aviator reached Madrid, Frenchman Jules Védrines, who instantly joined the thickening ranks of aviator-celebrities.

The start of another race that year, the Circuit-of-Europe, held in June, brought further carnage. Three pilots

> ## "Arriba, siempre arriba."
> ## (Higher, always higher.)
>
> **JORGE CHÁVEZ**
> CHÁVEZ'S REPORTED DYING WORDS –
> NOW THE MOTTO OF THE PERUVIAN AIRFORCE

suffered horrifying crashes on take-off: two were killed – one burnt to death in full view of the stands – and the other was crippled, losing both his legs. Yet the race was not cancelled, continuing for three weeks on a circuit that covered over 1,600km (1,000 miles) from Paris through Brussels and London and back to Paris again. It was won by another Frenchman, Jean Conneau. Huge excitement was generated by these contests between the "magnificent men in their flying machines". When a tightly fought encounter in July saw the popular Védrines finish runner-up to Conneau in the Circuit-of-Britain race, thousands of pounds were raised by public subscription to compensate Védrines for missing the winner's cash prize.

While these long, competitive flights did not necessarily confirm the potential practicability of

LINCOLN BEACHEY

IN THE PERIOD BEFORE WORLD WAR I, stunt flying turned Californian pilot Lincoln Beachey (1887–1915) into one of the most famous personalities in America. Beachey originally made his name piloting experimental airships, before graduating to aeroplanes in 1910. Taken on by Curtiss to create publicity for his aircraft,

Beachey proved both a great showman and a masterly aviator with an iron nerve. He was the first American flier to loop the loop and briefly held the world altitude record in 1911 – the same year in which he performed one of his most famous feats of bravado, by flying across the Niagara Falls. A series of staged races with champion automobile driver Barney Oldfield in 1914 was especially popular with the public. Although daredevil by nature, Beachey's shows were always based on meticulous technical preparation. However, in March 1915, his machine let him down and he crashed into the sea off San Francisco. He was just 28 years old when he died.

AMERICAN DAREDEVIL
Lincoln Beachey is shown here seated at the controls of his specially modified Curtiss Beachy Special. His breathtaking stunts drew crowds in their hundreds of thousands.

swoop down and read the name of the town. Some pilots were prepared to navigate across country by map and compass, but they often got hopelessly lost. As a last resort, a pilot could land his aeroplane in a field and ask some rural worker to tell him the way.

Sporting entertainment

One of the most impressive of all early aviation feats, Roland Garros' non-stop flight across the Mediterranean from Fréjus (Provence) to Bizerte (Tunisia) took place in 1913 – known to the French as the "glorious year" of aviation. But by then, the public had begun to grow accustomed to the miracle of flight. Flying displays and tournaments had quickly turned into a routine form of sporting entertainment. Teams of professional aviators toured internationally, putting on shows, while pilots mounted patchy displays on almost any weekend when the weather was tolerable, in aerodromes based in the suburbs of most major European cities.

Future aircraft manufacturer Anthony Fokker, has left a fascinating description of his life as a young pilot based at the Johannisthal airfield, outside Berlin, in 1912. On Saturday and Sunday afternoons a substantial crowd would gather, paying a small entrance fee to watch the aviators perform. The local pilots divided the gate money between themselves in proportion to the amount of time each had spent in the air. This meant that any flier who was

aircraft as commercial or military machines, it was an impressive sign of progress that, for example, in the Circuit-of-Europe race, a whole fleet of aircraft crossed the Channel twice without incident. But flying was still only viable in good weather. Wind and fog were especially dangerous, while in a rain shower pilots were soaked, and in the cold they froze. The high number of minor crashes and engine failures meant that any long-distance flight required a vast support system on the ground – hence the accompanying steam train or cavalcade of cars – to supply spare parts and technical assistance at frequent intervals.

Cal Rodgers' spectacularly accident-prone flight across the United States in the *Vin Fiz* was only an extreme example of a common phenomenon. When French aviator Roland

Garros contested the 1911 Paris-to-Rome race, he set off in one Blériot monoplane, wrecked it in the south of France, bought another off a local aviation enthusiast, wrecked that in Italy, and had to have a third machine sent to him from Paris by express train to finish the course. Even in decent weather, with good visibility, fliers had serious trouble finding their way from place to place. Most followed roads or rail lines – railway stations were especially useful, because a disoriented pilot could

> "An aeroplane in the hands of Lincoln Beachey is poetry. His mastery is a thing of beauty to watch. He is the most wonderful flier of all."
>
> **ORVILLE WRIGHT**

prepared to risk taking his machine up in bad weather, when the rest stayed grounded, could make good money. The pilots were a raffish lot. As

AERIAL ACROBATICS
Young French pilot Adolphe Pégoud – the first to perform "loop-the-loops" as part of his aerobatic display – is shown here mid-manouevre in a Blériot XI above a crowded airfield near Vienna. The first ever loop was completed by a Russian pilot, Lt Pyotr Nesterov, in August 1913. Expecting a rousing welcome for this daring feat, he was instead arrested for endangering government property.

Fokker writes, "sober, industrious pilots and designers were in the minority" among the "daring spirits, ne'er-do-wells, and adventurers". Their rewards were not only financial, as "beautiful women from the theatre and nightclubs hung around the flying field… unstinting of favours to their current heroes". Even higher class women might succumb to the rough glamour of aviation. In Paris, where Port Aviation at Juvisy was a similarly popular weekend draw, a diplomat referred sneeringly to a duchess associating with "the scum of the aerodromes".

Inevitably, as the spectacle of flight became more familiar, it was no longer enough for pilots simply to take off and fly around a field. The public demanded novel stunts and ever more risky manoeuvres to stimulate its jaded appetite. French pilot Adolphe Pégoud came to prominence in 1912 as the ace of aerial

acrobats. He was the second aviator to "loop the loop", a stunt that he made part of his repertoire in a series of lucrative appearances around Europe. That a pilot could fly upside down seemed a sheer miracle to the public and to most aviators, who were barely used to strapping

PURPLE LADY
Harriet Quimby, the first licensed female pilot in America (1911) and the first woman to cross the Channel (1912), is shown here in her trademark purple flight suit.

themselves to their seats during a flight. Once it became clear that the public would flock to see aerobatic stunts, they became de rigueur for pilots wanting to make money.

Despite their apparent frivolity, the development of airshow acrobatics represented a significant advance in flying techniques and aircraft control. The loops, tight-banked turns, high-speed dives, and other daredevil manoeuvres that began as high-risk entertainment would soon be used to deadly effect by World War I aces in dogfights over Flanders.

BELGIAN AVIATRIX
Hélène Dutrieu, a former Belgian champion cyclist, was admired by her contemporaries for her singular feats of aviation. In August 1910 she flew 45km (28 miles) from Ostend to Bruges in 20 minutes, and in December 1911 she smashed the world speed record, flying an average 80kph (50mph) over three hours.

The dangers of aviation

The aerobatic American pilot Lincoln Beachey always flew in a business suit, to emphasize what an easy, everyday matter flying was, but for most people, the drama and thrills of races, stunts, and shows only underlined the dangers of aviation. The rising death toll among aviators was something that eventually neither governments nor the fledgling aviation industry could ignore. A total of 32 pilots were killed in 1910, and another 30 died in the first six months of the following year – at a time when there were probably fewer than 600 fliers worldwide. In 1911 the French government set up a commission to enquire into measures for protecting aviators, resulting in the first flight regulations, including a ban on flying over towns and crowds.

Level-headed citizens were inclined to dismiss pilots as "flying fools" addicted to a dangerous sport. When Winston Churchill, then Britain's First Lord of the Admiralty, took up flying just before World War I, his wife and friends begged him to refrain from what they regarded as a suicidal obsession. Churchill eventually gave in after his wife told him: "Every time I see a telegram now I think it is to announce that you have been killed flying." Private fliers fell broadly into two categories: wealthy individuals who saw in flying a thrill or a challenge; and mechanically

A VIOLENT END
Human error was a major cause of crashes. In the early days, pilots had only a sketchy idea of the hazards involved. They often flew too low over the ground, creating the risk of hitting obstacles and giving no time for recovery if something went wrong.

THE FIRST PARACHUTE JUMP

IN 1912, HEAVIER-THAN-AIR AVIATORS began to experiment with parachutes, which were already in widespread use by balloonists. The first parachute jump from a powered aeroplane was made by an American, Captain Albert Berry, over Jefferson Barracks, Missouri. He was testing a "parachute carrying and dispensing means carried by an airplane", which later received a US patent for its inventors, Tom Benoist and Tony Jannus, the pilot. Asked if he would repeat the performance, Berry replied: "Never again! I believe I turned five somersaults on my way down… My course downward… was like a crazy arrow. I was not prepared for the violent sensation that I felt when I broke away from the aeroplane."

Parachuting was taken up by the French acrobatic pilot, Adolphe Pégoud, who made it a part of his aerial displays. Parachutes did not, however, become a standard item of a pilot's kit. Fliers regarded them as impossibly bulky and, in any case, of little use in most of the types of accident to which they were prone.

Parachute pulled from conical container attached to landing skid

DOWN TO EARTH
Captain Albert Berry (wearing a cap) made the first parachute jump from a powered aeroplane on 1 March 1912. Berry jumped from a height of 460m (1,500ft) over Jefferson Barracks, St Louis, Missouri, from a Benoist biplane (above).

gifted working men, who dreamed of making their fortune out of prize money or exhibitions. Pilots also included a few women, ready to brave not only the physical risks of flying but also the prejudices of their male-dominated society. Hélène Dutrieu, a Belgian aviatrix, and Harriet Quimby, the first American woman to gain a pilot's licence, won suitable renown for their aviation exploits, but interest in them often focused more on the elegance of their flying outfits than their evident skill and bravery.

Learning to fly

Flying schools proliferated, especially in France, as increasing numbers of people came forward to

LANDING TECHNIQUES

LANDING WAS A TRICKY MANOEUVRE and much harder to master than taking off. A smooth touchdown required the pilot to adopt a reasonable angle of descent and switch off the engine at the right moment to reach the ground at a suitable landing speed. (More experienced pilots would switch a rotary engine on and off several times in their descent as a way of reducing speed.) Old hands often enjoyed watching novices come down, either landing too steep and too fast and bouncing alarmingly across the field, or losing speed too soon and "pancaking", often with spectacularly destructive effect on the undercarriage.

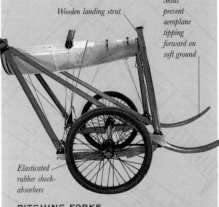

Wooden landing strut

Skids prevent aeroplane tipping forward on soft ground

Elasticated rubber shock-absorbers

PITCHING FORKS
Curved skids on the front of this 1909 Deperdussin monoplane helped to stop the aeroplane pitching forward when landing on soft ground – a common hazard in the early days of flying.

learn. Although some aircraft were built with dual controls, most flight training took place with the trainee in an aircraft on his or her own. The lessons began with sessions taxiing on the ground. The trainee would trundle backwards and forwards across the airfield, turned around at each end by a mechanic. This got them used to the "noisy, jarring vibration of the engine" and the castor oil spraying back from the rotary engine into the pilot's face. This was followed by flying in short hops a few feet off the ground before the great moment of taking to the air. Fortunately the main aircraft of the time were quite stable and fairly easy to fly in good weather as long as they did not attempt any manoeuvres. Grahame-White, writing in 1911, described the ease of

maintaining level flight on a calm day: "One's feet move just a little to and fro upon the rudder bar. This little 'joggling' of the rudder is sufficient to keep the machine on a straight course. As regards the elevator, one is moving the rod in one's hand a matter of an inch or so…"

Carelessness, foolhardiness, and inadequate landing grounds were all common causes of crashes. Early aeroplanes were fragile machines. Engine failure was common but did not necessarily lead to a crash. Aeroplanes could glide well, and an experienced pilot would expect to be able to nurse a powerless machine to a forced landing in a flat field. Structural failure, however, was a serious matter. If the wings or control surfaces

> ## "The danger? But danger is one of the attractions of flight."
>
> **JEAN CONNEAU**
> FRENCH AVIATOR, 1911

collapsed under the pressure of sudden manoeuvres or through the cumulative strain of use, a pilot was doomed. English car manufacturer Charles Rolls, for example, was killed in July 1910 when a rear elevator on his Wright biplane cracked as he came in to land, sending the machine diving into the ground.

Of course, much of the attraction of the early air shows was a ghoulish expectation of witnessing violent death. A journalist described the end of popular pilot Arch Hoxey, who lost control during an exhibition flight at the Los Angeles air meet in 1910: "The cracking of the spars and ripping of the cloth could be heard as the machine… came hurtling to the ground in a series of somersaults. When the attendants rushed to the tangled mass of wreckage they found the body crushed out of all semblance to a human being. The crowd waited until the announcer megaphoned the fatal news and then turned homeward." It had presumably been worth the price of the ticket.

ON THE FACTORY FLOOR
At first, airframes were mostly made using hand tools, often in workshops employing no more than a few dozen staff, and the production process was slow and laborious. More mechanized, larger-scale production only developed after 1911.

THE FLYING DUCK
On 28 March 1910, 28-year-old French engineer Henri Fabre made the first take-off from water in his canard (tail-first) seaplane, the Hydravion, *at Lake Berre near Martigues on the Mediterranean, despite never having flown before. The* Hydravion *had three floats and was powered by a 50hp Gnome engine.*

From 1909 onwards, there was a swift expansion in the range of aircraft designs. While great strides were being made in the new science of aerodynamics, it was not until World War I that such insights would seriously affect practical aircraft design. Successful aeroplanes evolved instead through the accumulation of a practical body of knowledge based on the experience of flying and of building flying machines. Small-scale manufacturing companies, often set up by experienced pilots, employed engineers and artisans who might have previously worked on anything from ship-building to furniture making. The production process was slow and laborious. More mechanized, larger-scale production only began to develop after 1911, when the first military contracts arrived.

Production line

The French led the way in aircraft production – France's lead in the conquest of the air became an important focus of national pride. By 1910, companies run by Farman, the Voisin brothers, Levavasseur, and Blériot had staked out the territory for which others had to compete. They were joined in 1910–11 by several other notable names. The Caudron brothers, Gaston and René, based in Picardie, became famous for their distinctive tractor biplanes. Louis Bréguet, descended from a family of wealthy Parisian clockmakers, opened a factory in Douai, northern France, where he pioneered the use of metal in airframe construction. Pilot and engineer Edouard de Nieuport, set up a company specializing in high-performance monoplanes which outlived its founder, who died in a plane crash in 1911. Designer Raymond Saulnier, who had worked for Blériot, joined the Morane brothers to create Morane-Saulnier.

Germany and Britain were the only other European countries to make significant steps towards building up an aviation industry before 1914, although many of their aircraft were copies of French models, and the British used almost exclusively French aero-engines. In Germany, companies such as Albatros, Rumpler, Aviatik, and Fokker were founded; while in Britain, famous names beginning to emerge included Short Brothers, Avro, Handley Page, Sopwith, and Bristol. In the United States, Wright and Curtiss, locked in their bitter patent dispute, remained the unchallenged market leaders. Between 1910 and 1914 the United States fell behind Europe in aviation and would not

THOMAS SOPWITH

BRITISH AIRCRAFT MANUFACTURER Thomas Octave Murdoch (TOM) Sopwith (1888–1989) had a relentless appetite for sport and machines. Learning to fly in 1910, he crashed on his first solo flight, but went on to become one of Britain's top sporting pilots.

In 1912, he set up the Sopwith Aviation Company, producing a string of successful aircraft including the *Bat Boat* (Britain's first successful flying boat), the Tabloid Seaplane (the first British aeroplane to win the prestigious Schneider trophy), and numerous fighter planes during World War I.

COMPANY FOUNDER
As a young man, Thomas Sopwith's taste for sport (ballooning and sailing) and his engineering skills (indulged on cars and motorcycles) led him to aviation.

take the lead again until the late 1920s. Americans did play a prominent part in taking aircraft to sea, but even then they cannot claim to have created the first seaplane.

Early seaplanes

That honour belongs to French engineer, Henri Fabre, whose Gnome-powered seaplane made the first flights from water in March 1910 (see below). The first truly practical seaplane, however, was tested by Curtiss in 1911, when the Curtiss-Ellyson "hydroplane" took off from San Diego Bay, California. Curtiss achieved another first in 1912 by designing a flying boat – that is, an aircraft resting in the water on a boat-like hull, rather than on floats. Manufacturers on the other side of the Atlantic, including Short Brothers and Sopwith, soon joined in with their own seaplanes and flying boats, and, in 1912, the first seaplane meeting was held at Monaco.

During this period, and for several decades to come, seaplanes and flying boats had several clear advantages over land-planes. The oceans provided an almost limitless space for take-off and landing at a time when airfields were limited. This not only meant that seaplanes and flying boats could operate where aerodromes did not exist, but also that they could potentially be larger and faster than land-planes.

Mail and passengers

One area in which the advantage of speed outweighed the disadvantage of a limited payload was in mail delivery. From 1911, exhibition airmail flights were frequently arranged in connection with an airshow or other event, and financed by the sale of souvenir postcards or franked envelopes. Surprisingly, the first such flight took place at Allahabad in British-ruled India in January 1911. Later in the year, exhibition airmail flights were authorized in Britain, France, Germany, and the United States. Claude Grahame-White's Aviation Company carried 130,000 letters and cards between Hendon and Windsor, in England, during

FIRST NAVAL TAKE-OFF
On 14 November 1910, American pilot Eugene Ely's demonstration sparks off the US Navy's interest in flight. Flying a Curtiss pusher, Ely successfully took off from a wooden platform on the cruiser USS Birmingham, in Hampton Roads, Virginia.

TO RUSSIA WITH LOVE
In the first deal of its kind, wealthy French industrialists Paul and Pierre Lebaudy built the airship Lebaudy 6, "La Russie" (above), for export to Russia in 1909. Seven years earlier the brothers had sponsored engineer Henri Julliot's construction of a semi-rigid airship, nicknamed "Le Jaune" (The Yellow One).

the celebrations for King George V's coronation. The following month, Earle L. Ovington carried mail from an aviation meeting on Long Island to Mineola, New York, dropping a mail sack over the side of his Blériot monoplane into a field behind the post office. Before World War I airmail services did not develop beyond this experimental stage.

The only sustained aeroplane passenger service at this time was provided by the Saint Petersburg-Tampa Airboat Line in Florida during the opening months of 1914. Some 1,200 passengers paid $5 for the 23-minute flight across Tampa Bay in a Benoist flying boat, but the service did not survive the end of the tourist season. The potential of aeroplanes as passenger or freight transports was

"Behind us lies the last period of German weakness and inferiority. The future of Germany is in the air!"

RUDOLF MARTIN
GERMAN AIRSHIP ENTHUSIAST, WRITING IN 1910

severely limited by their inability
to operate in poor weather conditions
and to carry substantial loads. However,
while aeroplanes struggled to get off
the ground with more than two people
on board, airships measured their
payloads in tonnes.

Aerial monsters

As far as most Germans were concerned,
the dominant form of aircraft before
1914 was the zeppelin, named after its
creator, Count Ferdinand von Zeppelin.
Although airships were also developed
in other countries, it was only in
Germany that they attained the status
of a national icon.

Zeppelins were "rigid" airships –
that is, the shape of the hydrogen-filled
envelope was maintained by a solid
framework rather than by the pressure
of the gas inside. The LZ 1, von
Zeppelin's first airship, tested in 1900,
was 128m (420ft) long. Eventually these
aerial monsters would grow to almost twice
that size. Inevitably, such dimensions made them
hard to handle on the ground, and expensive
to manufacture – it was reckoned that in 1914
you could make 34 aeroplanes for the cost
of one zeppelin.

Germany's love affair with the zeppelin really
began in 1906 when the Count first achieved
sustained flight with his LZ 3. Carrying out
flights of up to eight hours' duration, he won
official support from both the army and the
royal family. Patriotic fervour rose to fever
pitch when Zeppelin's next airship, LZ 4,
was destroyed on the ground by a storm
during a highly publicized journey up
the Rhine valley in 1908. The
German public spontaneously
subscribed the sum of six million
marks to allow von Zeppelin to
continue his work, and zeppelin
airships began passenger

AIRSHIP PASSENGERS
*The LZ 11 "Viktoria Luise",
produced at Zeppelin's
Friedrichshafen base, made its
first passenger flight on 4 March
1912. In total it made 1,000
trips, flying between Hamburg,
Heligoland, and Copenhagen.*

services in 1910. Under the direction of chief
engineer Ludwig Durr, zeppelin design steadily
improved. From 1912, goldbeater's skin – a fine
membrane from a cow's intestine – replaced
rubberized cotton as the material used to make
the gas cells inside the airship's envelope. It was
lighter and removed the risk of igniting
the hydrogen with static electricity that
could be generated by cotton surfaces
rubbing together. For the framework, a
new aluminium alloy called duralumin
was in use by 1914, offering the
strength of steel at one third of the
weight. These improvements, plus more
powerful engines, allowed the LZ 26 to
carry a 13-tonne (12.7-ton) load at
more than 80kph (50mph).

Instruments of war

Although zeppelins had carried more
than 37,000 passengers by 1914,
passenger transport in aeroplanes was
an idea whose day had yet to come.
They were to find their first
practical use in war. In the
summer of 1911, a stand-off
between France and Germany
over their interests in Morocco
brought Europe to the brink of
war. The crisis was resolved after
much sabre-rattling, but a major
conflict seemed likely – a question
not of whether, but of when.
Stirred up by a jingoistic press,
popular opinion demanded that

FLYING IN TO HAMBURG
*The LZ 13 "Hansa" is pictured looming over Hamburg harbour
in 1912 on the last leg of a journey from Scandinavia. From
1910, zeppelins made regular flights and carried over 37,000
passengers before World War I, mostly on sightseeing trips.*

COUNT FERDINAND VON ZEPPELIN

AN UNLIKELY AERIAL PIONEER, Ferdinand
von Zeppelin (1838–1917) was born
into the military aristocracy, and served
as a cavalry officer until his fifties. His
interest in airships was inspired by a visit
to the United States during the Civil
War, where he witnessed the use of
tethered balloons as military
observation posts. From 1891, he
devoted his personal fortune to the
development of powered rigid
airships. Despite numerous
setbacks, his first airship, LZ 1,
made its maiden voyage on 2 July
1900. When the LZ
4 was destroyed by
a storm in 1908,
the popular response revived von
Zeppelin's fortunes. He lived long enough
to see his airships being used as bombers
during World War I.

NATIONAL HERO
*The success of Count
von Zeppelin's airships
revived national pride
and made him a celebrity.*

AIRSHIP DISPLAY
*This German poster of 1913 depicts a heaving
crowd attending a zeppelin airshow. From
1908, zeppelins made routine flights, carrying
mail and passengers throughout Germany.*

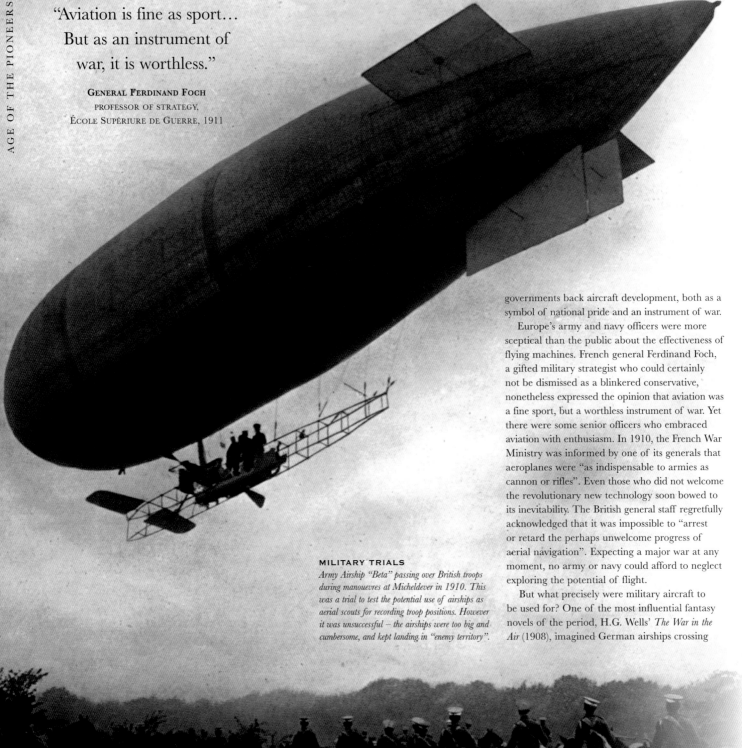

"Aviation is fine as sport…
But as an instrument of
war, it is worthless."

GENERAL FERDINAND FOCH
PROFESSOR OF STRATEGY,
ÉCOLE SUPÉRIURE DE GUERRE, 1911

MILITARY TRIALS
*Army Airship "Beta" passing over British troops
during manouevres at Micheldever in 1910. This
was a trial to test the potential use of airships as
aerial scouts for recording troop positions. However
it was unsuccessful – the airships were too big and
cumbersome, and kept landing in "enemy territory".*

governments back aircraft development, both as a
symbol of national pride and an instrument of war.

Europe's army and navy officers were more
sceptical than the public about the effectiveness of
flying machines. French general Ferdinand Foch,
a gifted military strategist who could certainly
not be dismissed as a blinkered conservative,
nonetheless expressed the opinion that aviation was
a fine sport, but a worthless instrument of war. Yet
there were some senior officers who embraced
aviation with enthusiasm. In 1910, the French War
Ministry was informed by one of its generals that
aeroplanes were "as indispensable to armies as
cannon or rifles". Even those who did not welcome
the revolutionary new technology soon bowed to
its inevitability. The British general staff regretfully
acknowledged that it was impossible to "arrest
or retard the perhaps unwelcome progress of
aerial navigation". Expecting a major war at any
moment, no army or navy could afford to neglect
exploring the potential of flight.

But what precisely were military aircraft to
be used for? One of the most influential fantasy
novels of the period, H.G. Wells' *The War in the
Air* (1908), imagined German airships crossing

the Atlantic to attack New York. Wells' vision of an air raid was apocalyptic: "As the airships sailed along they smashed up the city… Below, they left ruins and blazing conflagrations and heaped and scattered dead." But although the Germans hoped to create a zeppelin bombing fleet, cooler heads realized that aircraft of the period were neither reliable enough nor capable of carrying a large enough bombload to wreak such devastation on enemy cities. While experiments were carried out by dropping bombs and firing guns from aeroplanes – the majority of which were aimed at enemy troops – such offensive uses of aviation made relatively little progress before 1914. Instead more modest, but still vital, military roles were emphasized, centred on reconnaissance, message carrying, and artillery spotting – that is, helping gunners hit their targets by telling them where their shells were landing. By 1912, aircraft fulfilling these roles had become a standard feature of military manoeuvres, and experiments had begun in air-to-ground communication (including the use of radio) and aerial photography.

FLYING SOLDIERS
German military officers – among the first of such to take to the sky – pose in flying gear beside a Harlan Eindecker monoplane during an instruction course (c.1910–13).

Military demand

Although airships were added to the resources of both navies and armies, aeroplanes generally proved themselves more useful and reliable. They were also far cheaper to produce – a very important consideration. It was only in Germany that zeppelin advocates held their ground, diverting major resources away from aeroplane production.

Between 1911 and 1914, European military establishments became major buyers of aeroplanes and the main influence on the development of the air industry. Military competitions set manufacturers targets to aim at, with lucrative contracts at stake. The lure of profits brought substantial investment from the likes of German banker Hugo Stinnes, arms manufacturer Gustav Krupps, and Russian industrialist Mikhail Shidlovski. Some private firms experienced rapid growth. Henri Farman (see page 32) was employing around 1,000 workers by 1914, and the Gnome aero-engine company operated on a similar scale. Governments also set up their own establishments to encourage aircraft development – notably Britain's Royal Aircraft Factory at Farnborough.

However, the situation in the United States was strikingly different. America was not preparing for a major war. Its armed forces were under no pressure to embrace cutting-edge technology, and its politicians were reluctant to vote funds for military hardware. By the summer of 1913, when the biggest European military air arms were already numbered in hundreds, the US Army had 15 aeroplanes. Without substantial military contracts, the American air industry stagnated. In 1914, only 168 Americans were employed making aircraft.

Aircraft designs

The domination of European aviation by military contracts brought a distinct change in priorities. Since they did not yet take seriously the prospect of combat in the air, the armed forces demanded sturdy, reliable aircraft that could be flown in most weather conditions by average pilots and still carry a reasonable payload. Sporting pilots willingly risked their lives in treacherous high-performance machines built for speed or for stunting, but the military wanted stable aeroplanes that would survive prolonged use and keep their newly trained pilots alive. Although light monoplanes continued to be ordered for army use – for example, Taubes in Germany and Morane-Saulniers in France – there was a strong prejudice in favour of solid biplanes. A typical example was the two-seater B.E.2, designed by Geoffrey de Havilland for the Royal Aircraft Factory in 1912.

FIRST BOMBING RAID

IN THE AUTUMN OF 1911, Italy declared war with Turkey in a dispute over the territory now known as Libya, then part of the decaying Turkish Empire. The Italian army possessed a number of foreign aircraft – French Blériots, Farmans, and Nieuports, and German Taubes. An air flotilla, initially comprising just nine aeroplanes and 11 pilots, was sent off with the Italian force that embarked for the Libyan coast in North Africa. In the short but brutal war that followed, the aeroplanes performed creditably, carrying out reconnaissance missions, mapping areas of the desert, and dropping propaganda leaflets promising a gold coin and sack of wheat to all those who surrendered. On 1 November, Lieutenant Giulio Gavotti dropped four grenades over the side of his Blériot on to a Turkish military encampment at the Taguira oasis, in the first ever bombing raid by an aeroplane. Despite the fact that they faced little opposition, the aviators were hailed as heroes by patriotic Italians. Although the 1899 Hague Convention banned aerial bombing from balloons, Italy argued that this ban could not be extended to aeroplanes.

BATTLE OF DERNA
This propaganda poster shows three Italian monoplanes circling the battle of Derna during the Italo-Turkish war (1911–12).

Deperdussin 1913 Monocoque Racer

THE DEPERDUSSIN AVIATION COMPANY was set up by Belgian-born businessman Armand Deperdussin in 1910. Deperdussin had made his fortune as a silk importer and knew absolutely nothing about engineering or aeronautics. However, he could spot a good business opportunity, and in the wake of Blériot's Channel crossing and the first Reims air meeting, aviation was a tempting area for entrepreneurs. Deperdussin backed a gifted young engineer, Louis Béchereau, to develop new aeroplane designs. The Deperdussin 1911 Type C was already an elegant and successful variation on the basic pattern of the famous Blériot XI, but it was in 1912 that Béchereau came up with the revolutionary monocoque design that was to prove the fastest racing model of its time.

While other aircraft had a fuselage made of a framework of struts covered with varnished cloth, the Deperdussin fuselage consisted of a hollow wooden skin with no internal framework – similar to the fuselage of a modern aeroplane. It was

also light, streamlined, and, by the standards of the time, carried a powerful engine. In September 1912, Jules Védrines piloted one of the monocoque models to victory in the annual Gordon Bennett race, establishing a new world speed record of 174kph (108mph). The following year, a seaplane version, piloted by Maurice Prévost, won the first Schneider Trophy race at Monte Carlo, and an improved landplane model once again carried off the Gordon Bennett Trophy, as well as establishing a new world speed record of 204kph (127mph).

By that time, Deperdussin had run into financial difficulties and he was arrested on charges of fraud and forgery. A reorganized version of his company, taken over by his rival Blériot, went on to produce one of the most famous fighters of World War I, the SPAD XIII. In 1924, Deperdussin took his own life.

THE "DEP" SYSTEM
Deperdussin's racing monoplanes used an innovative flight control system – incorporating a wheel on top of the control stick – which is now standard on many aircraft. The "Dep" system controlled pitch by fore and aft movement of the column, but roll was controlled by wheel rotation rather than lateral stick movement.

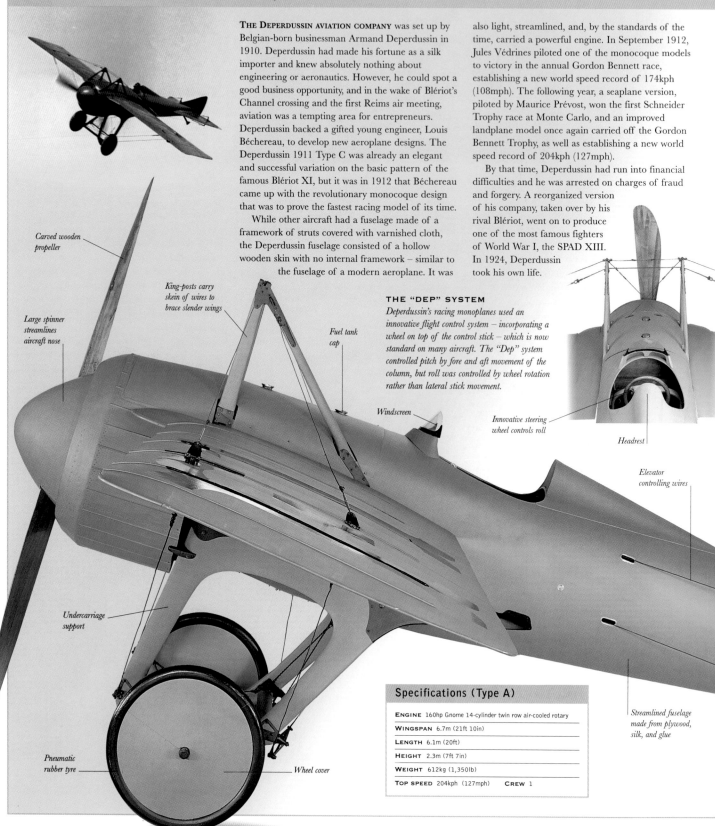

Carved wooden propeller

King-posts carry skein of wires to brace slender wings

Large spinner streamlines aircraft nose

Fuel tank cap

Windscreen

Innovative steering wheel controls roll

Headrest

Elevator controlling wires

Undercarriage support

Pneumatic rubber tyre

Wheel cover

Streamlined fuselage made from plywood, silk, and glue

Specifications (Type A)

ENGINE	160hp Gnome 14-cylinder twin row air-cooled rotary
WINGSPAN	6.7m (21ft 10in)
LENGTH	6.1m (20ft)
HEIGHT	2.3m (7ft 7in)
WEIGHT	612kg (1,350lb)
TOP SPEED	204kph (127mph) CREW 1

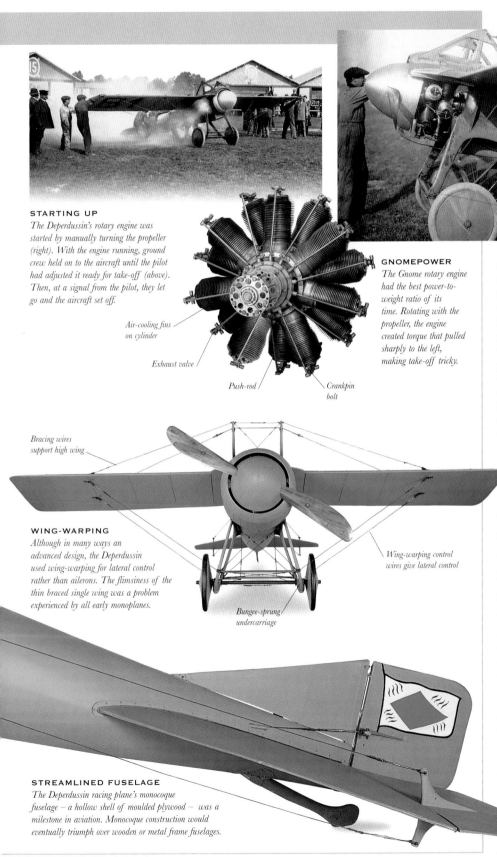

STARTING UP
The Deperdussin's rotary engine was started by manually turning the propeller (right). With the engine running, ground crew held on to the aircraft until the pilot had adjusted it ready for take-off (above). Then, at a signal from the pilot, they let go and the aircraft set off.

Air-cooling fins on cylinder

Exhaust valve

Push-rod

Crankpin bolt

GNOMEPOWER
The Gnome rotary engine had the best power-to-weight ratio of its time. Rotating with the propeller, the engine created torque that pulled sharply to the left, making take-off tricky.

Bracing wires support high wing

WING-WARPING
Although in many ways an advanced design, the Deperdussin used wing-warping for lateral control rather than ailerons. The flimsiness of the thin braced single wing was a problem experienced by all early monoplanes.

Wing-warping control wires give lateral control

Bungee-sprung undercarriage

STREAMLINED FUSELAGE
The Deperdussin racing plane's monocoque fuselage – a hollow shell of moulded plywood – was a milestone in aviation. Monocoque construction would eventually triumph over wooden or metal frame fuselages.

The record-breakers for speed around 1912–13 were the light monoplanes produced by French manufacturers Nieuport, Morane-Saulnier, and above all, Deperdussin (see left), all of which made an attempt at streamlining with a fully enclosed fuselage and engine cowling. In comparison, a biplane such as the Farman Shorthorn, used for military training, was described by a cynical trainee pilot as looking "like an assemblage of birdcages". But although the monoplanes were sleek and fast, their thin single wing generated inadequate lift for carrying much weight. It was also structurally frail, and was still braced by external wires attached to struts on the fuselage. Their control systems also made these aircraft difficult to handle.

At the time, a thin wing section was considered obligatory by aeroplane designers. In fact, as aerodynamic research would soon reveal, a thicker wing section provided improved lift, as well as a stronger structure. In 1910 a German high-school professor, Hugo Junkers, took out a patent for "an aeroplane consisting of one wing, which would house all components, engines, crew, passengers, fuel, and framework". This flying wing was never built, but the idea led the way to the cantilever wing, requiring no external struts or bracing wires, that Junkers would incorporate into aircraft design during World War I. The cantilever wing would eventually make the monoplane the aircraft of the future. But in 1913–14 the machine that established a new benchmark for performance was a biplane, the Sopwith Tabloid – the first British-designed aircraft to compete successfully for speed with the French. The Tabloid pointed forward to the leading fighter-aircraft design of World War I.

The question of size

Perhaps the greatest technical breakthrough immediately before World War I concerned size. While small, well-designed monoplanes and biplanes were breaking speed and altitude records, to be of any practical use, both in peace and war, aeroplanes simply had to get bigger. But no one had any clear idea of the feasibility of large flying machines. Greater size implied the use of more than one engine, yet many people doubted that a multi-engined aircraft could ever fly safely. Concern centred on how the aeroplane would behave in the extremely likely event of one of its engines failing. Would this throw the machine into a spin, with fatal consequences? The question was resolved by the young Russian designer and pilot Igor Sikorsky. In 1913–14, he repeatedly flew his large four-engined aeroplanes – first the Grand and then the *Il'ya Muromets* – proving that they could remain

THE SPECTACULAR FLIGHT OF THE IL'YA MUROMETS

ON 30 JUNE 1914, pilot and designer Igor Sikorsky took off on one of the most spectacular flights of the pre-war, pioneering era of aviation. In his four-engined *Il'ya Muromets*, Sikorsky intended to fly from St Petersburg over the forests and swamps of northern Russia to Kiev and back – a round trip of 2,600km (1,600 miles).

Working for his wealthy patron Mikhail Shidlovski at the Russo-Baltic Wagon Works in St Petersburg, Sikorsky had first designed a four-engined aeroplane, the Grand, in 1913. With a wingspan of 31m (88ft 7in), it was a giant aircraft for its day. Foreign aviation experts, convinced the monster plane would never fly, dubbed it the "Petersburg Duck". But in May 1913, fly it did, and the "Duck" soon proved itself airworthy in test flights with eight passengers aboard.

> ## "Aeronautics was neither an industry nor a science. It was a miracle."
>
> **IGOR SIKORSKY**

In October 1913, Sikorsky built the even larger *Il'ya Muromets*. Its wingspan was 37m (105ft), its fuselage was 27m (77ft) long, and fully loaded it weighed over 5,400kg (12,000lb). Even with four 100hp engines driving four tractor propellers, this was a lot of weight to get off the ground. The aeroplane could carry 16 people, including the crew, and by the spartan standards of the day, offered remarkable comfort. It had a heated passenger cabin with electric lights powered by a wind-driven generator, a bedroom, and the first airborne toilet. There was a balcony at the front, allowing passengers spectacular aerial views, and (for the brave) an observation platform on the rear fuselage.

The 1914 flight of the *Il'ya Muromets* was intended to demonstrate beyond any doubt that Sikorsky had created a truly practical large aircraft.

It took off at first light on 30 June carrying Sikorsky, three other crew, and substantial supplies of spare parts, food, and fuel. They were going to fly over a wilderness with no accompanying trains or cars.

After an uneventful eight hours, the aircraft made its first stop at a refuelling site at Orsha. The next leg of the journey, to Kiev, was, by contrast, dangerously exciting. An engine caught fire and two of the crew had to climb on to the wing to beat out the flames with their coats. Sikorsky made an emergency landing for repairs. Taking off the following morning, he soon ran into rain and low cloud. Turbulent air currents pitched the giant aircraft about, at one point throwing it into a spin from which it emerged only after dropping over 350m (1,000ft). It was

stable in the air with one or even two engines shut down. The path was open to the design of viable passenger-transport aircraft and heavy bombers.

Sikorsky's remarkable 2,600-km (1,600-mile) round trip from St Petersburg to Kiev and back in the *Il'ya Muromets* (see left) provided a fitting finale to an era in which heavier-than-air flight itself had travelled a vast distance from its tentative origins.

Legacy of the pioneers

Flying machines and the adventurers who flew them had conquered the hearts and imaginations of millions of people. Although those who had experienced what Grahame-White called the "great, curious sense of power" conferred by piloting an aeroplane still totalled only a few thousand, vast numbers of individuals from all sections of society had been caught up in the romance of flight. It was famously embraced by poets and painters, who adopted the aeroplane as a symbol of the modernism to which they aspired. Prominent Italian poet Gabriele d'Annunzio himself became a pilot after hymning the aviator as "the messenger of a vaster life". But the excitement about aviation stretched far beyond intellectuals. In 1911 a school in provincial England asked its pupils to state their greatest aim in life.

GIANT BIPLANE
The Sikorsky S–27 Il'ya Muromets Ye had an enclosed glass cockpit, which is clearly visible above. The postage stamp, right, was issued by the Soviet Union in 1976 to honour the ground-breaking flight of the Il'ya Muromets.

One seven-year-old, the future novelist Graham Greene, wrote: "To go up in an aeroplane".

By 1914 this era was drawing to an end. Aeronautics was rapidly becoming an industry, and, a little more slowly, a science. More disturbingly, to those who had hoped aviation might by its nature transcend national frontiers and bring different peoples together, aviation had become a branch of the armaments business and was about to turn into a major instrument of war.

APPROACHING KIEV
This painting shows the Il'ya Muromets *over Kiev during its remarkable round trip between St Petersburg and Kiev from 30 June–11 July. By the end of July, Russia was mobilizing for a war in which Sikorsky's giant aircraft would be put to work as heavy bombers.*

JOURNEY'S END
This photograph shows the Il'ya Muromets *landing at Korposnoi aerodrome, outside St Petersburg, after completing its epic round trip on 11 July 1914. Two figures can be seen standing on the outdoor observation platform.*

with immense relief that Sikorsky eventually brought the aircraft down from the cloud in sight of the golden domes of Kiev, landing to an enthusiastic reception. The return flight to St Petersburg was less dramatic and the total flying time for the remarkable 2,600-km (1,600-mile) round trip was 26 hours. Among the awards and acclaim showered upon Sikorsky came a personal expression of gratitude from Tsar Nicholas II.

IGOR SIKORSKY

BORN IN KIEV, Igor Sikorsky (1889–1972) grew up in a household where intellectual curiosity was encouraged. As a boy he developed an interest in flight through reading the science fiction of French novelist Jules Verne and accounts of Leonardo da Vinci's designs for helicopters. After studying engineering, he failed in his attempts to make a helicopter and turned to more conventional fixed-wing designs. In 1913, he constructed the world's first four-engined aeroplane to fly. Known as the Grand, it formed the prototype for the Il'ya Muromets, later adapted as a long-range bomber for World War I. In 1918, Sikorsky emigrated to the United States to escape the Bolshevik Revolution. After some years teaching, he founded his own engineering company, producing many successful flying boats. In the 1930s he returned to his original obsession, producing the prototype of the first mass-produced helicopter in 1939 (see pages 282–283).

RUSSIAN PIONEER
Igor Sikorsky stands in front of his S–22 Il'ya Muromets. Sikorsky's tried and tested designs proved that larger multi-engined aeroplanes were viable.

PRE-WAR PIONEERS (1908–14)

IN THE PIONEERING YEARS OF FLIGHT, wealthy engineers, sportsmen, and entrepreneurs vied with one another to come up with successful designs. The basis for a nascent aircraft industry lay in the craft skills of men who made furniture, or bicycles, using fabric, wood, and piano wire. Engines evolved from those designed for early automobiles and motorbikes. The Wright Type A was not as influential as might have been expected. Wheels were generally preferred to skids, and ailerons mostly won out over the Wright's wing-warping, for lateral control. The Blériot XI popularized another important configuration: the monoplane. But monoplanes were not considered sufficiently robust and suitable only for dare-devil sportsmen. The aircraft configuration that predominated in WWI, the tractor biplane, emerged only slowly to predominance. The 1909 Breguet established this configuration, evolving later into the more robust, high-performance Sopwith Tabloid.

CROSS-CHANNEL BLÉRIOT
A Blériot XI, like that used by Louis Blériot to cross the English Channel in 1909, flying over the pier at Nice, France.

Antoinette IV

The elegant, boat-shaped fuselage and advanced, lightweight engine showed the unconventional thinking and artistic background of the Antoinette's designer, Léon Levavasseur. On 19 July 1909, Hubert Latham made the first, unsuccessful attempt to fly the English Channel in this striking French monoplane, named after its sponsor's daughter.

ENGINE 50hp fuel-injected Antoinette	
WINGSPAN 12.2m (40ft)	LENGTH 11.3m (37ft)
TOP SPEED 72kph (45mph)	CREW 1
PASSENGERS None	

Avro Roe IV Triplane

In 1910, the BRITISH pioneer Alliot Verdon Roe created Avro, and this Avro IV, introduced in 1911, was the first successful triplane. The main advantage of triplanes over biplanes was that they could be built with a shorter wingspan to achieve the same lifting power. A shorter wingspan also gave greater manoeuvrability. A replica was flown in the film, *Those Magnificent Men in their Flying Machines.*

ENGINE 35hp Green 4-cylinder water-cooled	
WINGSPAN 9.8m (32ft)	LENGTH 9.2m (30ft)
TOP SPEED Unknown	CREW 1
PASSENGERS None	

Breguet Tractor Biplane

After an abortive attempt in 1907 to build a helicopter, Breguet built his first biplane in 1909, using steel tubing for its main structure. The configuration of the improved 1910 version would establish the sleek, "modern" design for all future tractor-engined biplanes.

ENGINE 50hp 8 Renault V8 cylinder	
WINGSPAN 13m (42ft 9in)	LENGTH 10m (25ft 9in)
TOP SPEED 70kph (44mph)	CREW 1
PASSENGERS None	

Bristol Boxkite

The Bristol Aeroplane Company's first successful product, the Standard Biplane, universally known as the Boxkite, was an improved version of a Henri Farman design. A total of 76 were built for military and civilian training schools around the world.

ENGINE 50hp Gnome rotary	
WINGSPAN 10.5m (34ft 6in)	LENGTH 11.7m (38ft 6in)
TOP SPEED 64kph (40mph)	CREW 1
PASSENGERS None	

Curtiss Model D

Prior to March 1909, Glen H. Curtiss was part of Alexander Graham Bell's Aerial Experiment Association (AEA) which built four increasingly successful biplanes. With the collapse of the AEA, Curtiss began to build what became known as the Curtiss "Pusher", based on the final AEA design. The early aeroplanes were all custom-built to order but, by mid-1911, the level of production and degree of standardisation led to the official designation of Models D (single-seat) and E (two-seat). The specifications below are for the Model D-4, which was advertised for sale at $4,500.

ENGINE 40hp 4-cylinder water-cooled Curtiss	
WINGSPAN 10.2m (33ft 4in)	LENGTH 7.9m (25ft 9in)
TOP SPEED 72kph (45mph)	CREW 1
PASSENGERS None	

Deperdussin Type C

Armand Deperdussin was a colourful entrepreneur who founded the Deperdussin Aviation Company in 1910. The company's aircraft (including the 1911 Type C) were designed by a 30-year-old engineer, Louis Béchereau. A series of advanced monoplanes ensued, which set a number of world speed records. The first in excess of 162kph (100mph) was achieved by Jules Védrines on 22 February 1912. When, in 1913, Deperdussin was jailed for fraud, his company was bought by Louis Blériot, with Béchereau retained as chief designer.

ENGINE	100hp inline		
WINGSPAN	8.8m (28ft 9in)	LENGTH	5.5m (18ft 6in)
TOP SPEED	162kph (100mph)	CREW	1
PASSENGERS	None		

Voisin Henri Farman H.F.1

The Voisin Henri Farman H.F.1 was a biplane with a boxkite arrangement and a forward biplane elevator. With castered wheels, it managed to hop on 30 September 1907 and flew for 30m (98ft) on October 7th. Modified over the course of the next few months, the H.F.1 reappeared with a monoplane forward elevator and covered nacelle, and tail with side curtains. With Henri Farman at the controls, on 13 January 1908 the H.F.1 made the first one-kilometre [0.62-mile] closed-circuit flight in Europe at Issy-les-Moulineaux. For this achievement he won a 50,000-franc prize and much acclaim for the aircraft.

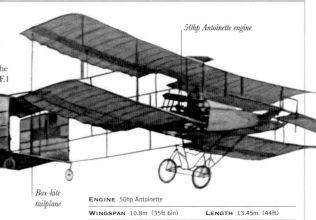

50hp Antoinette engine

Box-kite tailplane

ENGINE	50hp Antoinette		
WINGSPAN	10.8m (35ft 6in)	LENGTH	13.45m (44ft)
TOP SPEED	60kph (38mph)	CREW	1
PASSENGERS	None		

Farman (Maurice) 1912 Type Militaire

Although the Farman brothers became joint aircraft manufacturers in 1912, their designs were separate, with Maurice concentrating, from 1909 onwards, on improvements to the Voisin concept. It was a Maurice Farman design that was entered by Airco into the 1912 Military Aeroplane Competition, leading to a substantial order from the British Army for training and reconnaissance purposes. This type became known as the "Longhorn" after the extended front landing skids.

ENGINE	80hp Renault		
WINGSPAN	15.5m (50ft 11in)	LENGTH	11.5m (37ft 10in)
TOP SPEED	105kph (65mph)	CREW	2
PASSENGERS	None		

Goupy 1909 (No.2)

Although not particularly successful, the Goupy No.2, together with the Breguet Tractor Biplane, was to establish the standard configuration of all future front-engined biplanes. Designed by Ambroise Goupy and Lieutenant A. Calderara, the machine was built in the Bleriot factory and first "hopped" in March 1909. The more successful designs of one of Britain's aviation pioneers, A.V. Roe, were influenced by the Goupy No.2.

ENGINE	25hp REP		
WINGSPAN	6m (19ft 8in)	LENGTH	7m (23ft)
TOP SPEED	97kph (69mph)	CREW	1
PASSENGERS	None		

Santos-Dumont Demoiselle

The Santos-Dumont No.19 Demoiselle ("Dragonfly") made its maiden flight in November 1907, but it was the improved No. 20 version which went into commercial production in 1909. The smallest aeroplane of its time, it was the true ancestor of modern ultra-light aircraft, having a bamboo fuselage, a detachable single wing, and a tricycle landing gear. It was also one of the first "home-built" aeroplanes as Santos-Dumont freely gave the rights to the design. The lightweight 107kg (235lb) Demoiselle – selling for 5,000 francs (£300) – was more popular with the public than with pilots.

ENGINE	25hp 2-cylinder Dutheil-Chalmers		
WINGSPAN	5m (16ft 5in)	LENGTH	6m (19ft 8in)
TOP SPEED	100kph (62mph)	CREW	1
PASSENGERS	None		

Sopwith Tabloid

Although not the first single-seat scout aircraft, the Tabloid was the most successful British racing and scout aircraft prior to WWI. The Tabloid won the Schneider Trophy in 1914 and flew with both the Royal Flying Corps and the Royal Naval Air Service during the war.

ENGINE	80hp Gnome rotary		
WINGSPAN	7.7m (25ft 6in)	LENGTH	6.2m (20ft 4in)
TOP SPEED	148kph (92mph)	CREW	1
PASSENGERS	None		

Wright Type A

At Le Mans, on 8 August 1908, Wilbur Wright flew the Wright Type A in front of an astonished French audience. His obvious and apparently effortless mastery of the air during a brief 1 minute 45 second flight, was to revolutionize European aviation. A month later, on 3 September, Orville Wright began a series of tests at Fort Meyer near Washington DC, which silenced the brothers' American critics. A new era of practical aviation began, as the Type A was built under licence in France, Britain, and Germany, as well as by the Wrights themselves.

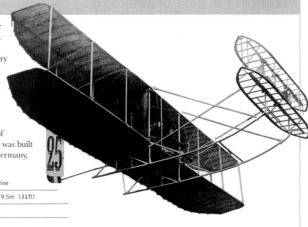

ENGINE	30hp 4-cylinder Wright water-cooled inline		
WINGSPAN	12.5m (41ft)	LENGTH	9.5m (31ft)
TOP SPEED	64kph (40mph)	CREW	1
PASSENGERS	None		

AIRCRAFT GO TO WAR 2

AIRCRAFT FOUND THEIR FIRST PRACTICAL USE as instruments of war. Between 1914 and 1918, aviation matured under the stress of combat. For the first time, aircraft were operated on a daily basis, with all that implies of regular servicing and a focus on reliability. More powerful engines and sturdier airframes brought a great leap forwards in overall performance. There was also a change of scale: aircraft had been manufactured in hundreds before the war; now they were produced in thousands. Militarily, the Great War saw the identification of the different roles aircraft could perform and the design of specialist aircraft to fulfil them – including bombers, which would develop after the war into the first airliners. The air aces who fought in the skies over the Western Front consolidated the tradition of pilots as popular heroes.

ON PATROL
A B.E.2c of the Royal Flying Corps flies over trenches in Belgium during World War I. The B.E.2c entered the War at a time when the future role of aircraft was unclear. Eventually it was used as a reconnaissance plane, a light bomber, and a Home Defence fighter against dirigibles.

"Men were going to die
in the air as they had for
centuries on the ground
and on the seas, by
killing each other. The
conquest of the air was
truly accomplished."

RENÉ CHAMBE
AU TEMPS DES CARABINES

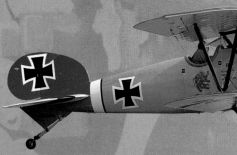

ELEGANT GERMAN FIGHTER
*Over 3,000 Albatros D.Vs were built in 1917–18,
providing a mount for many German aces. But despite its
streamlined monocoque fuselage and elegant lines, the D.V
was outclassed by Allied fighters. In July 1917, a German
pilot wrote: "The D.V is so antiquated and laughably
inferior that we can do nothing with it."*

WORKING FOR THE GENERALS

THE PRINCIPLE ROLE OF AIRCRAFT IN WORLD WAR I WAS TO SUPPORT THE ARMIES IN THE TRENCHES – AIRMEN GAVE THEIR LIVES FOR MEN ON THE GROUND

O**N THE OUTBREAK** of war in Europe in August 1914, aircraft did not seem set to play a serious part in the conflict. The ground forces of the major European armies were counted in millions; the front-line aircraft deployed by all combatants amounted to little over 500 fragile, unarmed monoplanes and biplanes. Caught up in the patriotic fervour of the moment, civilian pilots rushing to join up included such well-known stars of peacetime aviation as Roland Garros and Jules Védrines. But the military establishments initially had little use for the skills of the daredevil sportsman-aviators who had so recently enthralled the public.

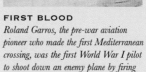

FIRST BLOOD
Roland Garros, the pre-war aviation pioneer who made the first Mediterranean crossing, was the first World War I pilot to shoot down an enemy plane by firing through his propeller blades (see page 73).

Aerial chauffeurs
Army pilots were essentially aerial chauffeurs. Their job was to ferry an observer – sometimes a senior officer – over the countryside to report on the movement of enemy troops. In the first months of the war there was plenty of movement to observe, with rapid advances, encirclements, and desperate retreats. In the west, the German forces overran Belgium and advanced on Paris, while in the east the Russians marched menacingly into East Prussia. Flying mostly from improvised airstrips (any unploughed field) close to the ever-shifting front line, pilots and observers roamed the thinly populated skies, seeking out bodies of enemy troops and recording their size, location, and direction of march in scribbled notes and hastily sketched maps. It was no easy task to locate the enemy

in unfamiliar territory while trying to avoid becoming hopelessly lost and coping with unpredictable weather. Low cloud hampered observation and the sheer flimsiness of the machines led to frequent accidents and forced landings.

The appearance of aircraft was greeted with volleys of rifle fire from friend and foe alike. And after undergoing these hazards, airmen often saw their reports simply disregarded by the crustier generals who distrusted information from such a novel source.

A vital role
Nonetheless, aerial reconnaissance made a decisive contribution to both fronts. In the east, the Russians failed to make effective use of the few aircraft they possessed, while the Germans employed their Taubes to crucial advantage. Ranging over the forests and lakes of East Prussia, German aviators located the advancing Russian armies, giving the high command time to move reinforcements to the front. When battle was joined at Tannenberg, information from aerial observers let the numerically inferior Germans concentrate their forces in the right place at the right time to carry off an epic victory.

In the west, French and British aviators were caught up in a rapid retreat across France as the grey columns of the German army swept towards Paris. Anticipating triumph, on 29 August 1914, a German pilot flew round the Eiffel Tower and dropped a single bomb on the city. But on 3 September, French aircraft assigned to the defence of their capital, reported that the enemy's armies had turned away from

A NEW KIND OF WAR
A squadron of French cavalry watches a biplane passing overhead in 1915. Aircraft largely replaced the cavalry in its traditional role of scouting, since they could cover more ground more quickly. Many of the best World War I pilots were men who transferred from the cavalry, sensing its irrelevance on the Western Front.

AERIAL VIEW

This photograph (below right) was taken on a reconnaissance mission over France. Once trench warfare set in on the Western Front at the end of 1914, army commanders became totally dependent on aerial reconnaissance for information on what was happening on the other side of no-man's land. "Photo-recce" was a risky business for airmen, requiring slow, straight-and-level flights over enemy positions, repeated many times to build up a complete picture.

Eyepiece

Lever for focusing

HAND-HELD BOX CAMERA

A Royal Flying Corps observer demonstrates a Thornton-Pickard "A" Type photo-reconnaissance camera (above). Cameras were initially hand-held, using straps or handles (right). Later they were mounted on the aircraft itself.

Handle to hold camera

Paris to the east. This information enabled General Joseph Gallieni, who was in charge of the defence of the city, to launch an attack on the exposed German flank on the Marne that turned the tide of the war. By the end of the autumn, the Germans had been driven back towards the Belgian frontier and the war of movement on the Western Front had come to an end. The armies dug in along a line from the English Channel to Switzerland, where they would stay for the next three and a half years.

During the long agony of trench warfare, with its monstrous artillery barrages and its massed infantry offensives, in which hundreds of thousands of lives were sacrificed for pitifully small territorial gains, aviation dutifully played the role assigned to it by army commanders. As far as they were concerned, the function of aircraft was to carry out reconnaissance and the closely related role of artillery spotting, as well as to inflict damage on the enemy's soldiers and material through tactical bombing and ground attack.

Building a picture

From 1915, photography replaced sketches and notes as a technique for aerial reconnaissance. Aircraft with unwieldy box cameras were dispatched day after day over the front to build up an exact picture of the enemy's trench systems and gun emplacements. Initially the cameras were like those found in photographers' studios, with large glass plates that had to be changed by hand after every shot. This was ghastly work for observers with freezing fingers operating in the gale of the aircraft's slipstream. Later, cameras with a mechanically operated plate change made the observer's job more practical, but photo-reconnaissance remained as hazardous as it was unglamorous. An aircraft held steady and straight for photography presented an inviting target

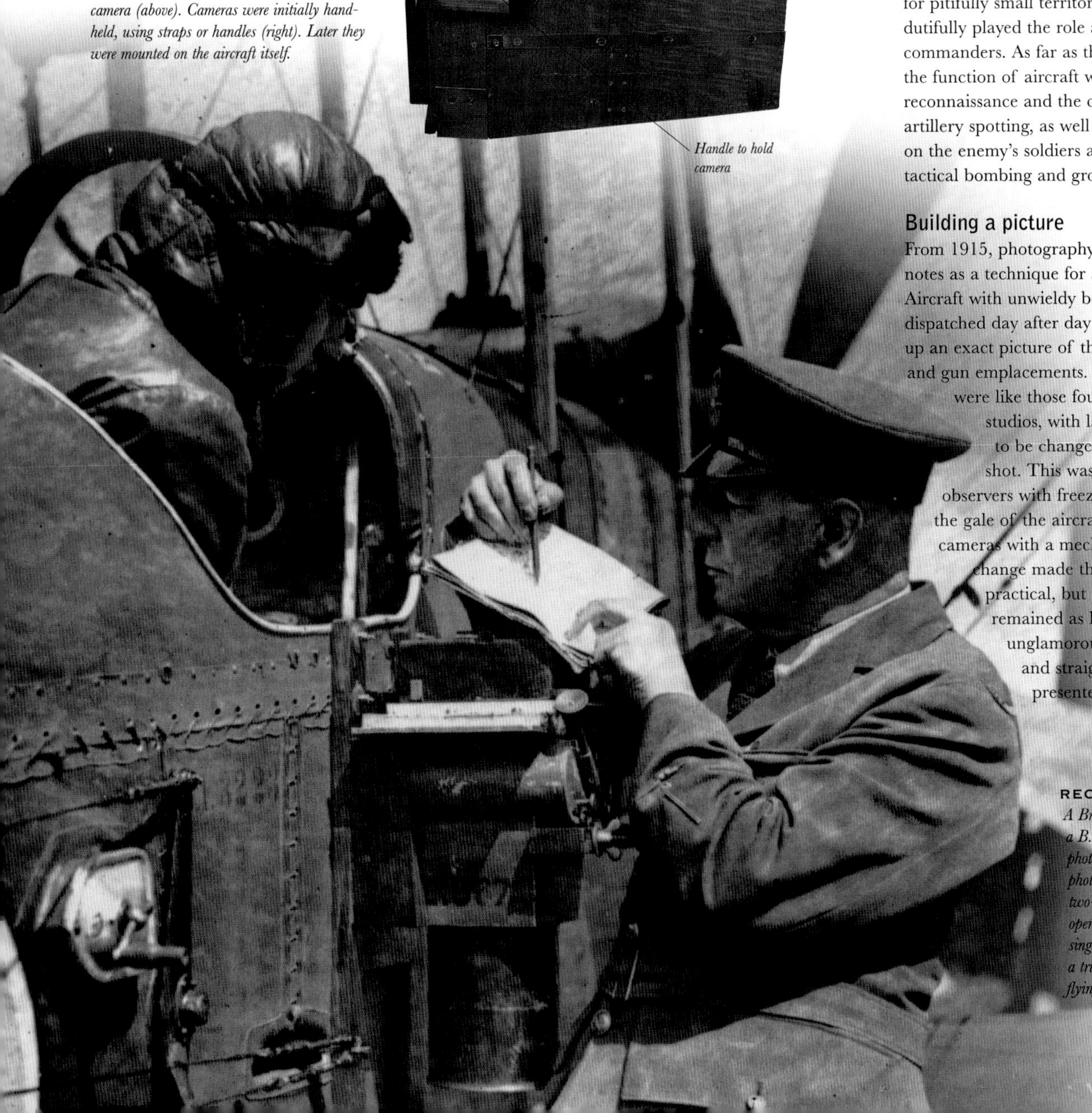

RECONNAISSANCE BRIEF

A British officer briefs the pilot of a B.E.12 on the areas to be photographed on his mission. Most photo-reconnaissance was entrusted to two-seater aircraft, with an observer operating the camera. The pilot of a single-seater B.E.12 would have had a tricky time taking pictures while flying in hostile airspace.

WWI RECONNAISSANCE AIRCRAFT

AT THE START OF WWI, the German General Staff stated that "the duty of the aviator is to see, not to fight". Given that reconnaissance and artillery spotting were seen as the central purpose of military aviation, it is perhaps surprising that the aircraft dedicated to this role often had such poor performance – from frail monoplanes, such as the Taube, used at the outset of the war, to the many unexciting biplanes that trundled over the trenches. Stability was regarded as the chief virtue of reconnaissance machines, in order to provide a platform for observation and photography. But this meant they were slow and clumsy to manoeuvre, making them easy prey for enemy fighters and vulnerable to ground fire. Some, such as the Renault A.R. and the RAF R.E.8, had a particularly poor reputation. After a spate of accidents, the R.E.8 was temporarily withdrawn from service, but investigations revealed inadequate training rather than poor design to be the culprit. Late in WWI, progress came on the Allied side with the introduction of the Bristol F.2B fighter for reconnaissance. Improved cameras allowed the Germans to initiate high-altitude photo-reconnaissance outside the range of most fighter aircraft.

VULNERABLE DOVE
Already outdated by the outbreak of WWI, the Rumpler Taube only had a top speed of 97kph (60mph) and was extremely vulnerable in the air.

Caudron Type G.IV

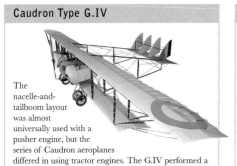

The nacelle-and-tailboom layout was almost universally used with a pusher engine, but the series of Caudron aeroplanes differed in using tractor engines. The G.IV performed a variety of roles in addition to that of reconnaissance, including bomber and trainer.

ENGINE 2 x 80hp Le Rhone 9-cylinder rotary

WINGSPAN 17.2m (56ft 5in)	**LENGTH** 7.2m (23ft 8in)
TOP SPEED 132kph (82mph)	**CREW** 2

ARMAMENT 2 x 7mm machine gun

Farman H.F.20

With the outbreak of war, the Farman brothers' factory received large orders for the Maurice-designed M.F.7 and 11 and for Henri's H.F.20. Its primary duty in France was in an observation role until mid-1915, when it became a trainer.

ENGINE Gnome 7A 7-cylinder rotary

WINGSPAN 15.5m (44ft 10in)	**LENGTH** 8.8m (27ft 9in)
TOP SPEED 100kph (62mph)	**CREW** 2

ARMAMENT None

RAF B.E.2c

The B.E.2 was an observation machine – slow, stable, and unarmed. It was immediately obvious that some form of defence was needed, so the B.E.2c, introduced in October 1914, was fitted with a machine gun in the observer's cockpit.

ENGINE 90hp Royal Aircraft Factory 1a V-8 air-cooled

WINGSPAN 11.29m (37ft)	**LENGTH** 8.3m (27ft 3in)
TOP SPEED 120kph (75mph)	**CREW** 2

ARMAMENT 1 x .303in Lewis machine gun in observer's cockpit (can carry up to four); 104kg (230lb) bombload (when flown solo)

RAF R.E.8

The R.E.8, known to its crews as the "Harry Tate" after a musical-hall comedian of the day, replaced the outmoded B.E.2c. Entering service in 1916, it became the most widely used reconnaissance and artillery spotting aircraft operated by the R.F.C. during WWI.

ENGINE 140hp Royal Aircraft Factory 4a air-cooled V-12

WINGSPAN 13m (42ft 7in)	**LENGTH** 8.5m (27ft 10in)
TOP SPEED 164kph (102mph)	**CREW** 2

ARMAMENT 1 x .303in Vickers machine gun; 1 x .303in Lewis machine gun in observer's cockpit; 101kg (224lb) bombload

Rumpler Taube

Instantly recognizable by its sweptback, birdlike wing tips, which warped for flight control, the Austrian Taube ("Dove") had its origins in the Etrich-Wels glider of 1907. Manufacture was initially licensed to Rumpler, and the design is now generally associated with that company. Although a pre-war design, its initial success as a reconnaissance machine on the Western Front led to it being built by Albatross, Gotha, and D.F.W.

ENGINE 100hp Mercedes D1, 6-cylinder liquid-cooled inline

WINGSPAN 14.5m (47ft 7in)	**LENGTH** 10m (32ft 10in)
TOP SPEED 97kph (60mph)	**CREW** 2

ARMAMENT None

EINDECKER IN ACTION
This rare aerial photograph captures a Fokker Eindecker in action over France. The aircraft had a metal tubular frame that gave it strength in a high-speed dive, but in other ways it was not of an advanced design. It was one of the last important aircraft to use wing-warping, rather than ailerons, for lateral control.

for ground fire, and the underpowered obsolescent aircraft usually thought suitable for reconnaissance were easy prey for enemy fighters. But, at the cost of heavy loss of life, comprehensive photomontages of trench systems were built up and used for selecting targets for the artillery.

When the heavy guns opened up, again the aircraft came into play. The gunners needed observers to tell them where the shells were landing so they could correct their range and direction. In decent weather, airborne observers could usually see where a shot was falling but there was no efficient way of communicating this to the ground until 1916, when some new aircraft were capable of carrying radio transmitters. Combining the use of radio with the "clock system" – a code

of number and letter co-ordinates that identified where a shell had fallen in relation to the target – created a reasonably efficient spotting technique. At the static battle of the Somme in 1916, the British found that no amount of aerial observation could make the artillery barrages actually work against troops in deep fortifications. In the mobile battles of 1918, however, thousands of airmen sacrificed their lives in an artillery-bombardment strategy that,

alone, would have failed, but employed with other arms and methods was very effective.

Towards the end of the war the Germans at last introduced a reconnaissance aircraft that gave its crew a reasonable measure of safety. The Rumpler C.VII could fly at 6,000m (20,000ft) and had an automatic camera that took a series of pictures when triggered. The downside was the lack of protection for high-altitude flight. Rumpler crews suffered from freezing cold (they were in open cockpits), lack of oxygen, and the bends.

Arming the aeroplanes

For the generals, the main reason for putting guns in aircraft was to protect their own reconnaissance aircraft and shoot down the enemy's. But the initial impetus towards arming aircraft came from pilots and observers who simply wanted to "have a go" at the opposition. Firing pistols and carbines at passing aircraft had limited effect, while attempts at dropping grenades on them from above were a total failure. Machine guns were what were needed. But carrying such a weapon was a considerable burden for the lightweight, underpowered aircraft of 1914. It was also hazardous: there was a serious risk of blowing bits off your own

FACING THE FLAK

GERMAN ANTI-AIRCRAFT FIRE, known to the Royal Flying Corps as "Archie", took a heavy toll on reconnaissance aircraft. Pilot Lt William Read wrote: "I wonder how long my nerves will stand this almost daily bombardment by 'Archie'... I would not mind quite so much if I were in a machine that was fast and that would climb a little more willingly. Today... some of the shells burst much too near and I could hear the pieces of shell whistling past... Well, I suppose the end will be pretty sharp and quick." Most anti-aircraft commanders believed shrapnel stood the best chance of bringing down an aircraft. Others preferred high explosive or incendiary shells.

GOOD ODDS
German anti-aircraft gunners like these had a fair chance of hitting observation aircraft travelling below 1,500m (5,000ft).

machine, with its array of struts and wires. The first recorded aerial victory is credited to a French aviator. On 5 October 1914, observer Louis Quénault shot down an Aviatik with a Hotchkiss machine gun mounted on a Voisin 8 – a pusher aircraft (with the propeller at the rear). Affording a clear field of fire to the front, pushers were one option for air-combat machines. They proved especially attractive to the British, who introduced the Vickers "Gun bus" in 1915 and the F.E.2 and single-seat D.H.2 pushers the following year. But while pusher machines were by no means ineffective, tractor machines (propeller at the front) were faster and more manoeuvrable.

What the more skilful and adventurous pilots instinctively yearned for was a gun they could aim simply by pointing their aircraft at the target. Before the war, French and German designers had discovered that it was feasible to create an interrupter gear that would pause a machine gun each time a propeller blade was in its line of fire. Raymond Saulnier, designer of the Morane-Saulnier monoplane, was one of those who experimented with interrupters, but he had not been able to make one work in practice. So it was Dutch designer Anthony Fokker who fitted the first effective interrupter gear to one of his Eindecker monoplanes. The Germans went on to use guns firing through the propeller arc on all their fighters for the rest of the war. Interrupter gears and other forms of synchronizing mechanism tended to reduce the rate of fire of the machine gun, but in later German aircraft, such as the Albatros D.V and Fokker D.VII, the use of twin guns compensated for this drawback.

The Allies' first effective riposte to the Eindecker's interrupter gear was to mount a machine gun on the upper wing of a biplane so that it fired over the top of the propeller. Even after the Allies developed their own synchronizing mechanisms to allow firing through the propeller arc, they remained attached to the concept of the wing-mounted gun. Successful solo fighters such as the Nieuport 17 and the S.E.5a were usually fitted with both. In tandem with new armaments, new tactics were also being developed.

Early in the war individual fighters prowled the skies as lone hunters in search of unsuspecting enemy aircraft. By 1916, fighter aircraft were being grouped in squadrons as tactics were developed for fighting in formation. During the titanic battles of Verdun and the Somme, Allied and German airmen fought for air superiority; losses on both sides were heavy in an aerial combat that mirrored the war of attrition on the ground. Numerically inferior,

INVENTING THE FIGHTER PILOT

ON 1 APRIL 1915, FRENCH PILOT Roland Garros positioned his Morane-Saulnier Parasol monoplane behind the tail of a German observation aircraft and fired a burst from his machine gun through the propeller arc. As the German machine plummeted to earth, Garros could claim to have become the first solo fighter pilot. The secret of his success – the ability to fire forwards through the propeller – had been achieved by fitting metal plates to deflect any rounds that struck the blades.

Garros had shot down three aircraft by 18 April, when engine failure forced him to land behind German lines. His exploits had been highly publicized and the Germans rushed to examine his downed plane. Dutch designer Anthony Fokker was called to Berlin and told to imitate the metal deflectors. Instead, he fitted one of his Eindecker monoplanes (an unarmed reconnaissance aircraft) with an 08/15 Maxim ("Spandau") machine gun and an interrupter gear copied from a pre-war German patent design. The interrupter allowed the pilot to fire through his propeller with much less risk to the machine and himself.

German commanders were slow to realize that an important new weapon had been placed in their hands. Eindeckers were introduced in small numbers and were initially spread out in ones and twos, supporting reconnaissance units, which limited their effectiveness. Some German pilots instantly recognized the potential of the new machine. Through the winter of 1915–16, using the simple tactic of swooping down on their enemy from behind in a steep dive, the Eindecker pilots shot down unprecedented numbers of Allied aircraft. The British called it the "Fokker Scourge". Yet the Eindecker was in fact seriously flawed. It was underpowered, not especially nimble, and had structural

Swivel

Cord

Ammunition drum

Foresight

7.62-mm barrel fires 600rpm

Mounting pivot

WING-MOUNTED GUN
A Lewis Aerial Gun mounted on the top wing allowed pilots to shoot in the line of flight, over the propeller. The gun was fired by pulling on a cord that led down into the cockpit. The pilot's problem with this arrangement was changing the ammunition drum in flight, a perilous moment requiring him to take his hands off the controls.

weaknesses. With careful handling it was effective, but it could be a death-trap even for an experienced pilot if he put too much stress on the airframe.

The Allies responded with their own solo fighter aircraft. In July 1915, France introduced the small Nieuport 11, affectionately known as the "Bébé" ("Baby"). Originally designed for racing, this light biplane was fast and extremely manoeuvrable. Its only major weakness was the single-spar lower wing, which allowed the wings to twist in a dive. Although it lacked a synchronized machine gun, it carried a wing-mounted Lewis machine gun, and virtually drove the Eindeckers from the skies. The scourge had finally been scourged.

FOKKER SCHWERIN

BOUNCING BULLETS
A Morane monoplane (far left) displays metal plates on its propeller blades. When its machine gun fired, about one in ten bullets hit the deflectors, bouncing off. This poster shows a German ace in an Eindecker firing through his propeller. The image captures the essence of the solo fighter pilot – the welding of man and aircraft into a single fighting machine.

the German aircraft tended to stay on their own side of the trenches and concentrate their resources in ever larger units capable of winning local air superiority on crucial sectors of the front.

Rapid expansion

A battle for production intensified in step with the struggle at the front. The growth in volume of engine and airframe output was spectacular. Quite early in the war, military contracts allowed small manufacturers to grow into major industrial concerns. The French company Nieuport was not untypical in seeing its turnover grow from 285,000 francs in 1914 to 26.4 million francs in 1916. New players entered the aero-engine and aircraft industries, notably automobile manufacturers such as Renault and Fiat. Expansion was most rapid in Britain, which had entered the war with a weak aircraft industry almost entirely dependent on imported French engines. By the end of the war, Britain had the largest aircraft industry in the

> "Aviation has assumed a capital importance... It is necessary to be master of the air."
>
> GENERAL HENRI PÉTAIN, MAY 1917

world, employing an estimated 270,000 workers.

Plagued by shortages of skilled labour and of vital materials, Germany critically lost out in the battle for volume production. In 1917, the Germans undershot a production target of 1,000 aircraft a month – at a time when the British and French between them were manufacturing about 30,000 aircraft a year.

Quantity was not, of course, the same as quality. Delays and bureaucratic incompetence sometimes led to aircraft being manufactured that were obsolescent before they were ever flown – the notorious R.E.8s and Renault ARs delivered to the front in 1917 were cases in point. The twin aims of maximizing output and improving aircraft performance often proved contradictory. Aircraft that had been good in their day were kept too long in production so that the demand for numbers could be met.

Necessity breeds invention

But there was also a built-in conservatism through the need to exploit existing resources and tried-and-tested techniques. Throughout the war the vast majority of Allied aircraft remained strut-and-wire biplanes, with a fabric skin stretched over a wooden frame. They achieved improved performance largely through the use of more powerful and reliable engines. The Germans were more innovative in their use of materials, partly because of shortages of good quality wood and of skilled workers required to build wooden-frame aircraft. For their fabric-skinned machines, the Germans mostly adopted welded steel-tube

POWERING UP

THE POWER OF ENGINES USED IN combat aircraft grew from around 80hp at the start of the war to a maximum of 400hp by 1918. The two main families of World War I power plant were rotary and in-line water-cooled engines. Rotary engines were lighter and more compact, but ran into problems when required to deliver over 150hp. With their cylinders whirling around a fixed crankshaft they created a powerful gyroscopic effect that made an aircraft tricky to fly, but they worked well on manoeuvrable dogfighters such as the Sopwith Camel and Fokker Triplane. In-line engines powered stronger, faster aircraft.

The Allies won the battle for engine development because they had a wider range of suppliers, mostly French. The Americans produced the most powerful engine of the war – the 12-cylinder, 400hp Liberty engine – in 1918.

Cylinder

Induction pipe

Connecting rod (con-rod)

Spark plug

LE RHÔNE 9B ROTARY ENGINE
Many of the rotary engines used by the Allies were produced by the Gnome and Le Rhône companies. Despite some disturbing characteristics, including spraying the pilot with castor oil, they were light and powerful. The best tribute to their quality is that the Germans often fitted captured Allied rotary engines in place of their own.

Cooling jacket around cylinder

Exhaust

Laminated wooden propeller

Brass-sheathed leading edge

"HISSO" IN-LINE ENGINE
The in-line Hispano-Suiza V8 is often regarded as the outstanding engine of World War I. It was powerful, compact, durable, and light for its size. The engine was used in aircraft like the SPAD XIII and the S.E.5.

SUPERIOR S.E.5A

When the British-built S.E.5a entered the war in 1917 it proved itself superior to all its German opponents. Faster than the Sopwith Camel and easier to fly, it developed a formidable reputation in the hands of celebrated aces like Edward Mannock. Designed by H.P. Folland at the Royal Aircraft Factory in Farnborough, it combined a strong airframe with good, solid performance. Over 5,000 were produced.

Wing-mounted Lewis gun

Faired headrest behind cockpit

Elevator

Metal engine cowling

Laminated wooden propeller

SKILLED HANDS

Workers construct S.E.5s at the Royal Aircraft Factory, Farnborough, Britain. Despite mass production, making aircraft remained a labour-intensive job, as it had been in the artisan workshops of the pre-war era. Processes such as attaching the wire rigging needed a high level of skill and were very time-consuming.

frames, which were strong, light, and easier to make. On aircraft such as the Albatros D-series fighters, they took plywood and wrapped it in strips around an inner framework to create wooden-skinned monocoque fuselages.

Although the Albatros fighters were fine machines, their evolution showed the increasing difficulties the Germans ran into, trapped between the demands of quantity and quality. Aiming to achieve its 1,000-aircraft-a-month target for 1917, Germany opted for mass production of the Albatros D.V, a variant of the D.III. But by the summer of that year, the Albatroses were being outclassed by a new generation of Allied fighters, especially the S.E.5 and the SPAD XIII. Yet dedicated to meeting their production targets, German factories went on churning out Albatroses into 1918.

The skies over the Western Front were an essentially Darwinian environment in which aircraft constantly evolved to survive – sometimes by straight imitation. When the French Nieuport 17 scout threatened the predominance of the German Albatros D.II fighter in 1916, for example, the Germans simply copied the Nieuport's single-spar lower wing – and with it, its tendency to twist and fail – to create the Albatros D.III. Similarly, after the Sopwith Triplane flown by Britain's Royal Naval Air Service shocked German fliers with its agility in 1917, Fokker copied it to create its own Dr.I triplane as a mount for Baron Manfred von Richthofen.

The entry of the United States into the war in April 1917 inspired Germany with a desperate urge to achieve victory before the overwhelming might of American manpower and industry could be brought to bear. The German high command planned an ultimate offensive on the Western Front for the spring of 1918. Evaluation trials were held to find the aircraft that would win the war. Both Claudius Dornier and Hugo Junkers put forward radical designs that looked to the future – aircraft with metal skins and, in Junkers' case, a cantilever wing that required no external struts or bracing. But again conservatism prevailed. The aircraft adopted for mass production were the Fokker D.VII and Pfalz D.XII, superb fighting machines but representing only a limited degree of innovation.

Racing to catch up

The expectation that America's entry into the conflict would swiftly swing the air war in the Allies' favour failed to take account of the degree to which its aviation had fallen behind that of Europe. The sole viable combat aircraft under

Bristol F.2B Fighter

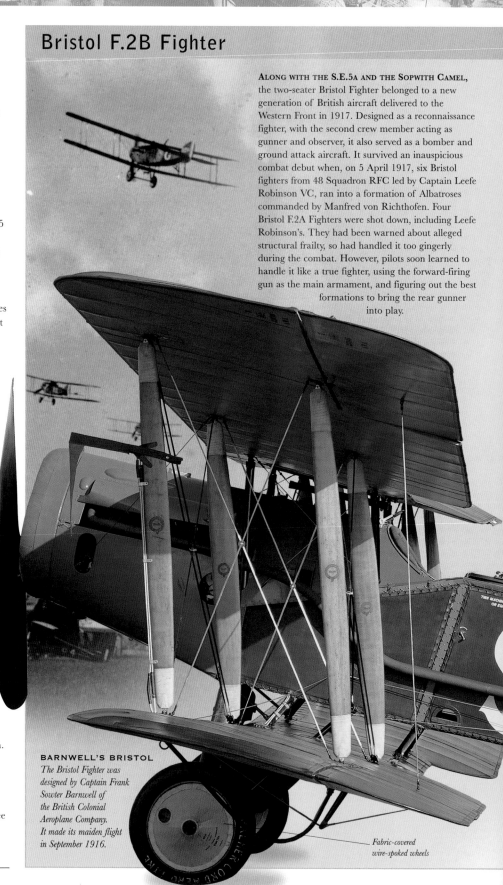

ALONG WITH THE S.E.5A AND THE SOPWITH CAMEL, the two-seater Bristol Fighter belonged to a new generation of British aircraft delivered to the Western Front in 1917. Designed as a reconnaissance fighter, with the second crew member acting as gunner and observer, it also served as a bomber and ground attack aircraft. It survived an inauspicious combat debut when, on 5 April 1917, six Bristol fighters from 48 Squadron RFC led by Captain Leefe Robinson VC, ran into a formation of Albatroses commanded by Manfred von Richthofen. Four Bristol F.2A Fighters were shot down, including Leefe Robinson's. They had been warned about alleged structural frailty, so had handled it too gingerly during the combat. However, pilots soon learned to handle it like a true fighter, using the forward-firing gun as the main armament, and figuring out the best formations to bring the rear gunner into play.

BARNWELL'S BRISTOL
The Bristol Fighter was designed by Captain Frank Sowter Barnwell of the British Colonial Aeroplane Company. It made its maiden flight in September 1916.

Fabric-covered
wire-spoked wheels

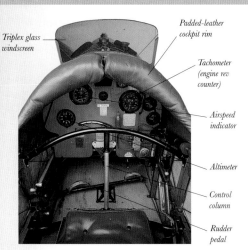

Triplex glass windscreen

Padded-leather cockpit rim

Tachometer (engine rev counter)

Airspeed indicator

Altimeter

Control column

Rudder pedal

SPARSE INSTRUMENTATION
The simplicity of the Bristol Fighter cockpit was typical of a World War I aircraft. The rev counter was the most useful instrument, but generally pilots flew more by feel than by dials.

"If you cannot fly a Bristol Fighter you must resign yourself to remaining an indifferent conductor of B.E.s, F.E.s, D.H.6s, for you will never be any kind of pilot."

MAJOR G. ALLEN
ON HOW PILOTS SHOULD FLY A
BRISTOL FIGHTER (1918)

A POPULAR MOUNT
The crews of two Bristol Fighters discuss tactics for their next mission. By October 1918 there were more than 1,500 F.2Bs in service.

Specifications

ENGINE	275hp Rolls-Royce Falcon III water-cooled V12
WINGSPAN	12m (39ft 3in)
LENGTH	7.9m (25ft 10in)
HEIGHT	3m (9ft 9in)
TOP SPEED	198kph (123mph)
ARMAMENT	2 x .303in machine guns; 12 x 11kg (25lb) bombs

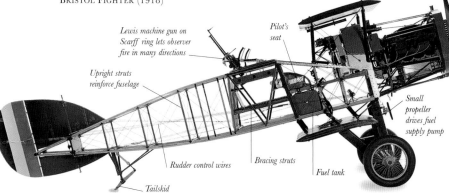

Lewis machine gun on Scarff ring lets observer fire in many directions

Upright struts reinforce fuselage

Pilot's seat

Small propeller drives fuel supply pump

Rudder control wires

Bracing struts

Fuel tank

Tailskid

STRIPPED SIDE VIEW

Oval radiator

Aileron

Struts linking upper and lower wing

FRONT VIEW

Bracing wires linking upper and lower wing

Axle

Two-spar fabric-covered wooden wings

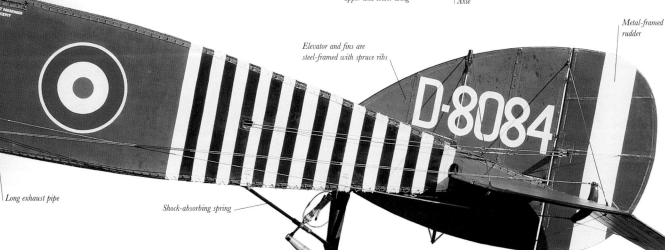

Rear-mounted Lewis machine gun

Metal-framed rudder

Elevator and fins are steel-framed with spruce ribs

Long exhaust pipe

Shock-absorbing spring

D-8084

production in the United States in 1917 was the Curtiss flying boat. The US Army had about 50 obsolete combat aircraft. But if America lacked an aircraft industry, it did have an automobile industry that was using assembly-line techniques to transform output. Confident that if America could make cars it could make aeroplanes, Congress enthusiastically voted funds for the mass manufacture of aircraft. Optimistic plans saw the United States equipping not only its own air service but those of Britain, France, and Italy. That is not how it worked out. Assembly-line methods proved hard to apply and differences between European and American standard measurements posed retooling problems. By the war's end, US factories had delivered only around 1,400 combat aircraft, mostly versions of the D.H.4 bomber.

Similar frustrations were experienced in engine manufacture. Seven automobile manufacturers were contracted to make the Liberty aero-engine, but an original eight-cylinder design was declared obsolete before it went into production and had to be replaced by a heavier 12-cylinder model. Delays meant that only 1,300 Liberty engines had been delivered by June 1918. At 400hp they were the most powerful engines in the war – in fact far too powerful for existing airframes. American pilots who took delivery

MODERN WARFARE

During the Second Battle of the Marne in July 1918, a German biplane patrols the trenches as a British tank looms into view. The ubiquity of aircraft and tanks at this time was a foretaste of the type of mobile warfare that would predominate in World War II.

of Liberty-powered D.H.4s complained that if they ran the engine at full throttle it would shake the aircraft to pieces.

The final push

Germany launched its final offensive of the war, the *Kaiserschlacht*, on 21 March 1918. Using ground-attack aircraft in support of small groups of "shock troops", the Germans punched holes in the Allied lines, ending the stalemate of trench warfare. In a repeat of 1914, Allied troops fell back towards Paris. The battle in the air was every bit as intense as on the ground. The Germans even threw their heavy Gotha bombers into the fray, attacking ammunition dumps behind the lines. The arrival of the superb Fokker D.VII at the end of April meant that German pilots had their best fighter of the entire war.

Yet with victory apparently in sight, the German war effort began to crumble. By June, flying missions were being cut back due to lack of fuel. While the Allies were mostly able to replace their losses of aircraft and pilots, the Germans could not. American fliers were arriving in their thousands and their number would continue to increase. In July 1918 the Allies went on the offensive. In the air,

"What is the point of shooting down five out of 50 machines? The other 45 will… bomb as much as they want. The enemy's material superiority was… dooming us to failure."

LIEUTENANT RUDOLF STARK
GERMAN PILOT DESCRIBING THE BATTLES OF 1918

they swamped the Germans with their sheer weight of numbers. France had grouped together aerial divisions of 700 bombers and fighters. By the time of the St Mihiel offensive in September, US General William "Billy" Mitchell commanded a force of 1,500 French, British, and American aircraft in his sector of the front. Exhausted, short of fuel and spares, and with their airfields increasingly exposed to air attack, the German airmen never gave up the fight and, indeed, inflicted heavier losses than they suffered. But it was a struggle driven by despair rather than hope.

World War I ended with the armistice of 11 November 1918; the Germans surrendered without truly accepting defeat. The war had cost an estimated nine million lives. Of those, probably some 15,000 were airmen. This may seem a relatively small figure in absolute terms, but individually a pilot probably stood no better chance of surviving the war than an infantryman in the trenches. Aviation had come of age in a war of mass slaughter driven by industrial technology. Although many people saw airmen soaring above that impersonal butchery, they had fully played their part as victims and killers. Flight had lost its innocence.

FALLEN WARRIOR

Allied soldiers examine the wreckage of a German aircraft after one of the last battles of the war. Although the German pilots were often better trained than their opponents and had at least marginally superior aircraft, they could not cope with the number of Allied machines filling the skies.

BOMBS FROM THE AIR

DROPPING EXPLOSIVES ON PEOPLE on the ground was one of the first conceived uses for aircraft. During World War I this primitive urge was refined into strategic bombing (such as factories and cities); tactical bombing of targets behind the front line (including railroads or supply dumps); and front-line attack. The earliest bombs were artillery shells dropped over the side of aircraft by hand.

By 1917, sturdier, specialized bomber aircraft had appeared with bomb racks, bomb sights, and release systems. Raids were carried out by day when air superiority allowed it, or by night with much reduced accuracy. By the final stages of the war, aircraft roamed the front, bombing bridges and airfields, and strafing troops and trucks, proving beyond a doubt the vital importance of air superiority.

FLECHETTE

Feathered flight helps guide 12-cm (5-in) dart

MARTEN HALE BOMB

Shell contains 2kg (4½lb) of explosive

Propeller guides bomb

INCENDIARY BOMB

Perforated "carcass" casing helps bomb catch fire on impact

BOMBS AND ARROWS

The bombs of both sides at the start of the war were small and basic. The British also used flechettes – dart-like weapons released in boxes of 500 from around 1,500m (5,000ft) – but they were ineffective against infantry and cavalry concentrations.

NIGHT BOMBARDMENT

This painting shows a night bombardment by a Voisin biplane. Night operations made bombing less accurate, but in daylight, bombers were easy targets for enemy fighters.

KNIGHTS OF THE AIR

WHILE BRUTAL TRENCH WARFARE BOGGED DOWN THE ARMIES, THE MYTH OF FIGHTER PILOTS AS "KNIGHTS OF THE AIR" SUPPLIED A PUBLIC NEED FOR HEROES

"I hate to shoot a Hun down without him seeing me, for although this method is in accordance with my doctrine, it is against what little sporting instincts I have left."

JAMES McCUDDEN VC, 1917

AMERICA'S TOP ACE
A former racing driver and military chauffeur, Captain Eddie Rickenbacker was the most successful American fighter pilot of World War I, with 26 kills.

WORLD WAR I was the first total war, in which the entire human resources of industrialized societies were mobilized in the drive for victory. The mass slaughter in the trenches put an immense strain upon social solidarity and morale. Even with deep reserves of patriotism to draw upon, political and military leaders recognized that popular support for the war might evaporate. Fighter pilots offered a welcome supply of heroes to be used as a focus for patriotic enthusiasm.

The French and Germans created a formalized system for allotting "ace" status to a flier based on a certain number of confirmed kills – a number that had to be raised in the course of the war as air combat intensified. The British high command never formally accepted the existence of aces but awarded a few of the highest scoring pilots the Victoria Cross, their most coveted military decoration. Aces were turned into celebrities by patriotic publicity machines. Their faces decorated the front pages of newspapers; they were

filmed for movie newsreels; they were showered with honours; and their funerals were occasions of national mourning. When French ace Georges Guynemer was killed in 1917, his name was inscribed on the walls of the Panthéon, alongside France's greatest philosophers and poets.

Soaring above the trenches
The propaganda worked because it fed a widely felt nostalgia for a cleaner, nobler form of warfare. The idealism felt by some at the start of the war – poet Rupert Brooke described entering the conflict as "like swimmers into cleanness leaping" – found no satisfaction in the squalid attrition of the trenches. Airmen seemed, morally as well as literally, to soar above the cratered mud of Flanders. This was war as it was meant to be – an opportunity to demonstrate the

RED BARON'S MOUNT
Anthony Fokker based his highly manoeuvrable 1917 Dr.I triplane, of which the above is a modern replica, on the Sopwith Triplane, whose three wings gave it extra lift and agility. It gained notoriety as the mount of Baron Manfred von Richthofen, who claimed 19 of his 80 victories in it.

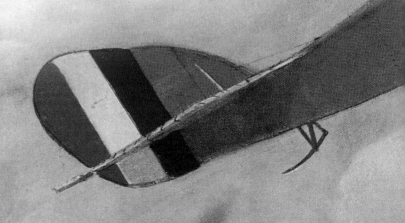

DOGFIGHT OVER THE WESTERN FRONT
This painting shows a de Havilland D.H.4. Fighter taking on an Albatros D.II biplane, possibly from Baron von Richthofen's "Flying Circus". In a tradition dating from 1917, all pilots from the Flying Circus displayed some red colouring to show their solidarity with their leader, whose triplane was completely red.

GERMANY'S FIRST ACES

MAX IMMELMANN (1890–1916) and Oswald Boelcke (1891–1916) were members of Flight Section 62 stationed at Douai, France, in August 1915. They were among the first pilots issued with the Fokker Eindecker, and they used it to deadly effect. As their victories mounted, they were trumpeted in German propaganda as examples of fearless devotion to the Fatherland. They were courted by the aristocracy, lauded by journalists, and deluged with fan mail.

Immelmann, a fitness fanatic and teetotaller, was killed in June 1916, probably by a faulty interrupter gear in his Eindecker – causing him to blow off his own propeller. His legacy was a tricky manoeuvre, the Immelmann turn, which consisted of pulling upwards out of an attacking dive, performing a half-roll, and dropping on the enemy from above a second time.

PENETRATING STARE
People who met Boelcke were struck by the intensity of his gaze. As squadron leader, he trained many of Germany's greatest fliers.

THE EAGLE OF LILLE
Max Immelmann's death was commemorated in a variety of ways, including this specially composed march for the piano dedicated to "the memory of our flying hero".

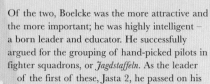

Of the two, Boelcke was the more attractive and the more important; he was highly intelligent – a born leader and educator. He successfully argued for the grouping of hand-picked pilots in fighter squadrons, or *Jagdstaffeln*. As the leader of the first of these, Jasta 2, he passed on his knowledge to many of Germany's greatest fliers, including Manfred von Richthofen. Boelcke also set out the principles of air combat – known as Boelcke's Dicta – that were taught to all German pilots. He said, among other things, that pilots should attack from behind and out of the sun; fire only at close range; and when attacked from above, turn to face the enemy instead of trying to escape. In autumn 1916, Jasta 2 was thrown into the intense air combats over the Somme. In two months, Boelcke downed 21 Allied aircraft, to give him a total of 40 victories. On 28 October, during a fierce dogfight, he collided with one of his own colleagues and spiralled to earth, fatally fracturing his skull.

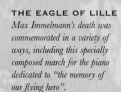

THE BLUE MAX
The Pour le Mérite *was the highest German military decoration. The British nicknamed it the "Blue Max" after the first airman to win it, Max Immelmann.*

SECTION 62
Flight Section 62 at Douai, France, in January 1916, including Boelcke (front, fourth from left) and Immelmann (front, third from right). German airmen were infused with a formal military discipline that contrasted strongly with the individualism of French elite pilots.

masculine virtues of physical courage, skilful aggression, chivalry, and noble sacrifice.

Britain's prime minister David Lloyd George told a wartime audience that airmen were "the knighthood of the war, without fear and without reproach". Airmen were by no means immune to this elevated view of their own activities. There are many recorded instances of self-consciously chivalrous behaviour – for example, pilots who declined to finish off a brave and skilful enemy whose gun had jammed. Gestures of respect for the enemy were common. When the German ace Oswald Boelcke was killed in 1916, British fliers dropped a wreath at his burial site inscribed to "the memory of Captain Boelcke, our brave and chivalrous foe". But gestures of that kind were

only a veneer on the ruthlessness of aerial warfare. The aces did not spend much time fighting aerial duels with skilful opponents, choosing to prey on lone reconnaissance aircraft instead. During dogfights they picked on the least experienced pilots, exploiting their errors for an easy kill. Top British ace Edward "Mick" Mannock once chanced upon six German aircraft on a training-school flight; he shot down the instructor and then picked off the five defenceless pupils one by one.

Souvenir hunting

Mannock was noted for his hatred of Germans and contempt for gestures of chivalry – his comment on the death of Germany's most

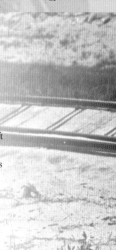

PUBLIC IDOL
Albert Ball was a fearless pilot and excellent marksman who became the first British ace to be idolized by the public. He was still only 20 when his plane crashed under mysterious circumstances in May 1917.

There is no doubting the adrenalin rush that many individuals experienced through combining the thrills of fighting and flying – even though "laughing in the face of danger" is an attitude found in postwar memoirs rather than in contemporary letters or diaries. Many soldiers stuck in damp verminous trenches undoubtedly looked on airmen with envy. A flier had a warm, dry, lice-free bed ten or 15 miles behind the lines; there was never any shortage of volunteers for the air services. Yet the war in the air had more in common with the war in the trenches than is often recognized. There was a gruelling attrition of pilots and aircraft. Freshly trained British pilots arriving at the front in 1917 had an average life-expectancy of little over a fortnight. Like "shell-shocked" infantry, airmen were prone to nervous breakdowns as the strain of combat intensified and losses mounted. Every ace was first and foremost a survivor.

Pilot material

Fighter pilots were of varied origins. A good number, like Richthofen, transferred from the cavalry, which had lost its function in the face of barbed wire and the machine gun. Some, like the British ace James McCudden or the German Werner Voss, were drawn to aviation because of an interest in machines and worked their way up from ground crew to

pilots. Some of the best fliers, including the indomitable French ace Charles Nungesser, had been sportsmen. Pilots in general were extremely young. British ace Albert Ball was a squadron leader at the age of 19. Many of them were also quite short– cockpits were small and weight was a prime factor in aircraft performance. Guynemer was a case in point: he weighed less than 60kg (132lb) and had been rejected as too frail for service in the infantry.

Many would-be fighter pilots never made it as far as the front line. The air services were largely unprepared for the challenge of training thousands of new pilots. The result was a great waste of young lives. Almost 500 American Air Service volunteers died learning to fly, more than twice the number killed in combat. Most cadets were introduced to the controls of an aircraft either through being taken up by an instructor or, in France, "flying" on the ground in flightless "aircraft" known as Penguins. But at some point they had to take the controls themselves. Although training aircraft were chosen for their inherent stability, one moment of panic could be fatal. Many novices forgot the simplest principles

famous ace, Baron von Richthofen, was, "I hope he roasted the whole way down." But even those who were less savage in their rejection of the ethic of chivalry took keen pleasure in a "kill". Fighter pilots drove to view the wreckage of aircraft they had shot down over their own lines, to examine the bodies and collect souvenirs. Richthofen himself was a renowned collector of momentos of his victories.

SHOOTING PRACTICE
An airman undergoes weapons training as his "cockpit" moves along rails in a primitive attempt to simulate the difficulty of hitting enemy aircraft in a dogfight. Many pilots were sent to the front without ever having fired a gun.

CHARLES NUNGESSER

FRENCH ACE CHARLES NUNGESSER (1892–1927) was a fearless individualist in love with danger, and perhaps more than half in love with death. A champion boxer and swimmer, he began the war in the cavalry but soon transferred to the air service. Flying for the N65 squadron based at Nancy, his bravery and flare soon became legendary, achieving ten victories

during the battle of Verdun. In 1916 he crashed while trialling a new aircraft, breaking both legs; the joystick smashed into his face, breaking his jaw and perforating his palate. Two months later, still walking on crutches, he was back in the air. More crashes followed, and more injuries. By the summer of 1917 he was so ill he had to be carried to his cockpit. His decorations included the Military Cross and Legion d'Honneur. Yet he never gave up trying to improve his score and ended up with 43 victories.

His death in 1927 has remained a mystery, after his biplane, L'Oiseau Blanc, was lost at sea between Paris and New York during an attempted non-stop Atlantic crossing.

SKULL AND CROSSBONES
Nungesser decorated his Nieuport with a macabre array of symbols of mortality, including skulls, coffins, and candlesticks. He miraculously survived World War I, only to disappear, along with his aeroplane, in 1927.

that had been drummed into them. For instance, every trainee was told that if his engine failed after take-off, he should under no circumstances attempt to turn back to the airfield. But hundreds did just that, going into a fatal spin.

When losses at the front were heavy, replacements were sent to combat units with ten hours or fewer of flying time to their credit. This was not a certain death sentence, but came close. Any aircraft of the time was difficult to fly. Basic errors in landing and take-off cost hundreds of lives. And in the air you needed to learn quickly what manoeuvres your aircraft was able to stand without falling apart. Many fighter pilots started their combat careers with a stint as an observer or reconnaissance pilot. Those sent directly into fighter units at first had little idea what was going on around them. Some testified to going through their first dogfight without seeing the enemy at all – everything happened too fast. The lack of radio contact between aircraft meant that once a pilot was in the air, even flying in formation, he was essentially on his own. Some squadrons protected novices, but others developed a ruthless attitude towards them, regarding them as disposable. They were given the unit's worst machines and left to fend for themselves.

Combat style
As the war went on, all sides got better at readying pilots for the shock of the war, but it was the Germans who made most effort to prevent deaths

LICENCE TO FLY
This pilot's certificate was issued to a British officer by the Royal Aero Club in 1916. Pilots fresh from flight training had a low survival rate at the front – at the height of the fighting their average life-expectancy could be as low as two to three weeks.

MEN AND MACHINES
A Royal Air Force squadron is pictured at its base in northern France in 1918. The airmen are in flying kit; as usual, the ground crew are relegated to the background. The aircraft are Royal Aircraft Factory S.E.5as, probably the best British fighter of the war.

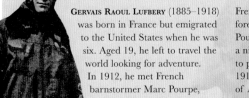

in training and who prepared their trainee pilots best for combat. They disseminated knowledge of the principles of air combat worked out by Boelcke in 1916 (see page 82), and built up a body of highly skilled fighter pilots through careful selection of suitable individuals and their integration into fighter squadrons that fought in formation. The system did not begin to break down until the end of the war, when heavy losses forced Germany to throw thousands of inexperienced pilots into the fray.

There was always a tension in fighter units between the practical advantages of fighting in formation and the individualist temperament of most of the best pilots. Aces were often loners who developed their own secrets for success in combat that they were not keen to share with their colleagues. Even when they were squadron leaders, many found it hard to abandon the role of "lone wolf" hunter. Werner Voss, for example, was a noted success as leader of Jasta 10 in 1917, but he would still head off on his own at dawn or dusk to track down enemy observation aircraft. Canadian pilot Billy Bishop for a time commanded Britain's 85th Squadron yet still fought his own personal war, recording most of his victories on solitary missions.

RAOUL LUFBERY

GERVAIS RAOUL LUFBERY (1885–1918) was born in France but emigrated to the United States when he was six. Aged 19, he left to travel the world looking for adventure. In 1912, he met French barnstormer Marc Pourpe, became his mechanic, and toured India and the East with Pourpe's exhibition flying show. When war broke out, Lufbery enlisted in the

AMERICAN ADVENTURER
Raoul Lufbery was revered by his men, who called him the "ace of aces"; his official tally was 16 victories.

French Foreign Legion – the only unit that foreigners could join – before teaming up with Pourpe again. When Pourpe was killed attempting a night landing, Lufbery graduated from mechanic to pilot, seeing his first service in 1915. In May 1916, he joined the Lafayette Escadrille, composed of American volunteers who had joined the French air force. He soon became its leading ace, as well as devising fighter tactics such as the Lufbery Circle, in which fliers formed a circle with each aeroplane protecting the one in front. In 1918 he joined the US Army Air Service as a combat instructor. He was killed in May 1918 after rushing to the aid of an inexperienced pilot fighting a German Rumpler. Lufbery's Nieuport caught fire in sight of the squadron airfield and he jumped to his death rather than burn alive.

The French were the first to create self-consciously elite formations of fighter pilots, the *Cigognes* (Storks). Originally a single squadron, N.3, fighting at Verdun in 1916, the *Cigognes* had expanded to five squadrons by the following year. Despite being grouped together, however, top French pilots were often reluctant to fight as part of a team. One observer, Jean Villars, wrote that "the veterans want to hunt individually, through overconfidence and a desire to work on their own; the novices imitate them through vanity and ignorance".

High living
Fighter squadrons were never noted for their respect for formal discipline. During any lull in the action at the front, French elite pilots were in the habit of flying off to Paris, where they would be familiar figures in the best restaurants, always with attractive women in attendance and a smart automobile parked outside.

British fighter squadrons were more noted for their drunken sprees and riotous behaviour in the mess, a traditional way of coping with the fear and personal loss that were inseparable from combat. Yet under a firm squadron leader, such behaviour would be kept carefully within limits. Few pilots relished the idea of embarking on a dawn patrol with a stinking hangover – it tended to be an experience that a man would have once and never again.

The relaxed style of the fliers was hard for more traditional army officers to accept. A US military intelligence report in 1917 identified indiscipline in the British flying corps as a problem caused by "the fact that the service, owing to its

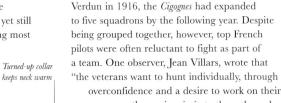

OUTER HELMET

Hole aids hearing

Turned-up collar keeps neck warm

Soft, supple leather

INNER HELMET

Sheepskin lining

LEATHER GLOVES

FLYING JACKET

Tinted, anti-splinter glass

SHATTERPROOF GOGGLES

Thick lining

Leather straps

FLYING BOOTS

FIGHTING THE COLD
Flying in open cockpits, cold was one of the major enemies all airmen faced. Wearing two layers of helmet and gloves, plus a leather coat and sheepskin boots, offered some protection. Goggles shielded the eyes against freezing wind, dust, and oil sprayed from the engine. Parachutes and oxygen equipment for high-altitude flight became available late in the war, but only for German airmen.

DUELS IN THE SKY

"**THE WHOLE SQUADRON WOULD ENTER** the fight in good formation," British pilot Lieutenant Cecil Lewis remembered, "but within half a minute the whole formation had gone to hell. Just chaps wheeling and zooming and diving. On each other's tails… a German going down, one of our chaps on his tail, another German on his tail, another Hun behind that… People approaching head-on firing at each other as they came and then just at the last moment turning and slipping away." This was the classic "dogfight", the result of the meeting of two fighter formations on the Western Front. Committed to a policy of "offensive patrols" over enemy lines, the pilots of the Royal Flying Corps directly invited their German enemies to such combat.

Typically a dozen aircraft of a fighter squadron flew towards the enemy lines in a well-rehearsed close formation, each pilot craning his neck and straining his eyes to pick out distant black specks or a flash of sunlight on a windshield that might reveal the enemy. Whichever side spotted the other first would, if they felt they had the

> ## "Fighting in the air is not sport, it is scientific murder."
>
> **CAPTAIN EDWARD V. "EDDIE" RICKENBACKER, USAS**
> *FIGHTING THE FLYING CIRCUS*

advantage, manoeuvre into a favourable position and dive. As Lewis described, the sky then filled with whirling aircraft as the engagement broke up into individual combats. Lieutenant Norman Macmillan recalled: "Our machines could turn in such tiny circles that we simply swerved round in an amazingly small space of air, missing each other sometimes by inches." In this chaos, pilots sometimes shot down or collided with their own colleagues. Guns jammed or ran out of ammunition at crucial moments.

Combat was close up and personal. Macmillan recalled how, as he closed in on a Fokker, the pilot looked around: "I was close enough to see his keen blue-grey eyes behind his goggle glasses… He saw I was dead on his tail and instantly banked and curved to the right… My tracers passed close over his central left wing, just outside his cockpit and in line with his head." Some aircraft returned to base with a film of blood on the

windscreen from an enemy shot at close range. Pilots also saw their own colleagues go down, at worst enveloped in flames. Lieutenant Ira Jones recalled being overcome by "a sudden feeling of sickness, of vomiting" as a comrade's machine blazed in the sky nearby.

At the end of a dogfight, pilots often found themselves heading back to their airfield alone, pursued by the enemy. An Australian flier, Lieutenant George Jones, remembered the experience of "being chased without any ability to retaliate" as the most nerve-shattering of all: "Every time I thought of it I could hardly hold a knife and fork if I was having a meal." But most pilots kept a stiff upper lip: according to Lieutenant John Grider, after evading enemy fighters, you "roll in derision as you cross the lines and hasten home for tea".

PULP HEROES
The dogfights of the Western Front were a popular topic for pulp magazines in the 1920s and 1930s. Colourful pilot memoirs were often combined with outright fiction. Ironically, many pilots later made a living as stunt fliers in Hollywood movies about the romanticized exploits of the aces.

FOKKER SCOURGE
This artwork depicts Lieutenant Oswald Boelcke in a Fokker Eindecker shooting down a British enemy in 1915 – the first victim of Anthony Fokker's synchronized forward-firing system, which allowed the pilot to fire though the propeller blades. It was a great success and allowed Boelcke to lead an aerial reign of terror known as the Fokker Scourge.

Fokker D.VII

"We got into a dogfight with the new brand of Fokkers… we put up the best fight of our lives, but these Huns were just too good for us."

LIEUTENANT JOHN M. GRIDER
BRITISH PILOT'S DIARY ENTRY ON FIRST
ENCOUNTERING THE FOKKER D.VII

BY 1918, GERMAN PILOTS WERE DESPERATE FOR A SINGLE-SEAT FIGHTER to replace their outdated Albatroses and Fokker Dr.I triplanes. After evaluation trials held at Adlershof, Berlin, at the end of January, the Fokker D.VII was selected for mass production, and the first models arrived at the front the following April.

Hard-pressed Jastas (fighter squadrons) greeted their new mounts with relief and enthusiasm. German pilot Rudolf Stark wrote: "The machines climb wonderfully and respond to the slightest movement of the controls." Their impact on the fighting peaked during the summer of 1918, by which time some 40 Jastas were flying D.VIIs, many of them with BMW engines that gave substantially better peformance than the original Mercedes power plants. Operating in skies crowded with Allied aircraft of all kinds, D.VII pilots achieved exceptional kill-rates. For example, one squadron, Jasta Boelcke, scored 46 confirmed victories in a month for the loss of only two of its own pilots.

The BMW-powered D.VII was especially effective at high altitude – its pilots were among the first to be issued with experimental oxygen equipment, as well as parachutes. Flying high gave the D.VII the initial advantage in encounters with Allied fighters and also allowed it to hunt down the Allied reconnaissance aircraft, which depended on altitude for safety. About 1,500 D.VIIs were delivered before the end of the war in November 1918.

Scalloped trailing edge formed by taut wire linking ends of ribs

Radiator filler tube

Aluminium panels cover nose of aircraft

Wire-spoked wheels with fabric covers

Louvres aid engine cooling

Fabric-covered plywood decking on top of rear fuselage

Fixed stirrup aids entry into cockpit

Coloured "lozenges"

COLOURFUL CAMOUFLAGE
Most D.VIIs were delivered with this imaginative camouflage scheme, printed with patterns of irregular, coloured "lozenges". When viewed from a distance these lozenges merged to form an effective camouflage.

Specifications

Engine	185hp BMW III six-cylinder water-cooled inline
Wingspan	8.9m (29ft 3in)
Length	7m (23ft)
Height	2.75m (9ft 2in)
Top speed	186.5kph (116.6mph)
Armament	2 x 7.92mm 08/15 Maxim machine guns

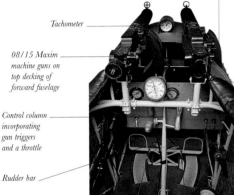

Tachometer

08/15 Maxim machine guns on top decking of forward fuselage

Control column incorporating gun triggers and a throttle

Rudder bar

EASY FLIER
The Fokker D.VII was considered a fairly easy aircraft to fly – an important consideration, since, by the summer of 1918, pilots were being rushed to the front after a bare minimum of training.

Streamlined-section steel-tube interplane "N" struts

Thick, semi-cantilever wing

FRONT VIEW

Aerofoil-shaped lifting surface encloses undercarriage axle

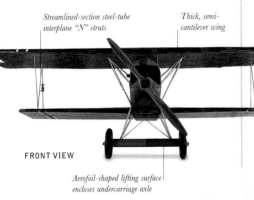

Sprung wooden tailskid with steel shoe

picturesque nature, is very likely to attract the wrong class of men". No doubt the author of the report would have included among "the wrong class of men" the courageous American volunteers who fought for France in the famous Lafayette Squadron.

Lafayette Squadron
It was in April 1916 that the French allowed New Englander Norman Prince to group seven American pilots into a high-profile squadron, reasoning that this might encourage the United States to enter the war on the Allies' side. The American volunteers earned a deserved reputation for high living and rough partying. But thrown into the thick of the fighting at Verdun in June 1916, they flew four or five patrols a day, until all seven original pilots were either dead or wounded. The Lafayette Squadron continued in existence, never numbering more than 20 members, with fresh entrants drawn from the hundreds of other American volunteers serving with more mundane French air units.

Almost a year after the US entered the war, the Lafayette Squadron was integrated into the US Army Air Service – with a sharp warning that discipline would have to be tightened. Other American pilots serving with the French were also invited to transfer to US units. An exception was African-American flier Eugene "Jacques" Bullard. Blacks were not accepted as pilots in the US force.

Despite their late arrival in the war, some Air Service pilots were enthusiastic in the pursuit of ace status. The most successful, Eddie Rickenbacker (a former "pupil" of Lufbery's), lent himself tirelessly to the demands of celebrity and publicity, even staging a fake dogfight for the cine cameras while the war was still on – a stunt that nearly turned sour when French pilots unaware of the filming tried to join in the fight.

Aces high
But beyond all the flim-flam of propaganda and publicity, fighter pilots had a real job to do, and the aces were the pilots who did it best. Broadly, the job was to win air superiority. At different times this might involve flying escort to reconnaissance or bomber missions; carrying out offensive patrols to challenge the enemy's forces; or picking off enemy reconnaissance aircraft or bombers. It was in the last of these roles that aces amassed most of their

ANTHONY FOKKER

FRIEND OF ACES
Fokker was a pilot before he became an aeroplane maker. He got on well with leading German pilots, who often called the shots in decisions over aircraft procurement.

Born into a wealthy Dutch family in Kediri, Java, Anthony Fokker (1890–1939) was caught up in the pre-war flying craze. He learned to fly in 1910 and soon designed his first mono-plane, setting up a factory at Schwerin in Germany in 1912. The war made his fortune. Wily and ambitious, he formed close ties with German bureaucrats and ace pilots alike. Orders flooded in for aircraft such as the Eindecker E.III and the D.VII.

Fokker had a pilot's feel for aeroplanes and a businessman's ability to organize large-scale production, but he was not an innovative designer. The Fokker trademark welded steel-tube frame was, in fact, dreamed up by his chief technician, Reinhold Platz. At the war's end, Fokker moved to the Netherlands, smuggling airframes, engines, and parts across the border. He produced successful civil and military aircraft in the postwar period, moving on to the USA, where he headed the Fokker Aircraft Corporation.

victories. The many qualities that made a successful fighter pilot included keen eyesight, fine reflexes, and perfect co-ordination; dedication to the task in hand, including a meticulous attention to the detailed preparation of their machine and guns; cool nerve when under fire; and utter ruthlessness in executing a kill. Air combat was about winning, not about giving the other side a fair chance. British ace James McCudden wrote that "the correct way to wage war is to down as many as possible of the enemy at the least risk, expense, and casualties to one's own side".

MANFRED VON RICHTHOFEN ("THE RED BARON")

AN OFFICER IN THE GERMAN CAVALRY when the war started, Baron Manfred von Richthofen (1892–1918) transferred to the air service in 1915. Having flown as an observer and bomber pilot on the Eastern front, he was selected by Oswald Boelcke to join his elite fighter group in France. He soon proved himself to be one of Boelcke's ablest pupils, and the deaths of Boelcke and Max Immelmann in 1916 cleared the way for Richthofen's emergence as Germany's most prominent ace. He was given command of his own squadron, Jasta 11, and then of the first *Jagdgeschwader*, grouping four squadrons in a fighter wing of about 50 aircraft. With its garishly coloured machines, this was the

THE HIGHEST ACE
In 1917 the Red Baron painted his planes red to mark him out to friend and foe alike. He was a ruthless hunter and, with 80 kills, achieved the highest tally of the war.

formation the British christened Richthofen's Flying Circus. Richthofen was a fine leader of men, but lacked personal warmth. His closest relationship was with his wolfhound, Moritz. Arrogant and ruthless, he showed few signs of chivalry or respect for his enemies – he was known to especially despise the French. Where many pilots used sport as a metaphor for air combat, Richthofen saw it in terms of hunting – his favourite leisure activity. He once wrote: "When I have shot down an Englishman my hunting passion is satisfied for a quarter of an hour."

On leave from the front, Richthofen was a celebrity, pursued by photographers and journalists, and dining with the Kaiser.

CLOSING IN
One of von Richthofen's hunting strategies was to let fly a short burst of fire while still far away from his target: "I did not mean so much to hit him as to frighten him… He began flying curves and this allowed me to draw near."

He even took time out to write his memoirs. Having survived a headwound during a dogfight in July 1917, Richthofen came under pressure from his superiors to withdraw from combat. It was felt that his death would be a severe blow to morale. However, during the 1918 spring offensive he headed back into the fray. On 21 April he was shot through the heart while pursuing a potential victim over enemy lines. Whether the fatal shot was fired by an Australian machine-gunner or by Canadian pilot Roy Brown is still a matter of dispute.

HONOURED DEATH
On 21 April 1918 Baron von Richthofen was shot down over a sector of the front manned by Australian troops. The Commonwealth troops buried the famous enemy with full military honours.

BARON'S MOUNT
Although it is the highly manoeuvrable "blood-red" Fokker Dr.I triplane with which Richthofen is associated, he spent most of his time flying biplanes like the Albatros D.II (right) and D.III.

BRAVE WARRIOR
Werner Voss, killed in 1917, was described by British ace James McCudden as "the bravest German airman whom it has been my privilege to see".

The ruthlessness of air combat was never better exemplified than by one of the very few recorded meetings of two aces in single combat. On 23 November 1916 British ace Major Lanoe Hawker, found himself isolated from his formation and targeted by the Red Baron, Manfred von Richthofen. The German was flying an Albatros D.I, Hawker a markedly inferior pusher D.H.2. The two machines, in Richthofen's words, "circled round and round like madmen". Hawker performed small miracles of agility, side-slipping his aircraft downwards each time the German was about to capture him in his sights. But, inevitably, he eventually ran out of height. Down to treetop level and over enemy lines, Hawker desperately headed for home, zigzagging all the time. But Richthofen's machine was faster. Unable to shake him off, Hawker made a last attempt to turn and face his pursuer, but Richthofen opened fire as the Englishman banked and shot him through the head. Recounted in Richthofen's memoirs, this story hardly conforms to the chivalric stereotype. In effect, Richthofen ruthlessly hunted down a pilot who, for all his courage and skill, had no chance because his machine was inferior. It reads more like an assassination than a duel.

Mental torture

Whether aces or not, fighter pilots were subjected to unbearable strain at peak periods of the war. When men were flying three or four missions a day, seeing friends and colleagues dying before their eyes, often in the most gruesome fashion, they would pray for the relief of a "dud" day, when weather conditions made flying impossible. If losses were high, there would be a sharp rise in the number of pilots aborting missions, alleging

that their magazines had jammed or their engine was playing up. Pilots suffered from repeated nightmares, began to behave oddly, and sometimes lost their nerve completely.

In this context it is perhaps just possible to understand the mentality of British commanders who opposed the development of suitable parachutes for pilots because it might encourage them to jump out unnecessarily – in effect deserting in the face of the enemy. Yet the lack of parachutes contributed to the worst moments of the air war. A pilot might spiral down for minutes, out of control but uninjured, with nothing to do but wait to hit the ground. Or, worse, be trapped in a burning aircraft with no better options than jumping to his death or blowing his brains out with a revolver.

The best fighter pilots were far from immune to such pressures of combat flying. Before they died – and most of them did die – many of the aces were harrowed men, flying too many missions and living on the edge of their nerves. Some deliberately drove themselves beyond the

> "Fight on and fly on to the last drop of blood and the last drop of petrol, to the last beat of the heart."
>
> **MANFRED VON RICHTHOFEN**
> DRINKING A TOAST TO HIS FELLOW PILOTS

limit. Guynemer, for example, feared that if he withdrew from combat, people would say it was because he had "won all the awards". Thus he died flying while patently unfit, physically ill, and racked by paranoia and insomnia.

A few aces died what could be called a hero's death. One such ace was German Werner Voss. On 23 September 1917, flying a Fokker triplane, Voss was bounced by a flight of S.E.5s led by James McCudden. As the S.E.5s dived down on his tail, Voss spun his aircraft to face them. McCudden wrote: "By now the German triplane was in the middle of our

TOP ALLIED ACE
With 75 victories to his name, French pilot René Fonck was the top Allied ace of the war. He twice shot down six German aircraft in a single day. A boastful individual, Fonck never won the popular affection granted to his compatriots Nungesser and Guynemer.

formation, and its handling was wonderful to behold. The pilot seemed to be firing at all of us simultaneously, and although I got behind him a second time, I could hardly stay there for a second. His movements were so quick and uncertain that none of us could hold him in sight…" Two British machines were forced to withdraw, shot up by Voss' bullets, but then the German's luck ran out. One pilot, Arthur Rhys Davids, latched on to his tail and raked the triplane with repeated bursts of fire. McCudden saw Voss' aircraft "hit the ground and disappear into a thousand fragments… it literally went to a powder".

Inglorious deaths

Such a death was rare indeed. McCudden himself died in a mundane flying accident when returning from a spell in Britain to take command of a squadron in France. Guynemer simply vanished on a mission – no trace of him or his aircraft was ever found. Boelcke's funeral oration declared that there could be for him "no more beautiful way to end his life than flying for the Fatherland", a rhetoric that contrasted sadly with the manner of his death, as victim of a collision with one of his own men.

But however much the reality of their mostly brief, brutal, nerve-racked lives might contrast with a romantic view of air war, the legend of the aces as "knights of the air" proved durable and an inspiration to a future generation of fliers. Charles Lindbergh, famed for the first solo, non-stop Atlantic crossing, recalled how, as a child during the war, he had "searched newspapers for reports of aerial combats – articles about Fonck, Mannock, Bishop, von Richthofen, and Rickenbacker", who were to him the modern equivalent of "King Arthur's knights in childhood stories". This myth was the aces' legacy.

WWI FIGHTERS

THE SPECIALIZATION OF MILITARY AIRCRAFT brought the emergence of the fighting "scout" from early in the war. The earliest effective fighters, the French Morane-Saulnier and the German Fokker Eindecker, were flimsy monoplanes which used outdated wing-warping, but benefitted from guns firing forward through a tractor propeller. The British still maintained an attachment to pusher propellers on aircraft with a forward-firing gun, in the one-seater Airco D.H.2 and the two-seater Vickers Gunbus. Tractor propellers gave superior performance, however, and by 1917, various single-seat tractor biplanes predominated. The two broad groups were fast machines with inline engines – like the SPAD XIII – which excelled at diving attacks and pursuit, and more manoeuvrable machines with rotary engines, such as the Nieuport 17 and Sopwith Camel. These were difficult to fly but in a dogfight they could turn on a sixpence. Most manoeuvrable of all were the rotary-engined triplanes such as the Fokker Dr.I. Probably the finest all-round fighter of the war was the inline-engined Fokker D.VII (see pages 88–9), introduced in 1918.

BRISTOL F.2B
Another late-war success story was the Bristol F.2B, which reintroduced the two-seat fighter concept. See pages 76–77.

Airco D.H.2

Introduced in 1915, the D.H.2 could only mount a forward firing gun by fitting a pusher engine. Being a single-seater, it was fast enough to keep up with its prey, although the pilot had to control the aircraft and fire at the same time, making it hard to fly.

ENGINE 100hp Gnome Monosoupape 9-cylinder rotary	
WINGSPAN 8.6m (28ft 3in)	LENGTH 7.7m (25ft 2in)
TOP SPEED 150kph (93mph)	CREW 1
ARMAMENT 1 x .303in Lewis machine gun	

Albatros D.III

The Albatros D series were part of the German effort to recover air superiority in 1917 after the eclipse of the Fokker E.III. During "Bloody April", the D.III inflicted heavy casualties on the RFC. However the improved D.V was still inferior to the SPAD and RAF SE.5a.

ENGINE 170hp Mercedes D.IIIa 6-cylinder water-cooled	
WINGSPAN 9.1m (29ft 8in)	LENGTH 7.3m (24ft)
TOP SPEED 175kph (109mph)	CREW 1
ARMAMENT 2 x 7.92mm Spandau machine guns	

Brandenburg D.I

Designed in Germany by Ernst Heinkel, the D.I was built in Austria and entered service with the Austrian Air Force in the autumn of 1916. However, poor flight characteristics and limited visibility caused a number of accidents, leading to its pilot nickname, "The Coffin".

ENGINE 160hp Austro-Daimler 6-cylinder liquid-cooled inline	
WINGSPAN 8.5m (27ft 11in)	LENGTH 6.4m (20ft 10in)
TOP SPEED 187kph (116mph)	CREW 1
ARMAMENT 1 x 8mm Schwarzlose machine gun	

Fokker Dr.I triplane

Fitting three wings allowed a lot of lift to be packed into a small area, giving the Fokker Dr.I a fast rate of climb and good manoeuvrability, which made it ideal for air fighting. During late 1917, it was feared along the Western Front, most famously in the hands of Manfred von Richthofen.

ENGINE 110hp Le Rhone 9-cylinder rotary	
WINGSPAN 7.2m (23ft 7in)	LENGTH 5.8m (18ft 11in)
TOP SPEED 165kph (103mph)	CREW 1
ARMAMENT 2 x 7.92mm Spandau machine guns	

Fokker E.III Eindecker

Fitted with a machine gun and an interrupter gear that allowed it to fire through the propeller, Anthony Fokker's pre-war monoplane design became the deadly E.III (replica shown), which cut a swathe through Allied observation aircraft in the summer of 1915. This was mainly due its innovative armament and the brilliance of its pilots, including Boelcke and Immelmann.

ENGINE 100hp Oberursel 7-cylinder rotary	
WINGSPAN 9.5m (31ft 3in)	LENGTH 7.2m (23ft 7in)
TOP SPEED 130kph (81mph)	CREW 1
ARMAMENT 1 x 7.92mm Spandau machine gun	

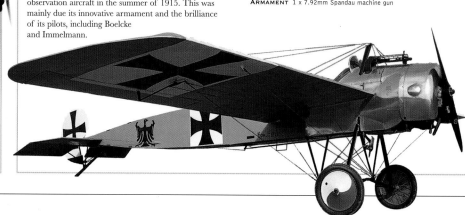

Morane-Saulnier Model N

French pilot Roland Garros was the first to fit a machine gun firing through the propeller to an aircraft in early 1915. He used steel deflector plates to prevent the bullets damaging the blades. Following his success, the manufacturers produced about 50 Model Ns with similar fittings. As it was one of the few armed fighters available, it was operated by several French and British units, but it was never popular, control being by old-fashioned wing-warping rather than ailerons, and it was phased out by 1916.

ENGINE	80hp Gnome 9-cylinder rotary		
WINGSPAN	8.1m (26ft 9in)	LENGTH	5.8m (19ft 2in)
TOP SPEED	144kph (89mph)	CREW	1
ARMAMENT	1 x .303in Lewis machine gun		

Nieuport Type 17

Developed from the highly agile Type 11 "Bébé" (Baby) racer, that helped defeat the "Fokker Scourge" in 1916, the Type 17 offered superior performance and a synchronized machine gun on the upper fuselage.

ENGINE	110hp Le Rhone 9Ja 9-cylinder rotary		
WINGSPAN	8.2m (26ft 10in)	LENGTH	5.8m (19ft)
TOP SPEED	165kph (102mph)	CREW	1
ARMAMENT	1 x .303in Vickers machine gun (firing through propeller); up to 8 Le Prieur rockets on interplane struts		

Pfalz D.III

The Pfalz factory was in Bavaria, which was still a kingdom in its own right within the German Empire and wished to retain some control over its armed forces. Hence in the early war years, Pfalz built German and French planes under licence for supply to Bavarian units only. In 1917, they brought out their own design, which was similar to the Albatros, with monocoque fuselage and Mercedes engine, but by no means a copy and in several respects superior.

ENGINE	160hp Mercedes D.III 6-cylinder water-cooled		
WINGSPAN	9.4m (30ft 10in)	LENGTH	7m (22ft 10in)
TOP SPEED	165kph (103mph)	CREW	1
ARMAMENT	2 x 7.92mm Spandau machine guns		

RAF S.E.5a

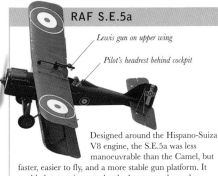

Lewis gun on upper wing

Pilot's headrest behind cockpit

Designed around the Hispano-Suiza V8 engine, the S.E.5a was less manoeuvrable than the Camel, but faster, easier to fly, and a more stable gun platform. It could also sustain more battle damage and was the preferred mount of the great British aces, including Ball, Mannock, and Bishop. Its only problem was engine unreliability, solved by the introduction of the Viper V8.

ENGINE	200hp Wolseley W.4A Viper water-cooled V8		
WINGSPAN	8.1m (26ft 7in)	LENGTH	6.4m (20ft 11in)
TOP SPEED	193kph (120mph)	CREW	1
ARMAMENT	1 x .303in Vickers machine gun (firing through propeller); 1 x .303in Lewis machine gun; 4 x 11kg (25lb) bombs		

Sopwith Camel

The Camel was a development of the Pup, so-called because the tight grouping of the engine, guns, and pilot gave it a humped appearance. Formidable at height, and in particular in the right-hand turn, when the torque of the engine pulled it round amazingly swiftly, it needed a skilful and experienced pilot to fly it without getting into difficulty. Crashes in training were frequent. However, during the war, Camels shot down more enemy aircraft (over 1,200) than any other type.

ENGINE	130hp Clerget 9B 9-cylinder rotary		
WINGSPAN	8.5m (28ft)	LENGTH	5.7m (18ft 9in)
TOP SPEED	182kph (113mph)	CREW	1
ARMAMENT	2 x .303-in Vickers machine guns		

Sopwith Pup

ENGINE	80hp Le Rhone 9C 9-cylinder rotary		
WINGSPAN	8.1m (26ft 6in)	LENGTH	6m (19ft 4in)
TOP SPEED	179kph (112mph)	CREW	1
ARMAMENT	1 x .303in Vickers machine gun		

Officially the Type 9901 in the RNAS and the Scout in the RFC, the "Pup" was universally known in both services by its nickname. This came about because it was a smaller version of the previous Sopwith model and because it was a delight to fly.

SPAD XIII

The great strength of this aircraft was in the powerful V8 engine designed by Marc Birgit of Hispano-Suiza, who was also responsible for the machine gun interrupter gear. During 1917, more were produced than any other Allied fighter type and it was extremely popular in the French fighting squadrons.

ENGINE	235hp Hispano-Suiza water-cooled V8		
WINGSPAN	8.1m (26ft 4in)	LENGTH	6.2m (20ft 4in)
TOP SPEED	222kph (139mph)	CREW	1
ARMAMENT	2 x Vickers or Marlin machine guns		

Vickers F.B.5 Gunbus

One of the first aircraft designed to fight; the F.B.5 used the layout of observation aircraft, to mount a forward firing machine gun. The less efficient pusher propellers combined with a low-powered engine and weight made for a slow and unresponsive aeroplane.

Movable .303in machine gun

ENGINE	100hp Gnome Monosoupape 9-cylinder rotary		
WINGSPAN	11.1m (36ft 6in)	LENGTH	8.3m (27ft 2in)
TOP SPEED	113kph (70mph)	CREW	2
ARMAMENT	1 x .303in Lewis machine gun		

ZEPPELINS AND BOMBERS

GIANT AEROPLANES AND AIRSHIPS BROUGHT THE TERROR OF WAR TO THE CITIES OF EUROPE

"Those deaths must have been the most dramatic in the world's history. They fell – a cone of blazing wreckage – watched by eight million of their enemies."

MURIEL DAYRELL-BROWNING
EYEWITNESS TO THE DESTRUCTION OF
ZEPPELIN SL 11 OVER LONDON, 1916

EVEN BEFORE World War I, the airship was fixed in the popular imagination as a symbol of terror. In his 1908 book The War in the Air, H.G. Wells had described an airship raid on New York leaving "ruins and blazing conflagrations and heaped and scattered dead".

It was a vision that appealed to some military commanders. Captain Peter Strasser, head of the German navy's airship fleet, believed that Britain could be "overcome by means of airships… through increasingly extensive destruction of cities, factory complexes, dockyards…".

The Germans had no monopoly on the intent to bomb enemy cities and factories. But they did have the lead in airship technology, and in 1914 airships were the only aircraft capable of carrying a significant bombload far enough to hit strategic targets.

AIRSHIP COMMANDER
Peter Strasser, Commander of the German Navy Zeppelin Fleet during World War I, was an ambitious officer who made the zeppelin an effective weapon of war. He died when the L 70 was shot down over England.

SHADOW OVER THE WESTERN FRONT
The Zeppelin Staaken R.IV, of which the above is a model, flew on the Western Front in 1917–18. This six-engined monster had a wingspan of 42.2m (138ft 6in) and, when fully loaded, weighed over 13 tonnes (almost 13 tons).

However, at this stage they were not capable of fulfilling apocalyptic visions of mass destruction. Whether metal-framed Zeppelins or plywood-framed Schütte-Lanzes, German airships revealed serious drawbacks early in the war. The army found that they could not survive over the battlefield. Travelling at under 80kph (50mph), they presented large, tempting targets for gunners on the ground; four army

airships were shot down in the first month of the war. They were also distressingly accident-prone, especially in bad weather.

Stealthy raiders

However, an answer to the airship's vulnerability in combat was sought in stealth. On a moonless night, despite their bulk, zeppelins could evade detection and pursuit. Out of necessity they became night raiders. The German army and navy used airships to bomb a variety of strategic targets under cover of darkness, including Paris and other French cities. But their most prized target was Britain, and above all London. In those bitter, hate-filled days of war, German schoolchildren learned to sing: "Zeppelin, fly! Fly to England! England will burn in fire!"

NIGHT BOMBER
Introduced in 1917, the four-engined Zeppelin Staaken R.VI was mainly used for night bombing raids on London. It was unusual in its day for having a fully enclosed cabin, but during bomb runs the commander stood in the open observation post in the nose.

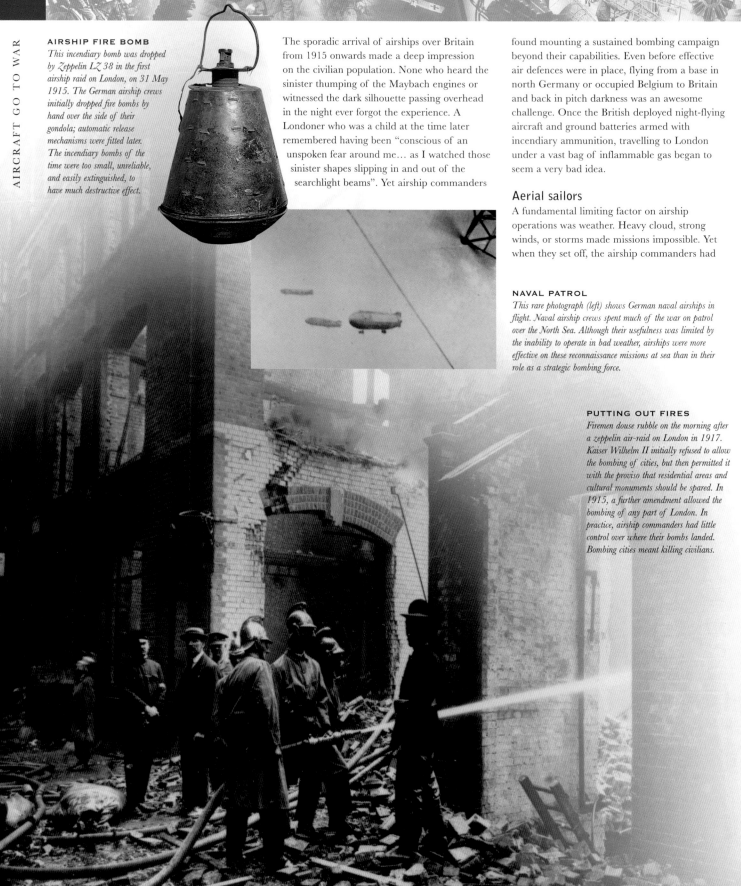

AIRSHIP FIRE BOMB
This incendiary bomb was dropped by Zeppelin LZ 38 in the first airship raid on London, on 31 May 1915. The German airship crews initially dropped fire bombs by hand over the side of their gondola; automatic release mechanisms were fitted later. The incendiary bombs of the time were too small, unreliable, and easily extinguished, to have much destructive effect.

The sporadic arrival of airships over Britain from 1915 onwards made a deep impression on the civilian population. None who heard the sinister thumping of the Maybach engines or witnessed the dark silhouette passing overhead in the night ever forgot the experience. A Londoner who was a child at the time later remembered having been "conscious of an unspoken fear around me… as I watched those sinister shapes slipping in and out of the searchlight beams". Yet airship commanders

found mounting a sustained bombing campaign beyond their capabilities. Even before effective air defences were in place, flying from a base in north Germany or occupied Belgium to Britain and back in pitch darkness was an awesome challenge. Once the British deployed night-flying aircraft and ground batteries armed with incendiary ammunition, travelling to London under a vast bag of inflammable gas began to seem a very bad idea.

Aerial sailors

A fundamental limiting factor on airship operations was weather. Heavy cloud, strong winds, or storms made missions impossible. Yet when they set off, the airship commanders had

NAVAL PATROL
This rare photograph (left) shows German naval airships in flight. Naval airship crews spent much of the war on patrol over the North Sea. Although their usefulness was limited by the inability to operate in bad weather, airships were more effective on these reconnaissance missions at sea than in their role as a strategic bombing force.

PUTTING OUT FIRES
Firemen douse rubble on the morning after a zeppelin air-raid on London in 1917. Kaiser Wilhelm II initially refused to allow the bombing of cities, but then permitted it with the proviso that residential areas and cultural monuments should be spared. In 1915, a further amendment allowed the bombing of any part of London. In practice, airship commanders had little control over where their bombs landed. Bombing cities meant killing civilians.

no idea what the weather was like over Britain. Time and again they set off in promising conditions only to encounter fog or strong head winds as they approached their targets.

Just flying the huge aircraft posed complex problems. They were more like warships than aeroplanes. The commander strode about his control cabin with binoculars around his neck while a coxswain steered the ship with a nautical-style wheel. Another coxswain monitored altitude and gas pressure. The engines were tended in flight by mechanics in the engine cars, and a

sailmaker checked for damage to the outer fabric. The commander and his officers were constantly engaged in complex calculations about the airship's altitude. A variety of factors made the craft rise or fall. For instance, when it rained, the water on the vast cover of the airship would increase its weight, making it lose height. Constant fine tuning of ballast and gas pressure was needed to maintain a steady height.

But this was nothing compared to the problem of navigation. On night raids, zeppelins often did well to find the right country, let alone a city the

size of London. The best means of navigation available was by a radio fix: the airship transmitted a signal to two ground stations, each then identified the direction of the signal, and the two direction lines allowed the airship's precise position to be calculated. But airship commanders mostly preferred to keep radio silence for fear of revealing their position to the enemy. They fell back on the age-old nautical technique of dead reckoning. If you knew how fast you had travelled, for how long, and in what direction, you could plot your position on a chart. Yet these

SHOOTING DOWN AN AIRSHIP

AT AROUND 2.30AM ON 3 SEPTEMBER 1916, a woman in north London, woken by explosions, looked out of her window and saw sailing close above "a cigar of bright silver in the full glare of about 20 magnificent searchlights". It was the Schütte-Lanz SL 11, one of 16 naval and army airships sent against London in the largest raid of the war. Also in the night sky was a B.E.2c biplane, piloted by Lieutenant William Leefe Robinson, part of the Home Defence Wing assigned to protect Britain against airship raids. Leefe Robinson and his colleagues had grown used to the unnerving experience of night flying. When an airship raid was detected, they took off

"I saw, far behind us, a bright ball of fire… Poor fellows, they were lost the moment the ship took fire."

ZEPPELIN CAPTAIN ERNST LEHMANN
RECALLING THE LOSS OF THE SL 11

from airstrips lit by flares and, once in the air, relied on the faintly visible line of the horizon to keep their sense of balance and hold the aircraft true.

Looking for airships in total darkness was, however, a frustrating experience. Leefe Robinson had been on patrol for three hours without seeing anything when the SL 11 was suddenly lit up for all to see. Armed with explosive and incendiary ammunition, he dived towards the airship and came up under its nose, raking it with fire from bow to stern. He had shot off three drums of ammunition before he saw it begin to glow: "In a few seconds the whole rear part was blazing," he wrote. "I quickly got out of the way of the falling, blazing zeppelin." The event was visible for miles around. Londoners cheered as the airship fell. The blaze was witnessed with very different sentiments by Captain Ernst

MACABRE MEMENTOS
Pieces of the SL 11, the first airship shot down over London, were made into souvenirs such as these cufflinks and pin, which were sold to raise money for the Red Cross.

Lehmann, commander of a zeppelin at that moment heading for home. "I saw, far behind us, a bright ball of fire," he wrote. "Poor fellows, they were lost the moment the ship took fire." The SL 11 was the first airship shot down over London. Leefe Robinson was awarded the Victoria Cross for this feat, only to be killed the following year on the Western Front.

ZEPPELIN DESTROYER
The Ranken dart was designed to be dropped on an airship from above and explode after penetrating its outer cover. Explosive and incendiary ammunition eventually made airships fatally vulnerable to attack.

BURNT-OUT SHELL
The Zeppelin L 33 was one of two German airships shot down over Britain on the night of 23–24 September 1916. It was destroyed by a combination of anti-aircraft fire and air attack.

ADAPTABLE TWO-SEATER
The B.E.2c was introduced in 1914. Originally intended for reconnaissance duties, it quickly expanded to other roles, such as light bomber and Home Defence fighter.

FLYING THE GIANTS

THE LARGEST AEROPLANES THAT FLEW in the Great War were the heavy bombers the Germans called *Riesenflugzeug* ("giant aircraft"). The most famous of these "R-planes", the Zeppelin Staaken R.VI, was introduced into Germany's arsenal in September 1917, to join the smaller Gothas in their mass raids on Britain. These huge machines had two pilots sitting side by side operating steering wheels like those on airships. They also carried mechanics on

board, who tended the engines in flight. Arthur Schoeller, commander of a R.VI, wrote a vivid account of a night raid on London.

He described how 40 ground crew prepared the "Giant" for action, loading the bomb bay, filling the vast fuel tanks, and tuning the four 250hp engines. After a light supper, the eight-man crew headed out to the aircraft, "whose idling engines sing a song of subdued power". Six R-planes taxied out on to the take-off strip. Their engines at full throttle emitting a deafening roar, the heavily laden machines slowly rose into the air and headed out across the sea into black nothingness. With only calculations of time, speed, and direction to tell them where they were, Schoeller and his observer had begun to suspect they might be lost when, to their relief, they saw searchlights probing the sky: "bright beams making glowing circles in the thin

overcast clouds". They must be over England. An airfield, lit by flares for use by British night fighters, appeared startlingly bright beneath them. They bombed it in passing, while their machine-gunners blasted away at the searchlights. Then, through a break in the cloud, they spotted the Thames. Soon the observer, who had moved to the open observation post in the nose of the aircraft, was pressing the bomb-release keys, hoping to hit the London docks somewhere below. After dropping their bombload, they turned for home along the Thames with bursts of anti-aircraft fire dangerously close. A shell splinter tore the fabric of the upper wing but caused no serious damage. Heading back across the sea their troubles were far from over. In sight of the Belgian coast, all four propellers stopped because of a frozen fuel line. They glided as far as dry land, where their flares lit up a cratered terrain. Schoeller stalled the aircraft just above the ground and pancaked, smashing the landing gear and one wing. But his crew was safely home.

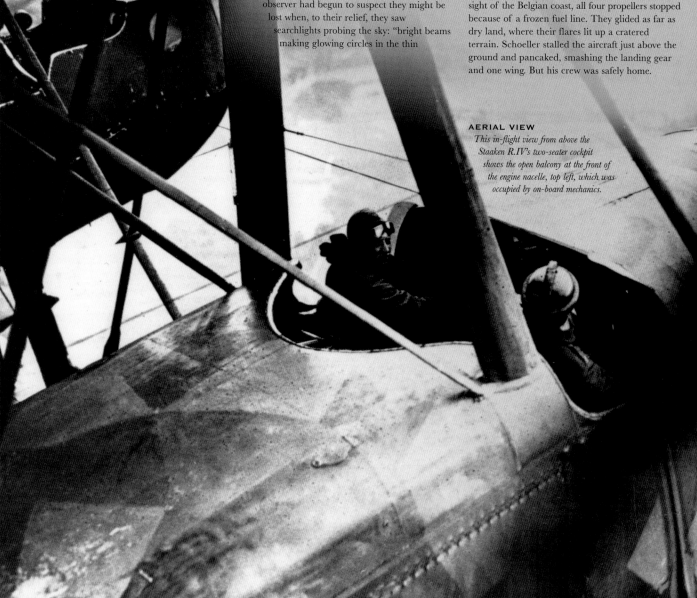

AERIAL VIEW
This in-flight view from above the Staaken R.IV's two-seater cockpit shows the open balcony at the front of the engine nacelle, top left, which was occupied by on-board mechanics.

"GIANT AIRCRAFT"
The massive scale of the "R-planes" is shown by this image of an R.IV surrounded by people. Its wingspan measured 42m (138ft) and it weighed over 13,000kg (28,600lb).

DAREDEVIL MECHANICS
If the aircraft's engines needed attention, mechanics would move around its airframe even in mid-flight.

calculations could be thrown completely by the effect of wind. If navigators failed to realize that they had been blown off course, they were lost and had no way to regain their bearings.

Of course, a fair number of airships heading for London did find it. The city's lights were hidden by a strictly enforced blackout, but they were able to use the river Thames as a navigational aid, since it offered a readily indentifiable shape for them to follow. They also had searchlight and gun batteries marked on their maps, not only so they could avoid them but also to use them for orientation around the darkened city.

Airships did not attack a target in formation. Each came from a different direction, at staggered intervals. Once through the ring of defences around London, the method of attack was simple – to fly as fast as possible in a straight line across the city, releasing their explosive and incendiary bombs as they went.

The first airship raids on Britain were carried out in January 1915; the first raid on London the following May. Up to the late summer of 1916, although sporadic, the raids could be counted a success. Some inflicted substantial damage – for example, on 8 September 1915, Captain Heinrich Mathy's L 13 killed 22 people and injured 87 in a single pass across London. Even wandering lost over England in the middle of the night was not necessarily a waste of energy. One of the most damaging raids of the war came in January 1916, when nine airships trying to bomb Liverpool became totally confused in the darkness and arrived instead over the cities of the Midlands, where no blackout was in force. 70 people were killed in the bombing.

The British were forced to divert valuable resources of aircraft, pilots, and guns from the Western Front as the public demanded better protection. But these resources, accompanied by the crucial introduction of explosive and incendiary ammunition, succeeded in tipping the scales fatally against the zeppelins. In late August and September 1916, the Germans launched their most ambitious airship bombing offensive. It was a disaster. In a series of raids, four airships were lost to ground fire or pursuit aircraft. Among those who died was the experienced Mathy; he jumped to his death from his burning zeppelin rather than roast alive.

High fliers

The Germans responded to this setback with a new type of airship that was bigger but lighter than previous ones. These "height-climbers" regularly flew at over 5,600m (16,000ft), out of reach of ground fire and difficult for aircraft to attack. But for the unfortunate crews, missions at high altitude were a severe trial. For hours on end they endured freezing cold; they found it hard to breathe in the thin air but using their crude oxygen equipment made them feel nauseous; rapid changes of pressure in ascending and descending gave them the bends. And for all this suffering they inflicted only limited damage.

However, the technical achievements of German airships were extraordinary. In November 1917, the L 59 flew a total of 6,760km (4,200 miles) non-stop on an abortive mission to Africa. In the same year, L 55 established an enduring altitude record of 8,400m (24,000ft).

But turning technical capability into military effectiveness proved impossible. To the end, naval airship leader Peter Strasser still dreamed of a decisive coup – an airship raid across the Atlantic to devastate New York. However, in August 1918 he was shot down on a final hopeless mission over the North Sea.

> "Inside the fuselage the pale glow of dimmed lights outlines the chart table, the wireless equipment, and the instrument panel… Under us is a black abyss."
>
> **HAUPTMANN ARTHUR SCHOELLER**
> COMMANDER OF AN R-PLANE, DESCRIBING A NIGHT FLIGHT TO BRITAIN IN MARCH 1918

SKY JUMPER
In this stereoscopic photograph, British soldiers look at the hole left in the ground by the body of a man who jumped from a burning zeppelin over Billericay, Essex, England. Using a stereoscopic viewer, it is possible to see the hole in 3-D.

Arrival of the Gotha

From the start of the war, the Germans had wanted to use heavier-than-air aircraft for strategic bombing, but lacked a suitable machine until the advent of the *Friedrichshafen* and Gotha twin-engined bombers. The Gothas began raids on Britain in the summer of 1917 with impressive effect. Both the air defences and the civilian population were unprepared for formations of bombers attacking in broad daylight. When 14 Gothas appeared over London for the first time on 13 June, crowds of people ran out into the streets to watch them. The bombs falling on unsheltered civilians killed 162 people.

The Gothas flew faster than zeppelins and were far harder to shoot down. When flying in formation, they could dish out heavy punishment to pursuit aircraft from the combined firepower of their machine guns. This relative invulnerability meant that they could operate initially in daylight and later, when air defences improved, on moonlit, rather than moonless, nights. This made it much easier for them to locate their targets. Also, because they could be produced far more quickly and cheaply than airships, they could be deployed in substantially greater numbers – 43 were used in the largest single raid on London.

Sporadic night attacks on London by Gothas and the even larger R-planes (see page 98) continued through the winter and spring of 1917–18. Paris also became a regular target for the German bombers. Parisians and Londoners grew used to a routine of air-raid warnings and all-clears, huddling through the long dangerous hours in the cellars and basements of their houses or in the tunnels of underground railway stations. The raids petered out in May 1918, since, from this time on, all Germany's resources were devoted to the desperate battle on the Western Front. But

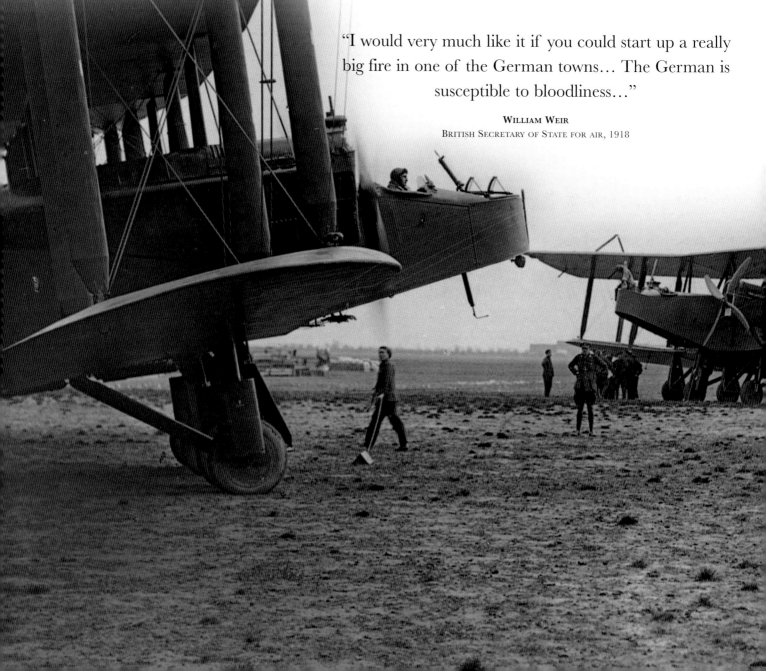

"I would very much like it if you could start up a really big fire in one of the German towns... The German is susceptible to bloodliness..."

WILLIAM WEIR
BRITISH SECRETARY OF STATE FOR AIR, 1918

aeroplanes had proved their relative effectiveness: in 51 airship raids on Britain, 557 people were killed and 1,358 injured; in 52 aeroplane attacks 857 were killed and 2,508 injured.

Hitting back

The damage inflicted by both airships and bombers was, in truth, a puny return for the investment of men and material. But its psychological impact was far-reaching. Shrill demands for a more effective defence of the civilian population caused a crisis in British policy on the air war. And the demand for revenge against the Germans gave fresh impetus to moves towards creating an Allied strategic bomber force.

Throughout the war, debate raged among both politicians and military commanders in the Allied countries as to the merits or demerits of bombing enemy cities and factories. The stalemate and heavy losses on the Western Front naturally encouraged speculation that bombers might be able to end the deadlock by breaking the enemy's will to fight or destroying his industrial capacity. It was certainly tempting to grasp at any alternative to yet another apparently futile infantry offensive. But those who saw the

GIANNI CAPRONI

GIANNI CAPRONI (1886–1957) WAS ITALY'S most prominent aircraft designer and manufacturer during World War I. After building his first aircraft in 1910, his efforts to win contracts from the Italian army were frustrated until he was befriended by the air-minded Colonel Giulio Douhet. By 1914 Caproni had already designed an innovative prototype monoplane fighter, the Ca 20, with a forward-firing gun mounted on top of the wing. Under the influence of Douhet, however, Caproni devoted himself primarily to the production of large bomber aircraft.

By the time Italy entered the war in 1915, Caproni three-engined biplane bombers (Ca 42s) were coming into service. By 1918 an estimated 70,000 workers were building Caproni biplanes and triplanes, not only in Italy, but also under licence in France and the United States.

Caproni joined Douhet in advocating a strategic air offensive that would have used thousands of heavy bombers to batter Germany into submission. This did not happen, but his arguments were influential, not least with American military aviation enthusiasts. After the war Caproni went on building civil and military aircraft.

BOMBER MAN
Caproni designed a number of heavy bombers, including the Ca 33, which played a major role in the allied bombing campaign.

CAPRONI CA.42
The Ca 4 series of triplanes were used on Italian bombing missions against Austria. Despite being less common than the Caproni biplanes, they were more powerful.

---

trenches of the Western Front as the place where, inevitably, the war would eventually be won and lost, argued against any diversion of aerial resources away from direct support for the hard-pressed armies on the ground.

At first the debate was largely theoretical, since the Allies simply did not possess the equipment to carry out a strategic bombing campaign. French airmen made a brave attempt at bombing Germany from 1915 using slow, clumsy Voisin 8s, first by day and then by night, but their losses were high and the damage they inflicted negligible. Strangely, at first those Allied countries who generally had less-strong air forces possessed the most powerful bombers. Russia deployed Sikorsky's four-engined Il'ya Muromets (see pages 62–63) as bombers and reconnaissance aircraft.

HEAVY BOMBERS
Entering service in the late summer of 1918, the Handley Page O/400 bomber was the key aircraft in the newly formed Independent Air Force. Large formations of O/400s – up to 40 bombers at a time – carried out night raids deep inside Germany. The aircraft's bombload could include a 750kg (1,650lb) "blockbuster" bomb.

Italy, which joined in the war on the Allied side in 1915, was at that time the only power to have an aircraft specifically designed for bombing, the Caproni Ca.1. Italy also had one of the most aggressive and influential advocates of strategic bombing in Colonel Giulio Douhet. His outspoken views and undisciplined behaviour brought him in conflict with his superiors, and he spent part of the war in prison. But Italy nonetheless used Caproni bombers to attack the cities of its nearest enemy, Austria-Hungary.

Call for action

By 1918 there was a solid weight of opinion in the Allied countries calling for a bombing offensive against Germany, and frustration among political leaders at the military's failure to deliver it. In April 1918, Britain created the world's first independent air force, the Royal Air Force, to replace the army's Royal Flying Corps and the navy's Royal Naval Air Service. It was intended, among other things, to help give Britain more effective air defences, and to promote a strategic air offensive against Germany. The Independent Force of bombers was set up in June to carry out this offensive. Meanwhile the French commander-in-chief, Marshal Henri Pétain, called for a fleet of heavy bombers to "paralyze

AIR WAR AT SEA

IN THE EARLY YEARS OF AVIATION, navies were, on balance, more aware of the potential of aircraft than were armies. This was especially true of Britain's Royal Navy, where the influence of an imaginative, progressive-minded First Lord of the Admiralty, Winston Churchill, made itself felt. But when war started, there was no effective way of taking aircraft to sea with the fleet. In September 1914, the Royal Navy converted three cross-Channel steamers into seaplane carriers. The seaplanes were winched off the ship to take off from the sea, and lifted back on board after their mission. It sounded simple and effective, but it was not. The seaplanes found taking off and landing at sea impossible, except under highly favourable conditions – they needed exactly the right degree of swell.

Before the war, both the US and British navies had experimented with launching aircraft off a platform on the deck of a ship. The Royal Navy resumed these experiments in earnest in 1917. A light battlecruiser, HMS *Furious*, had its forward guns removed and replaced by a take-off deck. The idea was that the latest land-based aircraft, superior in performance to seaplanes, would take off and land from the ship.

The manoeuvre was undertaken by Squadron Commander E.H. Dunning flying the highly agile Sopwith Pup biplane. Taking off was relatively easy if the ship sailed into the wind, but landing was another matter. Dunning managed it twice by matching his speed to that of the ship so he

PLATFORM LAUNCH
One system for launching aircraft at sea was to put them on a lighter – a sort of barge – towed behind a light, speedy vessel (left). When travelling fast enough, this would generate the windspeed, and hence the lift, needed for take-off.

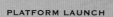

HOISTED OVERBOARD
Seaplanes offered a way of carrying aircraft with a fleet. Machines like this Short seaplane were winched overboard from a ship, took off from the sea, and might with luck land safely alongside and be lifted back on board.

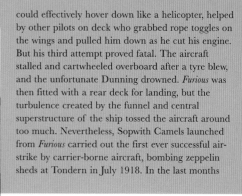

could effectively hover down like a helicopter, helped by other pilots on deck who grabbed rope toggles on the wings and pulled him down as he cut his engine. But his third attempt proved fatal. The aircraft stalled and cartwheeled overboard after a tyre blew, and the unfortunate Dunning drowned. *Furious* was then fitted with a rear deck for landing, but the turbulence created by the funnel and central superstructure of the ship tossed the aircraft around too much. Nevertheless, Sopwith Camels launched from *Furious* carried out the first ever successful air-strike by carrier-borne aircraft, bombing zeppelin sheds at Tondern in July 1918. In the last months the economic life of Germany and its war industries by methodical and repeated action against principal industrial cities…". In 1918 the Allies had the de Havilland-designed Airco D.H.9 and excellent Breguet Br.14 as day bombers and the Handley Page O/400 and Caproni biplanes and triplanes as night bombers. The Handley Page, although nothing like as big as the German R-planes, could carry a maximum bombload of 900kg (2,000lb), and formed the backbone of the Independent Force. Other heavy bombers, including the French Farman Goliath and the British Vickers Vimy, were under construction in 1918 but arrived too late to see service.

Lesson in terror

In the summer and autumn of 1918, formations of up to 40 Allied bombers flew raids deep into Germany. Predictably, bad weather and unreliable aircraft limited the effectiveness of the bomber offensive. But civilians in cities such as Frankfurt and Mannheim were taught the terror of air-raids that had already been experienced by inhabitants of Paris and London. Allied airmen were always under orders to aim for precise targets, such as factories or communications centres. But Allied political leaders were keen to affect civilian morale. The British Secretary of State for air, William Weir, told Hugh Trenchard, the commander of the Independent Force, not to be scrupulous in respect for civilian life: "If I were you," he wrote, "I would not be too exacting as regards accuracy in bombing railway stations in the middle of towns. The German is susceptible to bloodiness, and I would not mind a few accidents due to inaccuracy."

In fact, the air commanders were generally more sceptical about strategic bombing than the politicians. Trenchard knew he was supposed to use his force to bomb German cities and factories, but more often he directed it against tactical objectives such as airfields and communications centres behind the front. The same is true of US General Billy Mitchell. Both Trenchard and Mitchell later became advocates of strategic bombing, but they devoted themselves in the last months of the war to the tactical use of airpower.

The evidence of World War I was that, at current levels of technology, strategic bombing

FIRST INDEPENDENT AIR FORCE
This recruiting poster invites volunteers to join the Royal Air Force (RAF), the world's first independent air force. Created by combining Britain's army and navy air arms in April 1918, the RAF was intended to prioritize the protection of Britain against air attack and promote the strategic bombing of enemy factories and cities.

could neither seriously disrupt industrial production nor significantly weaken a population's will to fight. Bombing was costly and inaccurate. Its chief positive effect lay in forcing the enemy to divert resources to air defence.

The building and operation of large bomber aircraft was nonetheless an important step in the progress of aviation. Bomber aircrews had accumulated extensive experience of long-distance flight and night flying, and the large aircraft they flew carrying bombs could, with relatively small modifications, carry passengers or freight instead. Strategic bombing in the Great War helped pave the way for the development of commercial aviation – as well as the devastation of Dresden, Tokyo, and Hiroshima.

> "It is probable that future war will be conducted by a special class, the air force, as it was by the armored knights of the Middle Ages."
>
> BRIGADIER GENERAL WILLIAM "BILLY" MITCHELL
> *WINGED DEFENSE*, 1924

FIRST DECK LANDING
On 2 August 1917, pilot E.H. Dunning made the first successful landing on the deck of a moving ship. He was flying a Sopwith Pup, a popular, highly manoeuvrable aeroplane. An unsuccessful attempt five days later ended in his death.

of the war, an ocean liner was coverted into HMS *Argus*, the first true aircraft carrier. The funnel was hidden away at the back of the ship, allowing a long, unobstructed flight deck. British naval commanders planned to use *Argus* for a Pearl Harbor-style strike against the German fleet in port. Modern naval aviation had been born.

WWI BOMBERS AND GROUND ATTACK AIRCRAFT

WHILE **WWI** GENERALS welcomed the chance to launch air raids behind enemy lines, the effectiveness of tactical bombing was limited by the small bomb payload that bombers could carry and by their vulnerability to enemy fighters, which forced them to stick to inaccurate night raids for much of the war. Still, by 1918, tactical bombers such as the Breguet 14B2 and Airco D.H.4 were making a significant impact. While the British chiefly used fighters to attack enemy ground forces during land battles, the Germans used dedicated ground-attack aircraft, such as the Halberstadts and the all-metal Junkers J 4, able to take considerable punishment as its crew flew low over the trenches. Some very large, multi-engined aircraft were developed for strategic bombing. The biggest were the Zeppelin R-planes, which supported Gothas in the bombing campaign against England.

RUSSIAN GIANT
Sikorsky's Il'ya Muromets was the world's first four-engined bomber. During WWI, over 75 Il'ya Muromets were deployed in a special squadron on the Eastern Front for bombing and reconnaissance missions from 1915 onwards.

A.E.G. G.IV

Over 540 of the G series of excellent twin-engined medium bombers were built by Allegemeine Elektrizitats Gesellschaft from 1915. The most numerous (around 400) was the G.IV, which entered service in late 1916 as a short-range tactical bomber but was later relegated to photo-reconnaissance duties.

ENGINE	2 x 260hp Mercedes D.IVa 6-cylinder inline	
WINGSPAN	18.4m (60ft 5in)	LENGTH 9.7m (31ft 10in)
TOP SPEED	165kph (103mph)	CREW 3
ARMAMENT	2 x machine guns; 400kg (882lb) bombload	

Airco D.H.4

Designed in 1916 as a high-speed day bomber, the D.H.4 was first used as a bomber over the Western Front in March 1917. It was also used for reconnaissance, anti-submarine patrols, and even as a night fighter.

ENGINE	375hp Rolls-Royce Eagle VII water-cooled V-12	
WINGSPAN	12.9m (42ft 4in)	LENGTH 9.4m (30ft 8in)
TOP SPEED	230kph (143mph)	CREW 2
ARMAMENT	4 x machine guns; 209kg (460lb) bombload	

Breguet 14B2

On 21 November 1916, Louis Breguet personally flew the Breguet 14 prototype, and by 1926, over 8,000 had been built. As well as the bomber, reconnaissance and training versions were also produced. The type had a long postwar career with many foreign air forces.

ENGINE	300hp Renault 12 Fox V-12	
WINGSPAN	14.4m (47ft 2in)	LENGTH 8.9m (29ft 1in)
TOP SPEED	177kph (110mph)	CREW 2
ARMAMENT	3 x machine guns; 300kg (661lb) bombload	

Caproni Ca 42 (Ca 4)

Together with Sikorsky, Caproni pioneered the construction of giant aircraft. The three-engined Ca 30 appeared in 1913, and the series developed throughout WWI, with most variants employing the single pusher and double tractor engine layout.

ENGINE	3 x 270hp Isotta-Fraschini water-cooled V-6	
WINGSPAN	29.9m (98ft 1in)	LENGTH 13.1m (43ft)
TOP SPEED	126kph (78mph)	CREW 4
ARMAMENT	4 x machine guns; 1,450kg (3,197lb) bombload	

Gotha G.V

If to the British public every airship was a Zeppelin, then every German bomber was a Gotha. The G.IV first undertook daylight raids over southern England in May 1917, operating with virtual impunity at 4,575m (15,000ft). The improved G.V entered service in September and continued night raids to May 1918. Although a typical bombload was only six 50kg (110lb) bombs, the Gothas dropped nearly 85,000kg (187,435lb) of bombs on Britain for the loss of 24 aircraft.

ENGINE	2 x 260hp Mercedes D.IVa 6-cylinder inline	
WINGSPAN	23.7m (77ft 9in)	LENGTH 12.2m (40ft)
TOP SPEED	140kph (87mph)	CREW 3
ARMAMENT	2 x machine guns; 500kg (1,102lb) bombload	

Halberstadt CL.II

Although designed as an escort fighter, the CL.II was first used for close support in September 1917, when 24 aircraft attacked a British division. By the beginning of 1918, the CL.II was joined by the improved CL.IV. Their use in close support of the German infantry during allied counter-attacks was frequently crucial.

ENGINE	160hp Mercedes D III 6-cylinder water-cooled inline		
WINGSPAN	10.8m (35ft 4in)	LENGTH	7.3m (23ft 11in)
TOP SPEED	165kph (103mph)	CREW	2
ARMAMENT	2 x fixed and 1 x movable machine guns; 50kg (110lb) anti-personnel grenades or 10kg (22lb) bombload		

Handley Page 0/400

In comparison with Russia and Italy, Britain was slow to build a heavy bomber. A requirement for a "bloody paralyser of an aeroplane" was eventually met by the HP O/100, which entered service in November 1916. This led to the much more numerous O/400. Very large orders were placed with some 550 being built in Britain and over 100 manufactured in the US. The type began as a day bomber on the Western Front in April 1917 and continued in RAF service, as a transport, until 1920. Four were subsequently converted to civil airliners for overseas route-proving.

ENGINE	2 x 360hp Rolls-Royce Eagle VIII water-cooled V-12		
WINGSPAN	30.5m (100ft)	LENGTH	19.2m (62ft 10in)
TOP SPEED	156kph (97mph)	CREW	4
ARMAMENT	5 x .303in machine guns; 907kg (2,000lb) bombload		

Boxtail configuration

Forward machine gunner

Junkers J 4 (J.I)

When Dr Hugo Junkers built his first aeroplane, the J 1, in 1915, it appeared as a remarkable, pioneering, all-metal monoplane skinned with thin sheet iron. The success of this advanced design and its successor the J 2, led to an order for an armoured biplane specifically for low-level reconnaissance and close support of the army. The result was the J.I, built under the factory identity of J 4, which retained the all-metal construction but was skinned with a corrugated aluminium alloy, a manufacturing technique that continued with the WWII Ju 52. An ash tailskid was the only wooden component. The first angular J.I reached the squadrons in France towards the end of 1917. Although somewhat cumbersome and tricky to handle on the ground, the new Junkers were immediately popular for their strength and armour protection. A total of 227 J.Is were built and nearly 190 served on the Western Front.

Corrugated aluminium skin

200hp Benz Bz.VI inline

ENGINE	200hp Benz Bz.IV 6-cylinder inline		
WINGSPAN	16m (52ft 6in)	LENGTH	9.1m (29ft 10in)
TOP SPEED	155kph (97mph)	CREW	2
ARMAMENT	2 x fixed and 1 x movable machine guns		

Sikorsky Il'ya Muromets S-23 V

The Il'ya Muromets series of giant biplanes were developed from the world's first four-engined aircraft, *Le Grand*, which first flew in May 1913. During the war, nearly 80 were produced, in a number of variants, and served successfully with the Russian Air Force.

ENGINE	4 x 150hp Sunbeam V8 water-cooled inline		
WINGSPAN	29.8m (97ft 9in)	LENGTH	17.1m (56ft 1in)
TOP SPEED	121kph (75mph)	CREW	4–7
ARMAMENT	7 x machine guns; 522kg (1,150lb) bombload		

Zeppelin Staaken R.VI

The most remarkable aircraft built by the Germans during WWI were the "R" (Riesenflugzeug) type giant aeroplanes with four, five, or six engines. Of all the heavy bombers, those designed by Zeppelin at their Straaken works were the most remarkable. While not the largest, the R.VI was the only version to be produced in reasonable numbers. The first attack on England was on 17 September 1917, introducing the British public to an even more psychologically terrifying weapon than the Gotha.

ENGINE	4 x 260hp Mercedes D.IVa 6-cylinder inline		
WINGSPAN	42.2m (138ft 6in)	LENGTH	22.1m (72ft 6in)
TOP SPEED	135kph (84mph)	CREW	7
ARMAMENT	4–7 x Parabellum machine guns in nose, dorsal, and ventral positions; 2,000kg (4,409lb) bombload		

THE GOLDEN AGE

IN THE AFTERMATH OF WORLD WAR I, aircraft manufacturers struggled to survive as air forces were run down. But, defying the postwar recession and the Great Depression that followed, the 1920s and 1930s blossomed into a "Golden Age" of aviation. Pilots were among the most celebrated heroes of the day, and the public thrilled to the excitement of air races and the feats of record-breakers such as Charles Lindbergh and Amelia Earhart. Helped by pioneering long-distance survey flights, airlines began to stretch their networks across and between continents. Great airships brought unparalleled luxury to air travel, challenged only by the stately flying boats. The advent of sleek all-metal monoplanes led to radical advances in speed and range, while improved flight instruments and navigation devices made aircraft increasingly safe to fly.

STREAMLINED SUPERSTAR
This aerial view of a TWA Douglas DC-3 passenger aeroplane shows it flying over midtown Manhattan. The DC-3, which came into service in the mid-1930s, could carry 21 passengers and marked a turning point in aviation by making a profit on passenger services alone.

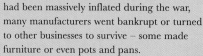

BLAZING THE TRAIL

PILOTS RISKED THEIR LIVES EXPLORING REMOTE PARTS OF THE GLOBE, CROSSING OCEANS, AND PUSHING AIRCRAFT TO THE LIMITS OF SPEED AND ENDURANCE

> "I have lifted my plane… for perhaps a thousand flights and I have never felt her wheels glide from the earth into the air without knowing the uncertainty and the exhilaration of first-born adventure."
>
> **BERYL MARKHAM**
> PILOT IN KENYA IN THE 1930s

THE PERIOD between the World Wars has often been called the Golden Age of aviation. But the first years after 1918 certainly did not look or feel like the beginning of a Golden Age to either pilots or aircraft manufacturers. The end of World War I was little short of catastrophic for the aircraft-manufacturing business. The market was awash with surplus military aircraft. In the United States, you could buy a Curtiss Jenny trainer for a knock-down $300. With demand collapsing, the whole United States aircraft industry manufactured just 328 new aircraft in 1920. In Europe, where aircraft-making capacity had been massively inflated during the war, many manufacturers went bankrupt or turned to other businesses to survive – some made furniture or even pots and pans.

Only a minority of the tens of thousands of military pilots stayed in the army or navy. For those who returned to civilian life still determined to make a living out of flying, the peacetime world offered mostly precarious employment. In Europe, fledgling passenger and airmail services absorbed a number of aircrew. Skywriting, first demonstrated by Major Jack Savage in Britain in 1922 and adopted with enthusiasm in the United States, gave some pilots a living as a branch of the burgeoning advertising industry. Other respectable employment for fliers and their aeroplanes was found in crop-dusting, also an innovation of the 1920s, and survey work using aerial photography. Hollywood, which responded to the appeal of aerial adventure with a clutch of movies, had a constant need for stunt pilots. Quite a few World War I fliers – including German ace Ernst Udet – found themselves recreating air combat for the movie cameras.

SOLO PIONEER
Amelia Earhart captured the imagination of America after she became the first woman to complete a solo transatlantic flight in 1932.

The barnstormers

The most prominent role of aviation in the United States was as a fairground sideshow or daredevil circus. At the bottom end of the scale, small groups of "gypsy fliers" toured the remote towns of rural America, putting on a show for local people and taking them up for a ride at a dollar a time. This was life on the breadline. Asked what was the most dangerous thing about his occupation, one gypsy flier in the early 1920s said, "The risk of starving to death". But at the other extreme, some gifted pilots

TINGMISSARTOQ
With this Lockheed Sirius seaplane, Charles Lindbergh and his wife Anne Morrow explored flight routes across both the Pacific and Atlantic Oceans. The aircraft was christened Tingmissartoq *(meaning "the man who flies like a big bird") by an Inuit boy in Greenland.*

grew famous and tolerably well off, taking aerial acrobatics to new heights in highly publicized "barnstorming" displays that drew huge crowds.

With no safety regulations in place, barnstormers developed an astonishingly risky repertoire, adding to the standard stunts of the aerobatic pilot a whole range of circus tricks such as wing-walking, crossing from one aeroplane to another in flight, hanging by a trapeze underneath the plane, or apparent death falls avoided by a parachute opening at the last moment. Since audiences were not prepared to wait for a chance accident to liven up the day, the pilots staged pleasingly spectacular crashes as part of the show. The risk of death or crippling injury was accepted as part of the show.

Setting new targets

Despite the rundown state of aviation, the public's fascination with flight remained intense. As before the war, cash prizes were made available by "air-minded" press magnates and other wealthy individuals for record-breaking long-distance flights. But now, after the rapid improvements in aircraft performance brought by the war, the flights were to far more distant destinations. No one was impressed any longer by flights between European cities or

DAREDEVIL STUNTS
This wing-walker is bracing himself to leap from one aeroplane to another during a 1926 barnstorming show. After the war, many unemployed pilots (including Charles Lindbergh) toured the US, thrilling small-town America with their daredevil stunts. Unfortunately the barnstormers also reinforced the public's belief that aeroplanes were dangerous, hindering the efforts of entrepreneurs trying to develop passenger airlines.

BESSIE COLEMAN

AFRICAN-AMERICAN PILOT Bessie Coleman (1892–1926) grew up in the Texas cotton fields, before moving to Chicago in 1915 and training as a beautician. There she decided that she wanted to become a flier, but she was refused entry to flight-training schools because of prejudice against her colour. Undeterred, Coleman went to France in 1920, inspired by her brothers' tales of women fliers and racial tolerance, returning to the USA in 1921 with an international pilot's licence.

"Queen Bess" went on to become one of the most famous barnstormers. The curiosity aroused by the novel spectacle of a black woman pilot undoubtedly helped her achieve celebrity status. Her ambition was to raise enough capital to open a flying school for African-Americans, but tragically she did not live to realize it. In 1926, while practising for a show in Jacksonville, Florida, her plane went into a tailspin and she was thrown to her death.

ONE-IN-THREE CHANCE?

Of the three US Navy Curtiss flying boats (designated NC-1, NC-3, and NC-4) that set out to cross the Atlantic, the NC-4 (below) was the only one to complete the 6,280-km (3,925-mile) journey. On landing in Lisbon, pilot Lieutenant-Commander Albert Read sent a radio message to his base: "We are safely on the other side of the pond. The job is finished."

FIRST ATLANTIC CROSSING

The US Navy Curtiss NC-4 flying boat made its first flight on 30 April 1919, just over a week before the start of its historic transatlantic crossing. This four-engined aircraft had a wingspan of 38.4m (126ft), was 20.8m (68ft 3in) long, and weighed over 7,000kg (16,000lb) when empty. Its maximum attainable speed was 146kph (91mph).

across the Mediterranean – these were routes now being targeted for regular commercial services. Whole continents and oceans had to be crossed. Pilots were to take aircraft to every remote corner of the world, traversing mountainous wastes and impenetrable jungle.

Transatlantic challenge

The most obvious unconquered space on the globe in 1918 was the Atlantic Ocean. Before war broke out in Europe in 1914, Glenn Curtiss had already been planning to send a seaplane across the Atlantic. With the end of hostilities the project was picked up again. In May 1919, three US Navy Curtiss flying boats set out from Newfoundland for Lisbon, via the Azores. Only one of the aircraft, NC-4 captained by Lieutenant-Commander Albert C. Read, completed the journey, and that after many interruptions. The trip took 19 days, including 42 hours flying time. Although the

NC-4 was the first to cross the ocean, there remained an unclaimed prize put up by Lord Northcliffe for the first non-stop transatlantic flight. Australian pilot Harry Hawker and navigator Kenneth Mackenzie-Grieve made an attempt in May 1919 that they were lucky to survive; after coming down in the ocean they were rescued by a passing steamer. The first successful non-stop Atlantic crossing was made the following month by two

British airmen, Captain John Alcock and Lieutenant Arthur Whitten Brown, in a modified Vickers Vimy bomber (see panel, right).

As with the long-distance races of the pre-World War I era, such flights could be taken either as revealing the potential of aviation or underlining its shortcomings. For instance, in 1919 the Australian government put up a prize of £10,000 for the first of their countrymen to fly from Britain to Australia in under 30 days. The prize was won by brothers Ross and Keith Smith, who, with two other crew members, flew a Vickers Vimy the distance of around 20,000km (12,000 miles) in 27 days 20 hours. They were congratulated by Winston Churchill, then Britain's Secretary of State for air, in a telegram: "Well done. Your great flight shows conclusively that the new element has been conquered for the use of man." But of the five aircraft that set out on the Australia flight, the Smiths' was the only one to arrive. Four airmen lost their lives in the attempt and another three narrowly escaped death. The Smiths themselves survived some uncomfortable moments, especially when their aircraft clipped the tops of trees taking off from a too-short racecourse in Rangoon, Burma. Clearly the "new element" was far from being conquered, although its challenge was being heroically confronted.

It was not surprising that many people still considered airships to have better potential for long-range passenger transport. In July 1919, for example, the British airship R 34, a close copy of a captured German zeppelin, flew across the Atlantic from Scotland to Mineola, New York, and back with 31 people on board. It showed that airships were still well ahead in load-carrying, if not in speed – the outward leg of the R 34's journey, against headwinds, took four and a half days.

RETURN TICKET
On 13 July 1919, the British military airship R 34 set off on the first two-way transatlantic crossing, carrying 31 people (including one stowaway) from Scotland to New York and back in a total flying time of under eight days.

NON-STOP ACROSS THE ATLANTIC

ON 14 JUNE 1919, two British airmen, Captain John Alcock and his navigator, Lieutenant Arthur Whitten Brown, took off from a grass airstrip at St John's, Newfoundland, in a modified Vickers Vimy bomber. Their aim was to win the *Daily Mail's* £10,000 prize for the first non-stop flight across the Atlantic. Some 3,040km (1,890 miles) of ocean separated them from Ireland, the nearest landfall to the east. Travelling in an open cockpit, deafened by the racket of the engines, and with only sparse and unreliable instruments, the journey was a test of endurance for the airmen as well as their machine. By ill chance, they also ran into some awful weather. At one point, flying through turbulent storm clouds with zero visibility, disoriented and blinded by lightning flashes, they span down to barely 35m (100ft) above the ocean before emerging from the clouds just in time for Alcock to regain control and pull up above the waves. Hail and heavy snow battered and froze the airmen; Brown had to climb on to the wings to de-ice the engines with a pocket knife. After 16 hours battling with the elements, Alcock and Brown were exhausted and short on hope when a dark line of shore grew faintly visible through the grey mist. They descended over Galway and landed inelegantly nose-down in a soft Irish bog at 9.40am on 15 June. The airmen were given a hero's welcome in London and were knighted for their achievement.

A HERO'S WELCOME
Alcock and Brown paraded through London in the back of an automobile after making the first non-stop transatlantic flight, from Newfoundland to Galway, Ireland, in 16 hours 27 minutes. For Alcock triumph was brief: he was killed in an air crash six months later.

SOFT LANDING
Alcock and Brown's Vickers Vimy ended its epic non-stop crossing by crash-landing in Derrygimla bog near Clifden, Galway, on 15 June 1919.

Flying around the world

It was indicative of the backwards state of US aviation in the early 1920s that the most prominent American long-distance flight was a propaganda effort to attract government funding and public support for the Army Air Service. The ever-publicity-conscious General Billy Mitchell had the newly formed Douglas aircraft company produce modified versions of the float planes it was manufacturing for the navy. Dubbed the Douglas World Cruisers, these four aircraft left Seattle to make the first round-the-world flight in April 1924. It was a hard and halting journey, but two of the World Cruisers eventually completed the circuit five months later, having spent a total of fifteen and a half days in the air.

In Europe, where commercial aviation was more advanced and governments committed to aviation as a focus of national prestige, long-distance flying in the 1920s focused on "trail-blazing" imperial routes, preparing the way for mail or passenger services that would link home countries to their far-flung colonies. Fliers were sent out to map remote territories, identify suitable landing sites, test weather conditions, and investigate ways over or around natural obstacles. This was the inspiration behind French pilots' pioneering flights across the Sahara desert to Dakar, the hub of French West Africa, and across Asia to Hanoi in Indochina; of Dutch aircraft opening up routes across the Middle East and southeast Asia to their Indonesian empire; and flights by British airmen via India to Australia, and via Cairo to South Africa.

Lacking an empire, Germany explored routes eastwards across the Soviet Union, with which it had formed a close relationship as a fellow "pariah state". In July and August 1926, the newly formed Deutsche Luft Hansa (DLH, later Lufthansa) airline sent two Junkers G 24s, state-of-the-art all-metal trimotor monoplanes, on a 10,000-km (6,000-mile) flight from Berlin to Peking via the Soviet cities of Moscow and Omsk. It was in its way a journey as adventurous as any in Africa or southeast Asia.

BYRD AND AMUNDSEN BATTLE TO THE NORTH POLE

IN THE 1920s Norwegian explorer Roald Amundsen (1872–1928) – the man who had beaten Captain Scott to the South Pole in 1911 – launched a bid to become first to fly over the North Pole. After Amundsen's first attempt failed in 1925, he found himself in a race with US Navy Commander Richard E. Byrd (1888–1957), who had the same ambition. Byrd signed up two pilots (a mail pilot called Charles Lindbergh applied for the job but was too late) and took a prototype trimotor Fokker to King's Bay, Spitzbergen, Norway. In May 1926, Amundsen also arrived, this time in an Italian airship, the *Norge*, piloted by Umberto Nobile.

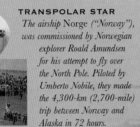

TRANSPOLAR STAR
The airship Norge *("Norway"),
was commissioned by Norwegian
explorer Roald Amundsen
for his attempt to fly over
the North Pole. Piloted by
Umberto Nobile, they made
the 4,300-km (2,700-mile)
trip between Norway and
Alaska in 72 hours.*

POLE PRETENDER?

*On 9 May 1926, Richard Byrd set out on a 2,455-km
(1,535-mile) circuit around the North Pole in this Fokker
F.VII, the* Josephine Ford, *piloted by Floyd Bennett. He
claimed to have accomplished his goal in 16 hours.*

On 9 May, while the *Norge* was still being readied for its attempt, Byrd flew off in the Fokker, piloted by Floyd Bennett, returning 16 hours later to announce that he had flown over the Pole. Somewhat downcast, Amundsen and Nobile set off two days later, successfully crossing over the Pole and flying on to Alaska between 11 and 13 May.

Whether Byrd actually reached the North Pole subsequently became a subject of controversy. It is now generally accepted that he did not, and that Amundsen thus deserves credit for the first transpolar flight. Byrd is still credited with the first flight over the South Pole three years later, and is recognized as a great Antarctic explorer. The story of Amundsen and the *Norge* had a tragic coda. In 1928, the airship crashed on a second Arctic flight under Nobile's command. Amundsen was among those who flew out to attempt to rescue the Italian and his crew. Amudsen and his pilot were lost without trace, though Nobile was later rescued.

THE TWO CONTENDERS
*Richard Byrd (left) shakes hands with Roald Amundsen in
Spitzbergen, Norway. Although Amundsen resigned himself
to having lost the race for the first transpolar flight, it was
later shown that Byrd never reached the North Pole.*

COAST-TO-COAST
This Fokker T-2, powered by a 420hp Liberty engine, was originally designed as a transport plane for the US Army Air Service. Modified with extra fuel tanks built into the wings, it made the first non-stop flight across the USA in May 1923 piloted by Lieutenants John Macready and Oakley Kelly.

The DLH pilots had to traverse the great empty spaces of Soviet Asia beyond the Urals with no accurate maps, no weather reports, no spare parts to be had, and often only primitive airfields. They then entered China, a dangerous and unstable country fought over by rival warlords whose attitude to the arrival of foreign aircraft was difficult to predict. After many difficulties, they not only made it to Peking but also flew back again to Germany.

Global imposition

The image of the aviator as a high-tech hero astonishing "primitive" imperial subjects was one of the most satisfactory aspects of aviation for the European public of the day. The book jacket of British aviator Alan Cobham's account of a flight from Britain to Cape Town and back, published in 1926, shows natives with spears and shields looking up in awe at his de Havilland D.H.50 passing overhead. But unfriendly locals could be a serious hazard for trail-blazing pilots in untamed areas of the globe. In the 1920s the French were fighting a Berber uprising in Morocco, while Britain tackled an Arab rebellion in Iraq. Since military aircraft were being used by the French and British against these rebels, their attitude to any aviators who crossed their territory was, not surprisingly, hostile.

During Cobham's flight to Australia and back, his mechanic, Arthur Elliott, was shot dead by a bullet fired at the aircraft from the ground near Basra in southern Iraq. French airmail pilots pioneering the route down the desert coast of North Africa from Casablanca to Dakar frequently

> "We have just witnessed a page of history being turned; a dream has come true."
>
> CALIFORNIA NEWSPAPER ON MACREADY AND KELLY'S FIRST NON-STOP US CROSSING

drew fire from camel-riding nomads. When a future hero of French aviation, pilot Jean Mermoz, came down in the desert in 1926, he was taken prisoner by tribesmen and held in a cave until a ransom was paid for his release. Later in the same year three of his colleagues were less fortunate after another forced landing – two were shot dead and the other, although ransomed, died later as a result of mistreatment in captivity.

BRITISH HERO
Celebrated British aviator Alan Cobham is shown landing his modified de Havilland D.H.50J biplane alongside the Houses of Parliament on 1 October 1926. Crowds lined the river Thames in London to applaud Cobham's completion of his three-month, 43,992-km (26,703-mile) England–Australia round-trip.

RECORD-BREAKERS
Lieutenants John Macready (left) and Oakley Kelly stand beside the tanks of fuel that powered their non-stop flight across America. They covered the 4,240km (2,650 miles) from Long Island, New York to San Diego, California in just under 27 hours.

In the 1920s, airmail pilots were an elite group flying regular services across some of the earth's most inhospitable terrain. Mail could be carried over many long-distance routes that were simply not safe enough for passenger transport. Subsidised by governments eager to promote aviation, airmail became a test-bed for scheduled commercial services. It allowed high-risk experiments in flying at night and in poor weather conditions, the application of new navigational techniques, and the development of support services and infrastructure, from airfields to weather forecasting. Airmail pilots were inculcated with a spirit of team discipline and professional dedication. They accepted high risks but were taught to minimize losses of aircraft or mail, while doing everything humanly possible to keep to a strict schedule.

Aéropostale

One of the first scheduled airmail services maintained under hazardous conditions was the organization for which Jean Mermoz and his colleagues flew, first known simply as *La Ligne* ("The Line") and later as Aéropostale. Founded by businessman Pierre-Georges Latécoère and based in Toulouse in southwest France, *La Ligne* began airmail flights in 1919, through Spain to Casablanca in Morocco. The extension of the postal service to Dakar on the coast of French West Africa followed in 1925. Two years later, Aéropostale boldly moved into South America, where the field had been left open for European competition. From March 1928, an airmail service of sorts linked France to Rio de Janeiro in Brazil – the mail crossed the Atlantic by boat from Dakar to the Brazilian port of Natal, from where it was flown on to Rio. In the 1930s, French pilots established regular flights across the South Atlantic in flying boats, as well as a network of airmail routes in Brazil, Argentina, and Chile.

JEAN MERMOZ

JEAN MERMOZ (1901–36) flew for the French army in Syria before joining the Latécoère line as an airmail pilot in 1924. After an adventurous period flying mail across North Africa, he was given the job of trail-blazing Aéropostale routes in South America in 1927. In 1928 Mermoz made the first night flight from Buenos Aires, Argentina, to Rio

FRENCH LEGEND
During the 1930s Mermoz became involved in French politics, exploiting his heroic status to drum up support for La Croix du Feu, an extreme right-wing organization.

WINGED HEAD
This stamp, bearing the portrait of French trail-blazer Jean Mermoz, was created to commemorate his feats of daring and bravery.

de Janeiro, Brazil. The following year, he embarked on the hazardous exploration of routes across the Andes to Chile. On one occasion, while making a forced landing on an Andean mountainside, he leapt from the aircraft and stopped the machine from rolling into a ravine by sheer muscle power. Such exploits – which lost nothing in the telling – along with a series of bold flights across the South Atlantic, made Mermoz a French national hero. On 7 December 1936, Mermoz crashed while piloting his flying boat, *La Croix du Sud*, across the South Atlantic. No trace of the crew or aircraft was ever found.

The exploits of the airmail pilots – not only Mermoz, but others such as Henri Guillaumet and the pilot-author Antoine de Saint-Exupéry – became legends of French aviation. The stories of some of their more extreme adventures, although well attested, read like fiction. In March 1929, for example, looking for a new route across the Andes from Chile to Argentina, Mermoz and his mechanic Alexandre Collenot crashed at around 4,000m (13,000ft) on a rocky slope surrounded by steep crevasses. For four days they worked in sub-zero temperatures to repair their Latécoère LAT 26, using what they had to hand – scraps of rubber, leather, fabric, wire, and glue. They then had to take off down a stony slope that was

fissured by two crevasses, which they hopped over as they gathered enough speed to lift off. The engine soon failed, but fortunately, once they had cleared the peaks, they were able to glide safely back down to the Chilean plains.

Army airmail

In the United States, airmail blazed the trail for commercial aviation from coast to coast, after a decidedly uncertain start. On 15 May 1918, with World War I still in progress, the Army Air Service was informed that it would be initiating the world's first regular airmail service, between Washington, D.C., and New York, via Philadelphia, until such time as the Post Office could procure its own pilots and aircraft. It was a task for which army pilots and their Curtiss Jennys were ill-prepared. The first mail to leave Washington had a high-profile send-off, with President Wilson himself in attendance, but the inexperienced pilot, Lt George L. Boyle, got lost immediately

POST OFFICE
The Aéropostale airmail company – one of the offices of which is shown above – was based in Toulouse, southwest France, and was the first to establish a regular mail service between South America and Europe.

NEW ROUTES
The French government hoped to use Aéropostale to help it conquer the commercial aviation market. Mermoz pioneered night flights over long distances to give airmail a clear advantage over ground mail.

SMALL-TOWN PARADE

A de Havilland airmail plane, decorated in bunting and streamers, is towed in a small-town parade celebrating the new transcontinental airmail service. During the 1920s, DH-4s like this one made up the bulk of the US Post Office fleet.

MAIL PICK-UP

A Bellanca CM monoplane is shown in low-level flight picking up a mailbag "on the hoof". Travelling at about 160kph (100mph), the mailbag shock cord can be seen catching on the Bellanca's pole hook.

after take-off and finished upside-down in a field a few miles from his point of departure.

After this inauspicious start, the army made a surprisingly successful job of flying the Washington-to-New York mail for the next three months. The US Post Office was sufficiently impressed to embark on an ambitious scheme for a transcontinental airmail service, using its own civilian pilots and specially commissioned aircraft. In the course of 1919 an airmail service was established first from New York to Cleveland and then on to Chicago. The following year it was extended to Omaha and over the Rocky Mountains to Salt Lake City, then on via Reno, Nevada, to San Francisco.

FIRST MAIL DROP

On 23 September 1911, Earle L. Ovington made America's first airmail delivery, carrying nearly 2,000 letters and postcards from Long Island to Mineola.

LONG-DISTANCE MAIL

This airmail bag, covered in signatures, was carried by Cal Rodgers (see page 47) on his 84-day transcontinental flight across the USA from New York to California, started in September 1911.

KNIGHT FLIGHT

IN 1921, WITH FUNDING for US airmail under threat, the Post Office decided to stage a daring night-flight experiment that would dramatically demonstrate the viability of a fully airborne transcontinental mail service. On 22 February, two DH-4s took off from New York with mail bound for San Francisco, while another two set off from San Francisco with mail for New York.

They were to cover the stretch between Cheyenne and Chicago in darkness. Disaster soon struck. One of the eastbound pilots crashed in Nevada and was killed. On the westbound journey, only one load of mail got as far as Chicago, where the onward flight was cancelled as a snowstorm set in. In the middle of the night the only mail still in transit was passed to pilot

Jack Knight who was to fly the segment from North Platte, Nebraska, to Omaha. Navigating by dead reckoning, and guided by bonfires that enthusiastic citizens on the ground below had lit along the way, Knight reached Omaha at 1am, only to find that the next pilot had failed to turn up. So Knight had a strong coffee, tore a road-map of the onward route off a wall, and set off for Chicago, 725km (435 miles) away.

Exhausted and numb with cold in his open cockpit, Knight reached Iowa City at 5am and, after a refuelling stop, flew on through snow to Maywood Airport, Chicago, arriving at 8.40am on 23 February to a hero's welcome. The westbound mail eventually reached New York 33 hours 25 minutes after leaving San Francisco.

TESTING, TESTING
Jack Knight, hero of the 1921 night-flight experiment, wears a radio microphone and helmet fitted with headphones to test radio air-to-ground communications.

The only way to make the service viable was to establish night flying as normal practice, so that mail could travel from coast to coast entirely by air. During the war, pilots had repeatedly shown that flying at night was feasible, but a regular and reliable scheduled night operation over long distances would require something more than reliance on the skills and instincts of experienced pilots. The way forward was indicated by experiments conducted by the army at McCook Field, near Dayton, Ohio, in 1923. They used rotating beacons and flashing markers to create a lighted airway between McCook Field and Columbus – a distance of over 112km (70 miles), along which army pilots proved capable of making safe and regular night flights. The Post Office set out to translate this experiment into a continent-wide system of lighted airways. Night-time airmail services on the first illuminated stretch, from Chicago to Cheyenne, began in July 1924. By 1925, when the Kelly

The US Air Mail pilots were necessarily a tough bunch of individuals. Armed with pistols to protect the mail, they attempted to maintain regular flights over mountains and deserts in inadequate aircraft with only primitive instruments. Without aerial maps or reliable compasses, the pilots found their way either by following railroad tracks or looking out for landmarks that they recorded in a notebook – a church steeple, the orange roof of a barn, a water tower. When cloud cover was low they had little choice but to fly underneath it, running the risk of crashing into a hillside or even tall trees. If they encountered fog or a storm – a frequent occurrence – they had no choice but to land wherever they could. Even in flat, open terrain, a forced landing could be hazardous, as on the occasion succinctly reported by pilot Dean C. Smith: "Dead-sticked – flying low – only place available – on cow – killed cow – scared me." The pilot would head off in search of

a farmstead, which might have a telephone, or a railroad, where he could flag down a passing train. Death was an accepted risk of the job. One pilot later recalled, "I would have been frightened if I had thought I would get maimed or crippled for life, but there was little chance of that. A mail pilot was usually killed outright."

Initially some senior officials in the US Post Office were inclined to dismiss airmail as a passing fad. As the aircraft could only fly by day, they had to work in conjunction with trains. They were loaded up with mail from the nearest railroad in the morning, flew it some way along the route, and then handed it back to the trains for transport during the night. Thus most mail crossing the United States made, at best, only a small part of its journey by air, achieving an almost negligible improvement in delivery time at very considerable expense to the Post Office and risk to its pilots.

Act was passed authorizing the Post Office to contract out mail services to private companies, mail was regularly being flown from coast to coast in around 30 hours, compared to three days by train.

A network of airways

The network of airways that the Post Office handed over to commercial operators in 1926–27 had, by the standards of the day, an impressive safety record. Ground crews ensured that aircraft were well maintained and emergency landing fields, established at roughly 48-km (30-mile) intervals along the transcontinental route, saved the skin of many a pilot in difficulty. But it was impossible at this stage to take the danger out of flying. Pilots still had to navigate by sight and had no adequate instruments for flying in poor visibility. The use of

radio was mostly restricted to transmitting weather reports from airfield to airfield along the route. This helped pilots about to take off to decide whether and where to fly, but once they were airborne they received no further information and were still at risk of flying into fog or violent storms.

One pilot who amply demonstrated the continued riskiness of airmail flying was the young Charles Lindbergh. A former army pilot and barnstormer, Lindbergh was flying the mail between St Louis and Chicago – one of the first routes exploited by a private contractor. In September 1926 he found himself trapped above a dense layer of ground fog outside Chicago. Unable to land, he eventually ran out of fuel and had to jump out, relying on his parachute. Unnervingly, Lindbergh heard the engine of his abandoned aeroplane start up again as he floated down through the fog, a little residual fuel having found its way in to the carburettor. The aircraft

LINDY'S LUCKY ESCAPE
Charles Lindbergh poses with a farmer next to the wreckage of his DH-4B biplane on the edge of a cornfield near Ottawa, Illinois, on 16 September 1926. He had managed to parachute to safety when the aircraft had run out of fuel.

crashed about a mile from where the pilot came down. Only six weeks later, Lindbergh again parachuted to safety, this time after being caught in snow and low cloud. It was a sufficiently notable repeat to catch the attention of safety-

LIGHTED AIRWAYS

TO COMPENSATE FOR THE ABSENCE of decent navigational aids or instruments for night flying, lines of beacons were installed across the United States during the 1920s, creating lighted airways. The beacons were steel towers about 15m (50ft) tall, supporting a rotating lamp and mirror that generated a powerful beam roughly equivalent to that of a lighthouse. Set roughly 16km (10 miles) apart, pilots were never out of visual range of a beacon in clear weather. Most of the lights were electric, though acetylene gas was used in remote areas. Installing beacons in mountains, swamps, and deserts was a herculean task for engineers and construction crews, especially since it was essential to site the beacons on the highest peaks.

PORTABLE FLOODLIGHT
Half-billion-candlepower floodlights were also used to mark out the airmail routes. By 1933 there were 28,800km (18,000 miles) of lighted airways across the United States.

LOADING THE MAIL
Ground crew load mailbags into the mail compartment situated in front of the pilot's cockpit in a Western Air Express (WAE) Douglas M-2 Mailplane. The WAE started Contract Air Mail Route 4 between Los Angeles and Salt Lake City, via Las Vegas, on 17 April 1926.

LINDBERGH'S NON-STOP NEW YORK-TO-PARIS FLIGHT

"A SLIM, TALL, BASHFUL, SMILING American boy is somewhere over the Atlantic ocean, where no lone human being has ever ventured before... If he is lost, it will be the most universally regretted loss we ever had," wrote humorist Will Rogers on 20 May 1927. The "smiling American boy" was 25-year-old Charles Lindbergh, and Rogers' anxious concern for his fate was shared by millions in America and Europe. Lindbergh was the latest pilot to attempt the first non-stop flight between New York and Paris, a contest that had already cost the lives of six airmen. He intended to fly the 5,760km (3,600 miles) alone in a single-engined monoplane. Lindbergh's quiet courage and good looks had won him an enthusiastic following, but few people rated his chances.

SPIRIT OF ST LOUIS

The Spirit of St Louis' *cramped cockpit, squeezed in behind the massive fuel tank, had no front view. Lindbergh had to turn the aeroplane to see ahead. He had bought the custom-built Ryan NYP with its 223hp Wright Whirlwind engine for $10,580, using money raised from St Louis businessmen.*

Lindbergh's Ryan NYP was carrying 2,000 litres (450 gallons) of fuel – a winged gasoline tank. He had jettisoned every luxury to lighten the aeroplane's load, carrying no radio and no sextant for navigation by the stars. His entire provisions for the journey were five sandwiches and two canteens of water. Yet still he did not know if the fuel-laden aircraft would get off the ground. Roosevelt Field was sodden and muddy and there was no headwind to aid take-off. As the monoplane accelerated through mud and puddles, it twice rose and bumped back to the ground before at last achieving sustained lift and rising clear of the telegraph wires at the end of the runway.

Once in the air, Lindbergh faced two crucial problems. One was navigation. He had to find his way by dead reckoning, using a clock and an airspeed indicator to measure his distance travelled, and two compasses to plot his direction. Once he was over the ocean he would have no visual reference to check the accuracy of his calculation. His other problem was tiredness. He had not slept the night before the flight and faced a further day and a half without rest. By the time he left North America behind, crossing the coast of Newfoundland, he had already been in the air for 11 hours and was racked by fatigue.

FLYING LOW

The main difficulty that Lindbergh faced was staying awake for the 33½ hours his transatlantic flight took. This photograph was taken during his US tour, undertaken on his return home.

The 15-hour ocean crossing was a severe test of the flying skills and the instinct for survival that Lindbergh had developed as an airmail pilot. He had to fly for two hours in total darkness between sunset and moonrise; he flew through storm clouds and freezing cold that iced his wings; after daybreak he ran into thick fog. Few pilots could have managed to fly for so long in such poor visibility with the minimal instruments that Lindbergh possessed – essentially a turn-and-bank indicator and an altimeter.

> "Here, all around me, is the Atlantic – its expanse, its depth, its power... If my plane can stay aloft, if its engine can keep running, then so can I."
>
> CHARLES LINDBERGH

The routines of the flight helped keep him awake: regularly switching from one gas tank to another, noting the readings from his instruments in his log. So did the instability of his aircraft, which was definitely not designed to fly itself and would soon give the pilot a jolt if he nodded off at the controls.

By the morning of 21 May, Lindbergh was hallucinating, creating mirages of land out of shapes in the fog. But first the sight of fishing boats and then a rugged coastline told him that he had made it across the ocean. Remarkably he was almost exactly on course, crossing the Irish coast

N-X-211

at Dingle Bay. The last 960km (600 miles) of the flight were relatively straightforward. Lindbergh entered France at Cherbourg and followed a lighted route towards Paris, which appeared to him as "a patch of starlit earth under a starlit sky".

News of the American's imminent arrival set Paris alight. A vast traffic jam developed as hundreds of thousands of Parisians headed for Le Bourget airfield. An American expatriate, Harry Crosby, described the chaotic scene: "Then sharp swift in the gold glare of the searchlights a small white hawk of a plane swoops hawk-like down and across the field – C'est lui, Lindbergh, LINDBERGH! and there is pandemonium [like] wild animals let loose and a stampede towards the plane…"

After a flight of 33 hours 30 minutes Lindbergh emerged from total solitude into the clutches of an hysterical mob. He had become the most famous man in the world.

conscious aviation adminstrator William P. MacCracken, who commented acidly: "He's not going to help commercial aviation if he keeps dropping these airplanes around the countryside." MacCracken must have been relieved when Lindbergh quit airmail flying to take up a more prominent challenge.

Non-stop challenge

French-born American hotelier Raymond Orteig had put up a $25,000 prize for a non-stop flight between New York and Paris in 1919, but it was only in the mid-1920s that interest in this challenge began to raise a buzz among fliers and the press. In 1926 French wartime ace René Fonck announced that he intended to mount the first attempt on the prize. Fonck and his team had Igor Sikorsky create a trimotor version of his two-engined S-35 transport plane. It was to take a crew of four across the Atlantic in some style – fittings included the latest radio equipment, seats with red leather upholstery, and a bed. But a rush to carry out the flight before the winter weather set in meant there was not enough time to test the aeroplane fully loaded. On 21 September, in front of a large crowd, Fonck attempted to take off from Roosevelt Field bound for Paris. The aircraft never left the ground. Spectators saw it plunge down an embankment at the end of the runway and explode in a fireball. Fonck and one crew member escaped, but two of the crew were killed.

Obviously, such drama only brought popular interest to a higher pitch. By the spring of 1927 more famous pilots were lining up for a crack at the prize. Commander Richard Byrd was planning to fly to Paris in a Fokker trimotor, while in France war aces Charles Nungesser and François Coli announced that they were preparing to fly to New York. But the prize seemed jinxed. Byrd's Fokker crashed on its first trial flight in April, virtually putting him out of the running. Ten days later, pilots Noel Davis and Stanton Wooster, who were being sponsored by the American Legion, crashed and died on their last test flight before attempting to fly to Paris.

On 8 May, Nungesser and the one-eyed Coli left Le Bourget aerodrome, Paris, in their single-engined Levasseur biplane *L'Oiseau Blanc* ("*The White Bird*"), confidently expecting to arrive in New York the following day. After leaving the coast of France, they were never seen again. The agonizing wait for news of the two French aces, and the slow ebbing of hope, stirred deep emotions in France and unquestionably contributed to the intensity of the reception

ILL-FATED WHITE BIRD
L'Oiseau Blanc was the Levasseur biplane in which Charles Nungesser and François Coli set out to fly from Paris to New York two weeks before Lindbergh. They disappeared without trace.

CHARLES A. LINDBERGH

SON OF A MINNESOTA CONGRESSMAN, Charles A. Lindbergh (1902–74) was a shy and solitary child, qualities that developed into mild-mannered self-reliance in his adulthood. Completely out of step with the clichéd image of youth in America's "Roaring Twenties", he did not smoke, dance, or drink alcohol. Obsessed with flying, he drifted from barnstorming to a spell in the Army Air Corps and, in 1926, a job as an airmail pilot.

At first he handled the extraordinary fame brought by his transatlantic flight remarkably well, impressing everyone with his charm and poise. He married the talented Anne Morrow and became the world's leading ambassador for aviation. But after the kidnap and murder of their infant son in 1932, Lindbergh was increasingly driven to escape the storm of publicity that surrounded his every action. In the late 1930s he developed an admiration for aspects of Nazi Germany and campaigned to keep the United States out of the war in Europe. Although he took an active part in the war against Japan after 1941, his reputation never fully recovered.

HOMECOMING HERO
After his record-breaking flight, Lindbergh returned from Paris to a hero's welcome. A ticker-tape parade through New York was witnessed by four million people. For his amazing endeavour, Lindbergh was promoted from captain to colonel in the US Army Air Corps Reserve. He also received the first Distinguished Flying Cross to be awarded and a cheque for $25,000.

LONE EAGLE
Charles Lindbergh's courage, daring, and endurance helped him to become the most famous aviator of his day and did much to inspire people's faith in flying. He is shown here posing by the Spirit of St Louis.

that Lindbergh received on successful completion of his New York-to-Paris flight on 21 May (see pages 118–9). But the extraordinary scale of the popular response to the young American's flight, both in Europe and on his return to the United States, has never been adequately explained. It was a considerable feat by any reckoning to have flown solo across the Atlantic, yet only a fortnight later Clarence Chamberlin and Charles Levine flew from New York to within 160km (100 miles) of Berlin in a Wright-Bellanca (incidentally the aeroplane Lindbergh had originally wanted to fly), beating Lindbergh both for distance and endurance. They made the headlines, but only for a few days. It was Lindbergh who was caught in the exhilarating but eventually destructive embrace of celebrity.

The "Lindbergh effect"

The only clear historical significance of Lindbergh's flight was as the turning point at which the United States took its place as the leader in world aviation for the first time since the Wright brothers. The "Lindbergh effect" would have given a mighty boost to America's struggling air transport and aeroplane-making businesses even without the new hero's enthusiastic active commitment to promoting aviation. In the event, Lindbergh followed his transatlantic flight with a Guggenheim-sponsored tour of every state of the Union, which led directly to airport building throughout the USA. In the following years he promoted passenger airlines, letting TAT (Transcontinental Air Transport, later TWA) advertise itself as the "Lindbergh Line", and carrying out wide-ranging route-survey and publicity flights on behalf of Juan Trippe and Pan Am.

Inevitably Lindbergh's Atlantic crossing inspired a new wave of transoceanic and transcontinental flights, keeping the press supplied with sensational copy and creating a new generation of pilot-celebrities. Each nation had to have its own heroes to compete with Lindbergh. France, for example, had Dieudonné Costes. In 1927 he crossed the South Atlantic, flew down the eastern seaboard of South America, and back up the Pacific coast as far as Washington state; and in 1930, with Maurice Bellonte, he made the first direct non-stop flight from Paris to New York in the Breguet 19 Point d'Interrogation. Italians worshipped Francesco De Pinedo, a naval officer who flew across the Atlantic in one Savoia-Marchetti flying boat, crashed it, and flew back in another. In 1930, Britain's Amy Johnson, a former stenographer from Hull, became the darling of the press after flying solo from England to Australia in a de Havilland Moth. Amelia Earhart, fulfilling America's need for a female Lindbergh, was lauded in 1928 as the first woman to fly across the Atlantic, although this was in a Fokker trimotor piloted by Wilmer Stultz. Four

LONG-DISTANCE AIRCRAFT

THE DEGREE OF PUBLIC attention focussed on long-distance flights between the wars often corresponded to the degree of danger in aircraft that often had only primitive navigational equipment. Given the problem of engine reliability, multi-engined aircraft were often preferred, as were seaplanes or flying boats. In the aftermath of WWI, most record-setting long-distance flights were made in converted military aircraft. A huge gulf separated their performance from that of 1930s long-distance racing aircraft such as the Lockheed Vega and D.H.88 Comet. And no aircraft could match the range of an airship like the *Graf Zeppelin*.

ROUSING RECEPTION
Alan Cobham prepares to land his D.H.50J by the Houses of Parliament, London, on 1 October 1926, after successfully completing a round-trip to Australia.

Breguet 19 *Point d'Intérrogation*

Introduced in 1922, more Breguet 19s were built than any other military aircraft between the wars. However it was a series of long-distance flights, helped by its large fuel-carrying capacity, that brought the Breguet 19 fame. Most notably, *Point d'Intérrogation* flew a world record 7,905km (4,912 miles) in September 1929.

ENGINE	650hp Hispano-Suiza 12Lb	
WINGSPAN	14.8m (48ft 8in)	LENGTH 9.5m (31ft 3in)
TOP SPEED	Unknown	CREW 2
PASSENGERS	None	

Curtiss NC-4

On 31 May, 1919, a Curtiss NC-4, flown by Lt.-Commander Read, became the first plane to fly across the Atlantic. One of three specially designed Curtiss flying boats commissioned by the US Navy, the NC-4 was the only one to complete the epic 6,319km (3,925 mile) flight, crossing from New York to Plymouth in stages, via Halifax, Nova Scotia, the Azores, and Lisbon.

ENGINE	4 x 400hp Liberty 12A	
WINGSPAN	38.4m (126ft)	LENGTH 20.8m (68ft 3in)
TOP SPEED	137kph (85mph)	CREW 5
ARMAMENT	Provision for 8 machine guns but none carried	

de Havilland D.H.88 Comet (Racer)

To mark the centenary of the State of Victoria in 1934, an England-to-Australia air race was planned. De Havilland designed an advanced two-seat racer specifically for the event and three were entered for the MacRobertson Mildenhall-to-Melbourne Air Race. The winning Comet *Grosvenor House* won in just under 71 hours.

Three fuel tanks in fuselage, ahead of cockpit

ENGINE	2 x 230hp de Havilland Gipsy Six R	
WINGSPAN	13.4m (44ft)	LENGTH 8.8m (29ft)
TOP SPEED	381kph (237mph)	CREW 2
PASSENGERS	None	

de Havilland D.H.60G Gipsy Moth

ENGINE	100hp de Havilland Gipsy I 4-cylinder water-cooled inline	
WINGSPAN	9.1m (29ft 8in)	LENGTH 7.3m (23ft 11in)
TOP SPEED	164kph (102mph)	CREW 2
PASSENGERS	None	

In the D.H.60 Moth, which first flew in 1925, de Havilland created the ideal, multi-purpose light aeroplane. Nearly 600 were built before production ceased in 1934, of which the most successful version was the D.H.60G which not only won the 1928 King's Cup air race, but also completed many long-distance flights.

Douglas World Cruiser

The Douglas World Cruiser (DWC) was specially commissioned by the US Army to attempt the first round-the-world flight. A flight of four aircraft left Seattle on 6 April 1924 and after an epic 175 days (365 hours flying time) covering 44,298km (27,553 miles), DWCs *Chicago* and *New Orleans* returned to Seattle on 28 September.

Liberty V 12 engine

ENGINE	420hp Liberty V 12 water-cooled	
WINGSPAN	15.2m (50ft)	LENGTH 10.8m (35ft 6in)
TOP SPEED	166kph (103mph)	CREW 2
PASSENGERS	None	

Fokker F.VII *Southern Cross*

One of the most successful commercial and long-distance aircraft of the 1920s and 30s, the trimotor F.VII was manufactured in the Netherlands, Britain, US, Belgium, France, Italy, and Poland, and equipped airlines and air forces worldwide. The F.VIIB-3m was used by pioneers and explorers such as Richard E. Byrd, Amelia Earhart, and, most famously, Charles Kingsford Smith. His *Southern Cross* achieved the first transpacific flight on 9 June 1928, after flying 12,555km (7,800 miles) in 88 hours.

ENGINE	3 x 237hp Wright J-5 Whirlwind 9-cylinder air-cooled radials	
WINGSPAN	21.7m (71ft 3in)	LENGTH 14.5m (47ft 7in)
TOP SPEED	185kph (115mph)	CREW 2
PASSENGERS	8–10	

Lockheed Model 5B Vega

Designed by John Northrop, the first Vega appeared in 1927 and was an immediate success for the young Lockheed Aircraft Company. A family of fast, small airliners and long-distance racers evolved, with 128 being sold by the mid-1930s. One of the most popular versions was the six-seat Vega 5B. In July 1933, the remarkable one-eyed Wiley Post flew his Vega 5C "Winnie Mae" in the first solo round-the-world flight. A Vega was also used by Amelia Earhart when she made her transatlantic solo flight in 1932.

ENGINE	450hp Pratt & Whitney Wasp C1 9-cylinder air-cooled radial	
WINGSPAN	12.5m (41ft)	LENGTH 8.4m (27ft 6in)
TOP SPEED	290kph (180mph)	CREW 1
PASSENGERS	6 (on commercial aircraft)	

Ryan NYP *Spirit of St. Louis*

In February 1927, Claude Ryan received an order for a special version of his monoplane, in which a young mail pilot, with the backing of a group of St Louis businessmen, aimed to win a $25,000 prize for the first non-stop flight between New York and Paris. The modifications included moving the engine forward to allow the installation of a huge fuel tank and increasing the wingspan. The pilot's name was Charles Lindbergh, who reached Paris on 21 May 1927, after 33.5 hours flying.

ENGINE	223hp Wright J-5 Whirlwind 9-cylinder air-cooled radial	
WINGSPAN	14m (46ft)	LENGTH 8.4m (27ft 7in)
AVE SPEED	173kph (108mph)	CREW 1
PASSENGERS	None	

R 34

The R 34 was almost an exact copy of the Zeppelin L 33 which had been brought down in September 1916. On 2 July 1919, it left East Fortune airfield near Edinburgh and reached New York 108 hours later, achieving the first East-West crossing against the prevailing winds. When it returned six days later in 75 hours, it completed the first ever double crossing.

ENGINE	5 x 240hp Sunbeam Maori 4	
LENGTH	196m (643ft)	DIAMETER 23.2m (76ft)
CAPACITY	55,224 cubic metres (1,950,000 cubic feet)	
SPEED	89kph (55mph)	CREW 30
PASSENGERS	None	

Vickers Vimy F.B.27

Designed as a long-range heavy bomber, the Vimy arrived too late for WWI. However, it achieved some renown pioneering long-distance flights, the most famous of which was the first non-stop transatlantic flight by Alcock and Brown in June 1919.

ENGINE	2 x 360hp Rolls-Royce Eagle VIII	
WINGSPAN	20.7m (68ft)	LENGTH 13.3m (43ft 7in)
TOP SPEED	166kph (103mph)	CREW 2
ARMAMENT	Provision for 2 machine guns	

LZ 127 *Graf Zeppelin*

The inventor of the rigid airship. Count von Zeppelin, was almost 62 when his LZ 1 first flew on 2 July 1900. When LZ 127 appeared in 1928, the new liner of the air was called *Graf Zeppelin* in his memory. For almost a decade, this giant carried passengers and cargo, crossing the Atlantic many times and, in 1929, flying around the world at an average speed of 113kph (70mph). A million miles in perfect safety. This all ended with the horrific loss of the even larger *Hindenburg* in May 1937.

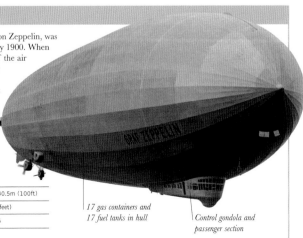

17 gas containers and 17 fuel tanks in hull

Control gondola and passenger section

ENGINE	5 x 550hp Maybach VL II V12	
LENGTH	236.6m (776ft 3in)	DIAMETER 30.5m (100ft)
CAPACITY	105,000 cubic metres (3,708,040 cubic feet)	
SPEED	115kph (72mph)	CREW 40–45
PASSENGERS	20	

years later, in 1932, on the fifth anniversary of Lindbergh's crossing – a timing carefully calculated by her publicists – Earhart did the real thing, flying the Atlantic solo in a Lockheed Vega.

The excitement surrounding these flights was above all fuelled by risk. Pilots could, and did, still die with appalling frequency, especially when crossing large expanses of ocean. The frisson of death kindled the emotions of the public. But at the same time patient efforts were being made by practical and ingenious inventors and designers precisely to take the risk out of aviation.

Flying blind

One of the most frequent causes of accidents – and the most inhibiting limit on the development of scheduled commercial flights – was the difficulty of flying in low cloud or fog. Poor visibility posed the problem not only of how to find your way, but also of how to keep control of the aeroplane. Pilots habitually flew by "the seat of their pants", relying on their sense of sight and their instinctive sense of balance. But in fog or dense cloud, with no visual feedback, pilots easily became disoriented. Typically they might misinterpret a banked turn as a dive, pull back on the stick, tighten the turn further, and end up in a fatal spin. In principle, the altimeters and turn-and-bank indicators available since World War I allowed pilots to fly even in zero visibility, but few fliers found it possible to trust these instruments when they contradicted their instincts.

Although the development of improved

BLIND-FLIGHT TECHNOLOGY

FOR THE FIRST EVER blind flight in 1929, test pilot James Doolittle's biplane was fitted with an altimeter 20 times more accurate than the standard devices then in use. To replace the turn-and-bank indicator, Elmer Sperry had developed an "artificial horizon". This ingenious instrument combined a bar representing the horizon and a small aeroplane symbol. When the aircraft banked, the horizon bar tilted, and if it changed pitch the bar rose or fell accordingly. Another of Sperry's contributions was a gyrocompass, which, unlike traditional compasses, held stable through turning manoeuvres. Doolittle's chief locating device was a radio. He was equipped to receive instructions from a ground controller and to orient himself on a radio beam.

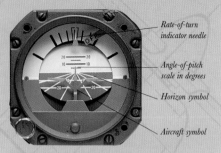

Rate-of-turn indicator needle

Angle-of-pitch scale in degrees

Horizon symbol

Aircraft symbol

MODERN ARTIFICIAL HORIZON
Originally conceived by Elmer Sperry in 1929, an artificial-horizon instrument shows the angle of an aircraft when banking, as well as its angle of pitch, in degrees.

instruments and instrument-flying was pursued in several countries, notably Germany, some of the most crucial progress was achieved in the late 1920s at the research laboratory established by the Guggenheim Foundation at Mitchel Field, Long Island, New York. Working in conjunction with Elmer Sperry, the world's leading expert on gyroscopic instruments, the Guggenheim Foundation employed racing pilot James Doolittle to experiment with blind flight. The holy grail of their quest was to take off, fly a specific course, and land, without reference to the world outside the cockpit. Doolittle was to attempt the feat in a Consolidated NY-2 modified with a new generation of flying instrumentation (see above).

On 24 September 1929 Doolittle taxied out on to Mitchell Field enclosed by a light-proof hood that cut him off from outside vision. For safety, a second pilot, allowed normal vision, was in the aeroplane but he did not touch his controls. Doolittle used the radio beam to find the correct line for take-off, lifted into the air, stayed in the air for a quarter of an hour, making two 180-degree turns, and landed somewhat roughly but safely. The first successful blind flight in history, it was immediately recognized as a major step forwards in air safety.

Autopilots

Shortly afterwards, Sperry went on to develop the first effective automatic pilot. This was a field in which he had much experience. In 1914, his son Lawrence Sperry had

demonstrated a gyrostabilizer at a French airshow. With the pilot's hands off the controls, the device kept a biplane in level, stable flight while a mechanic climbed first on to a wing and then on to the rear fuselage, a shift of weight that should have caused the aircraft to tilt or pitch. No practical use was found for the gyrostabilizer, partly because pilots did not need it and also because it was too unreliable due to the tendency of the gyroscope to "drift" out of alignment. In the 1930s, however, with airmen flying ever-longer distances in aeroplanes with increasingly complex instruments and radio equipment, the usefulness of a device that could at least temporarily take over control of the aeroplane became apparent, and Sperry set to work to remedy the defect of drift by linking the gyroscope to a pendulum. In 1933, when one-eyed American pilot Wiley Post made the first solo round-the-world flight, his Lockheed Vega, the *Winnie Mae*, was fitted with a prototype Sperry autopilot.

Design breakthroughs

Progress in instrumentation was matched by giant strides in aeroplane design. The aircraft used by adventurous pilots immediately after World War I – the Vickers Vimy, for example, or the Breguet 14 – were far superior to pre-war models in range, engine power, load-carrying capacity, and reliability. But they were still cumbersome strut-and-wire biplanes that looked as if they could have been designed specifically to maximize drag.

The most innovative airframes of the immediate postwar period came from German designers. Both Hugo Junkers and Reinhold Platz, the designer at Fokker, developed aircraft with a single strut-free cantilever wing. Junkers went for all-metal construction, making his monoplanes out of strong, lightweight Duralumin. The metal skin of Junkers aircraft was corrugated to give

JAMES "JIMMY" DOOLITTLE

BORN IN CALIFORNIA, James H. Doolittle (1896–1993) learned to fly with the Army Signal Corps during World War I. In 1922 he became the first man to cross the United States from coast to coast in under 24 hours. Doolittle was one of the most famous showmen of the 1920s, renowned both as a stunt pilot and a record-breaking racer – winner of the Schneider Trophy in 1925, the Bendix Trophy in 1931, and the Thompson Trophy in 1932. But there was also a more serious side to his flying. An exceptional test pilot, he became one of the first people to receive a doctorate in aeronautical engineering in 1925. During World War II, he returned to active service as a senior commander in the USAAF, notably leading a daring long-distance bombing raid on Tokyo in 1942.

PRECISION PILOT
An aviation pioneer who carefully calculated the risks for every stunt, Doolittle lived into his nineties and received almost every major aviation honour.

extra strength, at the expense of increasing drag. Platz stuck to wood as his prime material, although he used steel tubing for the internal structure of the fuselage and adopted a high wing in contrast to the low wing preferred by Junkers. Fokker monoplanes were much admired in the 1920s, especially the F.VII-3m (trimotor), flown by aviators such as Kingsford Smith and Byrd. However, both wooden construction and a high wing were to prove retrograde. When retractable undercarriages became common in the 1930s, high-wing aircraft had nowhere to retract the landing gear into. The Junkers corrugated metal designs also had limitations. By 1920 another German designer, Dr Adolph Rohrbach, had worked out the advantages of making wings and tail surfaces out of a smooth metal skin stretched over box spars. This "stressed skin" would bear some of the load previously borne entirely by the frame. Rohrbach's ideas for a smooth metal aeroplane did not come to the attention of American aircraft designers until 1926 but were to prove immensely influential on Jack Northrop and Boeing, and then on almost all manufacturers.

Reducing drag

Progress in the theory of aerodynamics, backed up by research at well-funded laboratories such as America's National Advisory Committee for Aeronautics (NACA) facility in Virginia, also began to have a profound effect from the second half of the 1920s. NACA's array of wind tunnels was especially important in studying airflow around different models of aeroplane. Some of its experiments fed directly into aircraft design. For example, air-cooled radial engines were being widely adopted in the 1920s because of their excellent weight-to-horsepower ratios, but their exposed ring of engine cylinders tended to have a serious adverse effect on streamlining. In 1928 wind-tunnel experiments conducted by NACA with various forms of cowling showed that a full cowling of the right design could eliminate 60 per cent of the drag from the engine while actually improving cooling. The "NACA cowling" became a standard feature of aeroplanes with radial engines, producing a significant increase in performance.

The Lockheed Vega, designed by Jack Northrop, was one of the first aeroplanes to adopt the NACA cowling in the late 1920s. With its Fokker-influenced high, strut-free single wing and streamlined monocoque fuselage, the Vega represented one of the most advanced designs of its day – not surprisingly embraced by the likes of Post and Earhart. The fact that its fuselage was made of two easily assembled prefabricated parts also looked forward to the increasing use of mass-production methods in the air industry in place of traditional craft skills. But the Vega was made of spruce wood and had a fixed undercarriage, two features that were soon to make it seem dated.

Established innovations

The 1930s brought to fruition the revolution in aircraft design begun in the previous decade. Monoplanes completed their triumph over high-drag biplanes, while all-metal stressed-skin construction became the rule, benefiting from improved metallurgy, especially lightweight aluminium alloys (aircraft manufacture was the first major use that had been found for aluminium). Engines continued to improve in power-to-weight ratio and reliability. Two types prevailed: air-cooled radials such as the Pratt & Whitney Wasp series, especially favoured in the United States; and liquid-cooled in-line engines such as the Rolls-Royce Merlin. The power of radial engines was increased by adding a second ring of cylinders, and the NACA cowling made them more efficient. In-line engines were

DAREDEVIL DOOLITTLE
James Doolittle, wearing a flying suit and parachute, refuels the centre wing tank of his Laird Super Solution racing plane. Doolittle went down in history for making the first blind flight. He flew a 24-km (15-mile) irregular course in a Consolidated NY-2 biplane before landing safely.

JACK NORTHROP

JACK NORTHROP (1895–1981) was born in Newark, New Jersey. Extraordinarily for a designer associated with pushing flight technology to its limits, he had no training as an engineer beyond high-school physics. He drifted into aircraft design in 1916, working for the Loughead brothers in their workshop in Santa Barbara, California. After a spell with Douglas, Northrop helped Allan Loughead found the Lockheed Aircraft Company in 1926. The following year, Northrop designed the Lockheed Vega, an aeroplane that exemplified his taste for radical and elegant design solutions. In 1929 he left to found the Northrop Aircraft Company, where he produced the Alpha, an aeroplane regarded as ahead of its time. Over the next 25 years he developed his pet "flying-wing" project. Although his flying-wing bombers such as the XB-35 were not adopted by the USAF – a personal catastrophe – their visionary design was later incorporated into stealth-aircraft designs.

DESIGN MEETING
Jack Northrop (right) is featured discussing designs with the Northrop Aircraft Company's assistant chief of design Walt Cerny (left) and project engineer Tom Quayle.

improved in particular by the adoption of ethylene glycol as a coolant – with a low freezing and high boiling point, it allowed radiators to be made smaller, reducing weight and drag. By the late 1930s, aircraft engines were capable of delivering well in excess of 1,000hp. Multi-engine aircraft were now seriously powerful machines.

Modern refinements

Experiments at the NACA wind tunnel revealed that a fixed undercarriage contributed an astonishing 40 per cent of the entire drag acting upon an aeroplane. Some designers responded by enclosing fixed undercarriages in streamlined "trousers", but retractable undercarriages inevitably became standard in the course of the 1930s. Initially, there were considerable safety concerns about the possibility of the undercarriage failing to extend for landing. This is why in the Douglas DC-3, the most successful aircraft of the decade, the retracted landing gear still protruded below the fuselage so that, if the undercarriage failed to extend, the aeroplane could still come down on its wheels.

Other improvements included variable-pitch and then constant-speed propellers. Before this innovation, the setting of the propeller had to be optimized for one stage of the aeroplane's flight at the expense of others. For example, one of Lindbergh's problems taking off for his transatlantic journey in the heavily laden Spirit of St Louis was that his propeller was set to be optimal for cruising, not for take-off. Another innovation was the use of flaps to temporarily change the shape of the aircraft's wing so that, for example, an aircraft designed for high speed would have enough lift to fly at lower speed when coming in to land. Safety was improved by fitting de-icers to leading wing edges – initially inflatable rubber devices that punched the ice so that it cracked.

WIND-TUNNEL TESTING
This image taken in the test chamber at the Langley Research Centre, Virginia, in 1932, shows the massive scale of their 9 x 18-m (30 x 60-ft) wind tunnel. An 9.4m- (31ft-) span Loening XSL-1 seaplane can be seen mounted on the test rack.

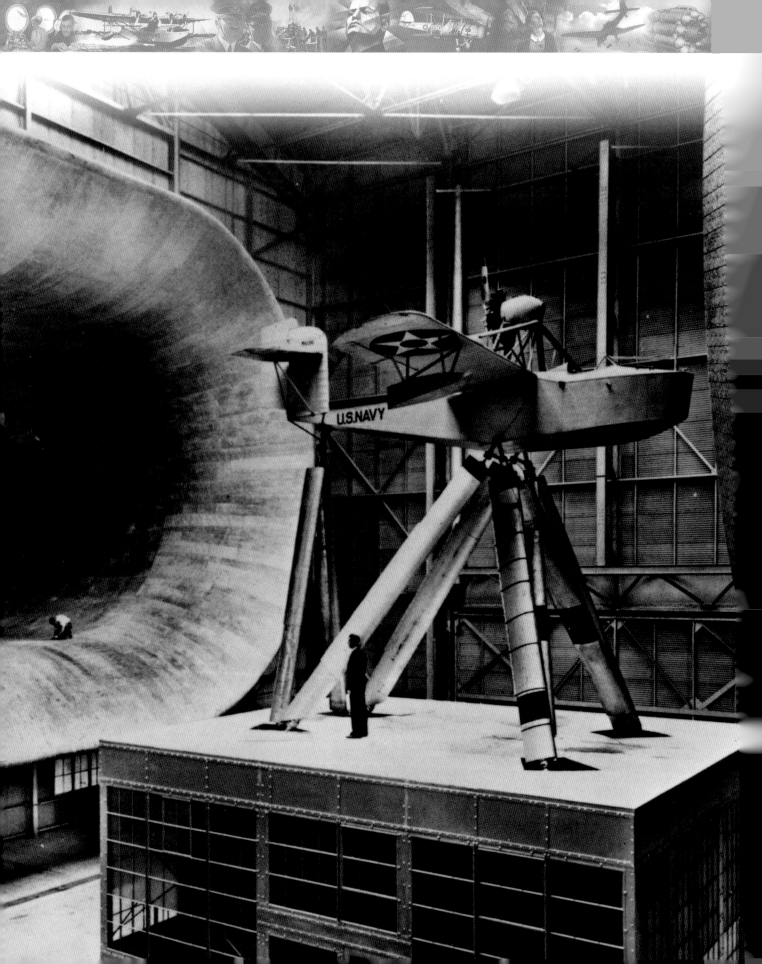

Lockheed Model 10 Electra

MAIL RUNNER
British Airways was one of the first foreign companies that bought Electras. This aircraft is serving as an airmail carrier, a suitable use for an aircraft that was fast but relatively low on payload. A British Airways Electra carried Prime Minister Neville Chamberlain to negotiate a flawed peace deal with Nazi dictator Adolf Hitler, at Munich in 1938.

STYLISH INTERIOR
The interior of the Lockheed 10 was small but stylish. Heating, comfortable seats, and sound insulation made it luxurious by the standards of the day.

DATING FROM 1934, THE LOCKHEED Model 10 Electra belonged to the same ground-breaking generation of airliners as the Boeing 247 and the Douglas DC-2 and DC-3 – twin-engined, all-metal, stressed-skin monoplanes incorporating the latest features to reduce drag, including retractable undercarriages. They could all fly passengers further and faster than their predecessors, but the Electra was the fastest and most stylish of the set. The Electra design team was headed by Hal Hibberd, although the distinctive double tail was down to a junior designer, Kelly Johnson, who had a great career ahead of him at Lockheed. They produced an aircraft that could carry ten passengers plus mail or freight, at an average speed of around 305kph (190mph). Although the DC-3 won the lion's share of the market because it could carry more passengers, the Electra pulled in sufficient orders from major US and foreign airlines to rescue the Lockheed company from potential bankruptcy. In all, 149 Model 10s were built.

The subsequent Model 14 Super Electra (shown above left), introduced in 1937, could cruise at a remarkable 370kph (230mph). In 1938, the eccentric millionaire Howard Hughes piloted one around the world in under four days.

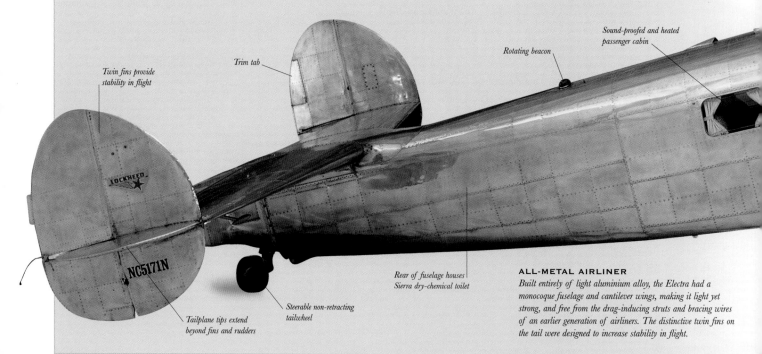

Twin fins provide stability in flight

Trim tab

Rotating beacon

Sound-proofed and heated passenger cabin

Tailplane tips extend beyond fins and rudders

Steerable non-retracting tailwheel

Rear of fuselage houses Sierra dry-chemical toilet

NC5171N

ALL-METAL AIRLINER
Built entirely of light aluminium alloy, the Electra had a monocoque fuselage and cantilever wings, making it light yet strong, and free from the drag-inducing struts and bracing wires of an earlier generation of airliners. The distinctive twin fins on the tail were designed to increase stability in flight.

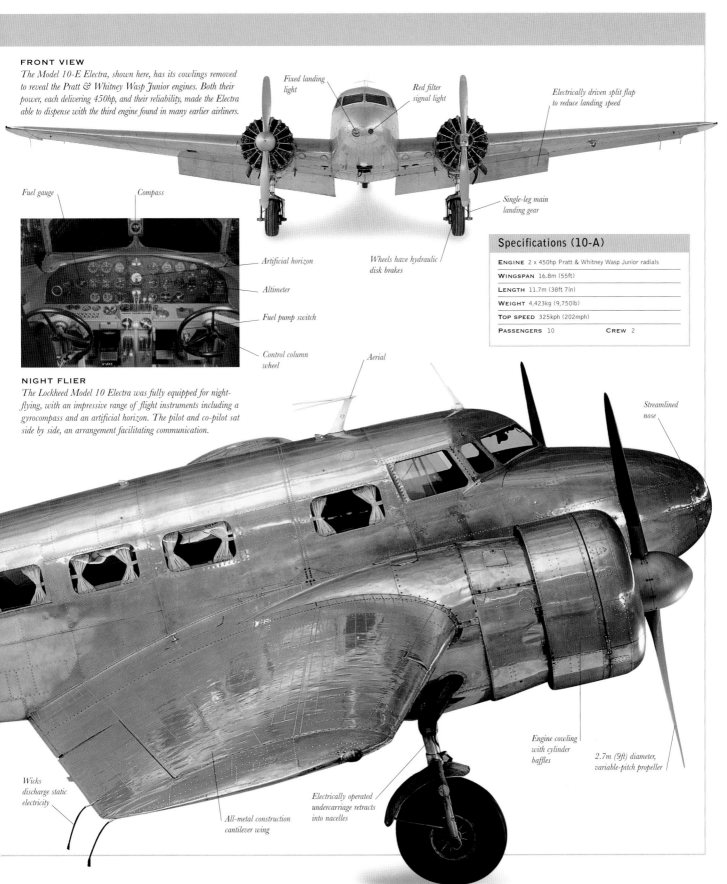

FRONT VIEW

The Model 10-E Electra, shown here, has its cowlings removed to reveal the Pratt & Whitney Wasp Junior engines. Both their power, each delivering 450hp, and their reliability, made the Electra able to dispense with the third engine found in many earlier airliners.

Fixed landing light

Red filter signal light

Electrically driven split flap to reduce landing speed

Fuel gauge

Compass

Artificial horizon

Altimeter

Fuel pump switch

Control column wheel

Single-leg main landing gear

Wheels have hydraulic disk brakes

Specifications (10-A)

ENGINE	2 x 450hp Pratt & Whitney Wasp Junior radials
WINGSPAN	16.8m (55ft)
LENGTH	11.7m (38ft 7in)
WEIGHT	4,423kg (9,750lb)
TOP SPEED	325kph (202mph)
PASSENGERS 10	**CREW** 2

NIGHT FLIER

The Lockheed Model 10 Electra was fully equipped for night-flying, with an impressive range of flight instruments including a gyrocompass and an artificial horizon. The pilot and co-pilot sat side by side, an arrangement facilitating communication.

Aerial

Streamlined nose

Wicks discharge static electricity

Electrically operated undercarriage retracts into nacelles

All-metal construction cantilever wing

Engine cowling with cylinder baffles

2.7m (9ft) diameter, variable-pitch propeller

INTERWAR RACING PLANES

DESIGNERS OF INTERWAR RACING PLANES concentrated exclusively on speed over short distances at low altitude. Concerns such as range or higher altitude performance were left to the designers of serious commercial and military aircraft. What was needed was an engine delivering maximum horsepower for its weight, allied to a lightweight airframe with minimum drag. Racing planes provided a testbed for high-octane fuels and high-performance engines, as well as for developments in streamlining. But racing remained a sport – with the usual mix of spectacular thrills and competition for international prestige. Throughout this period, seaplanes were often the fastest aircraft, because of the almost unlimited take-off run from open water, which made up for the drag associated with floats. Until 1931, the Schneider Trophy seaplane competition provided the focus for attempts on the world speed record, with competitors funded by their governments. Landplane racing flourished in the United States from the late 1920s, eliciting inspired designs from racing pilots like James Wedell or smaller planemakers like the Granville brothers.

SUPER SPORTSTER
A Gee Bee R-1 being flown by Jimmy Doolittle during the 1932 Cleveland Air Races. Doolittle's R-1 set a landplane world speed record of 473kph (294mph).

Caudron C.460

The C.460 was the sensation of the 1936 US National Air Race season. Designed by Marcel Riffard at the long-established Caudron company, it set an airspeed record of 506kph (314mph) in 1934. At the 1936 Thompson Trophy, it trounced the American opposition.

ENGINE 370hp 8-litre Renault Bengali 6-cylinder inline	
WINGSPAN 6.7m (22ft 1in)	LENGTH 7.1m (23ft 3in)
TOP SPEED 505kph (314mph)	CREW 1
PASSENGERS None	

Curtiss R3C-2 Racer

For the 1925 season, the Curtiss company (which had built one of the earliest seaplanes in 1911) produced the R3C-1 and the R3C-2 (with floats). On 25 October, Jimmy Doolittle flew the R3C-2 to a new world speed record of 395kph (246mph).

ENGINE 565hp Curtiss V-1400 12-cylinder liquid-cooled	
WINGSPAN 6.7m (22ft)	LENGTH 6.7m (22ft)
TOP SPEED 395kph (246mph)	CREW 1
PASSENGERS None	

Granville Model R-1 Super Sportster

The most extraordinary of all the early 1930s racing designs were the highly distinctive Gee Bee racers. In 1932, the Granville Brothers produced the ultimate racer – the Super Sportster – which came in two versions: the R-1 and the longer range 525hp R-2.

ENGINE 745hp supercharged P&W R-1340 Wasp 9-cylinder radial	
WINGSPAN 7.6m (25ft)	LENGTH 5.4m (17ft 9in)
TOP SPEED 473kph (294mph)	CREW 1
PASSENGERS None	

Hughes H-1 Racer

Conceived by the wealthy but eccentric Howard Hughes, the H-1 was designed to be the world's fastest landplane. That aim was achieved on 13 September 1935, when Hughes flew the H-1 at 566.7kph (352.2mph). Great attention was paid to streamlining, with retractable landing gear and all rivets and seams flush.

ENGINE 700hp supercharged Pratt & Whitney Twin Wasp radial	
WINGSPAN 9.6m (31ft 9in)	LENGTH 8.2m (27ft)
TOP SPEED 567kph (352mph)	CREW 1
PASSENGERS None	

Macchi M.39

In November 1926, the Italians won the Schneider Trophy for fastest seaplane at Hampton Roads, Norfolk, Virginia, with their remarkable M.39 monoplane racer. Designed by Mario Castoldi, its winning speed over the closed circuit was 396.6kph (246.5mph). The Americans had won the previous two years with converted Curtiss fighter planes and were hoping for a record third victory, so that they could keep the trophy. In the event, they were thwarted, and the US Navy withdrew from the Schneider Trophy competition altogether. On 17 November, the winning pilot, Mario de Bernardi, subsequently flew the M.39 to a new world air speed record of 416.6kph (258.9mph).

ENGINE 800hp Fiat AS-2 V12 liquid-cooled	
WINGSPAN 9.3m (30ft 6in)	LENGTH 6.7m (22ft 1in)
TOP SPEED 417kph (259mph)	CREW 1
PASSENGERS None	

Red racing colours of Italian team

Low-mounted propeller

Wires and struts brace submergeed floats

Macchi M.C.72

The beautiful M.C. (Macchi-Castoldi) 72 was built for the 1931 Schneider Trophy but was unable to compete due to the torque effect of its tandem-mounted engine, which made the aircraft uncontrollable on the water. The problem was solved by fitting contra-rotating propellers to cancel the effect. In October 1934, it set a new world speed record of 709.1kph (440.7mph).

ENGINE 2,800hp Fiat A S 6 tandem V-24 cylinder	
WINGSPAN 9.5m (31ft 1in)	LENGTH 8.3m (27ft 3in)
TOP SPEED 709kph (441mph)	CREW 1
PASSENGERS None	

Supermarine S.6B

The S.6B was the final link in a chain of Supermarine racing seaplanes that had started in 1925 with the unsuccessful S.4. In 1927, the S.5 won in Italy, and in 1929 the S.6 won in England. In 1931, the S.6B completed the course alone, winning the Schneider Trophy outright for Britain after three successive victories. On 29 September, the S.6B raised the world "absolute" air speed record to over 400 mph for the first time.

ENGINE 2,350hp Rolls-Royce R V-12 cylinder liquid-cooled	
WINGSPAN 9.1m (30ft)	LENGTH 8.8m (28ft 10in)
TOP SPEED 655kph (407mph)	CREW 1
PASSENGERS None	

Wedell-Williams Model 44

Three Model 44s were built by Jim Wedell between 1930 and 1932. They were among the most consistently successful radial-engined racers of the early 1930s, finishing in the first three in both the Bendix and Thompson Trophy races from 1931 to 1935.

ENGINE 525hp supercharged Pratt & Whitney Wasp Junior	
WINGSPAN 7.9m (26ft 2in)	LENGTH 7.1m (23ft 4in)
TOP SPEED 491kph (305mph)	CREW 1
PASSENGERS None	

One aspect of aircraft performance that especially fascinated the public in the 1920s and 1930s was pure speed. Achievements in this dimension were certainly an impressive token of the general progress of aviation technology. At the Reims meeting in 1909 the top speed of any aircraft had been 77kph (48mph). In 1920 French pilot Sadi Lecointe set a world speed record of 275.2kph (171.1mph) in a Nieuport. In 1928 Italian Major Mario de Bernardi piloted a Macchi M.52 seaplane at 512.7kph (318.6mph). Only three years later, a British Supermarine S.6B seaplane raised the record to 652kph (407mph). Maximum speed had more than doubled in little over a decade, but further advance slowed as designers ran into technological barriers. By 1939, the top speed had progressed to 751.7kph (469.2mph), a record set by German pilot Captain Fritz Wendel in a Messerschmitt Me 209. Much beyond this point piston-engined aircraft could not go. Still, a mere 30 years on from Reims, record speeds had increased almost tenfold.

Popular air races

Not surprisingly in the context of this rapid increase in speeds, air races became more and more popular. The annual Schneider Trophy seaplane race became a focus for intense national rivalries, drawing substantial investment for engine and airframe development from official sources. Intended by its founder Jacques Schneider as a competition to promote practical reliable seaplanes, it turned into a testbed for single-point designs aimed at speed to the exclusion of all other qualities. Seaplanes were the fastest aircraft of the time, largely because they enjoyed a "runway" of unlimited length for take-off and landing.

Between 1920 and 1931 the competition was dominated by three companies: Curtiss for the United States, Macchi for Italy, and Supermarine for Britain. The Curtiss R3C, piloted to victory by Doolittle in 1925, was the last biplane to win the Schneider Trophy. After that the Macchi and Supermarine monoplanes predominated, the Supermarines eventually winning the trophy outright for Britain after a "contest" in 1931 in which they were the sole contestants. These light, streamlined metal monoplanes, with their powerful liquid-cooled inline engines, obviously showed the path to the fighter aircraft of World War II, even if the lineage was not as direct as has sometimes been assumed.

In the United States, where private individuals and small companies built and

SCHNEIDER TROPHY
The Schneider Trophy was set up in 1912 by Jacques Schneider – a wealthy industrialist and keen balloon pilot – for a seaplane race of at least 150 nautical miles (278km). He hoped the trophy would inspire practical improvements in seaplane design, but it developed into a contest for pure speed.

flew their own aircraft for competitions, air racing was not sharply distinguished from barnstorming. Roscoe Turner, for example, one of the most successful racing pilots, was also a consummate showman, famous for stunts such as flying with a lion cub (wearing a parachute) in the passenger seat. Events such as the National Air Races attracted huge crowds who expected thrills and spills. The top races of the 1930s were the Thompson Trophy, a speed contest flown on a closed course around pylons, and the Bendix Trophy, a race across America from coast to coast. Famous winners included James Doolittle, Roscoe Turner, Jacqueline Cochran, James Wedell, and millionaire Howard Hughes.

Freakish designs

Some custom-built racing aircraft reached extremes of freakish design, none more so than the Gee Bee racers produced by the Granville brothers of Springfield, Massachusetts. These were little more than engines with wings attached, and were regarded as among the most dangerous aircraft ever built. But they were impressively fast. The Gee Bee Model R-1, piloted by Doolittle, won the 1932 Thompson Trophy with a speed of 404.8kph (252.7mph). In a straight line it could reach almost 480kph (300mph). Other victorious aircraft were closer to the mainstream of engine and airframe design, for example the stylish Northrop Gamma or Howard Hughes' record-breaking H-1 racer. But however much design ingenuity and inventiveness went into speedsters, air racing remained essentially a peripheral if exciting spectacle, not a vital testbed for military or commercial airplane development.

NATIONAL AIR RACE TICKET
This ticket admits one to the 1929 National Air Races, held in Cleveland, Ohio. The race around pylons was won by Douglas Davis in the Travel Air Mystery Ship – giving birth to the annual Thompson Trophy Race.

The longest air race of all was held in 1934. The MacRobertson race required competitors to fly from Mildenhall, England, to Melbourne, Australia – a distance of 18,100km (11,300 miles) across 19 countries and seven seas. Adventurous aviators from around the world rose to the challenge, although stringent entry conditions saw a field of only 20 entrants set off for Australia. Interestingly, aircraft in the race included both dedicated racers, such as the Granville "Gee Bee" flown by Jacqueline Cochran and Wesley Smith, and standard passenger-transport aircraft, such as the Boeing 247D piloted by Roscoe Turner and Clyde Pangborne.

In the event, the race was won by an aeroplane specifically designed for the occasion, the de Havilland D.H.88 Comet, flown by British airmen C.W.A. Scott and T. Campbell Black. Its performance was an extraordinary tribute to the progress that had been made in engines and airframes, in instrument flying, and also in the provision of facilities such as navigational aids and aerodromes along long-distance routes. The Comet reached Darwin, its first port of call in Australia, in 2 days, 4 hours, and 38 minutes. Fifteen years earlier the same journey had taken the Smith brothers 27 days 20 hours. But as striking as the winning time – 70 hours 54 minutes to

Melbourne – was the fact that the Comet was hard pressed by two passenger aircraft – a Douglas DC-2 entered by Dutch airline KLM and Turner and Pangborne's Boeing 247D. The DC-2 reached Melbourne only seven hours behind the Comet.

End of an era

By the time commercial aeroplanes were able to fly halfway across the world in a matter of days, the age of the celebrity pilot-heroes was patently drawing to a close. In 1938, only 11 years after Lindbergh's celebrated Atlantic crossing, a Lufthansa Focke-Wulf Fw 200 Condor, a commercial aeroplane built to carry 26 passengers, flew non-stop from Berlin to New York in 24½ hours. This did not exhaust the Condor's range – from

AMELIA EARHART

KANSAS-BORN Amelia Earhart (1897–1937) was an average amateur pilot when she was invited to become the first woman to fly across the Atlantic. Her publicist (later her husband) judged her personality and looks suitable for promotion as a female Lindbergh. With no experience of instrument-flying, she could only travel as a passive third member in the Fokker trimotor that crossed the ocean in June 1928, but this did not prevent her from achieving celebrity status.

Although she continued to benefit from a powerful publicity machine, Earhart subsequently built up a list of impressive achievements that justified her star status. In May 1932, she flew solo across the Atlantic on the fifth anniversary of Lindbergh's New York-to-Paris flight, and in January 1935 she flew solo over the Pacific from Hawaii to California. Earhart now began to

"LADY LINDY"
The press dubbed Earhart "Lady Lindy", an association with Lindbergh that her publicists encouraged. During the 1930s she built up an impressive list of achievements to justify her fame.

formulate plans for a round-the-world flight in a Lockheed Electra with former Pan Am navigator Fred Noonan as her co-pilot. In March 1937, they set out westwards from Oakland, California, but after crashing on take-off, they decided it would be safer to fly eastwards. On 2 May 1937, they set out again, reaching Lae, New Guinea, by the end of June, having flown some 35,200km (22,000 miles). On 2 July 1937, a visibly exhausted Earhart took off on the 4,000-km (2,500-mile) leg to Howland Island, the last stop before California. She never arrived.

ALOHA FROM HAWAII
Amelia Earhart stands in the cockpit of her Lockheed Vega 5C after becoming the first person to fly solo from Hawaii to California on 11–12 January 1935.

Germany it could even reach Tokyo, a two-day journey, stopping three times for fuel. And most of these long-distance flights could be made with minimal risk. Transcontinental and transoceanic flight had ceased to be an heroic enterprise.

The romance of the "heroic age" of aviation had required an awesome waste of young lives. The list of those lost pushing the limits of distance or speed included the famous – Jean Mermoz, Wiley Post, Charles Kingsford Smith, Amelia Earhart, James Wedell, Bert Hinkler, Harry Hawker – and hundreds whose names were never known to the public or have long been forgotten. Without the high risks, the drama of flight could never have reached the pitch that

enthralled the public and made the likes of Lindbergh and Earhart into legends of solitary endeavour.

But while the deeds of individual pilots occupied the front of the public stage, flight was being adopted by governments and large corporations as a business interest and a projection of national power and prestige. It was becoming a serious matter of concern for government bureaucracies, official research institutes, commercial interests, and military establishments, and as a result both a safe and reliable form of transport and an increasingly effective weapon of war. By the late 1930s, the image of flight as a realm in which heroic individuals pitted their courage and skill against the elements was going out of date.

WORLD'S LONGEST AIR RACE
Miss Clara Johnson, a United Air Lines stewardess, points to an artwork on the Boeing 247D that took part in the 1934 MacRobertson London-to-Melbourne Air Race. The Boeing came in third; the race was won by a de Havilland Comet.

FLIGHT OF THE CONDOR
This four-engined Focke-Wulf Fw 200 Condor transport plane is shown being welcomed home at Templehof airport, Berlin. The Condor's non-stop flight from Berlin to New York in 24½ hours on 11 August 1938 gave proof of the capabilities of Germany's renascent air industry.

MISR AIRLINES CAIRO

"First Europe, and then the globe, will be linked by flight, and nations so knit together that they will grow to be next-door neighbours… What railways have done for nations, airways will do for the world."

CLAUDE GRAHAME-WHITE, 1914

COMMERCIAL BREAKTHROUGH
The Douglas DC-2 was introduced in 1934. An instant hit, it established 19 American speed and distance records in its first six months. For the first time, the American business traveller could fly from coast to coast without losing a day.

PASSENGERS NOW BOARDING

BETWEEN THE WARS, PASSENGER AIR TRAVEL RAPIDLY DEVELOPED FROM A PRIMITIVE AND HAZARDOUS ADVENTURE INTO A REFINED, TIME-SAVING SERVICE

THE FIRST sustained passenger aeroplane services were established in Europe in the immediate aftermath of World War I. They required a sturdy breed of customer. When the first daily international scheduled air service, between Hounslow, London, and Le Bourget, Paris, began in August 1919, the passengers travelled in open cockpits and wore protective clothing against the cold. The aircraft did not operate in bad weather and frequently failed to complete the three-hour journey non-stop, making a forced landing in farmers' fields for repairs or refuelling. For this dubious service passengers paid £42 return – equivalent to six months' pay for an average British worker.

Conditions on European passenger lines soon became less challengingly spartan, yet even in enclosed cabins passengers were subjected to deafening noise, sickening turbulence, bone-shaking vibration, and either stifling heat or freezing cold. Although forced landings soon became less common, cancellation of flights due to bad weather did not. The speed advantage aircraft enjoyed over the frequent and punctual trains that linked European cities was, to a large extent, undermined by the time it took to travel to and from aerodromes. What air transport mostly had to offer the traveller was novelty and excitement, a new view of the earth, and the sense of adventure and superiority that came from experiencing the

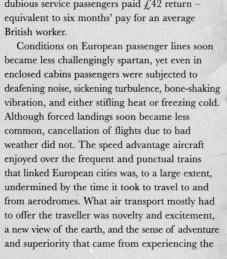

BUSINESS MOGUL
In 1916 William Boeing founded Pacific Aero Products. The next year, the company was renamed the Boeing Airplane Company, and between the wars it produced fighter planes for the US army and navy, as well as commercial passenger airliners. Boeing retired in 1934.

world's most modern technology at first hand. For a long time flight would remain an uncomfortable, expensive, and unreliable way to travel. It was also unprofitable. No aeroplane could carry enough passengers to cover costs. When European passenger air services developed after the war, it was less in response to public demand than to the needs of aircraft manufacturers, which, faced with the collapse of the military market at the end of the war, had to find another use for the aircraft they produced. Passenger transport offered a practical alternative to bankruptcy. It was logical that the French airline Compagnie des Messageries Aériennes should have been created by the joint action of France's leading aeroplane makers

– Blériot, Farman, Caudron, Morane, Renault, and Breguet. Similarly, in Germany, manufacturers such as Junkers and Albatros were involved in setting up airlines. The ready availability of ex-military pilots and cheap war-surplus aeroplanes also encouraged adventurous entrepreneurs to try establishing passenger services run on shoestring budgets.

JOY RIDERS

Suitably decked out in goggles, flying helmets, and protective overalls, passengers prepare for a trip in a Standard J-1 at Long Island, New York, in 1927. At that time in the US there was little passenger aviation beyond this primitive level.

Whatever their origin, early ventures into passenger transport only survived because European governments were prepared to promote air travel. Contrary to the dream of the early aviators that flight would transcend borders and make redundant the division of the world into nation states, the 1919 Convention of Paris decreed that there was to be no "free and universal thoroughfare" of the air. This first attempt to lay down rules for international air traffic also stated clearly that each country had "complete and exclusive sovereignty over the airspace above its territory". Indeed, the development of flight was entirely conditioned by national and imperial rivalries. European governments were persuaded of the need to actively encourage commercial aviation because they saw it as part of the struggle for national prestige and national defence. They recognized the need to maintain air technology and manufacturing capacity for possible future wars. And they saw in aircraft a means to bind together their far-flung empires.

European governments provided airlines with open or concealed subsidies – for example, through profitable airmail contracts – and introduced supportive regulations to encourage safety. They also intervened to protect airlines from competition, granting monopoly rights to fly certain routes and enforcing company mergers in line with their perception of the national interest.

The country that emerged as the leader in European – and therefore world – passenger air transport in the 1920s was Germany. This was an especially impressive achievement since, for years after the war, the Allies continued to impose restrictions on German civil aviation, alongside the total ban on military aviation, under the terms of the Versailles Treaty. Germany responded with a steely determination to maintain its aircraft industry. To avoid restrictions – for a time German aircraft production was banned completely – manufacturers such as Junkers and Dornier relocated outside Germany. Fokker had already hastily shifted operations back to his native Netherlands at the war's end.

TRANSPORT PIONEER
This AEG J.II biplane is one of the aircraft with which Deutsche Luft-Reederei (DLR) began the first postwar European passenger air service in 1919. DLR was one of the forerunners of Lufthansa, which adopted the distinctive crane logo seen here on the tail.

GERMAN NATIONAL AIRLINE
After the Nazis came to power in 1933, all Lufthansa airliners displayed the swastika. The aircraft shown in this poster, a Junkers Ju 52 was used as both a civil airliner and a military transport.

First passenger airlines

The very first passenger service of the postwar era was initiated by German airline Deutsche Luft-Reederei on 5 February 1919 – three days before the French Farman company began a tentative service on the Paris–London route. Deutsche Luft-Reederei flew between Weimar, where the constituent assembly of the new German Republic was sitting, and the capital, Berlin. From these small beginnings, through the 1920s Germany developed a network of commercial air routes, stretching north into Scandinavia, east through Poland into the newly formed Soviet Union, and south into the Balkans and the Mediterranean. By 1923 these routes were being operated by just two airlines, one owned by Junkers and the other financed by shipping companies and bankers. In 1926 the German government, which was subsidizing both, forced them to merge into a single national airline, Deutsche Luft Hansa (DLH, known as Lufthansa after 1934).

When it was formed, Luft Hansa was by far the world's largest airline – it was reckoned to be responsible for 40 per cent of the world's passenger air traffic. It certainly operated with the most advanced aviation technology. In imitation of the US Air Mail, airways lighted by beacons were created for night flying. This allowed, for instance, a direct air service from Berlin to Moscow, travelling the lighted section from Berlin to Konigsberg by night. By 1929, Luft Hansa pilots were routinely trained in instrument-flying, and aircraft were linked to air-traffic controllers by radio. By 1931, the airline was able to run a scheduled passenger service to Italy over the formidable barrier of the Alps.

HUGO JUNKERS

PROFESSOR HUGO JUNKERS (1859–1935) was almost 50 years old when he first took an interest in aviation. A professor at Aachen High School, he began exploring the aerodynamic possibilities of metal cantilever wings and, in 1910, patented a revolutionary design for an all-metal aeroplane without a fuselage or tail, housing the engines, crew, and passengers in the wing. This "flying wing" was never built, but during World War I, other Junkers all-metal designs were adopted by the German air forces. At the end of the war, Junkers told his team to turn their efforts to civil air transport.

Junkers Flugzeugwerke went on to produce outstanding aircraft, from the F 13 to the Ju 52, while Junkers also established a short-lived but highly successful airline business. When the Nazis came to power in 1933, the independent-minded Junkers was one of their first targets, and he was bullied into handing over his company and his personal patents to the state. He was put under house arrest and died on his 76th birthday in 1935.

INSPIRED DESIGNER
At a time when most of his competitors were creating strut-and-wire, fabric-covered biplanes, Junkers was producing all-metal, cantilever-wing monoplanes that pointed the way forwards for aircraft design.

The French government saw the growth of German civil aviation as a direct threat to its interests. As a consequence, it poured money into its own passenger network. By 1920 eight French airlines were operating, each with its own monopoly route and subsidy. Britain lagged behind in the promotion of passenger transport, but in 1924, after a series of air-transport ventures had struggled to survive without subsidy, the British government belatedly promoted a merger to form Imperial Airways, a private company officially backed as the nation's "chosen instrument". The logic of the situation eventually led to the creation of a single state airline in both France (Air France, founded in 1933) and Britain (BOAC, founded in 1939).

Even the smaller European countries had to have their own airline, and at least one of these "flag carriers" was outstandingly successful. Founded in 1919, the Dutch airline Koninklijke Luchtvaart Maatschappij (KLM) was the brainchild of army pilot Albert Plesman. Working closely with Fokker, who supplied KLM with some of the best passenger aircraft around, Plesman made his company a notable presence in world aviation despite the lack of a significant domestic network. By 1929 KLM was regularly flying an eight-day route from the Netherlands to Batavia (now Jakarta) in the Dutch East Indies – the longest scheduled service in the world.

Converted warplanes

The first passenger services after the war used converted bombers or reconnaissance aircraft. It was these aeroplanes that gave passengers the

CROSS-CHANNEL SERVICE
During the General Strike of 1926, cross-Channel air traffic greatly increased due to the cancellation of ferry services. Imperial Airways took advantage of this with its daily London–Paris service, here using a Handley Page W.10.

rawest experience of flight, since they were simply occupying seats that would once have held an observer or bomb-aimer. For instance, in the Handley Page O/11, a converted O/400 bomber, two of the passengers had balcony seats in the open bow, combining an astonishing view with total exposure to the elements. As custom-built passenger aircraft emerged, Britain and France largely stuck with biplanes, such as the de Havilland D.H.34, while the German and Dutch airlines flew more modern monoplanes such as the Junkers F 13 and Fokker F.III. Through the 1920s, Fokker and Junkers evolved these single-engine aeroplanes into multi-engine models that remained world leaders until the United States muscled into commercial aviation in the following decade.

As air travel became better organized through the 1920s, travelling in a purpose-built passenger aeroplane became a passably civilized experience. Climbing into an F 13,

RECORD-BREAKING FOKKER
The Fokker F.XVIII airliner, introduced to KLM airlines in 1932 and seen in this promotional poster of the time, cut down the travelling time to the Dutch East Indies to nine days. A year later, an airmail-only trip was achieved in a record time of four days.

TICKETS PLEASE
*America's largest airline in the early
1930s was United Air Lines, which
incorporated four different carriers. As the
tickets reproduced above show, it also carried mail.*

for example, you would find you were one of four
or five passengers in an enclosed cabin, sitting on
a cushioned seat with a picture window. You could
settle yourself for the flight with a fair expectation
of arriving at your destination approximately on
schedule. But first-time air travellers were still
stunned by the deafening engine noise, and once
in the air passengers often found themselves
vibrated through an extraordinary fairground ride
of bumps and drops as the aeroplane rode out
turbulence. They were always issued with paper
sick bags on boarding. Cold was also a problem
for passengers until heated cabins became the
norm – by 1934, an International Air Guide was
able to reassure its readers that "no special
clothing is required" for flight.

Every effort was made to promote the image of
flight as a luxury experience, for it was, after all,
suitably expensive. For example, publicity material

IMAGE OF LUXURY
*Airline publicity photos, such as this staged image of coffee
time on a DC-3, were understandably designed to represent
flight as comfortable, even luxurious. They also stressed the
presence of women on flights, to counter the popular prejudice
that flying was too risky for the "weaker sex".*

emphasized the serving
of champagne lunches
on some Paris-to-London services. The
first in-flight movie was projected during a
Luft Hansa flight in 1925 – a silent movie, of
course, so the deafening engine noise did not
matter. But such refinements were untypical.

The overwhelming majority of European air
passengers in the 1920s were male, chiefly
government officials or businessmen in a hurry.
Many men considered flying too risky for women
or children (traditionally categorized together).
European airlines in fact killed remarkably few of
their customers in those early days, but it would
have been an unnerving experience to find your
aircraft diving down almost to treetop height to
fly under low cloud or carefully sticking to a road
or rail line to avoid getting lost – a practice that
led to a fatal collision on the Paris–London route
in 1922 between a British and a French aircraft
that were following a road in opposite directions.
Not surprisingly, flying remained very much a
minority experience. By 1929 only 25,000 French

citizens had flown as aeroplane passengers –
less than one in a thousand of the population.

The Southern hemisphere
Although Europe was the centre of progress in
passenger air transport for most of the 1920s, air
travel also prospered in some less populated areas
of the world where other forms of transport were
undeveloped. Australia was a good place for air
travel because it was mostly flat, had huge
distances between settlements and an inadequate
railroad system, but had plenty of fine weather.
The Queensland and Northern Territory Aerial
Services Limited (Qantas) was founded in 1920 by
two pilots back from the war in Europe, Hudson
Fish and Ginty McGinnis. The key to its initial

FIRST IN-FLIGHT MOVIE

The first in-flight movie was shown to passengers on a Deutsche Luft Hansa flight on 6 April 1925. Due to weight limitations, single-reel shorts were generally shown; the fact that the films were silent made them ideal for showing in a noisy airliner. Note the wicker chairs, standard in most passenger aircraft in the 1920s.

success was that it served two towns, Charleville and Cloncurry, that were 960km (600 miles) apart and had no other viable means of communication. But Qantas could no more survive without government handouts than could any other airline, depending on a generous contract to carry airmail. From 1928, Qantas also provided the aircraft for the first "flying doctor" service in the Australian outback.

South America was another area of the world where air travel could radically cut journey times between towns and cities with otherwise inadequate or circuitous transport links. Despite the daunting challenge posed by jungle and mountain terrain, in the early 1920s air transport made more progress in South America than North America.

SITTING COMFORTABLY

Passengers on a Deutsche Luft Hansa flight of the late 1920s, travelling in a Junkers G 24 trimotor, enjoyed reasonable comfort with no frills. Notice that everyone has kept their coat on since the cabins were not heated. The Germans were world leaders in passenger aviation until well into the 1930s.

PARIS TO LONDON: A PILOT'S EXPERIENCE

IN THE EARLY 1920s, pilot Frank Courtney described the experience of flying a D.H.34 on a passenger service from Le Bourget, Paris, to Croydon, London, in poor weather conditions: "While taxiing out to take-off, I am getting wet and uncomfortable because of the open cockpit. Once airborne, though still comparatively close to the ground, we are just below cloud. I start on the usual compass course with a mentally calculated allowance for drift… but after six miles I find there are tree-covered hills shrouded in wisps of cloud. It is now obvious that a compass course is impossible, so I turn left and pick up the main road from Paris to Boulogne… [I] am compelled to stick to this road as completely as a motor-car, for if I lose sight of it I am to all intents and purposes lost…" Eventually he has to abandon following the road to Boulogne because it leads over hills where cloud is

at treetop height. Even then he does not dare cut across country, but has to trace the road back to a junction. There he follows an alternative road to Amiens. He finally crosses the Channel, reaches England's white cliffs and finds a circuitous route to Croydon that avoids cloud-covered hills. Courtney's concluding remarks were not reassuring: "Completion of the journey has been entirely dependent on what risks the pilot was prepared to take, and there is the added fact that on such a flight… the avoidance of a collision with a machine coming in the other direction is frequently a matter of luck."

BRAVING THE WEATHER

While the passengers travelling from Paris to London enjoyed relative comfort inside the cabin, the pilot remained largely exposed to the elements.

Pilot sits in open cockpit

Scadta, the first permanent airline in the Americas, was founded in Colombia by German expatriates in 1919 and began service in 1920.

Slow start in the USA

So why were travellers in the United States so slow to take to the air? Although it had an excellent railroad network, the country was built on the right scale for air transport, its continental expanse offering aeroplanes the chance to deliver major time-saving, as the US Air Mail quickly proved. But it was hard to persuade Americans to take flying seriously. Mesmerized by the antics of the barnstormers, the public failed to see the aeroplane as a straightforward way of getting from place to place. Most of the passenger services that did operate in the immediate postwar years were an amusing start to a vacation break – for example, trips from New York City to resorts on Long Island or in New Jersey, run by Aeromarine. Aviation in America seemed stuck with an image of frivolity and danger that deterred potential investors and customers alike.

Relatively free of the international rivalries and national-defence concerns that motivated

European governments, in the early 1920s the US federal government adopted a hands-off approach, neither supporting nor regulating commercial aviation. But a turning point came in 1925, with the decision to make the Post Office hand over the thriving US Air Mail network to private companies. Since no such companies yet existed, the federal authorities were in effect committed to creating the conditions in which they could flourish. Like elsewhere in the world, commercial aviation in the United States would be less the product of buccaneering free enterprise than of government handouts, public investment in infrastructure, and strict regulation to uphold standards and protect the new airlines from the cold winds of competition.

The first necessity was to persuade the public that flying was safe. Under the 1926 Air Commerce Act, all commercial aircraft and engine types had to be checked over by the Aeronautics Branch of the Department of Commerce and certified safe according to stringent standards. All pilots and mechanics had to apply for a licence. Pilots were subjected to a flying test and a physical examination; they also had to be attested as of

"good moral character". Crucially, pilots could have their licence revoked if they were considered to have flown dangerously – anything from flying while drunk to the frequent stunt of "buzzing" crowds at public events.

Concern for the safety of air passengers even led to the suggestion that they should all be issued with parachutes. This idea was hotly debated but never taken up, partly because it would have drawn attention to the dangers of flying, which was exactly the opposite of what the air industry and the federal aviation authorities wanted.

The Lindbergh factor

It was fortunate that the final handover of the airmail routes to private operators coincided with Lindbergh's sensational solo flight across the Atlantic in 1927 (see page 118). Suddenly there was a surge of enthusiasm for aviation among the public and business investors alike. While flying schools flourished, money poured in for aircraft manufacturers, engine and propeller makers, and air-transport companies, while city governments rushed to build new airports. But this wave of investment was based on optimism for the future, not current returns. Payment for airmail, calculated by weight carried, proved an unreliable source of income. Despite sharp practices such as air companies flying stacks of mail addressed to their own offices or bumping up weight by slipping bricks into the mailbags, it was hard to turn a profit. Yet airmail carriers had little incentive to develop passenger transport, because mail paid better.

The highest profile passenger service of the late 1920s was established by Transcontinental Air Transport (TAT), which took on Lindbergh as a consultant and marketed itself as the "Lindbergh Line". As its name suggested, TAT was set up specifically to run a coast-to-coast service. Because of the dangers of night flying, this had to involve a mix of air and rail travel. Starting in July 1929, passengers could travel overnight by train from New York o Columbus, Ohio, board one of TAT's Ford Tri-Motors for a day's flight to Waynoka, Oklahoma, carry on by overnight train to Clovis, New Mexico, and fly the last leg into Los Angeles. The two-day air-rail journey

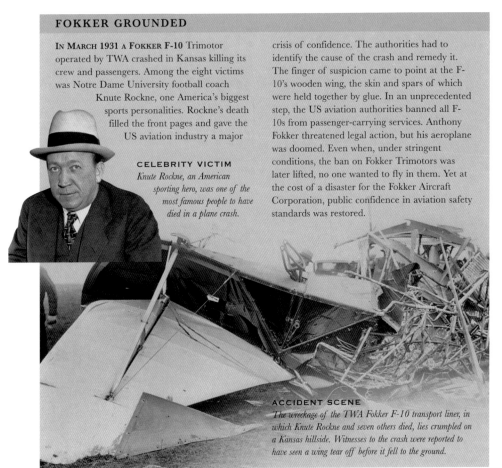

FOKKER GROUNDED

IN MARCH 1931 A FOKKER F-10 Trimotor operated by TWA crashed in Kansas killing its crew and passengers. Among the eight victims was Notre Dame University football coach Knute Rockne, one America's biggest sports personalities. Rockne's death filled the front pages and gave the US aviation industry a major

CELEBRITY VICTIM
Knute Rockne, an American sporting hero, was one of the most famous people to have died in a plane crash.

crisis of confidence. The authorities had to identify the cause of the crash and remedy it. The finger of suspicion came to point at the F-10's wooden wing, the skin and spars of which were held together by glue. In an unprecedented step, the US aviation authorities banned all F-10s from passenger-carrying services. Anthony Fokker threatened legal action, but his aeroplane was doomed. Even when, under stringent conditions, the ban on Fokker Trimotors was later lifted, no one wanted to fly in them. Yet at the cost of a disaster for the Fokker Aircraft Corporation, public confidence in aviation safety standards was restored.

ACCIDENT SCENE
The wreckage of the TWA Fokker F-10 transport liner, in which Knute Rockne and seven others died, lies crumpled on a Kansas hillside. Witnesses to the crash were reported to have seen a wing tear off before it fell to the ground.

LOADING THE GOOSE
Cargo handlers load air-express packages on to a TWA Ford Tri-Motor at Kansas City in the 1930s. The "Tin Goose" is instantly recognizable from its thick corrugated metal wing and its underwing radial engine.

was 20 hours faster than the trip by express train alone. However, this was not enough time-saving to attract many customers, especially after a TAT Tri-Motor flew into Mt Taylor, Arizona, just two months after the service started, killing everyone on board. The air-rail experiment ended in 1930.

Night flights

The development of the American air-transport network in the 1930s soon took it far beyond anything that had been achieved in Europe. One of the conditions of the contracts awarded

by the postmaster general in 1930 was that airlines had to be experienced in operating by night as well as day. By 1933 there were 28,800km (18,000 miles) of lighted airway in the United States, and airlines were flying passengers coast-to-coast without recourse to overnight train travel. Provision for bad-weather flying also made giant strides. Radio navigation stations – transmitters of radio "beams" – were established at 320-km (200-mile) intervals along US airways. Already by 1929 it was possible to fly "on the beam" from Boston to Omaha via New York

OVER THE SIERRA MADRE
In 1929, TAT opened up an air-rail coast-to-coast route from Los Angeles to New York. Charles Lindbergh piloted the first leg of the inaugural flight in the Ford Tri-Motor City of Los Angeles. Here he flies along the Sierra Madre mountains.

Ford 5-AT Tri-Motor

SLICK IMAGERY
A publicity shot emphasises the comfort and smoothness of flight onboard a Ford Tri-Motor, qualities that were in short supply on most flights. Vibration was a problem experienced by passengers on all propeller-driven airliners.

AUTOMOBILE MANUFACTURER Henry Ford, creator of the Model T, had a reputation second to none as a hard-headed businessman. When he announced an interest in aeroplane manufacture in 1924, the whole American business community sat up and took notice. Ford declared that he looked forward to a time when aircraft could be mass-produced like automobiles "by the thousands or by the millions". This was overambitious, but at least with the Tri-Motor – affectionately known as the "Tin Goose" – Ford made the first American passenger-carrier to be produced in the hundreds.

In the 1920s, Europeans were world leaders in aeroplane design. To come up with a successful commercial aeroplane, Ford's team put together a hybrid of features from the Fokker and Junkers stables. Ford incorporated a small company run by engineer Bill Stout, who was making all-metal aircraft with a corrugated tin skin in the Junkers style. The "tin" was aluminium and aluminium alloy. The story goes that after Richard Byrd's attempted flight to the North Pole in a Fokker, Byrd made a stop at Ford's Dearborn Field. Here the plane was secretly measured up by Ford engineers so that they could copy it for their prototype.

Ford's Tri-Motor 4-AT (Air Transport), which was introduced in 1926, was succeeded in 1928 by

> "I should like one thousand dollars, and I can only promise one thing. You'll never see the money again."
>
> **AEROPLANE BUILDER WILLIAM STOUT**
> TO HENRY FORD'S SON, EDSEL

IN-FLIGHT MEALS
Chicken salad was the staple centrepiece of the in-flight meals served during the pioneering years of passenger flight in the late 1920s. Air stewardesses became a feature of passenger travel from the early 1930s onwards.

Specifications (5-AT-B)

ENGINE	3 x 420hp Pratt & Witney Wasp radials
WINGSPAN	23.7m (77ft 10in)
LENGTH	15.2m (49ft 10in)
WEIGHT	5,897kg (13,000lb)
TOP SPEED	196kph (122mph)
PASSENGERS 15	**CREW** 2

Aircraft registration code

N9651

Ford
TRI-MOTOR

Elevator control cables

TRANS WORLD AIRLINE
TAT
TRANSCONTINENTAL AIR TRANSPORT. INC.

Swivelling tailwheel with shock absorber

Corrugated Alclad skinning on duralumin channel-section framework, riveted together

Entrance door to passenger cabin

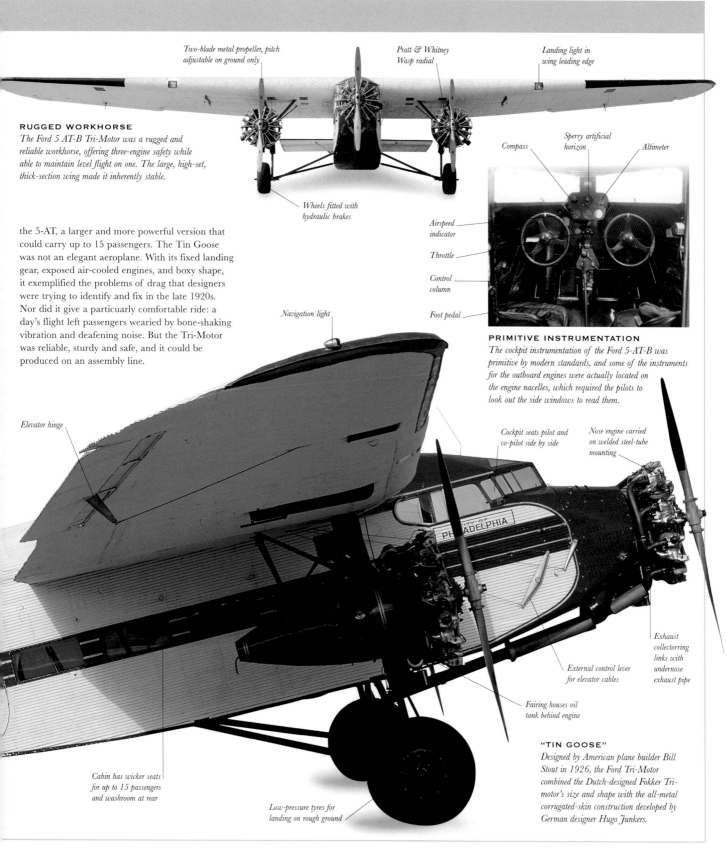

Two-blade metal propeller, pitch
adjustable on ground only

Pratt & Whitney
Wasp radial

Landing light in
wing leading edge

RUGGED WORKHORSE
The Ford 5 AT-B Tri-Motor was a rugged and
reliable workhorse, offering three-engine safety while
able to maintain level flight on one. The large, high-set,
thick-section wing made it inherently stable.

Wheels fitted with
hydraulic brakes

Compass

Sperry artificial
horizon

Altimeter

Airspeed
indicator

Throttle

Control
column

Foot pedal

the 5-AT, a larger and more powerful version that
could carry up to 15 passengers. The Tin Goose
was not an elegant aeroplane. With its fixed landing
gear, exposed air-cooled engines, and boxy shape,
it exemplified the problems of drag that designers
were trying to identify and fix in the late 1920s.
Nor did it give a particularly comfortable ride: a
day's flight left passengers wearied by bone-shaking
vibration and deafening noise. But the Tri-Motor
was reliable, sturdy and safe, and it could be
produced on an assembly line.

PRIMITIVE INSTRUMENTATION
The cockpit instrumentation of the Ford 5-AT-B was
primitive by modern standards, and some of the instruments
for the outboard engines were actually located on
the engine nacelles, which required the pilots to
look out the side windows to read them.

Navigation light

Elevator hinge

Cockpit seats pilot and
co-pilot side by side

Nose engine carried
on welded steel-tube
mounting

Exhaust
collectorring
links with
undernose
exhaust pipe

External control lever
for elevator cables

Fairing houses oil
tank behind engine

Cabin has wicker seats
for up to 15 passengers
and washroom at rear

Low-pressure tyres for
landing on rough ground

"TIN GOOSE"
Designed by American plane builder Bill
Stout in 1926, the Ford Tri-Motor
combined the Dutch-designed Fokker Tri-
motor's size and shape with the all-metal
corrugated-skin construction developed by
German designer Hugo Junkers.

and Chicago. Through the following decade the system was extended throughout the United States. Instrument-flying and navigation by radio beam became standard skills for commercial pilots. By the mid-1930s, passenger aircraft were also almost universally equipped with two-way radios for communication with ground controllers.

Air-traffic control

The control of air traffic around airports became an increasingly urgent concern as traffic levels increased. Arrangements for avoiding collisions were, at first, extremely primitive. Usually, a controller, positioned at a highly visible point in the airport, waved a green or red flag to tell aircraft whether or not it was safe to take off or land. The next refinement involved replacing the flag with a light gun, firing a red or green flare, but these were ineffective in poor visibility.

In 1930 the first air-traffic control tower, equipped with radio, was built at the busy Cleveland Municipal Airport. The pilots of approaching aircraft radioed information on their position to their airline representative at the airport. Controllers used this information to update a map showing where all the aircraft in the vicinity were positioned and radioed pilots if there seemed any risk of a collision. Permission to land or take off was also given by radio. By 1935, there were some 20 airports in the United States operating similar systems.

But traffic kept getting heavier. By the mid-1930s, busy airports such as Newark and Chicago would be handling 60 landings and take-offs an hour. Aircraft were travelling faster and routinely flying by instruments in poor visibility. Since there was no control over aircraft until they approached an airport, a number of aeroplanes would arrive at similar altitude in zero visibility, jostling for a chance to land as overburdened controllers strove to avert catastrophe. The situation was not helped by the fact that, at some underfunded airports, the

RIDING THE BEAM

A PRACTICAL PASSENGER air-transport system required something close to all-weather operation by day and night. Lighted airways were all very well to guide aircraft on clear nights, but they were useless in cloud or fog. The answer had to be some form of radio beacon that would remain "visible" to a receiver in the aeroplane cockpit at all times. But the transmitters available in the 1920s could not generate a beam-like radio signal for an aeroplane to follow – that would have to wait for the introduction of VHF. The solution was found in the use of the loop antenna, an

electrical circuit set upright from the earth. It could function as a direction-finding device because a receiver would pick up a strong signal when facing the edge of the loop, but almost none when at right angles to it. Radio engineers experimented with pairs of loop antennae set at right angles to one another, in the shape of a cross. If signals were transmitted alternately from the two antennae, they merged into a single unbroken tone in the equisignal zone – in effect, a radio beam. Pilots could tell whether they were "on the beam" when the humming was steady and continuous. They could also use the "cone of silence" directly above the radio transmitter to fix their position.

It would be an understatement to say that this navigational system was imperfect. The transmissions were sensitive to many forms of interference that could bend or shift the beam, and they were liable to be drowned out by static, especially in the bad weather conditions when they were most needed. But "riding the beam" was still a vast improvement on what had gone before and became standard in the airline business.

Direction-finding loop aerial

Radio tower

Loop antennae set at right angles to one another

GERMAN RADIO CONTROL
Germany was at the forefront of the development of radio navigation devices. This Lufthansa Ju 52 airliner has a direction-finding loop antenna and two-way radio to communicate with the control tower.

NIGHT SERVICE
In the 1930s, Britain's Imperial Airways and France's Air Union (later Air France) operated night passenger services between Le Bourget, Paris, and London's Croydon Airport. Imperial Airways only used British-made aircraft, such as this Handley Page H.P.42. Although slow and ungainly, the H.P.42 afforded a fairly safe and comfortable ride.

KEEPING TRACK
In a United Air Lines flight-dispatch room in the late 1930s, a woman updates information on the progress of flights on a status board. Airline dispatchers kept air-traffic controllers informed of aircraft's whereabouts by telephone.

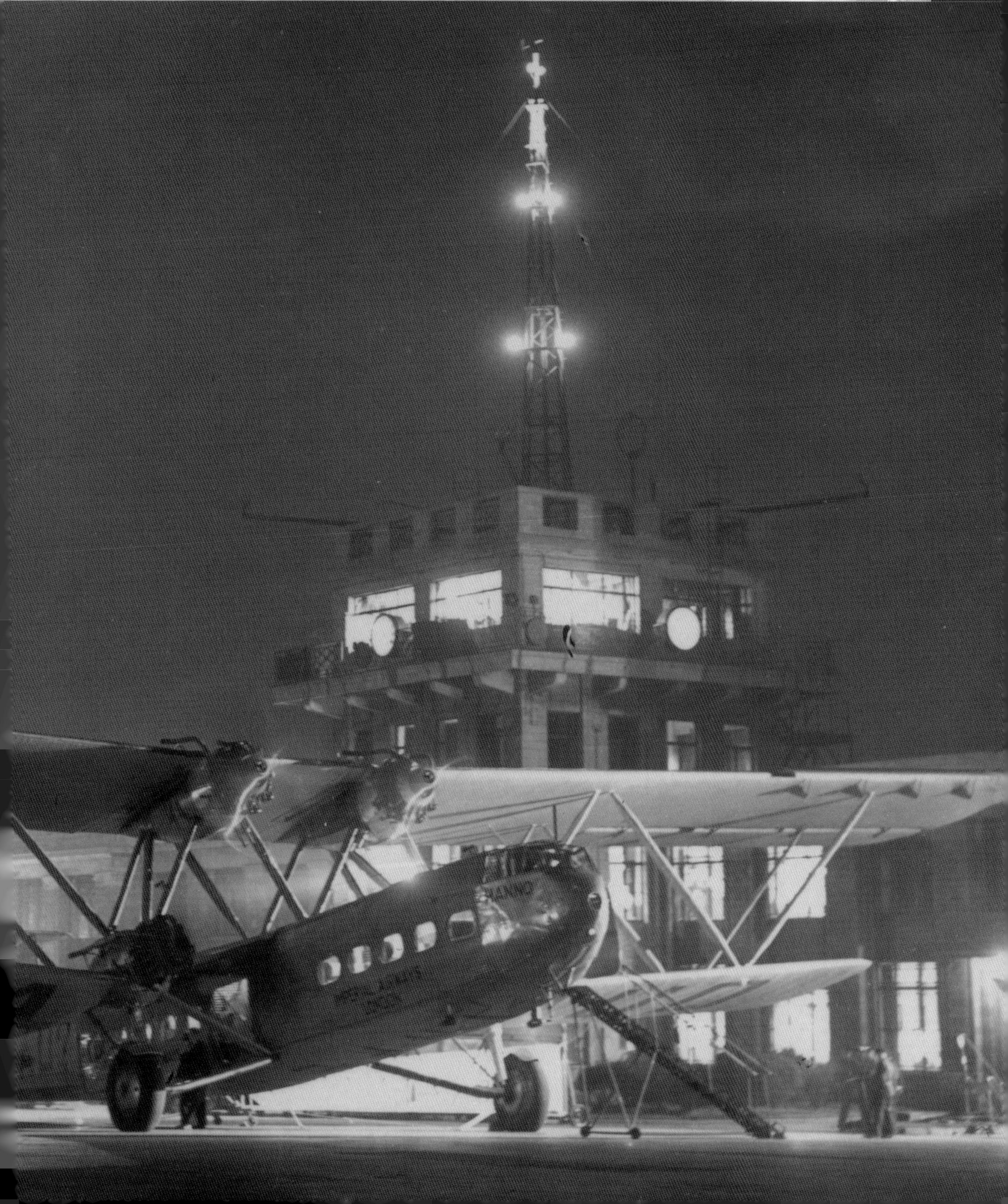

THE AIR HOSTESS ARRIVES

A United Air Lines stewardess, modelling the new summer uniform, stands in salute beside a Douglas DC-3. The first stewardesses were introduced by Boeing Air Transport in 1930, and most other airlines followed suit. In 1935, when TWA switched from male flight attendants, it dubbed its new female staff "air hostesses". First seen as a reassuring presence, stewardesses soon became part of the glamour of flight.

controller might have to double as switchboard operator or baggage handler.

An answer was found in federal control of the airways. From 1936, aircraft using the airways under instrument-flying conditions had to file a flight plan with federal airway-traffic controllers. The pilots then had to report their time at various checkpoints along the route, allowing controllers to plot their courses by shifting markers on a map. The controllers issued instructions to ensure that the aircraft arrived in the vicinity of airports at different times and altitudes.

Pilots in the 1930s did not necessarily accept these new disciplines with a good grace. It was by no means unknown for a pilot to decide he could not be bothered to wait any longer and simply take off or land without clearance. But, in time,

DUTIES OF A STEWARDESS

A manual prepared for air stewardesses in 1930 included the following instructions:
- Remember at all times to retain the respectful reserve of the well-trained servant.
- Captains and cockpit crew will be treated with strict formality while in uniform. A rigid military salute will be rendered the captain and co-pilot as they go aboard.
- Punch each ticket at each point passed.
- Tag all baggage and check it on board.
- Use a small broom on the floor prior to every flight. Check the floor bolts on wicker seats to ensure they are securely fastened down.
- Swat flies in cabin after take-off.
- Warn passengers against throwing lighted smoking butts or other objects out of the windows, particularly over populated areas.
- Carry a railroad timetable in case the plane is grounded somewhere. Stewardesses are expected to accompany stranded passengers to the railroad station.

the necessity for order in the skies became universally appreciated, even if a certain edginess between fliers and ground controllers was installed as part of the tradition of flight.

The first stewardesses

The effort to improve the image of flying extended from safety regulations and the avoidance of accidents to upgrading the in-flight experience of air passengers. One notable innovation came in 1930 at the initiative of a young nurse from Iowa, Ellen Church. Keen on flying, she persuaded Boeing Air Transport, then operating a mail-and-passenger service between San Francisco and Chicago, to hire her and seven other nurses as stewardesses. At that point all aircrew, as well as the vast majority of passengers, were men. Church argued that the presence of women on aircraft would encourage people to regard flying as safe, while a trained nurse was just what a man needed when faced with the rigours of a long air journey. The young women were not exactly welcomed by pilots, described by one of the first stewardesses as "rugged and temperamental characters who wore guns to protect the mail". But the idea of women looking after passengers quickly took hold and spread to other airlines.

The air passenger's need for comfort and reassurance in the early 1930s is easy to understand. The first air stewardesses, for example, worked on Boeing 80s. The latest in trimotor design, the Boeing 80 was comparable in comfort to the Ford "Tin Goose". The passenger cabin was fitted out to look like a luxury Pullman railroad car, with stylish wood panelling, plushly upholstered seats, and tasteful shaded lights. Chicken salad

STEWARDESS SERVICE
One of the first air stewardesses serves coffee on board a Boeing 80 in 1930. All the early stewardesses were qualified nurses and, when serving food, they donned a light grey nurse's uniform. Passengers found it comforting to be looked after by medically trained staff, especially given the prevalence of air sickness.

PAN AM WINGS
Pan American Airways hired its first stewards in 1929 and only employed men as cabin attendants until 1944, when it added its first female crew members. The Pan American brand represented adventure and glamour and enjoyed worldwide recognition.

CABIN CREW
In 1936 Eastern Airlines reverted to male stewards as an economy measure. Stewardesses had to leave their jobs if they married, which most did. A steward gave longer service in return for training.

and coffee were served in-flight with elegant china plates, cups, and saucers. An airspeed indicator and altimeter mounted on the front cabin wall kept passengers informed about their progress. The stewardess pointed out landmarks along the route, and provided blankets and pillows on request.

But despite this surface slickness, much of the experience of flight remained stubbornly discomforting. The noise from the three engines was as deafening as roadworks – every passenger was issued with earplugs on boarding. Chairs without shock-absorbers offered no protection against vibrations. The cabins were in principle heated, but the heating system was inefficient and passengers still often wore

overcoats. Toilet facilities were initially crude, as one stewardess described:
"The toilet was a can set in a ring and a hole cut in the floor, so when one opened the toilet seat, behold, open-air toilet!"

Such minor indignities were nothing compared with the effects of air-sickness. Unable to fly above the weather, aeroplanes frequently gave their passengers a bumpy ride. One of the stewardess's prime tasks was to care for people emptying their stomachs into the coyly named "burp cups" placed under every seat. And at times, having china plates and cups did not seem such a bright idea. Sitting among vomiting passengers and disintegrating crockery, a "bad flight" was as hellish an experience as the worst sea crossing. If the weather was really bad, it was still common in the early 1930s to make emergency landings in cow pastures or on remote emergency airfields. The stewardess's duties might extend to clearing obstacles out of the way for take-off from an improvised airstrip. Sometimes the stewardess

BILL BOEING

BORN IN DETROIT, William E. Boeing (1881–1956) studied engineering at Yale before following his father into the lumber business. In 1914 he bought himself a seaplane, but decided he could build a better one himself. With his friend US Navy Commander Conrad Westervelt, Boeing created the B&W seaplane and, in 1916, set up a company to manufacture it – Pacific Aero Products in Seattle. Relying on his timber and furniture business to keep him going during hard times, Boeing expanded his aviation interests throughout the 1920s, not only building aeroplanes but also creating the Boeing Air Transport Company to carry US airmail and passengers. The conglomerate of aircraft manufacture, air transport, and related companies that Boeing put together became an obvious

target for New Deal trust-busters after President F.D. Roosevelt's election in 1933. Boeing became heatedly involved in political disputes and retired in 1934 in protest at the decision to forcibly separate aircraft manufacturers from airlines. The Boeing Company has continued to combine the pursuit of cutting-edge technology with acute business acumen.

FOUNDING FATHER
Boeing founded what was to become, after his retirement, the world's leading aircraft business.

herself was regarded as an obstacle. The priorities on early 1930s flights were, from top to bottom: mail, passengers, stewardesses. If the pilot decided his aeroplane was overweight for take-off, he would dump the stewardess and continue without her.

From 1930, travellers tired of the discomforts of Fokker, Ford, and Boeing trimotors found relief on some routes in the Curtiss Condor. Flown first by Eastern and then by American Airways, it was

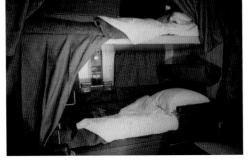

UNPROFITABLE SLEEPER
American airline companies wanted to offer sleeper accommodation on coast-to-coast flights to compete with the railroad's Pullman cars. These bunks are on a Douglas Sleeper Transport (DST). Introduced in 1935, the DST could not operate profitably because it carried only 14 passengers. The 21-seat DC-3 (opposite), the "day coach" version of the DST, did make a profit.

a design that, like most British interwar airliners, sacrificed looks and performance in the cause of passenger comfort. An old-fashioned strut-and-fabric biplane with a tendency to ice up and engines disturbingly prone to catch fire, the Condor could hardly be termed a technical success. Yet its wide fuselage offered a new level of luxury, including 12 sleeping berths for use on overnight journeys. In publicity for the Condor, much was made of the fact that a radio set installed in the passenger cabin relayed music and news during the flight. The selling point was not so much the in-flight entertainment as the fact that passengers could actually hear the radio, because the soundproofed Condor was far quieter to travel in than contemporary trimotors. In keeping with the luxury image, the women employed to look after Condor passengers were dubbed "air hostesses", a term that was to enjoy a long life.

Boeing versus Douglas

But the Condor was essentially a retrograde aircraft. The future lay with all-metal stressed-skin monoplanes reflecting the latest research in aerodynamics and streamlining, and powered by increasingly efficient air-cooled radial engines. In the 1930s the competition for the airliner market turned into a head-to-head between

Douglas and Boeing, both of which produced aircraft that revolutionized expectations of passenger-carrying performance.

Boeing set the ball rolling when it began developing its two-engine Model 247 in 1931. Here was an aircraft that was high-powered and streamlined, and capable of carrying 10 passengers at 250kph (155mph). It also mitigated the rigours of flight with a soundproofed cabin to cut down on engine noise – already reduced by doing without the third motor – and well-upholstered seats to reduce vibration.

As Boeing was tied to United Air Lines, it meant initially to keep the aircraft exclusively for United's use. This provoked TWA vice-president Jack Frye into asking the Douglas company to tender for a rival to the 247. The result was the Douglas Commercial DC-1. In February 1934, in a highly publicized stunt designed as a protest against Roosevelt's decision to transfer airmail to the army, Frye flew the prototype DC-1 from Burbank, California, to Newark, New Jersey, in 13 hours, despite running into a snowstorm. The model that went into production was the slightly longer DC-2. Like the 247, the DC-2 was a sleek and powerful two-engined all-metal monoplane with the latest features such as NACA engine cowlings and retractable undercarriage. But the DC-2 was faster, had longer range, and, crucially, could carry 14 passengers, offering its operators a potential 40 per cent extra revenue compared with the 247. Introduced in 1933, the 247 cut scheduled journey times from coast to coast from 27 hours in the old trimotors to 20 hours, with six refuelling stops in place of 14. In service a year later, the DC-2 shaved another two hours off the journey time and cut the number of intermediate stops down to three.

The holy grail

The obvious time for a busy man to travel 18 hours from New York to Los Angeles was overnight, but the DC-2 was not spacious enough to function comfortably as a sleeper. American Airlines' chief C.R. Smith was convinced that an aeroplane capable of carrying 14 passengers in bunks or 21

FIRST MODERN AIRLINER
The ten-passenger Boeing 247 was the first air transport to reflect the progress made in engines and streamlining in the early 1930s. It combined a sleek, all-metal, cantilever wing design with retractable landing gear and pneumatic de-icing. But the 247 was not a commercial success. Introduced in 1933, it was upstaged by the DC-2 the following year.

FLYING HIGH

A Pacific Northern Airlines Douglas DC-3 soars over the mountains of Alaska. The runaway success of the DC-3 made Douglas the world's leading manufacturer of civil aircraft – a lead that it maintained until the advent of the Boeing 707 in the late 1950s.

passengers seated would achieve the holy grail of commercial aviation – a profit on passenger operations alone. Smith persuaded Douglas to build a bigger version of the DC-2 to meet this specification. The result, the Douglas Sleeper Transport (DST), entered service in December 1935 and became one of the most successful aeroplanes in aviation history in its better-known "day coach" version, the DC-3.

With the advent of the DC-3, air travel had come of age. Progress in instrument-flying and radio navigation meant that emergency landings and cancelled flights had become uncommon, while night flying was a standard feature of airline schedules. Round-the-clock operations with an aeroplane carrying 21 passengers generated a profit despite falling ticket prices, which of course encouraged more people to fly. By the end of the 1930s, US airlines were carrying three million passengers a year, 90 per cent of them travelling in DC-2s or DC-3s.

DONALD DOUGLAS

BORN IN BROOKLYN, NEW YORK, Donald Douglas (1892–1981) spent two years at the US Naval Academy before switching to study aeronautic engineering at MIT, completing the four-year course in two years. In 1915, at the age of 23, he was taken on by the Martin company as chief engineer, helping to design America's first two-engined bomber, the MB-1. In 1920, Douglas moved to California hoping to set up his own aircraft company. He worked out of an office in a barber's shop until rich sportsman David Davis gave him $40,000 to build an aeroplane that could fly non-stop coast-to-coast. The aircraft Douglas produced, the Cloudster, never made it across the continent, and Davis later drifted away from aviation, but the Douglas Company was firmly established. It began making torpedo planes for the US Navy, and when four of these were adapted as Douglas World Cruisers for the first

global circumnavigation flight in 1924, the company's reputation was made. By 1928, the Douglas Company was worth $28 million and employed the best designers Douglas could find. During the 1930s they created some of the world's finest piston-engined aircraft, including the DC-1, DC-2, and DC-3, one of the most successful aircraft ever built.

SCOTS PATRIARCH

Of Scottish ancestry, with a fondness for Robbie Burns, Douglas ran his company in a distinct patriarchal style. He remained in control well past retirement age, until financial difficulties forced him to sell to McDonnell in 1967.

Douglas DC-3/C-47

"It was the first airplane that could make money just by hauling passengers."

C.R. SMITH
LONG-TIME PRESIDENT OF AMERICAN AIRLINES

THE DOUGLAS C-47 (known as the Skytrain in the US and the Dakota in Britain) was the military transport version of the DC-3, the most successful passenger aircraft of the 1930s. The DC-3 was apparently the result of a two-hour phone conversation between the president of American Airlines and Donald Douglas, founder of the Douglas Aircraft Company. Douglas was persuaded to produce a larger version of the DC-2, adapted for use as a 14-berth sleeper. This plane, the Douglas Sleeper Transport (DST) was soon eclipsed by the success of the 21-seat DC-3 version of the same aircraft, which first flew in 1935.

The robust virtues of the DC-3 were legion. It was reliable and easy to service – an engine could be changed in under two hours. It could operate equally well off dirt, grass, or concrete airfields. And it was considered virtually indestructible. The story is told of a DC-3 that had a wing shot off on the ground in China during the war with Japan. After being fitted with a spare wing from a DC-2 – considerably shorter than its own – it flew successfully to Hong Kong. The military transport version entered service in 1942 and soon became the universal workhorse of the Allies in WWII.

Variable-pitch propeller

Pratt & Whitney radial engine

ENDURING APPEAL
By the war's end, neither the C-47 (shown here) nor the DC-3 were state-of-the-art, but the aeroplanes just kept flying. In 1958, on the threshold of the jet age in commercial aviation, there were still more DC-3s in operation in the United States than any other commercial aeroplane.

US Civil Registration Number shows this to be a C-47

Fabric-covered metal rudder

All-metal wing

Specifications (C-47)

ENGINE	2 x 1,200hp P&W R-1830 Twin Wasp air-cooled radial
WINGSPAN	29m (95ft)
LENGTH	19.7m (64ft 6in)
WEIGHT	7,700kg (16,976lb)
CRUISING SPEED	298kph (185mph) CREW 3
PASSENGERS	27 troops

N147DC

N147DC

Tailwheel (non-retractable)

Fabric-covered metal aileron

COMFORTABLE RIDE
One of the features of the DC ("Douglas Commercials") line was its comfortable interior and smooth ride, enjoyed by pilots and passengers alike.

Windscreen wipers

Control yoke

Rudder pedals

Throttle quadrant

Pilot's seat

Co-pilot's seat

HEAVY WORKLOAD
A C-47 military transport carries US troops, along with a jeep and howitzer, into action during WWII. In addition to its roles as a heavy cargo and paratroop carrier, the C-47 was also used as a glider tug, a glider, a seaplane, and as the AC-47 "Spooky" gunship during the wars in Korea and Vietnam.

MASS PRODUCTION
The DC-3/C-47 was perfectly adapted for mass production, shown here in a production line in an Oklahoma City plant. Altogether, over 10,000 were built, with at least another 2,500 produced under licence in the USSR and Japan.

Radio antenna mast

Hamilton Standard propeller

Oil cooler

Main landing gear (semi-enclosed when retracted)

BIGGER AND BETTER
The Douglas C-47 (shown here) was the military version of the DC-3, which was a larger, modified version of the DC-2. It was slightly longer than the DC-2, had a larger wingspan, and was able to carry heavier loads.

The air passengers of the late 1930s travelled in properly heated cabins, sound-proofed to reduce engine noise to a loud drone rather than a nerve-shattering roar. Sitting in padded seats that reduced vibration, they were served hot in-flight meals brought on board in giant thermos flasks. To pass the time they played games with the free packs of playing cards handed out in the cabin, or wrote cards or letters in flight that would then be mailed by the airline. DST sleeper services attained a more lavish style. Passengers changed into their nightwear in a luxurious lounge while their beds were made up. In the morning, stewardesses served breakfast in bed, and during the day the extra space afforded to the smaller than usual complement of passengers was exploited to provide linen-covered tables with fresh flowers in vases.

Into the stratosphere
But the problem of turbulence remained. Many flights were still a nightmare of air-sickness for those with a tender stomach. The only solution would be to fly above the clouds in the weather-free stratosphere. Here came a chance for Boeing to steal back the lead from Douglas. By the end of 1938 they had developed the four-engined B-307 Stratoliner, the first commercial airplane capable of operating at stratospheric height. A pressurized cabin protected the crew and passengers from the effects of high altitude, while turbo-superchargers allowed the engines to function efficiently in the thin upper air, their performance enhanced by new high-octane fuel.

The Stratoliner was not a commercial success, but it pointed the way forward for passenger aviation as the United States headed into World War II. No fundamental technical obstacle stood in the way of land planes developing non-stop transcontinental and transoceanic services, as the Focke-Wulf Condor had shown with its celebrated Berlin–New York flight in 1938. Once peace returned, propliners cruising the clear blue heights of the stratosphere would whisk globetrotters from city to city across the world.

STRATOLINER
Entering airline service in 1940, the Stratoliner – shown here on a publicity flight – gave passengers a smoother, faster ride than they had ever known before.

INTERWAR AIRLINERS

THE FIRST PASSENGER TRANSPORT aircraft appeared in
Europe in the aftermath of WWI. Initially they were mostly
converted bombers, although Junkers and Fokker were very
quick to begin production of specialist aircraft for passenger
services. A wide diversity of types were flown, including
both monoplanes and biplanes of either wood or metal
construction. By the second half of the 1920s, safety
considerations had led to a prejudice in favour of tri-motors
– it was widely felt that if one engine failed, you would be
safe with the two remaining. After the launch of the Boeing
247 in 1933, however, twin-engined all-metal aircraft
dominated the airline market, at least in the United States,
with the Douglas DC-3 outselling any other type. By the
end of the 1930s, more powerful four-engined airliners were
beginning to appear, with increased range and payload. The
introduction of the pressurized cabin in the Boeing 307
Stratoliner pointed the way forward to a future of more
comfortable high-altitude flight above the weather.

POPULAR DC-3
*Luggage is unloaded from a United Airlines Douglas DC-3
(see pages 148–9). By the end of the 1930s, nine
out of ten air passengers in the United States
were travelling on DC-2s or DC-3s.*

Boeing 247

The era of the modern airliner began on 8 February
1933. With the first flight of the Boeing 247, the world's
airline fleets were made to look slow, old-fashioned, and
cumbersome, since it was some 113kph (70mph) faster
than its competitors. However, the success of the 247
was to be its commercial
undoing. With no less

than 70 on order for United Air Lines, none
were available for their competitors, who
turned to Douglas Aircraft. Their DC-1
led to the immortal DC-3, limiting
total production of the Boeing
Model 247 to just 75 aircraft.

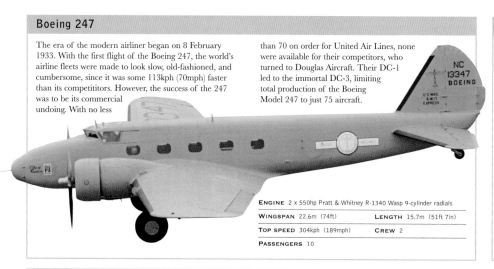

ENGINE	2 x 550hp Pratt & Whitney R-1340 Wasp 9-cylinder radials		
WINGSPAN	22.6m (74ft)	LENGTH	15.7m (51ft 7in)
TOP SPEED	304kph (189mph)	CREW	2
PASSENGERS	10		

Boeing 307 Stratoliner

Conceived as a commercial development of the B-17
Flying Fortress, the Stratoliner used the bomber's wings,
engines, and tail, married to a new fuselage, which,
uniquely for its time, was pressurized. For the first time
passengers were carried above the bad weather that had
made flying so unpleasant. Ordered by both TWA and
Pan American, the first were delivered in 1940.

ENGINE	4 x 1,100hp Wright GR-1820 Cyclone 9-cylinder radial		
WINGSPAN	32.7m (107ft 3in)	LENGTH	22.7m (74ft 4in)
TOP SPEED	357kph (222mph)	CREW	5
PASSENGERS	33		

Curtiss Condor T-32 (Condor II)

While bearing the same name and superficial appearance
as the Curtiss Condor 18, the T-32 was a new design that
first flew in January 1933. As a biplane, the aeroplane was
already an anachronism. Early operators included Eastern
Air Transport and American Airways. An improved
version, the AT-32, appeared in 1934, and most T-32s
were upgraded.

ENGINE	2 x 710hp Wright Cyclone SGR-1820 9-cylinder radials		
WINGSPAN	25m (82ft)	LENGTH	14.8m (48ft 7in)
TOP SPEED	269kph (167mph)	CREW	3
PASSENGERS	12 (sleeper version)		

de Havilland D.H.34

When the first D.H.34 entered service with Daimler
Airway in 1922, the passengers on the Paris route
experienced new standards of comfort. Refreshments
were served by a steward, and there was even a separate
compartment at the rear for baggage. The crew,
however, remained exposed to the elements. Apart
from one Russian sale, 11 of the 12
aircraft built flew with
British operators.

ENGINE	450hp Napier Lion 12-cylinder		
WINGSPAN	15.6m (51ft 4in)	LENGTH	11.9m (39ft)
TOP SPEED	169kph (105mph)	CREW	2
PASSENGERS	9		

Douglas DC-2

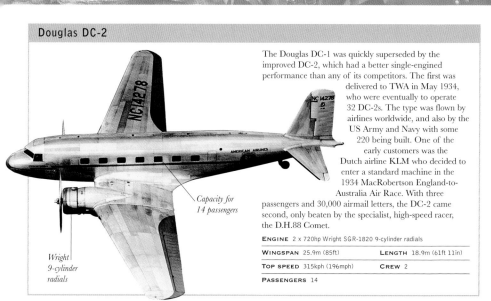

The Douglas DC-1 was quickly superseded by the improved DC-2, which had a better single-engined performance than any of its competitors. The first was delivered to TWA in May 1934, who were eventually to operate 32 DC-2s. The type was flown by airlines worldwide, and also by the US Army and Navy with some 220 being built. One of the early customers was the Dutch airline KLM who decided to enter a standard machine in the 1934 MacRobertson England-to-Australia Air Race. With three passengers and 30,000 airmail letters, the DC-2 came second, only beaten by the specialist, high-speed racer, the D.H.88 Comet.

Capacity for 14 passengers

Wright 9-cylinder radials

ENGINE	2 x 720hp Wright SGR-1820 9-cylinder radials	
WINGSPAN	25.9m (85ft)	LENGTH 18.9m (61ft 11in)
TOP SPEED	315kph (196mph)	CREW 2
PASSENGERS	14	

Farman F.60 Goliath

Like many of the early airliners, the Goliath was originally designed as a bomber but appeared too late for operational service. Servicing many European airlines, the 12 passengers were carried in two compartments, four forward of the open-crew cockpit and eight behind. In 1919, a Goliath set a world load-to-height record reaching 5,100m (16,732ft) with 25 passengers.

ENGINE	2 x 260hp Salmson 9CM 9-cylinder radial	
WINGSPAN	26.5m (86ft 10in)	LENGTH 14.3 (47ft)
TOP SPEED	120kph (75mph)	CREW 2
PASSENGERS	12	

Focke-Wulf Fw 200 Kondor

Designed in 1936 by Professor Kurt Tank to meet a Lufthansa specification for a long-range airliner, the prototype V1, powered by four Pratt & Whitney radial engines, first flew in July 1937. The two subsequent prototypes, V2 and V3, were fitted with BMW engines. The V1 prototype, named *Brandenburg*, flew non-stop from Berlin to New York in August 1938 in 24 hours 55 minutes, returning in less than 20 hours. Although used commercially by both Lufthansa and the Danish airline DDL, the Kondor is better known as a Luftwaffe maritime reconnaissance bomber.

ENGINE	4 x 720hp BMW 132G-1 9-cylinder radial	
WINGSPAN	33m (108ft 3in)	LENGTH 23.9m (78ft 3in)
TOP SPEED	325kph (202mph)	CREW 4
PASSENGERS	26	

Fokker F-10 Super Tri-motor

By the late 1920s, Fokker was the world's largest aircraft manufacturer, and its name was synonymous with safety and reliability until one accident in 1931 changed all that. The first F-10 Super Tri-motor appeared in April 1927 and was operated by many leading airlines, including Pan American and TWA. On 31 March 1931, a TWA F-10 crashed in a thunderstorm. The accident gained widespread publicity since one of the victims was the renowned coach of the Notre Dame football team. The F-10 was immediately grounded, and Fokker's reputation was ruined.

ENGINE	3 x 425hp Pratt & Whitney Wasp C radials	
WINGSPAN	24.1m (79ft 3in)	LENGTH 15.2m (49ft 11in)
TOP SPEED	198kph (123mph)	CREW 2
PASSENGERS	14	

Handley Page H.P.42

One of the largest biplanes ever built, the four-engined H.P.42 already looked old-fashioned when it joined Imperial Airways in June 1931. Compared with the competition's sleek monoplanes, the H.P.42 was derided for having its own "built-in headwind". Only eight were built, but their reputation for comfort, reliability, and, above all, safety, meant that only the outbreak of war curtailed their use. During their service life on European and Empire routes, the fleet flew over 10,000,000km (6,200,000 miles).

ENGINE	4 x 550hp Bristol Jupiter X(FBM) 9-cylinder radials	
WINGSPAN	39.6m (130ft)	LENGTH 27.4m (89ft 9in)
TOP SPEED	169kph (105mph)	CREW 3
PASSENGERS	38	

Junkers F 13

Using the Junkers all-metal corrugated structure with a cantilever monoplane wing, the F 13 was a very advanced design when it first flew in June 1919. It was the first commercial aircraft to be fitted with passenger seat belts. When production ceased in 1932, around 450 had been manufactured in some 60 different variants, including float and ski versions. The aircraft was flown by many major airlines worldwide.

ENGINE	185hp BMW.IIIa 6-cylinder inline	
WINGSPAN	17.8m (58ft 3in)	LENGTH 9.6m (31ft 6in)
TOP SPEED	140kph (87mph)	CREW 2
PASSENGERS	4	

28
HYDRAVIONS TYPE 'EMPIRE'
IMPERIAL AIRWAYS
EUROPE   AFRIQUE   INDES
EXTREME-ORIENT   AUSTRALIE

> "Half boat, half aeroplane, taking off in a tumult of spray – the flying boat was a journey of a lifetime."
>
> GRAHAM COSTER
> FROM CORSAIRVILLE: THE LOST DOMAIN OF THE FLYING BOAT

FLYING BOATS AND AIRSHIPS

AS THE ERA OF LONG-DISTANCE PASSENGER AIR TRAVEL TOOK SHAPE, AIRSHIPS AND FLYING BOATS COMPETED FOR TRANSOCEANIC ROUTES

IN OCTOBER 1928, at a time when every shaky, perilous crossing of the Atlantic by aeroplane was still headline news, the German airship *Graf Zeppelin* carried a score of passengers on a non-stop flight from the Alps to Lakehurst, New Jersey. They travelled in style, with private cabins, well-appointed bathrooms, gourmet food, and a carpeted dining lounge from which they enjoyed the breathtaking views. This was a real ship of the air, designed to compete with the Blue Riband liners of the ocean below. Its proponents believed fervently that the *Graf Zeppelin* represented the future of long-distance passenger air travel.

The performance of zeppelins during World War I, although militarily ineffectual, had made German airship technology an object of fear and envy for the country's enemies. At the end of the war, the victors were determined to procure zeppelins for themselves and deny them to the Germans. In 1919 a promising attempt to resurrect DELAG, the company that had operated an airship passenger service in Germany before the war, was nipped in the bud when its two zeppelins were seized as reparations. Further airship production was banned and the Zeppelin works at Friedrichshafen was marked down for destruction. Now under the dominant influence of wartime airship commander Hugo Eckener, the Zeppelin company won a stay of execution by offering to build an airship for the US Navy. Called the USS *Los Angeles*, it took until 1924 to produce, by which time wartime animosities were fading and Zeppelin was reprieved.

AERIAL AIRCRAFT CARRIER
Shown emerging from the clouds, the short-lived US Navy USS Akron (ZRS-4) was a product of the Goodyear-Zeppelin works and one of the world's strangest aircraft – an aerial aircraft carrier.

The Zeppelin company also sought survival through an alliance with the United States. In 1923, Goodyear formed a joint corporation with Zeppelin, obtaining rights to its patents. The United States had one major advantage over Germany – access to helium. During the war this had been identified as a suitable lifting gas because, unlike the hydrogen used in German airships, it was not flammable. Almost all the world's supply of this rare gas came from a single small area of Texas.

The sole customer for American airships was the US Navy, but its experience with helium-filled airships suggested that they were too fragile for long-term regular use. The *Shenandoah*, a US-built version of a wartime zeppelin, was torn in half when caught in a violent storm in Ohio in 1925; the *Akron*, the first product of the Goodyear-Zeppelin works, went down in the ocean off New Jersey in 1933; its sister airship, the *Macon*, met the same fate in 1935. Only the *Los Angeles* fulfilled its service life without disaster.

Imperial airships

The British experience of rigid airships was even less encouraging. The postwar period started promisingly with an impressive flight by the R 34, a British copy of the German L 33 that had been shot down over England during the war. With 31 people, including a stowaway, on board, the R 34 flew non-stop from Scotland to Mineola, New York, in July 1919, in four and a half days and then flew back. But attempts to build on this success led to disaster. In 1921 the British-built R 38 snapped in two under the stress of tight manoeuvres during flight trials over the port city of Hull. Both halves caught fire and exploded, killing 44 of the crew. Nevertheless, Britain eventually pursued the idea

SUPERB SUNDERLAND
The Short Sunderland – shown here with beaching gear, used after removing the craft from water – was a military version of the Empire-class flying boats used by Britain's Imperial Airways. First used by the RAF in 1938, the Sunderland gave outstanding service throughout WWII.

THE MIGHTY HINDENBURG

The LZ 129 Hindenburg *sits in its hangar at Rhein-Main, Frankfurt, Germany. Designed to revolutionize passenger travel, this mighty airship was the largest man-made object ever to fly. Measuring 245m (804ft) long and 41m (135ft) in diameter, it could carry up to 72 passengers and around 60 crew members. With its fiery destruction in May 1937, the age of the zeppelins came to an end.*

LEAVING THE HANGAR

A group of workmen walk out the German airship LZ 127 Graf Zeppelin from its hangar. Having first flown on 18 September 1928, it went on to become the most successful passenger airship ever built, clocking up 590 flights.

COVERING THE SKELETON

Here the outer fabric of the Graf Zeppelin is being fitted on to its framework at the Zeppelin works at Friedrichshafen, Germany. The huge gas cells inside the metal-and-fabric frame held 199,857 cubic metres (7,062,100 cubic ft) of hydrogen.

of operating rigid airships as passenger and mail carriers on imperial routes to Canada, India, and Australia. Two airships were built, the R 100 and R 101. Although the R 100 was an airworthy craft that flew to Canada and back in the summer of 1930, the government-built R 101 was a thoroughly bad design – overweight, leaky, and short of lift. But Britain's Secretary of State for air, Lord Thomson, insisted that the airship was "safe as a house" and commanded that its maiden flight take place in October 1930. The R 101 was to carry Thomson and other senior officials from England to India. It got no further than Beauvais, in northern France. Out of control on a night of wind and rain, the airship crashed into a hillside, killing all but six of the 54 people on board, including Thomson.

The tragic fiasco of the R 101 ended Britain's airship programme and the R 100 was broken up for scrap. France, whose inventors and sportsmen had once led the world in lighter-than-air flight, had also dropped out after the *Dixmude*, an adaptation of the L 72, broke up in a storm over the Mediterranean in 1923. Italy also abandoned airship development after Umberto Nobile's *Italia* was lost during an Arctic expedition in 1928.

The Graf Zeppelin

In the end, the Germans were the only Europeans with the knowledge and experience required to design and operate airships. And the German people continued to see their zeppelins as proud symbols of their country's technological prowess. In 1925, when Hugo Eckener decided to build an airship capable of operating a transatlantic passenger service, he raised the money by public subscription, appealing to national sentiment exactly as Count von Zeppelin had done after the crash of the LZ 4 back in 1908.

The result was the *Graf Zeppelin*, the most successful airship of all time. First flown in September 1928, it was one of the largest

AROUND THE WORLD IN 21 DAYS

IN THE SUMMER OF 1929, Zeppelin boss Hugo Eckener staged a sensational demonstration of the *Graf Zeppelin*'s potential by making the first passenger-carrying flight around the earth. The 19th-century fantasy writer Jules Verne had imagined a race to circle the globe in 80 days; the airship would do it in three weeks. Early in the morning of 8 August, the *Graf Zeppelin* left Lakehurst, New Jersey, carrying 16 passengers and 37 crew. Heading east across the Atlantic, it reached the Zeppelin base at Friedrichshafen in southern Germany at lunchtime on 10 August. After a rest and some sightseeing, the passengers re-embarked for a 11,247-km (7,029-mile) non-stop flight to Tokyo, crossing the sparsely populated wastes of Siberia. The captain would occasionally drop the airship down for a closer look at places of interest along the route. The inevitable tedious stretches were alleviated by a constantly changing menu that matched the countries traversed – Rhine salmon over Germany, beluga caviar over Russia.

After a few days in Tokyo, where the passengers and crew enjoyed a tumultuous welcome, the *Graf Zeppelin* set off across the Pacific to Los Angeles. After a one-night stop in California, the airship was on its way east again, reaching Lakehurst on the morning of 29 August. The journey had taken 21 days, 5 hours, and 31 minutes. Eckener was invited to the White House to meet President Hoover and the whole crew was treated to a ticker-tape procession down Broadway.

ARRIVAL IN JAPAN
The Graf Zeppelin *is viewed by curious visitors in its hangar in Tokyo, Japan, after completing the second leg of the world tour – a non-stop, 102-hour, 11,247-km (7,029-mile) flight.*

machines ever to fly – almost four times the length of a jumbo jet. It had the latest in aviation instruments, including radio direction-finding, but most striking was the quality of travel experience it offered its 20 passengers. The passenger section of the gondola was quiet and reasonably spacious, and the temperature was comfortable. Eckener did all he could to stress the luxury and glamour of the *Graf Zeppelin*. In the spring of 1929, a clutch of German dignitaries were taken on a magical voyage over the Mediterranean and the Near East, lapping up costly wines and haute cuisine meals while the sights of Rome, Capri, Crete, Cyprus, and Palestine filled the picture windows.

But the emphasis on glamour was a deliberate distraction from some of the Zeppelin's serious

drawbacks. It was a vast vehicle to transport only 20 paying customers; travelling at about 100kph (60mph), it was also fairly slow; and it could not comfortably cope with the weather in the North Atlantic. Instead of linking Germany to New York, the *Graf Zeppelin* settled into a scheduled service to Brazil, where a German expatriate community provided a sufficient pool of customers.

Despite the drawbacks of the *Graf Zeppelin*, in the early 1930s, faith in the future of airship travel was riding high. Germany embarked on the construction of an

HAUTE CUISINE
The Hindenburg *dining tables were set with linen tablecloths, china crockery, silverware, and glassware. Stewards served passengers gourmet meals and fine wines from the galley.*

SIGHTSEEING
This poster shows the Hindenburg *over Manhattan. Good views of sights like the Empire State Building were expected by passengers.*

IN 2 DAYS TO NORTH AMERICA!
DEUTSCHE ZEPPELIN-REEDEREI

HINDENBURG DISASTER

DEADLY SPARK
On 6 May 1937 the Hindenburg, *on approaching its mooring mast at Lakehurst, New Jersey, suddenly burst into flames and fell from the sky. This dramatic photograph shows the fireball rising high above the tail of the airship as it sinks to the ground. No one knows for sure what caused the tragedy, but the most likely suspect is a spark from a build-up of static electricity igniting the flammable skin.*

"It's burning, bursting into flames… this is one of the worst catastrophes in the world… Oh, the humanity, and all the passengers screaming around here!"

HERBERT MORRISON
RADIO REPORTER WITH CHICAGO'S WLS

SOARING OVER MANHATTAN
The LZ 129 Hindenburg *flies over Manhattan's famous skyline. At this time only the rich could afford airship travel. A one-way trip over the Atlantic could cost as much as a new car. A round-trip fare was equivalent to the cost of an average house.*

TRAGIC SEQUENCE

At 7.25pm, 13 hours behind schedule, the 244m- (800ft-) long Hindenburg *was approaching the mooring mast when there was a muffled bang and flames could be seen just forward of the rear fin. Within seconds nearly half the hull was ablaze and the zeppelin rapidly sank to the ground, stern first (see below). Some passengers tried to escape by jumping from windows while others slid down ropes. Just 34 seconds later all that remained was a glowing, red-hot skeleton.*

THE ARRIVAL OF THE ZEPPELIN *HINDENBURG* AT Lakehurst, New Jersey, on 6 May 1937 was not expected to be much of a headline event. The *Hindenburg* had already made 10 round trips to the United States the previous year. Still, it was the first in a new series of scheduled transatlantic flights and radio reporter Herbert Morrison was sent down to Lakehurst with a sound recordist to await its arrival. At 3.30 in the afternoon, strollers on the Manhattan streets looked up to goggle at the giant swastika-marked airship passing low over the Empire State Building – normal practice, both to give passengers a spectacular view and to publicize the airship service to New Yorkers. The *Hindenburg*'s captain Max Pruss, then directed his ship southwards to Lakehurst. But there were electrical storms around and he held off from landing, circling in wait of clearer weather. By the time Pruss finally began his approach to Lakehurst it was 7.00 in the evening. A light rain was falling. At 7.25, the *Hindenburg* put down mooring lines to the ground and was ready to be brought to the docking mast when a strange light appeared towards the stern of the ship. In seconds it had burst into a vast tongue of flame, and within 30 seconds, fire had engulfed the entire zeppelin.

Herb Morrison recorded his impressions of the scene with an emotional immediacy that would engrave itself on the hearts of American radio listeners when broadcast a few hours later:

"Get out of the way! Get out of the way... It's burning, bursting into flames... this is one of the worst catastrophes in the world. Oh! The flames are climbing four or five hundred feet into the sky and it's a terrific crash, ladies and gentlemen. There's smoke and there's flames, now, and the frame's crashing to the ground... Oh, the humanity, and all the passengers screaming around here!" Remarkably, 62 of the 97 people on board escaped the inferno, though some of the survivors, including Commander Pruss, were severely burned.

The cause of the disaster has never been established. It was once generally held that static electricity had ignited hydrogen leaking from the canopy. The crew of the airship always believed it had been sabotage – a time bomb planted in the rear of the airship by anti-Nazi saboteurs wishing to embarrass Hitler's regime. The latest theory is that the sealant used to coat the skin of the airship's gas bag – a mixture of iron oxide and aluminium powder – was inflammable and lit by a static spark caused by a build-up of electrostatic charge from the rainstorm.

international airport at Frankfurt-am-Main, from which a pair of even larger zeppelins was at last to provide the long-dreamed-of scheduled service to the United States.

The Hindenburg

The first of these airships, the *Hindenburg*, came into service in 1936, sporting Nazi swastikas. Powered by four 1,100hp diesel engines, it could carry 50 passengers at over 128kph (80mph) in unparalleled luxury. Where the *Graf Zeppelin* had crammed its passengers into a gondola, the *Hindenburg* used part of the massive hull for passenger accommodation on two decks. The upper deck had promenades on each side where passengers could stroll and gaze through panoramic windows. There was a dining room with linen-covered tables, a writing room, and a lounge with a baby grand piano. On the lower deck were bathrooms, a shower room, the crew's quarters, the kitchen, and a smoking room.

The latter drew attention to the *Hindenburg*'s one fatal flaw: its vast envelope was filled with inflammable hydrogen. The zeppelin's designers had intended to use helium, but the United States refused to supply it. Nevertheless, German engineers were sure that the *Hindenburg* was safe. A refit in the winter of 1936–37 upgraded it to accommodate 72 passengers, and in May 1937 it was ready to resume scheduled services between Frankfurt and Lakehurst. Ambitious plans were afoot for a German–US consortium to run a transatlantic service that would employ four airships. And then came the inferno of 6 May (see panel, left). As the *Hindenburg* met its mysterious and fiery end, international airship travel also went up in flames.

Whatever the specific reasons for the *Hindenburg* disaster, statistics suggest that airships were never really safe enough for passenger transport. Out of 161 airships built over three decades, 60 were destroyed in accidents, either through fire or structural failure. They were in any case too expensive to build and too slow to have provided the kind of mass international air travel that exists today. But for the *Hindenburg* disaster, giant rigid airships might just have found a niche as the cruiseliners of the sky. As it is, they have become no more than a remembered curiosity.

Age of the flying boat

The failure of airships left air-passenger transport over the world's oceans exclusively to flying boats. In the 1930s these glamorous machines enjoyed a brief golden age as the aristocrats of long-distance aeroplane travel. The ascendancy of flying boats made sense in the conditions of the time. Although they were not necessarily capable of alighting

READY FOR TAKE-OFF
The thrill of a take-off from water was one of the high points of clipper travel. Here, a Pan American Airways Sikorsky S-40 is shown taxiing for take-off from Miami, Florida, on its way to Port-au-Prince, Haiti.

safely on the open sea, they were, understandably, considered safer for transoceanic flight than land-planes. They could operate services to far-flung exotic locations without the need to build and maintain a chain of airfields. And their boat-like hulls lent themselves to more spacious and luxurious accommodation than contemporary land-based airliners. This allowed the flying boats to come closer in style to the luxury ocean liners with which they competed on most routes. It was no accident that flying-boat crews dressed in nautical fashion, or that those of the most famous fleet, operated by Pan American, were dubbed "clippers" after the fastest ships of the age of sail.

As in other areas of aviation, the United States lagged behind the Europeans in the development of commercial flying-boat services in the early 1920s, even on what the Monroe Doctrine had defined as its home turf – Central and South America. In August to September 1925, two German Dornier Wal flying boats carried out a demonstration flight from Colombia across the Caribbean to Miami, Florida, and offered to set up an airmail service linking the US with the Caribbean islands, and coastal states with South America. Stung in its national pride, the United States refused to play ball, but the incident highlighted the need for America to develop an overseas airmail system. The result was the Kelly Foreign Air Mail Act of 1928. As with US domestic commercial aviation, government-awarded airmail contracts were the springboard for the development of US international passenger services.

Empire building

Thanks largely to the extensive business contacts and lobbying skills of youthful entrepreneur Juan Trippe, his company Pan American, which had begun airmail services to Havana under a contract with the Cuban government in 1927, won a monopoly of US government contracts for routes throughout the Caribbean. These routes were operated by flying boat – initially Igor Sikorsky's eight-seat S-38. In September 1929, Trippe and his wife Betty, with Pan American's technical adviser Charles Lindbergh and his wife Anne Morrow Lindbergh, made a spectacular island-hopping flight across the Caribbean and around Central America. This publicity event

CARIBBEAN CLIPPER
Handlers offload cargo and mailbags from an S-40 Clipper in Miami. Three S-40s were built, carrying up to 40 passengers each on Pan American's Caribbean and South American routes.

JUAN TRIPPE

ONE OF AVIATION'S VISIONARY empire builders, Juan Trippe (1899–1981) was a Yale alumnus and scion of a well-connected New York family. After leaving college he entered the domestic airmail business, setting up Long Island Airways before moving on to Colonial Airways, operating the Boston–New York airmail route. But Trippe's ambitions drew him further afield. In 1925, he secured sole landing rights in Havana from the Cuban dictator General Machado. When a company called Pan American won a contract to carry mail between Key West and Havana in 1927, Trippe had to be cut in on the deal. He quickly took charge of Pan American, and with excellent connections in Wall Street and Washington, a showman's flair for publicity, and a gift for the cut-and-thrust of business competition, Trippe built the company into the first worldwide airline. To open up transoceanic routes, he promoted the development of bigger, better flying boats, culminating in the huge Boeing 314. He went on to lead Pan Am through the prop-liner era into the jet age. One of Trippe's last actions before his 1968 retirement was to commit the company to the Boeing 747, a bold decision that defined the shape of modern passenger transport – and set Pan Am on the path to bankruptcy.

DYNAMIC DUO
In 1927 Juan Trippe hired Charles Lindbergh (above right) to pioneer airline routes to every South and Central American country. Trippe had a gift for spotting new markets. On 28 March 1949 he was featured on the cover of Time *magazine.*

helped stamp the equation "flying boat equals glamour" on to the public consciousness.

Over the next few years, through a combination of shrewd company takeovers and intensive lobbying in Washington, Trippe extended Pan American's control of routes down both coasts of South America and established the company as the single airline representing the United States abroad – the "chosen instrument" of foreign policy.

With longer routes and expanding business, Trippe sought flying boats with a larger payload and greater range. The next to be introduced, developed directly in collaboration with Pan American, was the 40-passenger Sikorsky S-40. The S-40s were the first Pan American aeroplanes to be called clippers. They attempted to match the romantic name with elegant style, boasting spacious compartments, upholstered chairs, backgammon tables, and hot meals served by a uniformed steward. The romance of the flying-boat service to Brazil was celebrated by

Hollywood in the 1933 movie *Flying Down to Rio* – in which the S-40 appears occasionally but is upstaged by the first screen teaming of new dancing stars Fred Astaire and Ginger Rogers.

The first S-40s came into service in 1931, but by then Trippe was already looking towards transoceanic passenger routes. Commercial logic suggested starting with the Atlantic crossing, but Pan American ran into stiff resistance from European governments determined that their

national airlines should take an equal share in any transatlantic service. While negotiations about reciprocal landing rights became bogged down, Pan American looked to the Pacific. In 1931 Charles and Anne Morrow Lindbergh, flying the Lockheed Sirius Tingmissartoq, carried out a route-survey flight to Japan and China via Alaska. But although the Lindberghs showed that the Alaskan route was technically feasible, political instability ruled it out after Japan invaded the

PASSENGERS BOARDING
A Pan American Sikorsky S-42 takes on passengers at Pan Am International Airport, Dinner Key, Miami, Florida. Probably Sikorsky's finest flying boat, the S-42, introduced in 1934, cut the travel time from Miami to Buenos Aires to five days.

AFRICAN ADVENTURE

AUSTRALIAN WRITER Alan Moorehead travelled as a passenger in an Empire flying boat through East Africa. He recalled his exotic encounter with an undeveloped continent: "There was no flying after dark and the machine put down at some fascinating places… There were no familiar airport buildings, no advertisements, no other traffic of any kind; just this rush of muddy water as you lighted down on a river or a forest lake… On the Zambezi river… they had to run a launch up and down the water a few

minutes before the plane came to clear the hippopotami away. I remember too, with particular vividness, a little place called Malakal on the White Nile in the Sudan, where the women of the Dinka tribe… walked gravely along the riverbank and turned their heads away from the great flying boat on the water…"

GO SOUTH!
The poster above – featuring a Handley Page H.P.42, also right – advertised the Imperial Airways's services through Africa and Asia. The section through East Africa was made by flying boat.

Chinese province of Manchuria. Pan American had to settle for a route across the central Pacific via Hawaii and the Philippines to Hong Kong. This posed a severe technical challenge. The distance from California to Hong Kong was 12,000km (7,500 miles), including a 3,900-km (2,400-mile) stretch without landfall to Hawaii. However, no existing flying boat had that range.

Trippe ordered two powerful new flying boats: the Sikorsky S-42 to pioneer and survey the route, and the Martin 130 to follow up with scheduled services. Once again the Pan American boss's excellent contacts in Washington stood him in good stead. He not only won the transpacific airmail contract but also induced the US government to put the tiny Wake Island under US Navy administration, so that it could act as a stepping stone for flying boats crossing the Pacific, along with Midway and Guam.

Finding these specks in the ocean was another challenge. The traditional ocean voyager's techniques of dead-reckoning and celestial navigation – using a sextant to take a fix on the sun or stars – were useful but not sufficiently

OCEAN SURVEYOR
Sikorsky S-42 flying boats, such as this one, were used by Pan American for survey flights across the Pacific and Atlantic oceans. The Martin 130, which had much greater range, took over the transpacific route when scheduled flights were initiated in 1935.

CHINA CLIPPER
The Pan American Martin 130 China Clipper *rests at a mooring station off Manila, the Philippines, after completing its first scheduled transpacific mail flight from San Francisco in November 1935.*

reliable. Pan American's chief communications engineer, Hugo Leuteritz, developed a long-range radio direction finder to supplement traditional navigation methods. The aircraft transmitted a signal that allowed a ground station on one of the islands to establish its position, which was transmitted back to the aircraft's navigator.

Pacific crossings

Fitted with the new radio-navigation equipment, the S-42 with Pan American's star pilot Ed Musick at the controls set off from San Francisco in April 1935 on the first experimental flight to Hawaii. To maximize its range, the S-42 had been stripped of all superfluous weight and packed with extra fuel tanks. Even so, disaster was only narrowly averted. The outward flight went smoothly, reaching Honolulu in 18 hours 37 minutes. But on the return journey Musick had to battle headwinds, which added five hours to the flight time – putting it well beyond the S-42's theoretical endurance limit of 21 hours. By some miracle, there was still a little fuel in the tanks when the aeroplane reached California. Despite such teething troubles, the S-42

survey flights were successfully completed, and on 22 November 1935, the Martin 130 *China Clipper* set out from San Francisco on the first scheduled Pacific airmail flight to Manila in the Philippines. It was given a tumultuous send-off and an equally tumultuous reception at its destination, where more than 300,000 people gathered.

Passenger flights had to wait until crews had built up more experience of flying the route, and until hotels had been built at the island stopovers. But finally, on 21 October 1936, the Martin 130 *Philippine Clipper* set off from California with 15 paying passengers on board, arriving in Hong Kong three days later. The same journey by boat usually took three weeks.

Of course, the Americans had no monopoly of flying-boat operations. In the 1930s, France's Latécoère flying boats

operated in the Mediterranean and across the South Atlantic, while from 1934 Lufthansa flew a scheduled service from Berlin to Rio de Janeiro via Spain and West Africa. Britain used flying boats extensively on its imperial routes to South Africa and Australia. At first these journeys

FINE DINING
The Sikorsky S-42 Clipper *was fitted with a galley, allowing Pan American to serve high-quality meals to its passengers. As well as serving refreshments, it was also the steward's job to point out scenic attractions through the windows.*

VARIETY OF DESTINATIONS
The cover of a Pan American timetable advertises a wide range of destinations. Regular passenger services between the US and the Philippines started in October 1936, charging passengers $799 each for the 13,200-km (8,200-mile) flight.

SWEET DREAMS
Passengers on the Martin 130 Clipper flights across the Pacific had access to sleeping berths. However, Pan American built hotels at island stopover points along the route, so that travellers could have a proper night's rest away from the drone of the engines and constant vibration.

required a clumsy mix of land-plane, flying boat, and train travel, but in 1937, Britain's Imperial Airways introduced the Short Empire flying boats, which flew the Southampton to Sydney and Cape Town routes in stages.

Image of luxury

The Empire boats rivalled the American clippers for luxury. Passengers could recline in a spacious cabin filled with the rich smell of leather upholstery, as stewards recruited from the Cunard shipping line served lobster and caviar. Or they might stand at the windows of the observation deck taking in the panoramic views of exotic landscapes passing a few thousand feet below. Passengers on the Africa route were issued with guidebooks that suggested sights to look out for, including places where basking crocodiles or herds of elephant were to be seen. Some of the stops along the way were in remote locations, bringing encounters with

exotic animals and mud-hut-dwelling villagers.

There was no denying the glamour of the image of flying-boat travel. Pan American's Pacific passenger operation, for example, was an unashamedly exclusive service, with a round ticket from San Francisco to Manila costing over $1,400 – about equal to an average American worker's annual pay. The Pacific flying boats were even given the accolade of Hollywood treatment, with the release of the movie *China Clipper* starring Humphrey Bogart as a flying-boat pilot.

But the intended luxury of flying-boat travel was never quite matched by performance. Like 1930s land-planes, none of them was pressurized, and so they often flew through desperately sick-making weather. Speeds were quite slow, making journeys long and gruelling – the Empire flying boats took nine days from London to Sydney. The overnight stops at hotels did not necessarily help much, especially since passengers were usually required to board in the early hours of the morning, when cool conditions were ideal for take-off. One woman passenger wrote of a London-to-Cape Town flight: "It's a tremendous strain going on and on, being

relentlessly woken any time from 3am to 4.15am and flying and flying and flying and flying."

Nor was full reliability ever achieved. Many journeys were interrupted by bad weather, and navigational errors remained a source of frequent anxiety to crews, since even the best available radio-navigation devices were fallible. And flying boats often had difficulty in alighting if sea conditions were unfavourable. Safety concerns became acute in 1938, when Pan American lost two of its flying boats in the Pacific. In January, shortly after the opening of a second route to New Zealand, an S-42B exploded in the air near Pago Pago, killing the famous Captain Musick and the rest of its crew. The following July, another Pan American flying boat disappeared between Guam and Manila.

Bigger and better

Undeterred by these setbacks, Juan Trippe pressed ahead with introducing a fleet of six Boeing 314s. Twice the size of the Martin 130s, these extraordinary aircraft were the biggest airliners to fly until the age of the jumbo jet, and probably the most luxurious fixed-wing passenger aircraft ever built. They boasted a dining lounge and seven passenger compartments, one of which, the Deluxe Compartment in the tail, was described in the company's promotional literature as "corresponding roughly to a ship's bridal suite". They carried a maximum of 74 passengers seated, or 40 in sleeping berths. With these magnificent

The most luxurious Flying-Boat in the world

IMPERIAL AIRWAYS

IMPERIAL LUXURY
This poster for Imperial's Short flying boat reveals a complex interior layout. Introduced in 1936, the S.23 C-Class, or Empire, carried 17 passengers and five crew on stages of long-distance routes to Australia, South Africa, and India.

aeroplanes Pan American was at last able to initiate a transatlantic passenger service in May 1939, when the *Yankee Clipper* took off from the Marine Terminal in New York for Lisbon and Marseilles.

It was unfortunate for the Boeing 314 that war intervened just as the service to Europe was becoming established. Yet by this time it was in

any case already clear, even to Trippe, that the future lay with a new generation of land-planes. Lufthansa's FW 200 had already flown direct from Berlin to New York, and the German airline was only prevented from starting a scheduled service by the Americans' refusal to grant landing rights. Furthermore, the pressurized Boeing Stratoliner was pointing the way forwards to faster, smoother air travel above the weather. Trippe's next large-scale investment after the 314 was an order for 40 Lockheed Constellations.

The stately elegance of the flying boats, like the extravagant luxury of the *Graf Zeppelin* and *Hindenburg*, belonged to an era in which long-distance air travel was the preserve of the wealthy, and style seemed a better selling point than functionality. Quickly outmoded, they understandably remain a focus of nostalgic fascination for many modern air passengers who, trapped in the cramped blandness of a contemporary airline interior, like to imagine a time when a journey by air was an experience to be savoured and an adventure to be remembered for a lifetime.

TUCKING IN
The Pan American Boeing 314 Yankee Clipper, *which flew the first scheduled transatlantic service in 1939, was the most spacious and luxurious flying boat of all. Meal times in the dining lounge featured full waiter service and the finest haute cuisine.*

INTERESTING EXPERIMENTS

THE COMPLEX EQUATIONS OF POWER, payload, and fuel often did not add up for flying boats on long-distance routes. One failed attempt at resolving the dilemma of taking off with enough payload to be worthwhile and enough fuel to cross an ocean was the extravagantly powered Dornier Do X of 1929, which had no fewer than 12 engines. One Do X took up 170 passengers, but the aeroplane was plagued with problems and never entered service. It has been generally dismissed as, in the words of one expert, "an ambitious freak".

Another solution to the power–payload–fuel problem was some form of assisted take-off. The Germans experimented with catapulting flying boats and floatplanes into the air. In the Short Mayo Composite (below), the British used a flying boat with a light fuel load to lift a heavily loaded seaplane into the air, which was then released to continue its flight. The idea worked – from 1938–39, the *Mercury* made a number of long-distance mail flights – but was not very practical and remained an aviation curiosity.

COMPOSITE SOLUTION
Invented by the technical manager of Imperial Airways, Major R. Mayo, the Composite flying boat – first flown in July 1938 – involved a Short S.20 seaplane, Mercury, *riding piggyback on* Maia, *a Short S.21 flying boat.*

Seaplane carried on strut system

Mother-plane Maia *carries* Mercury *to cruising height*

FLYING BOATS

THE SLOW RATE OF DEVELOPMENT of flying boats in the 1920s and 1930s reflected a lack of funding for development of new aircraft. The ability to take off and land using any reasonably smooth stretch of water made flying boats an obvious choice for exploratory flights or airline services outside the technologically advanced areas of Europe and North America. They were also a reassuring form of transport for passengers worried about flying over oceans. The large "boat" hull could provide spacious accommodation, offering a chance to compete with ocean liners for the luxury market. The big four-engined monoplanes that flourished on long-distance routes in the late 1930s – such as the Boeing 314, Martin 130, Sikorsky S-42, and Short Empire boats – were fine machines, even if they proved unable to match the next generation of landplanes.

SHORT C-CLASS
The Cassiopeia is shown being loaded at Southampton for the first passenger and airmail flight service from England to Africa.

Boeing Model 314 Clipper

The Boeing Model 314 Clipper was arguably the finest flying boat ever built and for over 30 years the largest commercial aircraft. Based on the giant but unsuccessful XB-15 bomber, the Model 314 featured four passenger cabins on two different levels, some of which could be sleeping areas. With flush toilets and even, if required, a bridal suite, the Model 314 was the epitome of luxury. Pan American commenced mail and transatlantic passenger services in May 1939. Of the 12 Model 314s built, three were used by the British Overseas Airways Corporation (BOAC), and the rest were requisitioned off Pan Am by the US military, for wartime operations as far afield as North Africa and Southeast Asia.

ENGINE	4 x 1,200hp Wright Double Cyclone 14-cylinder radial	
WINGSPAN 46.3m (152ft)		**LENGTH** 32.3m (106ft)
TOP SPEED 294kph (183mph)		**CREW** 6–10
PASSENGERS 40–74		

Caproni Ca 60 Transaero

One of the most extraordinary aircraft ever built, the philosophy behind the Transaero's design was the more wings the better. Described as a Triple Hydro-Triplane, this amazing machine had no less than three sets of triplane wings and eight engines in tractor and pusher modes.

ENGINE	8 x 400hp Liberty V 12
WINGSPAN 30.5m (100ft)	**LENGTH** 23.5m (77ft)
TOP SPEED 130kph (81mph)	**CREW** up to 4
PASSENGERS 100	

Dornier Do 26

The last and most graceful of the long line of Dornier flying boats, the advanced Do 26 first flew in May 1938. Ordered as mailplanes by Lufthansa, two were used briefly on the South Atlantic route before war intervened. Only six were built, and these were converted into Do 26D military transports.

ENGINE	4 x 880hp Junkers Jumo 205D diesel	
WINGSPAN 30m (98ft 5in)		**LENGTH** 24.6m (80ft 9in)
TOP SPEED 323kph (201mph)		**CREW** 12
PASSENGERS 12 fully equipped soldiers		

Dornier Do J Wal

With the 1919 Versailles peace treaty restrictions on German aircraft manufacture, Professor Claudius Dornier built his Type J Wal ("Whale") in Italy. The first example flew in 1922, and over 300 were built. The Wal was used for many long-distance pioneering flights.

ENGINE	2 x 360hp Rolls-Royce Eagle IX mounted in tandem	
WINGSPAN 22.5m (74ft)		**LENGTH** 24.6m (80ft 9in)
TOP SPEED 324kph (201mph)		**CREW** 2
PASSENGERS 9–14		

Latécoère "Late" 521 *Lieutenant de Vaisseau Paris*

Originally conceived in 1930 for a transatlantic passenger service, the sole example of the "Late" 521 eventually appeared in 1935. Both technical and diplomatic problems delayed the inaugural flight, but finally in August 1938, the imposing 44 tonne (43 ton) flying boat, named *Lieutenant de Vaisseau Paris*, arrived in New York after a 22 hour 48 minute flight from the Azores. Although achieving several load-to-altitude records for flying boats, war would intervene before a French transatlantic service could be established.

ENGINE	6 x 650hp Hispano-Suiza 12Nbr	
WINGSPAN	49.3m (161ft 9in)	LENGTH 31.6m (103ft 9in)
TOP SPEED	213kph (132mph)	CREW 6
PASSENGERS	30–70	

Boat-shaped hull

Martin Model 130 *China Clipper*

During the 1930s, Pan American Airways (PAA) was seeking new routes, particularly across the Pacific but lacked a suitable aircraft. This need was met by the Model 130. *China Clipper*, the first of three built, inaugurated the transpacific service on 22 November 1935 – initially with mail – followed by the *Philippine Clipper* and the *Hawaii Clipper*. To the public, *China Clipper* became a generic term for all Pacific flying boats.

ENGINE	4 x 950hp P&W R-1830 Twin Wasp 14-cylinder radial	
WINGSPAN	39.7m (130ft)	LENGTH 27.7m (90ft 11in)
TOP SPEED	290kph (180mph)	CREW 6
PASSENGERS	30–70	

Savoia-Marchetti S.55X

The most spectacular achievement of the S.55 took place nine years after the type's first flight. In July 1933, Italy's General Italo Balbo led a formation of 24 S.55X flying boats from Italy to the Century of Progress Exposition in Chicago in just over 48 hours. With its twin hull, the aircraft itself was far from an orthodox design, but 170 were built, and it had a long, successful career with the Italian Navy.

ENGINE	2 x 750hp Issotta-Fraschini Asso 750R	
WINGSPAN	24m (78ft 9in)	LENGTH 16.8m (55ft)
TOP SPEED	279kph (173mph)	CREW 5–6
ARMAMENT	4 x machine guns; 1 x torpedo or 2,000kg (4,409lb) bombload	

Short S.23 C-class (Empire Flying Boat)

The unprecedented order from Imperial Airways for 28 advanced giant flying boats from Short Brothers was based on a forecast income from airmail. *Canopus*, the first C-class, made its inaugural scheduled flight on 30 October 1936. The Empire Postal Service started in June 1937, when *Centurion* flew 1,589kg (3,500lb) of mail to South Africa. By October, Imperial Airways could claim to be the world's largest carrier of mail.

By 1938, Australia was nine days away for £274 return. Despite wartime losses, the last Empire boat service was flown by *Caledonia* in March 1947.

ENGINE	4 x 900hp Bristol Pegasus 9-cylinder radial	
WINGSPAN	34.7m (114ft)	LENGTH 26.8m (88ft)
TOP SPEED	322kph (200mph)	CREW 5
PASSENGERS	17–24	

Sikorsky S-38

Although not the most elegant flying boat, being variously described as "Ugly Duckling", "Flying Tadpole", and, "A collection of aeroplane parts flying in formation", the S-38 was the first commercial success for the small Sikorsky Corporation. Being amphibian, the design was very versatile, and 111 were built. Pan American operated the S-38 primarily on routes around the Caribbean.

ENGINE	2 x 415hp Pratt & Whitney Wasp C radial	
WINGSPAN	21.9m (71ft 10in)	LENGTH 12.3m (40ft 4in)
TOP SPEED	177kph (110mph)	CREW 2
PASSENGERS	8	

Sikorsky S-42

The success of the S-38 led to the larger S-40, which was the first to carry the famous Pan American "Clipper" name. The first, even larger, S-42 flew on 29 May 1934 and set innumerable long-distance records before entering service with Pan Am in August 1934.

ENGINE	4 x 700hp Pratt & Whitney Hornet 9-cylinder radial	
WINGSPAN	34.8m (114ft 2in)	LENGTH 21.1m (69ft 2in)
TOP SPEED	274kph (170mph)	CREW Up to 5
PASSENGERS	32	

Supermarine Southampton

Entering service in August 1925, the Southampton served the RAF for 12 years, a flying boat record that only the Sunderland (see page 221) would surpass. Sixty-eight were built in two versions: the Mk.I with a wooden hull, and the more numerous duralumin-hulled Mk.II. The type became famous for long-distance formation flights.

ENGINE	2 x 502hp Napier Lion V	
WINGSPAN	22.9m (75ft)	LENGTH 15.6m (51ft 2in)
TOP SPEED	174kph (108mph)	CREW 5
ARMAMENT	3 x Lewis machine guns; 499kg (1,100lb) bombload	

THE SHADOW OF WAR

DURING THE 1920S AND 1930S, AVIATION DEVELOPED INTO A POTENTIALLY DEVASTATING WEAPON THAT WOULD CHANGE THE NATURE OF WAR

Deutschlandflug 1937

> "...would not the sight of a single enemy airplane be enough to induce a formidable panic? Normal life would be unable to continue under the constant threat of death and imminent destruction."
>
> **GENERAL GIULIO DOUHET**
> *THE COMMAND OF THE AIR*, 1921

IMPORTANT AIRCRAFT
Introduced in 1937, the Seversky P-35 was a vital stepping stone in the development of American fighter technology. It was the first single-seat, all-metal pursuit plane with retractable landing gear and enclosed cockpit to go into service with the US Army Air Corps.

THE EXPERIENCE of World War I set in motion an idea that was to have a potent influence on the future of warfare: the notion that wars could be won by air power alone. The appalling casualties endured by the infantry in the long stalemate in the trenches provided a powerful motive for seeking some other way of fighting a war. And the example of aerial bombardment, especially by airships and Gotha bombers on London, suggested what that new way of fighting might be.

In his book *The Command of the Air*, first published in 1921, Italian general Giulio Douhet argued that in future wars, armies and navies would be relevant only as defensive holding forces while large fleets of heavy bombers delivered massive attacks on enemy cities and industrial centres. Since civilian morale would soon crack, the war would end quickly and with relatively little loss of life.

Although Douhet's writings were little known outside his own country, they expressed an attitude shared by many leading figures in military aviation. They included Sir Hugh Trenchard, Britain's chief of air staff, who had commanded the unified bombing force intended to launch a major aerial offensive on Germany in 1919, before peace intervened. Trenchard ensured that the bombing of cities was the central plank in British air strategy between the wars. In the United States, the most strident propagandist for military air power was General Billy Mitchell, commander of the US air forces in Europe in 1918 and postwar assistant chief of the Army Air Service. To Trenchard and Mitchell, the great appeal of heavy bombing – whether

"IL DUCE"
Italy's fascist dictator Benito Mussolini was keen to identify himself with the modernity and dynamism that aircraft symbolized. This 1933 portrait by Gerardo Dottori flatters the dictator by composing his image of aeroplanes.

used for destroying cities or, as Mitchell advocated, to sink enemy ships approaching America's shores – was that it provided a rationale for a powerful air force, independent of, and with equal status to, the other armed services. Trenchard was fortunate in already having the world's only independent air force, the RAF, although in the 1920s he could not win it more than the meanest funding. Mitchell had to go out and campaign for the force that he felt destined to lead. It was a campaign that brought him into political entanglements and a conflict with his superiors that eventually led to his court martial for insubordination in 1925.

Disappearing funds

The pressing problem for military aviation in the immediate postwar period was to persuade tight-fisted governments to fund it adequately. The US air service contracted from 190,000 men in 1918 to 10,000 in 1920; the RAF shrank from a force of almost 300,000 to under 40,000 in the year after the war. The French, worried about Germany, kept a larger air arm – enabling Trenchard to use the "threat" of French aerial strength as an argument for building up the RAF in the 1920s.

In the postwar period, the RAF had a chance to practise the use of air power in a series of small-scale

LUFTWAFFE FLYPAST
In an ominous show of strength, a formation of German Dornier Do 17 bombers fly over a Tag der Wehrmacht rally in 1938. Germany's Nazi regime took every opportunity to show off its bomber force, seeking to intimidate Britain and France with the threat of aerial destruction.

ITALO BALBO

LIKE THOUSANDS OF OTHER disillusioned young men who fought for Italy in World War I, Italo Balbo (1896–1940) joined Mussolini's violent Fascist Blackshirts in the early 1920s, playing a prominent part in Mussolini's rise to power. In 1929, he was appointed head of Italian aviation. Only then did he learn to fly. In 1928, Balbo became famous beyond Italy's frontiers for the mass-formation flights he staged – 60 seaplanes across

DASHING PIONEER
As minister of aviation from 1929–33, Italo Balbo helped modernise Italian aviation, gaining it an international reputation with his mass-formation flights.

ITALIAN GLORY
This booklet, dedicated to the "Brave and Intrepid Italian Aviators… and Italo Balbo… who led the glory of Italian wings…", celebrates their internationally acclaimed flight from Rome to Chicago.

southern Europe in 1928, 10 seaplanes across the southern Atlantic to Brazil in 1931, and 24 seaplanes to Chicago in 1933. Jealous of the acclaim Balbo received for the Chicago flight, Mussolini then packed him off to govern the Italian colony of Libya. Balbo subsequently opposed Mussolini's alliance with Nazi Germany in 1939, proposing instead a rapprochement with Britain. He was killed in 1940 when his aeroplane was mistakenly shot down by an Italian anti-aircraft battery.

colonial conflicts. Faced with a rebellion in Afghanistan in 1919, the British bombed Kabul and Jalalabad, also dropping leaflets warning Afghani troops and tribesmen of the dire fate that awaited them if they did not surrender. This had no appreciable effect. But the use of air power against rebels in Iraq between 1922 and 1925 was judged a major success, showing that the RAF could act as a cheap imperial policeman. France and Spain both used air power against Berber rebels led by Abd al-Karim in their respective colonies in Morocco in the 1920s – the French achieving particular success employing their aircraft in tactical support of mobile columns of troops in trucks and armoured cars.

Symbols of power

Whatever the tactics employed and the practical results achieved, there was something profoundly satisfactory to the European psyche in the deployment of aircraft against "primitive" peoples. At a period when the unbridled assertion of white racial superiority and Western technological prowess was starting to be challenged by anti-colonial movements, aircraft stood as a

FASCIST FRIENDS
Italo Balbo (left) was the closest ally and heir-apparent of Benito Mussolini (right) until the acclaim following his transatlantic flights aroused Mussolini's jealousy. This photograph was taken in Italy in the mid-1930s.

IMPERIAL POLICING
A local soldier stands guard in front of an RAF Hawker Hardy in northern Iraq in the 1930s. British aircraft were sent to Iraq to protect the Kirkuk oilfields and pipelines from hostile tribes, a cheaper option than sending imperial troops.

comforting symbol of the dominance of the "civilized" peoples over the "uncivilized". Not surprisingly, air power appealed especially to the militaristic right-wing nationalist movements that came to power in much of Europe between the wars, for which Italian dictator Benito Mussolini's Fascists established the pattern.

When Mussolini's black-shirted followers bullied their way to power in 1922, several Italian World War I air aces were prominent in their ranks. Mussolini himself idolized aviation, revering it as a symbol of power and modernity. "Not every Italian can or should fly," he declared, "but all Italians should envy those who do and should follow with profound feeling the development of Italian wings."

In 1923, an independent Italian air force was created, the Regia Aeronautica, and the scale of Italy's air power rapidly expanded. Mussolini's dictatorship gloried in the incredible stunts that drew world attention to Italian aviation, from the individual achievement of naval officer Francesco de Pinedo's flight from Rome to Tokyo in 1925, to the mass flying-boat spectaculars staged by Italo Balbo between 1928 and 1933. The Italian air force got to carry out its own colonial campaign against rebels in the deserts of Libya and, in 1935, was used against the forces of Emperor Haile Selassie when Italy invaded the independent African state of Ethiopia.

German rebirth

Germany was another country where no questions were raised about the importance of air power. Formally denied the right to maintain an army or naval air force by the terms of the Treaty of Versailles, throughout the 1920s German military leaders, aviators, and aircraft makers worked, often covertly, to maintain pilot training and keep up with advances in military aviation technology and tactics. This was partly achieved through the development of German civil aviation, in which former World War I air commanders and pilots

WINGS OVER CHICAGO

ON THE EVENING OF 15 JULY 1933, a mass formation of 24 Italian Savoia-Marchetti S.55X flying boats, commanded by General Italo Balbo, flew over Lake Michigan towards Chicago after a 15-day, 9,200-km (5,750-mile) transatlantic flight from Orbetello, Italy, via Iceland.

That month Chicago's population was swollen with visitors to the Century of Progress World's Fair, staged to celebrate the city's first centenary. Hundreds of thousands of spectators lined the lakefront for the arrival of the Italian squadron. In perfect formation, three by three, the twin-hulled Savoia-Marchettis circled over the city and then descended gracefully on to the lake, while an escort of 43 American fighter aircraft spelled out the word "Italy" in the sky.

Balbo received a hero's welcome, fêted with celebratory dinners and parades – he even had a major avenue named after him. The Chicago flight was a sensational propaganda coup for the Italian Fascist dictatorship, helping to project an international image of an efficient and technologically advanced regime.

RECORD-BREAKING FLYING BOATS
Originally designed in 1925 as a torpedo-bomber, the huge 24m- (79ft-) wingspan Savoia-Marchetti S.55 was later adapted for passenger service. Identified by its back-to-back engines above the wing, it broke 14 flying-boat records for speed, altitude, distance, and load-carrying.

TRANSATLANTIC TRIUMPH
This postcard, commemorating the 50th anniversary of Italo Balbo's transatlantic flight, depicts Savoia-Marchetti S.55 flying boats approaching the New York skyline with the Statue of Liberty looming.

FORMATION FLYING
General Balbo's S.55 flying boats alight in Jamaica Bay, New York, during their epic 48-hour transatlantic crossing to Chicago. The 24 S.55s completed the entire flight in a tight V formation. Even today, pilots refer to a large formation of aircraft as a "Balbo".

inevitably played a major role. As director of Luft Hansa from 1925, ex-squadron commander Erhard Milch was in constant contact with top officers in the Reichswehr, ensuring that they were fully informed of the latest navigation techniques and flight instrumentation. Air clubs and flying schools also played their part, acting as a training network for future military pilots under the cover of sports aviation.

A major German military-aviation programme took place in the newly established Soviet Union. In 1922, the Germans and Soviets signed the Rapallo Treaty, finding common ground in their status as pariah states in postwar Europe.

Under secret military provisions of the treaty, Germany was allowed to carry out army and air-force training in Russia, in return for providing Soviet forces with training and the latest military technology. A substantial German base was established at Lipetsk, 350km (220 miles) outside Moscow, where, from 1925 to 1933, German pilots were secretly trained to fly state-of-the-art military aircraft, practising bombing, fighter tactics, and manoeuvres.

CIVIL AND MILITARY
Erhard Milch, a World War I fighter pilot, was head of Deutsche Luft Hansa from 1925 until 1933, when he was given the task of rebuilding the Luftwaffe.

DEMONSTRATION OF AIR POWER
In July 1921, General Billy Mitchell used Martin MB-2 bombers, similar to the Martin MT (above), to sink the captured German battleship Ostfriesland, *in the Chesapeake Bay, off the Virginia Capes. In September that year he followed up this demonstration by bombing the USS* Alabama *(below).*

Junkers and Rohrbach set up factories in the Soviet Union where they produced military aircraft prototypes in defiance of the peace treaty. At the same time, the transfer of technology from Germany boosted the development of Soviet military aviation, which became especially dependent on German aero-engines.

American isolationism

In the United States, the 1920s were a lean period for military aviation. In a reaction against America's involvement in World War I, public sentiment was overwhelmingly "isolationist". Determined to keep out of foreign quarrels, Americans saw their military needs as purely defensive. Since the only credible threat to the United States was an attack from the sea, the navy received the lion's share of a much reduced military budget. Advocates of a powerful independent air force with equal status to the army and navy had a hard furrow to plough, taking on both the chiefs of the established services – keen to keep control of their own air forces – and the general perception of America's defence needs.

The leading advocate of an independent US air force, General Billy Mitchell, made what was under the circumstances a pretty effective job of advancing his cause. The argument for aviation as the offensive arm that would win a major war had little impact, since that was not the sort of conflict the United States intended to get involved in. So Mitchell took it upon himself to demonstrate that aircraft could take over responsibility for the defence of America's coastal waters. The idea won some backing in Congress after Mitchell pointed out that a large fleet of bombers could be built for the price of a single battleship. The navy was forced to allow Mitchell the chance to prove his point.

Mitchell's demonstrations

In July 1921 Mitchell assembled some Martin MB-2 bombers and, in front of naval observers, undertook to sink three German warships that had come into American hands at the end of the war. Armed with 270-kg (600-lb) bombs, the Martins made short work of a destroyer and a light cruiser. The key test was whether they could sink the third vessel, a captured, heavily armoured battleship, the *Ostfriesland*. Their first attempt failed. The following morning, a series of attacks with 500-kg (1,100-lb) bombs left the battleship damaged but still afloat. Finally, a strike by seven bombers carrying 900-kg (2,000-lb) bombs sent the battleship to the bottom of the sea. Mitchell was triumphant and some naval observers reportedly watched with tears in their eyes as the

WILLIAM "BILLY" MITCHELL

WILLIAM "BILLY" MITCHELL (1879–1936) earned rapid promotion in the US Signal Corps in the years leading up to World War I. In 1912, aged 32, he became the youngest officer ever posted to the General Staff. Mitchell was an early enthusiast for aviation and in 1917 emerged as the leading combat commander of the US Army's air forces on the Western Front. He struck up a close relationship with RAF chief Sir Hugh Trenchard and became, like Trenchard, an advocate of independent air power. As assistant chief of the Army Service, Mitchell campaigned tirelessly for a well-funded US air force, lobbying Congressmen, writing popular books and articles, and staging publicity stunts such as the sinking of the *Ostfriesland* and the flight of the Douglas World Cruisers.

But Mitchell's vigorous self-promotion and his public denigration of senior officers who disagreed with him went far beyond what was acceptable from a serving officer. In 1925 he was courtmartialled and suspended from duty, after accusing his superiors of "incompetence [and] criminal negligence". Mitchell has remained a controversial figure in America, regarded by many aviation enthusiasts as an inspired prophet of air power.

HIGHLY DECORATED COMMANDER
The decorations and medals of General William "Billy" Mitchell, recognized as the top American combat commander of World War I, included the Distinguished Service Cross and Medal, several foreign decorations, and a posthumous Congressional Medal of Honor for outstanding services to military aviation.

pride of the oceans succumbed to air power. Others must have reflected on whether it was in fact so impressive to sink a tethered and undefended warship at the third attempt.

Mitchell's efforts may have had some effect

> "In the development of air power, one has to look ahead and not backward and figure out what is going to happen…"
>
> **WILLIAM "BILLY" MITCHELL**

in advancing the cause of naval aviation, but his advocacy of an independent air force came to nothing. The Army Air Service was officially upgraded to the Army Air Corps in 1926, but remained in practice an underfunded, subordinate branch of an underfunded army – during the 1930s, America's land forces were smaller than those of Poland or Romania.

The deficiencies of the Air Corps were publicly revealed when the army briefly took over flying US airmail routes in the winter of 1934 (see pages 114–15). The army aircraft – mostly

small, open-cockpit biplanes – were inferior to civil aircraft, and army pilots generally had little or no experience of flying in bad weather or at night. In short, American military aviation had fallen years behind the most advanced civil aviation. Air Corps pilots averaged more than one accident for every 1,000 hours of flying time. It was quite normal for them to die at the rate of about one a week. By bringing the spotlight of publicity to bear on these deficiencies, the attempt to fly the mail led to some improvements in equipment and training.

Seaborne aviation

Naval aviation in the United States made better progress, despite financial stringencies. The example of Britain suggests that this progress may have been partly due to the absence of an independent air force. For, at the end of World War I, Britain had led the way in the development of carrier-borne aviation. HMS *Eagle*, which joined the Royal Navy in 1924, set the template for future aircraft carriers, with an "island" set to one side of the flight deck incorporating a bridge and funnel. But until 1937 the aerial element of the Fleet Air Arm was under the direct control of the RAF and consistently starved of funds and first-rate aircraft by air commanders for whom naval aviation was a peripheral concern.

In the United States, the pivotal debate was not about the virtues of independent air power as opposed to an air force under naval command, but rather about the relative importance and roles of aircraft carriers and battleships. As early as 1921, one American naval commander, Admiral William Sims, predicted that "the airplane carrier, equipped with 80 planes," might be "the capital ship of the future". The Navy's Bureau of Aeronautics, led by Admiral William Moffett, worked

vigorously to establish the importance of naval aviation. But less air-minded admirals, although aware of the usefulness of aircraft, believed they should be employed in support of battleships, in a reconnaissance or air-defence role, and tended to dismiss the idea of using seaborne aircraft as a prime attack force. The first US aircraft carrier, USS *Langley*, entered service in 1922. It was not an

A PACKED FLIGHT-DECK
The flight-deck of the USS Saratoga *(CV-3) could carry up to 81 aircraft. Here it is shown crowded with Vought O2U-1 Corsair reconnaissance-fighters, Boeing F2B-1 fighters, and long-range Martin T4M-1 torpedo-bombers.*

PIONEER CARRIER
The USS Saratoga *was one of the US Navy's first fleet carriers, with a top speed of almost 34 knots and a capacity for 81 aircraft. A converted battle cruiser, she was commissioned in 1927 and participated in numerous task-force exercises that helped to develop American carrier strategy and doctrine.*

AIRCRAFT FOR CARRIERS

DURING THE INTERWAR YEARS, US naval air tacticians worked out a basic mix of aircraft for carriers. The three main types were fighters, for the defence of the fleet and to achieve air superiority; dive-bombers to attack enemy ships from above; and low-flying torpedo aircraft.

On the whole, naval air forces lagged behind land-based forces in making the transition from slower biplanes to higher performance monoplanes. The Royal Navy was especially archaic in introducing the Fairey Swordfish biplane – top speed 222kph (138mph) – as its latest torpedo-bomber in 1934, but the US Navy still had the Boeing F4B biplane as its main fighter through most of the 1930s. There were some good reasons for sticking to biplanes: they could land at lower speeds, a useful attribute at sea, and tended to take up less deck space than monoplanes (which needed a greater wingspan to achieve the same lift). However their slow speeds made them vulnerable to anti-aircraft fire during their long approaches on target when delivering their torpedoes.

LEGENDARY "STRINGBAG"
Introduced in 1934, the Fairey Swordfish was a three-man torpedo-bomber that could also be used for anti-submarine warfare. Remarkably, this open-cockpit biplane performed admirably in World War II.

impressive vessel – a converted collier with a top speed of 14 knots – but it was a start. The next two were much larger and faster. USS *Lexington* and USS *Saratoga* were originally meant to be battle cruisers. But at the Washington Conference in 1922 the leading naval powers agreed limits on warship numbers. The battle cruisers could not be built – but two fleet carriers could. *Saratoga* and *Lexington* were each capable of carrying up to 81 aircraft and had a top speed of 34 knots – faster than any warship of comparable size. In naval exercises from 1929 onwards they proved their ability to play a key role as an offensive strike force. Notably, in 1932 more than 150 aeroplanes from the two carriers executed a mock attack on the naval base at Pearl Harbor, which achieved total surprise and would have had a devastating effect if really carried out by an enemy power – the Japanese, for example.

Carriers were not the only means of providing aerial support at sea that were explored by the United States. In the early 1930s, experiments were also conducted with the naval airships *Akron* and *Macon*. They were designed to carry fighter planes, which they would launch and recover in the air – the returning aeroplane had to adjust its speed to that of the airship, position itself below the airship's hull, and then fly upwards to hook itself on to a support sticking out from the hull. These airborne aircraft carriers could accompany the fleet, acting as command and control centres for their aeroplanes, which would carry out reconnaissance missions and provide air defence. This bizarre-seeming idea might have worked but for the vulnerability of airships, not only to enemy action but also to the weather. The *Akron* was lost at sea in 1933 and the *Macon* suffered a similar fate in 1935. The only lighter-than-air aircraft the US Navy continued to use were non-rigid blimps.

Carrier fleets

Despite financial stringencies, the US Navy achieved a respectable development of its carrier force in the 1930s, with the addition of the USS *Yorktown* and *Enterprise* thanks to money from Roosevelt's New Deal programmes. But the exact role of carriers remained undecided, with many senior commanders still convinced that traditional warships held the key to sea victory. The Japanese, already identified as America's most likely enemy in a naval conflict, also prevaricated about the role of carrier aviation. But they accepted that aircraft would have to be used to weaken the US fleet before Japan's heavy warships could deliver a knockout blow. On this basis the Japanese developed the world's most effective carrier force in the 1930s, with better aircraft and better-trained aircrews than their American counterparts.

BIPLANE LANDING
Sailors standing in the USS Langley's safety nets watch as a US Navy Aeromarine 39-B biplane successfully lands on the carrier's deck in October 1922.

INTERWAR MILITARY AIRCRAFT

A GENERATION OF MILITARY aircraft developed between the end of WWI and the mid-1930s was obsolescent by the time war broke out again in 1939. These were primarily open-cockpit biplanes that achieved better performance than their WWI predecessors – chiefly through improved engines – without any dramatic change in technology. The technical revolution that brought the absolute dominance of streamlined, cantilever-wing monoplanes – often of all-metal construction with retractable undercarriages and closed cockpits – was already apparent in the 1932 Martin B-10B bomber. Some of the biplanes were used in small colonial conflicts, and others saw action in the Spanish Civil War – for example, the Fiat CR.32 and the Nieuport Type 52. A few soldiered on gallantly in the early stages of WWII, including the Gloster Gladiator and Fiat CR.32.

HART FORMATION
Hawker Harts were the RAF's standard light bombers during the 1930s. Their top speed of 296kph (184mph) was considered fast at the time.

Arado Ar 68

The Ar 68 was destined to be the Luftwaffe's last biplane, when it entered service in 1936, replacing the He 51. However, by the time it reached the squadrons, its successor, the Messerschmitt Bf 109 was already being tested.

ENGINE 690hp Junkers Jumo 210 Da	
WINGSPAN 11m (36ft 1in)	LENGTH 9.5m (31ft 2in)
TOP SPEED 335kph (208mph)	CREW 1
ARMAMENT 2 x 7.92mm MG17s; 60kg (132lb) bombload	

Boeing F4B-4

The F4B was Boeing's fourth and most significant member of a family of biplane fighters that had their origins in the 1923 US Army Air Corps' PW-9. The first F4Bs appeared in 1928 and served on the USS *Lexington*. Evolving through a number of variants, the F4B remained in service until the early 1940s.

ENGINE 550hp Pratt & Whitney 9-cylinder radial	
WINGSPAN 9.1m (30ft)	LENGTH 6.12m (20ft 1in)
TOP SPEED 302kph (188mph)	CREW 1
ARMAMENT 2 x machine guns	

Bristol Bulldog IIA

In its Mk.II version, the Bulldog joined the RAF in June 1929. Over 300 were built, equipping ten fighter squadrons and providing 70 per cent of Britain's air defences in the early 1930s. Exports included the air arms of Denmark, Australia, and Finland.

ENGINE 490hp Bristol Jupiter VIIF 9-cylinder radial	
WINGSPAN 10.3m (33ft 10in)	LENGTH 7.7m (25ft 2in)
TOP SPEED 280kph (174mph)	CREW 1
ARMAMENT 2 x machine guns; 35kg (80lb) bombload	

Dewoitine D.500

All-metal construction

Fixed undercarriage

The D.500 joined the French air force in 1935. Despite its open cockpit and fixed undercarriage, the D.500 provided the Armée de L'Air with its first modern all-metal monoplane fighter. The fighter was so important that production was shared between three manufacturers, including Dewoitine. The D.500, or its cannon-armed variant the D.501, were also exported to China, Lithuania, and Venezuela. A total of 308 D.500/501s were constructed, and the last 501s were withdrawn from service in 1941.

ENGINE 600hp Hispano-Suiza 12Xbrs V-12	
WINGSPAN 12m (39ft 8in)	LENGTH 7.7m (25ft 5in)
TOP SPEED 359kph (223mph)	CREW 1
ARMAMENT 4 x machine guns	

Farman F.222

A four-engined bomber was a rare sight during the 1930s, the Farman F.220 series and Tupolev TB-3 among the few. Farmans were initially used in night bombing raids over Germany, before being converted to military transports (civilian version shown here).

ENGINE 4 x 950hp Gnome-Rhone 14N 14-cylinder radials	
WINGSPAN 36m (118ft 1in)	LENGTH 21.4m (70ft 5in)
TOP SPEED 320kph (199mph)	CREW 5
ARMAMENT 3 x machine guns; 4,200kg (9,260lb) bombload	

Fiat CR.32

From the first all-Italian fighter, the Celestino Rosatelli-designed CR.1 of 1923, came a series of excellent combat machines. Perhaps the best was the CR.32, which entered service in 1934. The aircraft was the main fighter of the Nationalist forces during the Spanish Civil War, in both Italian and Spanish Hispano-Suiza licensed versions. Over 1,100 were manufactured and exported widely.

ENGINE	600hp Fiat A30 RA V-12	
WINGSPAN	9.5m (31ft 2in)	LENGTH 7.5m (24ft 5in)
TOP SPEED	375kph (233mph)	CREW 1
ARMAMENT	2 x machine guns	

Gloster Gladiator

The last Royal Air Force biplane fighter and the first with four forward-firing machine guns, the Gladiator entered service in 1937, less than a year before Hurricanes began to replace them. Twenty-four squadrons were equipped with them until 1941. Two squadrons sent to France in 1939 were overwhelmed by the German attack in May 1940. By 1941, all Gladiators had been withdrawn from front-line duties.

ENGINE	830hp Bristol Mercury IX 9-cylinder radial	
WINGSPAN	9.8m (32ft 3in)	LENGTH 8.2m (27ft 5in)
TOP SPEED	414kph (257mph)	CREW 1
ARMAMENT	4 x .303in Browning machine guns	

Grumman FF-1

The relationship with the US Navy that made the name Grumman synonymous with naval aircraft, began on 2 April 1931, when a contract was signed for a tubby biplane with unique features – a retractable undercarriage and an enclosed cockpit. Although the navy only acquired 27 FF-1s and 33 SF-1s (the reconnaissance version), the link was forged to such immortal aircraft as the Wildcat, Hellcat, and ultimately the Tomcat of *Top Gun* fame.

ENGINE	700hp Wright Cyclone 9-cylinder radial	
WINGSPAN	10.5m (34ft 6in)	LENGTH 7.5m (24ft 6in)
TOP SPEED	333kph (207mph)	CREW 2
ARMAMENT	3 x machine guns	

Hawker Hart

525hp Rolls-Royce Kestrel V12 engine

Sleek fuselage design

The 1930s Hawker family of sleek biplane fighters and bombers built around the renowned Kestrel engine were probably the most attractive British aircraft of their era. The first of that family was the superb Hart light bomber. First flown in June 1928, it was faster than contemporary RAF fighters and was quickly put into production. Over 500 Harts were manufactured and exported widely, and the Hart continued to serve the RAF until 1938.

ENGINE	525hp Rolls-Royce Kestrel 1B V-12	
WINGSPAN	11.4m (37ft 3in)	LENGTH 8.9m (29ft 4in)
TOP SPEED	296kph (184mph)	CREW 2
ARMAMENT	2 x machine guns; 226kg (500lb) bombload	

Heinkel He 51

Designed and built in secret, since Germany was not permitted to build fighters, the first He 51 flew in 1933 and deliveries to the Luftwaffe started in early 1935. Constantly improved, over 700 were built, of which the final variant was the He 51C, designed for a ground attack role. During the Spanish Civil War, He 51s were flown by the Nationalists and German Condor Legion.

ENGINE	750hp BMW VI V-12	
WINGSPAN	11m (36ft 1in)	LENGTH 8.4m (27ft 6in)
TOP SPEED	330kph (205mph)	CREW 1
ARMAMENT	2 x machine guns	

Martin B-10B (Model 139)

The B-10B first flew in 1932, during a time of rapid aeronautical development. It was the first US all-metal bomber to go into production, had the first gun turret, and was faster than the Army's fighters. However, the production of more advanced bombers, such as the Boeing B-17, limited its military role, although it continued to be manufactured for export until 1939.

ENGINE	2 x 775hp Wright R-1820 Cyclone 9-cylinder radials	
WINGSPAN	21.5m (70ft 6in)	LENGTH 13.6m (44ft 9in)
TOP SPEED	343kph (213mph)	CREW 4
ARMAMENT	3 x machine guns; 1,025kg (2,260lb) bombload	

Nakajima Ki-27 "Nate" (Type 97)

First flown in July 1936, the Ki-27, or Army Type 97, was an exceptionally agile monoplane fighter. It entered operational service with the Japanese Army in March 1938. A total of 3,999 were built between late 1937 and the end of 1942, with many ending their careers as kamikaze bombers.

ENGINE	500hp Nakajima-built Bristol Jupiter 9-cylinder radial	
WINGSPAN	11m (36ft 1in)	LENGTH 7.3m (23ft 10in)
TOP SPEED	299kph (186mph)	CREW 1
ARMAMENT	2 x machine guns	

Nieuport Type 52

Similar to the Type 62 C1, which served with the French air force, the Type 52, chosen by the Spanish air force from 1929 to 1936, had an all-metal construction. At the outbreak of the Spanish Civil War, it fought on both Nationalist and Republican sides.

ENGINE	580hp Hispano-Suiza 12Hb inline	
WINGSPAN	12m (39ft 5in)	LENGTH 7.7m (25ft 1in)
TOP SPEED	249kph (155mph)	CREW 1
ARMAMENT	2 x machine guns	

THE GOLDEN AGE

During the 1930s, the world shifted from a period of aspiration towards disarmament into widespread rearmament. By the second half of the decade, an arms race was under way, with the militarist governments of Germany and Japan and their potential enemies pumping money into military aviation. Advances in aircraft design, engines, and general aviation technology already seen in commercial and racing aeroplanes were applied to a new generation of military aircraft, while wars in China and Spain gave some air forces the chance to try out their new aeroplanes and tactics for real.

The Luftwaffe returns

After Adolf Hitler's rise to power in 1933, Nazi Germany began what would soon became a general rearmament in Europe. Like his fellow dictator Mussolini, Hitler found in aircraft an image of dynamism, modernity, and power that reflected his own vision of the Nazi state, as well as a practical tool for achieving his military ambitions. He ordered an immediate and massive programme of expansion in military aviation that was already well under way by the time the recreation of the Luftwaffe was publicly announced in 1935. The

> "Germany is once more a world power in the air. Her air force and her air industry have emerged from the kindergarten stage. Full manhood will still not be reached for three years."
>
> MAJOR TRUMAN SMITH
> US MILITARY ATTACHE IN BERLIN, 1936

official head of Nazi German aviation was Hermann Göring, but the true mastermind behind the rapid resurgence of the Luftwaffe was former Luft Hansa director Erhard Milch. Despite all that had been done to keep the "shadow Luftwaffe" in existence, Milch faced a daunting task in creating the large air force Hitler demanded. Between 1933 and 1936, he expanded Germany's aircraft production by a staggering 800 per cent, as well as training an entire new generation of pilots.

Although Milch's achievement was impressive – especially when new designs such as the

"OUR LUFTWAFFE"
Secretly reformed in 1923 in defiance of the terms of the Treaty of Versailles, the Luftwaffe's resurgence after the Nazis came to power in Germany in 1933 was masterminded by Erhard Milch. In 1935 its existence was publicly announced with posters such as this (left), promoting "Our Luftwaffe" in German.

Messerschmitt Bf 109 fighter and the Junkers Ju 87 Stuka dive-bomber began to roll off the production lines in the second half of the decade – the Luftwaffe was never quite as strong as its potential enemies thought it to be. Nazi propagandists ensured that the image of German air power was stamped on the imagination of foreign peoples and their leaders, undermining the will to resist Hitler's ever-escalating demands. But after 1936, the pressure to create a numerically massive air force, for which sufficient resources were not really available, led to chaotic inefficiency and disorganization – characteristics in any case typical of all parts of the Nazi system.

Former World War I ace and stunt flier Ernst Udet, appointed as Luftwaffe technical director and later head of aircraft production, became a wild card within the system. Among the many decisions by Udet that caused consternation, the most famous is probably his demand that the impressively fast Ju 88 medium bomber, ready to enter production in 1938, should be modified so it could also act as a dive-bomber. The result was a redesign that cut the bomber's speed from 500kph (310mph) to 300kph (185mph) and delayed its introduction by two years. With Udet and Milch at loggerheads and Göring pursuing his own erratic and self-serving course, it is a tribute to the abilities of German designers, scientists, and fliers that the Luftwaffe still proved such an impressive force.

PROPAGANDA PLANE
A Tupolev ANT-20 "Maksim Gorkii" flies over a 1935 May Day parade in Moscow's Red Square. The propaganda plane was fitted with loudspeakers, a printing press, and a pharmacy.

Allied developments

From 1935 onwards, Britain and France were acutely aware of the threat posed by the resurgence of the Luftwaffe. It coincided with the realization that their existing air fleets were rapidly becoming obsolescent because of technological developments. The RAF's new fighter ordered into production in 1935 was the Gloster Gladiator biplane (which won fame during the defence of Malta in 1940–41), but fortunately a prototype of the Hawker Hurricane flew in the same year and the Supermarine Spitfire was taking shape on the drawing board. More sensible, if less dynamic, than the Nazis, the liberal democracies embarked on a longer term but relatively slow development of new models and the industrial set-up to manufacture them. By 1938, state-of-the-art monoplanes were reaching RAF squadrons in growing numbers and production accelerated rapidly as an uneasy peace turned to war. Unfortunately for the French, their progress was slower and new models, such as the

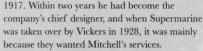

Dewoitine D.520 fighter, had only just begun to come into service when the German *Blitzkrieg* struck in 1940.

Soviet innovators

Hitler had made it clear throughout his political career that the "Jewish Bolsheviks" of the Soviet Union were intended for destruction. His rise to power brought a definitive end to the links between the Soviet and German aviation establishments that had served both so well in the 1920s and early 30s. Even in that decade, the Soviet Union had put considerable resources behind the creation of its air force and nascent air industry. Many of the leading talents in Russian aviation, including Igor Sikorsky, Alexander Seversky, and Alexander Kartveli, had gone into exile after the revolution, contributing instead to the progress of aviation in the United States – Kartveli, for example, designed the Republic P-47 Thunderbolt, one of the great American aircraft of World War II. But a tradition of aircraft design and aerodynamic research was firmly implanted. Andrei Tupolev, who as an engineering student in pre-war Russia had been arrested by the tsarist police for his revolutionary activities, emerged in the 1920s as the prime mover in Soviet aviation. Other talented individuals who rose through the Soviet system included

REGINALD MITCHELL

IN 1933 REGINALD MITCHELL (1895–1937), chief designer at the Supermarine aircraft company, took a holiday in Europe to convalesce after undergoing surgery for cancer. A conversation with some German aviators convinced him that war was on its way, and from that moment he devoted himself single-mindedly, and against medical advice, to the creation of the fighter that would be called the Spitfire.

Born in Stoke-on-Trent, in England's industrial Midlands, Mitchell was an apprentice railway engineer before joining the Supermarine Aviation Works in Southampton in

1917. Within two years he had become the company's chief designer, and when Supermarine was taken over by Vickers in 1928, it was mainly because they wanted Mitchell's services.

He made his reputation designing seaplanes for the Schneider Trophy; his 1925 S.4 gave the world its first view of the kind of fast, streamlined monoplane that was to be his speciality. Known for his attention to detail, he went on to design the Schneider-winning S.5 and S.6 – one version of which became the first aircraft to top 400mph (640kph) in 1931. Exhausted by his work on the Spitfire, Mitchell died at the age of 42, shortly before it went into production, but he was already sure of the aircraft's success. His only regret was the name "Spitfire", dreamed up by Vickers.

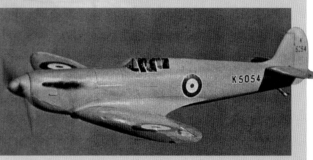

SUPER SPITFIRE
Mitchell's design for the Spitfire was revolutionary in its use of an elliptical wing, designed for maximum aerodynamic efficiency. The Spitfire evolved through many versions, late models being substantially different from the 1937 prototype (right).

Sergei Ilyushin (eventually assigned exclusively to the development of long-range bombers), Nikolai Polikarpov, Alexander Yakovlev, and Syemyen Lavochkin, creator of the Lavochkin La-5 fighter.

Through the 1930s, these designers had to cope with working under the increasingly paranoid rule of Soviet dictator Joseph Stalin. As early as 1929, Polikarpov was arrested for "sabotage" when his department's development of a new fighter fell behind schedule. Along with his entire design team, he was consigned to the prison section of a state aviation factory (the use of prison labour was an important element in the Soviet economy), where they designed the Polikarpov I-5 fighter, earning their release in 1933. Tupolev himself was one of the thousands of prominent individuals "purged" by Stalin in the second half of the 1930s, spending six years working on aircraft design in one of the "special camps" of the Gulag. He was only released in 1943.

Yet within this bizarre system some excellent innovative aircraft design was achieved. Polikarpov, for example, was responsible for the I-16 which, along with the Messerschmitt Bf 109, was one of the very first single-seat, low-wing monoplane fighters. And Stalin's ruthless drive to industrialize the Soviet Union in the 1930s, although carried out at the expense of great

human suffering, did create the basis for an effective mass-production aircraft industry. But the impact of the Stalinist purges of the late 1930s on the Soviet air force was devastating. About three quarters of senior officers were either executed or sent to the Gulag labour camps. The effects of this blow were still being felt when Germany invaded the Soviet Union in 1941.

New generation of fighters

In all the major air forces, the 1930s saw the same progress from biplane fighters – the sort of aeroplanes seen attacking King Kong on the top of the Empire State Building in the famous 1933 movie – to sleek cantilever-wing monoplanes such as the Spitfire, Bf 109, or the Fiat G.50 Freccia. The new generation of fighters consisted mostly of metal aeroplanes (although the Hurricane, one of the most successful, had a fabric-covered fuselage supported in part by wooden strips). They mounted a powerful engine in a lightweight frame and were designed with a scrupulous eye to reducing drag, not only abolishing the old biplane struts and wires but also having a retractable undercarriage and guns that were built into the wings or fuselage. From a traditional pilot's point of view, their most controversial aspect was an enclosed cockpit and an implied dependence on instrumentation. When

Udet first sat in the cockpit of a Messerschmitt Bf 109, he is said to have remarked that "this would never be a fighting aeroplane" because "the pilot has to feel the air to know the speed of the plane". It was a prejudice shared by many old-school pilots, brought up on "flying by the seat of your pants". But they could not argue with the speed – typically 480–560kph (300–350mph) – or rate of climb of the new models, which was combined with a breathtaking capacity to dive, spin, and roll that made them among the most exciting aircraft to pilot that have ever been created.

Heavy bombers

For most American and British air commanders, however, the crucial aeroplanes in their force were not the fighters but the heavy bombers. The US Army Air Corps and the RAF held that strategic bombing could be a war-winning use of air power, given the right aircraft to do the job. In Britain, Trenchard and other commanders drew support for this view by arguing that a bomber fleet would allow the British to fight a war in Europe without sending an army across the Channel – and a repeat of the trench warfare

WILLY MESSERSCHMITT

WILLY MESSERSCHMITT (1898–1978) built gliders as a teenager before World War I. Exempted from war service due to ill health, in the 1920s he began designing powered aircraft. From 1926 he had his aircraft built by BFW (Bayerische Flugzeugwerke), and he subsequently took over the company. During the vicious infighting of the Nazi regime after 1933, Messerschmitt had to cope with the hostility of the powerful state-aviation boss Erhard Milch and the bitter rivalry of designer and manufacturer Ernst Heinkel.

The adoption of the Bf 109 by the Luftwaffe in 1935 made Messerschmitt's reputation, and he renamed the BFW as Messerschmitt AG in 1938. He experimented with mixed success at the cutting edge of aviation technology, producing, among other models, the Komet rocket aircraft and the Me 262 jet fighter. After the defeat of the Nazis, Messerschmitt took refuge in Argentina, but in the 1950s he returned to Germany to continue his career.

INSPIRED DESIGNER
An inspired designer, Willy Messerschmitt was also capable of basic errors and miscalculations.

Messerschmitt Bf 109

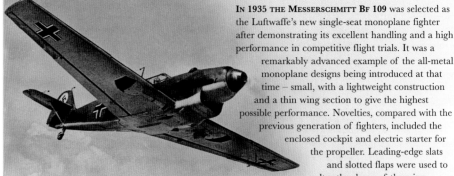

IN 1935 THE MESSERSCHMITT BF 109 was selected as the Luftwaffe's new single-seat monoplane fighter after demonstrating its excellent handling and a high performance in competitive flight trials. It was a remarkably advanced example of the all-metal monoplane designs being introduced at that time – small, with a lightweight construction and a thin wing section to give the highest possible performance. Novelties, compared with the previous generation of fighters, included the enclosed cockpit and electric starter for the propeller. Leading-edge slats and slotted flaps were used to alter the shape of the wing (optimized for maximum speed) to perform adequately at slow speed for landing.

The Bf 109 unfortunately also had serious drawbacks. Its cockpit was cramped and gave the pilot poor vision, especially when taxiing. Its thin wing had serious difficulty accommodating machine guns or cannon and was too weak to support the aircraft's weight on the ground. As a result the undercarriage had to be positioned close to the fuselage rather than out under the wings. This narrow undercarriage, along with difficulties in handling at slow speed, meant that the aircraft was prone to accidents on take-off and landing, especially when obliged to operate from improvised airstrips under wartime conditions. The Bf 109 was blooded in the Spanish Civil War, where it won air

"The new Bf 109 simply looks fabulous. The take-off is certainly unusual but... its flight characteristics are fantastic."

JOHANNES TRAUTLOFT
CONDOR LEGION PILOT

LIFE-JACKETS REQUIRED
Luftwaffe fighter pilots wore life-jackets during the Battle of Britain. The limited range of the Bf 109 meant that it was a common experience for pilots to run out of fuel over the Channel (on the way home). This necessitated "ditching" in the sea.

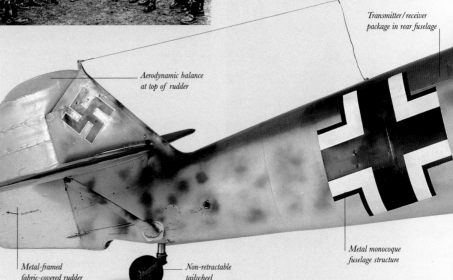

Transmitter/receiver package in rear fuselage

Aerodynamic balance at top of rudder

Metal-framed fabric-covered rudder

Non-retractable tailwheel

Metal monocoque fuselage structure

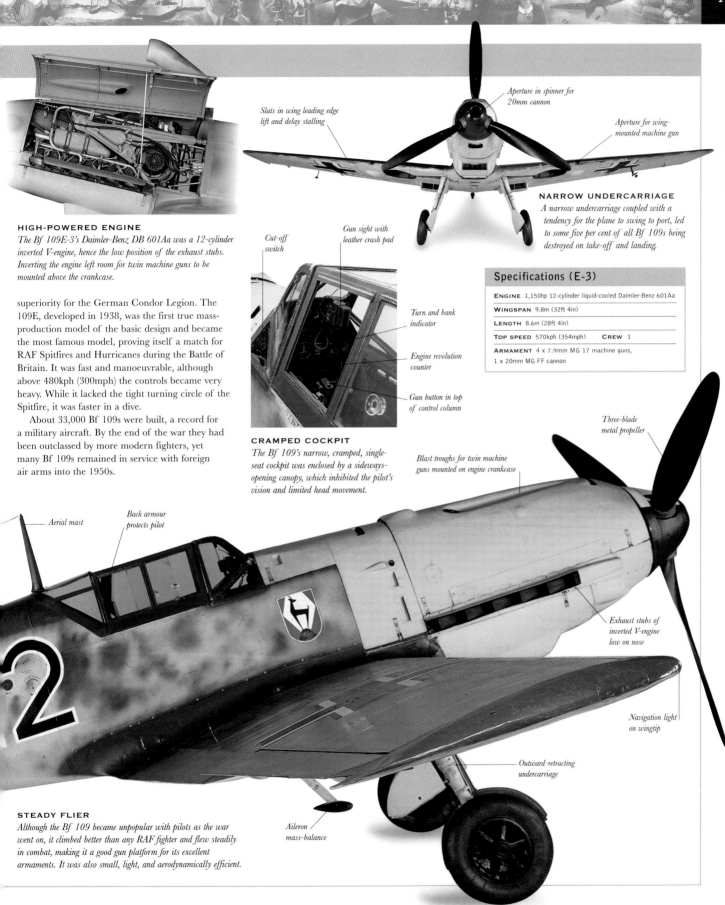

HIGH-POWERED ENGINE

The Bf 109E-3's Daimler-Benz DB 601Aa was a 12-cylinder inverted V-engine, hence the low position of the exhaust stubs. Inverting the engine left room for twin machine guns to be mounted above the crankcase.

superiority for the German Condor Legion. The 109E, developed in 1938, was the first true mass-production model of the basic design and became the most famous model, proving itself a match for RAF Spitfires and Hurricanes during the Battle of Britain. It was fast and manoeuvrable, although above 480kph (300mph) the controls became very heavy. While it lacked the tight turning circle of the Spitfire, it was faster in a dive.

About 33,000 Bf 109s were built, a record for a military aircraft. By the end of the war they had been outclassed by more modern fighters, yet many Bf 109s remained in service with foreign air arms into the 1950s.

Slats in wing leading edge lift and delay stalling

Aperture in spinner for 20mm cannon

Aperture for wing-mounted machine gun

NARROW UNDERCARRIAGE

A narrow undercarriage coupled with a tendency for the plane to swing to port, led to some five per cent of all Bf 109s being destroyed on take-off and landing.

Cut-off switch

Gun sight with leather crash pad

Turn and bank indicator

Engine revolution counter

Gun button in top of control column

CRAMPED COCKPIT

The Bf 109's narrow, cramped, single-seat cockpit was enclosed by a sideways-opening canopy, which inhibited the pilot's vision and limited head movement.

Specifications (E-3)

ENGINE	1,150hp 12-cylinder liquid-cooled Daimler-Benz 601Aa
WINGSPAN	9.8m (32ft 4in)
LENGTH	8.6m (28ft 4in)
TOP SPEED	570kph (354mph) CREW 1
ARMAMENT	4 x 7.9mm MG 17 machine guns, 1 x 20mm MG FF cannon

Blast troughs for twin machine guns mounted on engine crankcase

Three-blade metal propeller

Exhaust stubs of inverted V-engine low on nose

Navigation light on wingtip

Aerial mast

Back armour protects pilot

Outward-retracting undercarriage

STEADY FLIER

Although the Bf 109 became unpopular with pilots as the war went on, it climbed better than any RAF fighter and flew steadily in combat, making it a good gun platform for its excellent armaments. It was also small, light, and aerodynamically efficient.

Aileron mass-balance

of 1914–18 was what, above all else, the British wished to avoid. It was even argued that the existence of bombers might maintain peace by acting as a deterrent to would-be aggressors through a threat to attack their cities. Since, as British Prime Minister Stanley Baldwin said, "the bomber would always get through", the outbreak of war would be followed by the immediate destruction of cities – surely a prospect that would deter any country from breaking the peace.

Whereas the British concept of strategic bombing was essentially as a form of

psychological warfare, centred on terrorising civilians, American air commanders developed a notion of bombing as economic warfare. By precision bombing of factories and transport systems, the air force would undermine the enemy's capacity to continue a war. The USAAC was of necessity committed to the accurate bombing of precise targets because it chief function in the eyes of its political paymasters was to defend America's coasts. In other words, it had to claim to be able to sink ships – small, moving targets.

Bomber prototypes

If strategic bombing was to have any credibility, the American and British air forces needed the aircraft to do the job. They had to have the range to reach distant targets deep within enemy territory, the capacity to carry enough bombs to cause substantial damage, and the speed and firepower to brush aside air defences – there was no place in either British or American doctrine for the concept of an escort fighter.

The development of bombers was rapid. In 1932, the state-of-the-art machine was the Martin B-10, a twin-engined monoplane with a respectable top speed of around 320kph (200mph). But three years later, the Boeing company came up with the four-engined Model 299, the prototype of the B-17 Flying Fortress. The Model 299 crashed in October 1935, almost aborting the project and threatening

DIVE-BOMBING

IN THE 1930S, DIVE-BOMBERS attracted a lot of interest because they could deliver a bombload with far greater accuracy than normal bombers. This made them especially suitable for attacking ships and close air support of ground troops. Their effectiveness was first demonstrated by American pilots supporting the Nicaraguan government against leftist rebels in the late 1920s. They were then adopted by the US Navy as a key element of the carrier air force.

On a visit to America in 1934, leading Luftwaffe official and air ace Ernst Udet got a chance to fly a US Navy Curtiss Hawk dive-bomber, becoming a vigorous advocate of this novel form of air attack. Udet backed production of the Ju 87 Stuka and insisted that all new German bombers be capable of dive-bombing. But dive-bombers were slower and heavier than other warplanes of similar size because they had to be robust enough to withstand the stresses of a diving attack. This made them too easy a prey for enemy fighters. They were eventually upstaged by high-performance fighter-bombers that could hold their own in air-to-air combat.

WAILING DIVE-BOMBER
The Junkers Ju 87 resembled a bird of prey, with its inverted gull-wing and "jericho trumpet" sirens fitted to the landing gear. Its reputation was made during Germany's successful Blitzkrieg ("lightning war") campaigns of 1939–41. The "trousered" landing gear of early models can be seen below.

Boeing with bankruptcy, but the Air Corps ordered 14 of them anyway. Capable of flying at 480kph (300mph), the Flying Fortress was regarded by the US as the first credible strategic bomber. The RAF, after starting the war with twin-engined long-range bombers, followed on with the four-engined Sterling, Halifax, and Lancaster.

Strategic bombing

To the countries of continental Europe and Japan strategic bombing did not seem such an attractive use of air power. Whether primarily defence-oriented, as in the case of France and the Soviet Union, or bent on aggressive expansionism, as were Germany and Japan, they believed that in any war the crucial battles were going to be fought between armed land forces. Although they did not ignore the potential for direct air attacks

on enemy cities or economic targets, they felt that the essential role of air power must be to increase the chances of victory on the ground.

In the 1930s, Germany was ahead of any other country in its appreciation of the most effective use of air power. The Luftwaffe grasped the importance of seeking the destruction of the enemy's air forces – either in aerial combat or by attacks on airfields and aircraft factories – as a prelude to other air operations. It was trained to give close support to tanks and motorized infantry in mobile warfare, but also prepared for more independent operations, from attacks on tunnels and bridges along key road and rail routes behind the enemy front line, to the bombing of arms factories and fuel depots. Troop transport and logistical support were other areas to which the Luftwaffe paid close attention, in line with the

German armed forces' general preoccupation with mobility and shock tactics.

Far from ignoring strategic bombing, the Luftwaffe gave careful consideration to issues such as the guidance of bombers on to distant targets at night – a problem that the RAF, for all its obsession with bombing cities, had omitted to take seriously at all. But the Germans failed to develop a successful four-engined heavy bomber, largely because of the inability of German industry to produce adequate engines. For any strategic bombing campaign they would have to rely on twin-engined medium bombers such as the Dornier Do 17 and the Heinkel He 111.

BOMBER FORMATION
A formation of Keystone bombers flies down the Hudson Valley in a display of America's aerial force. At the start of the 1930s these lumbering biplanes represented virtually the entire bomber strength of the Army Air Corps.

Nazi Germany found a chance to try out its aircraft and tactics when a civil war broke out in Spain in 1936. Right-wing "Nationalist" Spanish officers, who had failed to overthrow the country's left-wing Republican government in an insurrection, asked for the Nazis' help to mount a sustained military campaign. The Luftwaffe immediately sent a score of Ju 52 transport aircraft to airlift soldiers from Spanish Morocco to Nationalist-controlled Seville in southern Spain. This was an unprecedented operation – the movement by air of a major military force. Between July and October 1936, some 20,000 soldiers with their equipment, including artillery, were airlifted into Seville, enabling the Nationalists to take the offensive.

Aerial artillery

While Nazi Germany and Fascist Italy provided air support for the Nationalists, the Soviet Union sent pilots and aircraft to fight for the Republican side. By the end of 1936, Soviet I-15 and I-16 fighters had won air superiority and were able to inflict serious damage on ground forces, notably with the destruction of an Italian motorized column at Guadalajara in March 1937. The Luftwaffe responded by sending in the Condor

Legion, a force of around 100 of its latest aircraft and best-trained pilots, with Colonel Wolfram von Richthofen, cousin of the Red Baron, as its chief of staff. Wherever the Messerschmitt Bf 109s appeared, Soviet aircraft were driven from the skies. With air superiority assured, the Condor Legion experimented with close air support, acting as "aerial artillery" to prepare the way for ground offensives, and interdiction – air attacks on enemy reserve troops and communications behind the front line. This use of air power proved decisive, allowing the Nationalists to achieve victory by 1939.

While the Luftwaffe was drawing invaluable insights and accurate conclusions from the actual experience of combined-arms operations, the attention of the world at large was fixed upon a single issue: the terror-bombing of civilians. This

Das ist das Heil, das sie bringen!

TERROR-BOMBING
A Spanish Republican propaganda poster makes a powerful attack on the German bombing of Guernica in 1937: "This is the health/salute [Heil] they bring".

was somewhat peripheral to air operations in the civil war but central to the fears and anxieties of the citizens of London, Paris, and even New York, who could not help but see events in Spain as prefiguring their own possible future fate. Both sides in the Spanish conflict at times bombed enemy-held towns and cities, but the German and Italian air forces had far more opportunity to do so and caused most loss of life. Apart from the devastation of the small Basque town of Guernica in April 1937, Republican-held Madrid came under intermittent aerial bombardment from 1936 onwards, and the Catalan city of Barcelona was heavily bombed by the Italians for three days in March 1938, killing around 1,300 people.

Civilian reaction

There was no question that air attack frightened people. Esmond Romilly, a British volunteer fighting with the Republican International Brigades, described being trapped in a metro station during an air-raid in Madrid: "A panic-stricken crowd made it impossible to move… women screamed and on the steps men were fighting to get into the shelter." US military attaché Stephen O. Fuqua, in Barcelona during the March 1938 raids, reported that economic and industrial life was "completely paralyzed" and "semi-panic permeated every form of city life". Fuqua graphically described civilians suddenly blown to pieces while sitting on buses or at the tables of sidewalk cafes – afterwards he saw the cafe waiters "sweeping up the human bits into containers".

Yet for all this horror, it was evident that civilians soon learned to cope psychologically with the threat of bombing. The Condor Legion's assessment of the effect of "destructive bombardments without clear military targets",

WATCHFUL CIVILIANS
Civilians in the Republican port of Bilbao, in Spain's Basque country, walk in fear as aircraft appear overhead. In April 1937 Bilbao was bombed by warplanes from the German Condor Legion and the Italian Avazione Legionaria, supporting the Nationalist side in the Spanish Civil War.

ITALIAN SUPPORT
During the Spanish Civil War, the Italian Savoia-Marchetti SM.81 bomber was used in support of Franco's Nationalist troops and in attacks on cities. The SM.81 shown here is escorted by Fiat CR.32 biplane fighters.

based on its experience in Spain, was that in the long run they were more likely to stiffen popular resistance than to undermine morale. This was not the conclusion drawn by most political and military leaders worldwide. What people expect is what they see, and Guernica in particular was widely interpreted as confirming the expectation of a swift devastation of cities in the early stages of any major war.

Theories of war
Meanwhile in 1937 Japan had invaded China, giving another demonstration of the effectiveness and frightfulness of air power. Cities such as Shanghai and Nanking were subjected to

THE DESTRUCTION OF GUERNICA

ON MONDAY 26 APRIL 1937 it was market day in the small but ancient Basque town of Guernica. At around 4.30 in the afternoon, when the town was at its most crowded, 43 aircraft of the Condor Legion – Heinkel He 111 medium bombers, He 51 fighters, and Ju 52 transport aircraft used as bombers – launched the first wave of an attack that lasted more than three hours. Strafing and dropping high-explosive bombs and incendiary devices, the German aircraft destroyed about half of the town.

A Basque Catholic priest, Alberto Onaindia, described the apocalyptic scene: "Five minutes did not elapse without the sky's being black with German planes. The planes descended very low, the machine-gun fire tearing up the woods and roads, in whose gutters, huddled together, lay old men, women and children… Fire enveloped the whole city. Screams of lamentation were

CONDOR LEGION
This postcard image shows the flags of Spain and the Condor Legion paraded side by side. The Condor Legion, which attacked Guernica, consisted of the best airmen from Hitler's developing Luftwaffe.

heard everywhere, and the people, filled with terror, knelt, lifting their hands to heaven…"

Foreign journalists were on the scene the following day and filed graphic descriptions of the carnage and destruction. The Nationalists and Germans for a long time denied that Guernica had been bombed at all. They later more plausibly argued that Guernica had been a valid military target, since it housed reserve troops and was a major crossroads. But the Republicans won the propaganda war, establishing Guernica as a symbol of the evils both of Nazism and of terror-bombing from the air. Artist Pablo Picasso, who had been commissioned to produce a large work for the Spanish Republic's pavilion at the Paris World Fair, created an iconic painting that was to prove the event's most lasting memorial.

SYMBOL OF DESTRUCTION
Guernica was reduced to ruins when 45,000kg (100,000lb) of explosives were dropped by German bombers in what was widely regarded as the deliberate terror-bombing of a civilian target. Up to 1,600 people may have perished.

bombing on a significant scale. This brought forth vigorous protests from the US government, which was inclined to head for the moral high ground as examples of the bombing of civilians multiplied. Secretary of State Cordell Hull said of the bombing of Barcelona: "No theory of war can justify such conduct". And in June 1938, the Senate passed a resolution condemning "the inhuman bombing of civilian populations". Yet the Roosevelt administration also sought increased funds for the American long-range bomber force, reasoning that the best way to stop an enemy bombing your cities was to credibly threaten to flatten his in reprisal.

The other answer to the threat of bombing was to prepare air defences to block an attack and civil defence to limit casualties. The concept of air defence was not popular with air commanders committed to war-winning strategic bombing – it was hard for them to argue that their bombers would "always get through" without accepting that enemy bombers would inevitably do the same. But the Germans had found in Spain just how vulnerable bombers would actually be – every bombing raid needed a fighter escort. In Britain, the government overruled the RAF and, in the second half of the 1930s, insisted on giving high priority to building fighters and preparing a co-ordinated air-defence system.

Defensive measures

One reason Douhet had given for believing that bombers could not be stopped was that their raids would come as a surprise, striking before fighters could be scrambled to intercept them. But in the 1930s, forms of radar were being developed and refined in all advanced countries – though not all were being applied to air defence – along with primitive IFF (Identification Friend or Foe) systems that allowed ground controllers to distinguish between

MONSTER EAR TRUMPETS
Emperor Hirohito inspects the huge trumpet-like aircraft detectors/audible rangefinders that were part of the air defence in Japan, and other countries, in 1935. Designed to pick up the low rumble of approaching enemy aircraft, they worked in conjunction with anti-aircraft guns, visible on the right.

enemy aircraft and their own. Controllers adapted the techniques developed during World War I – keeping track of aircraft movements by pushing models around on a chart and using two-way radio links to give instructions to pilots in the air – to direct interceptors against intruding bombers.

Britain was especially well placed to develop a radar-based defence system because its front line was the coast, and radar worked much better over the uncluttered sea than over land. The British

> "We thought of air warfare in 1938 rather as people think of nuclear warfare today."
>
> HAROLD MACMILLAN
> FORMER BRITISH PRIME MINISTER

were also highly motivated by fear of air attack. The age of the bomber had come as a far greater psychological shock to Britain than to any other country because, protected by the Royal Navy, its people had long thought themselves immune to attack from abroad. The British government was told by its aviation chiefs that it could expect 20,000 civilian casualties in London on the first day of a war with Germany, and 150,000 in the first week (in fact, there were 295,000 casualties from air attacks in the whole of Britain in six years of war

THE FIRST JETS

ON 27 AUGUST 1939, four days before the start of World War II, the first jet aircraft took off from Marienehe airfield in Germany. Test pilot Erich Warsitz kept the diminutive Heinkel He 178 in the air for just six minutes, but it was enough to open a new era in aviation. Aircraft manufacturer Ernst Heinkel commented: "The hideous wail of the engine was music to our ears."

The wailing turbojet engine was the brainchild of Hans von Ohain, a graduate of the University of Göttingen, Germany's most prestigious centre of theoretical aeronautics. It used a gas turbine to generate thrust in accordance with Newton's Third Law of Motion. The aircraft scooped up air as it moved along; this air was compressed, combined with fuel, and ignited; and the jet of hot gas forced out of the back of the engine propelled the aircraft forwards.

Ohain was only one of a number of researchers in the 1930s investigating the use of gas turbines to create jet propulsion. Another was Flight Lieutenant Frank Whittle of the RAF, who filed his first patent for a

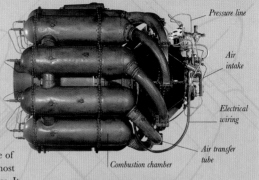

Pressure line
Air intake
Electrical wiring
Air transfer tube
Combustion chamber

JET PROPULSION ENGINE
Whittle's jet engine worked by sucking in air, then compressing and burning it with fuel, to create thrust. Faster than the propeller engine, it was also more economical on fuel.

jet engine as early as 1930. But whereas Whittle struggled to develop his engine in the face of financial difficulties and official indifference, Ohain and his assistant, Max Hahn, were backed both by Heinkel and the German Air Ministry. As a result, by the time Britain's first jet prototype, the Gloster E.28/39, flew in 1941, Germany was already developing practical jet-propelled warplanes.

FIRST JET AIRCRAFT
In August 1939 Erich Warsitz flew the world's first jet aircraft, an He 178, reaching a top speed of over 640kph (400mph).

from 1939 to 1945). Not surprisingly, faced with such alarmist predictions, the defenders of democracy in Europe were unnerved by fears of air attack as Hitler pressed for advantage in 1938. The last thing they wanted was a war that would begin with the immediate destruction of cities and hundreds of thousands of civilian casualties.

Looking back on that time, British politician Harold Macmillan commented: "We thought of air warfare in 1938 rather as people think of nuclear warfare today." In other words, it was expected to lead swiftly not only to mass slaughter but to a total breakdown of civilization. With war on the horizon, plans were finalized in 1939 for the distribution of gas masks, provision of bomb shelters, and evacuation of children. In the run-up to the war, filmgoers who saw the popular science-fiction movie *Things to Come* (1936), based on a 1933 novel by H.G. Wells, found in its image

of a world laid to waste by aerial bombardment a very plausible version of their own future. Watching warplanes manoeuvring over rural England in 1932, British poet Siegfried Sassoon had predicted that, one day, "fear will be synonymous with flight". For many, that day had come.

AERIAL APOCALYPSE
The film version of H.G. Wells' novel The Shape of Things to Come *vividly depicts the earth laid to waste by aerial bombardment in a 20-year war.*

WT 34

BATTLE FOR THE SKIES 4

WORLD WAR II WAS MOSTLY FOUGHT with aircraft that were at least on the drawing board before the war began. Although jet aircraft and missiles played a part in the later stages, the main technical developments of the war years centred on radar and other electronic devices. The real novelty was the sheer scale of the production and deployment of aircraft, with more than a thousand bombers sometimes sent on a single mission. Aircraft played a vital role in army operations, providing ground troops with mobility, supplies, and supporting fire. In the Pacific, naval war became a long-distance battle between aircraft carriers. But the most spectacular use of aircraft was in strategic bombing. Efforts to destroy the enemy's productive capacity and undermine the will to fight culminated in the dropping of the atomic bombs on Japan, which ushered in a new era of warfare.

DAYLIGHT BOMBER
*A Heinkel He 111 flies over London's Docklands on
7 September 1940, the day of the Luftwaffe's first deliberate
bombing raid on the British capital. The bombing of cities
from the air was to mount in intensity as the war progressed.*

COMMAND OF THE AIR

FROM THE START OF WORLD WAR II, AIRCRAFT WERE CRUCIAL TO THE SUCCESS OF ARMY OPERATIONS AS TROOPS WERE EXPOSED TO FIRE FROM THE SKY

"Anyone who has to fight…
against an enemy in
complete command of the
air fights like a savage
against modern European
troops… with the same
chances of success."

FIELD MARSHALL ERWIN ROMMEL
FROM THE POSTHUMOUS *ROMMEL PAPERS* (1953)

ON 1 SEPTEMBER 1939 Germany invaded Poland. Four weeks later the Polish forces surrendered and their country was divided between Germany and the Soviet Union. If anyone still had doubts about the importance of air power in warfare, this lightning campaign ended them. The Luftwaffe sent about 2,000 aircraft into Poland – a relatively small force compared with air operations later in World War II, but more than enough to overcome the few hundred aeroplanes of the Polish air force, despite the skill and gallantry with which its pilots fought. In command of the air, German aircraft battered Polish ground troops, clearing a path for advancing armour, shattered Polish rail and road networks, and, in a climactic gesture of terror and destruction, reduced much of Warsaw to burning ruins.

The Polish campaign was the first example of *Blitzkrieg*, the "lightning war" of short devastating mobile campaigns, which would give the Germans control of Denmark and Norway in April 1940, France and the Low Countries in the following

THUNDERBOLT FIGHTER-BOMBER
The Republic P-47 Thunderbolt was produced in greater numbers than any other American fighter in World War II. A large, heavy single-seater, the P-47 proved effective both in air-to-air combat and as a ground-attack aircraft. It was flying in Thunderbolts that Francis S. Gabreski, the war's top American ace in Europe, recorded his 28 kills.

May to June, Yugoslavia and Greece in the spring of 1941, and the Soviet Union as far as the outskirts of Moscow by the end of that year. These German victories were, to an important degree, triumphs of air power. It was not until later in the war that the awesome destructive capacity of strategic bombing, imaginatively envisaged in the 1930s, would become a reality. But from the outset air power held the balance between victory and defeat. A country beaten in the air found its army and navy fighting at a hopeless disadvantage. The only serious setback that the Germans experienced in the first two years of the war was the failure to subdue Britain – a direct result of the Luftwaffe's inability to establish air supremacy over southern England in 1940.

STUKA PILOT
A German dive-bomber pilot sits in the cramped confines of his two-man cockpit – the rear gunner covered his back, and the aircraft.

German dominance

The dominance of the German air force early in the war was not simply a result of overwhelming superiority of numbers or high quality of equipment. Luftwaffe methods could be surprisingly unsophisticated – during the bombing of Warsaw, Ju 52 crews scattered incendiaries over the city by shovelling them out of the aeroplane's side door. In the early days of the war, Messerschmitt Bf 109s were not even fitted with radios, so the pilots communicated with one another by waggling their wings. But the Luftwaffe pilots' training and numerical and technical superiority gave them a clear edge over their opponents. And, above all, German tactics were supremely well judged in the direction of air power to affect the course of battle.

The experience of the Battle of France, in May to June 1940, came as a profound shock to British and French airmen and their commanders. Expecting a long-drawn-out contest in the manner of the Western Front in 1914–18, they found themselves outfought and outthought, facing abject defeat in a matter of days. The Germans achieved air superiority from the outset, destroying Allied aircraft by attacks on airfields, the effective deployment of flak guns, and air combat. German commanders with a firm grasp of the principle of concentration of force assigned substantial numbers of aircraft to key points on the battlefield, so that British and French airmen, scattered in "penny packets" along the front, were overwhelmed. The Allies' problems were compounded by the inferior performance of many of their machines. Aircraft such as the RAF's Fairey Battle light bomber and the French Morane-Saulnier M.S.406 fighter proved little more than cannon-fodder for the Luftwaffe's

AGENTS OF BLITZKRIEG
During the lightning offensives of the early years of the war, Junkers Ju 87 Stuka dive-bombers like these operated as flying artillery in support of the German panzers. Despite their fearsome reputation, Stukas were slow and vulnerable to enemy fighters.

Messerschmitts. The Hurricanes and Spitfires that were a match for the German fighters soon had to be withdrawn to Britain because there were no airfields left from which they could operate.

Whereas the RAF and the Armée de l'Air had failed to establish any effective system for co-ordinating air and ground operations, the Luftwaffe was integrated into an overall strategy of shock and mobility. Avoiding sterile disputes about whether or not an air force should be "independent" or subordinate to the other arms, the Luftwaffe used its aeroplanes sometimes in direct support of army operations and sometimes on wide-ranging interdiction duties shading into strategic attacks on factories and cities.

Flying artillery

The most striking aspect of the *Blitzkrieg* strategy was the use of aircraft as "flying artillery" in support of armoured and motorized forces. On the ground Luftwaffe liaison officers were assigned to panzer units to co-ordinate air attacks with army operations, while the Luftwaffe's logistical organization kept air units supplied with fuel and munitions as they moved to keep up with the rapidly advancing front line. The Germans showed a keen appreciation of the psychological factor in warfare, especially the demoralizing effect of air attack on ground troops. The Ju 87 Stuka, the most famous and feared aircraft of the war's *Blitzkrieg* phase, was in some ways a backwards aeroplane for its time, a poorly armed two-seater with a fixed undercarriage. But the wailing soundtrack of its siren in the attack dive struck terror into its victims. Its bombing was accurate enough to destroy communications links such as bridges and rail junctions, and concentrated in tight formation it could deliver a devastating attack on ground troops or on ships – as during the Battle of Norway in April 1940.

The Stuka pilots typically attacked in formation, rolling over and peeling off into the dive in an accelerating cascade behind the group commander. The Stuka dived at near to vertical, accelerating to over 480kph (300mph) before the air-brakes activated to stop it reaching a velocity that would cause it to break up. Enemy pilots and flak-gunners soon noticed that the Stuka was especially vulnerable pulling out of its dive. At that moment its speed was at its slowest, and the pilot was preoccupied with resetting the machine for level flight. The aircraft was in any case vulnerable to enemy fighters, flying around 160kph (100mph) slower and with poor manoeuvrability. As the Luftwaffe lost its ability to guarantee air superiority, the Stuka's usefulness declined.

A general lesson of the first years of the war was the vulnerability of all bombers to attack by fighters. During the Dunkirk evacuation in June 1940, it was not only Stukas but also He 111s and Ju 88s – the latter fast and agile enough to later become a night fighter – that suffered heavily under attack from RAF Spitfires and Hurricanes. It was a lesson confirmed by the Battle of Britain. The single-seat fighter ruled the air (at least by day) and, as in World War I, dogfights pitted fighter against fighter in aerial combat where split-second reaction times determined life or death.

Airborne invasions

In its support of army operations, the Luftwaffe experimented with using aircraft to deliver troops to the point of battle, either by parachute or glider. One of the most successful

GROUND SUPPORT
German ground crew carry out maintenance and refuelling on a Dornier Do 17. The Luftwaffe was superbly organized on the ground to support operations from airstrips on a rapidly moving battlefield.

BREATHLESS COMBAT

AIR COMBAT IN WORLD WAR II monoplane fighters was conducted at a speed that pushed a pilot's reaction times to the limits. In 1941, future author Roald Dahl was flying with the RAF in Greece when the Luftwaffe arrived in strength:

"Over Athens on that morning, I can remember seeing our tight little formation of Hurricanes all peeling away and disappearing among the swarms of enemy aircraft. They came from above and they came from behind and they made frontal attacks from dead ahead, and I threw my Hurricane around as best I could and whenever a Hun came into my sights, I pressed the button. It was truly the most breathless and in a way the most exhilarating time I have ever had in my life. I caught glimpses of planes with black smoke pouring from their engines. I saw the bright red flashes coming from the wings of the Messerschmitts as they fired their guns, and once I saw a man whose Hurricane was in flames climb calmly out onto a wing and jump off…"

Landing at his airfield after surviving this experience, Dahl found he was perspiring so heavily the sweat was dripping to the ground, and his hand was shaking so much he couldn't light a cigarette.

HURRICANE FORCE
Designed by Sydney Camm, the Hawker Hurricane was the RAF's first monoplane fighter and the mainstay of Fighter Command in the Battle of Britain. Although slower than a Messerschmitt, it was tighter in the turn, an advantage in a dogfight.

"Wherever I looked I saw an endless blur of enemy fighters whizzing towards me from every side…"

ROALD DAHL
RAF PILOT AND WRITER

early examples of this novel military tactic was the taking of the apparently formidable Belgian frontier fortress of Eben Emael on 10 May 1940. Forty-one Ju 52s, each towing a glider with a contingent of troops on board, flew over German territory towards the Belgian border under cover of darkness, guided by beacons spaced out on the ground below. The gliders were released at very first light and most landed on top of the fortress or alongside nearby strongpoints. Taken completely

by surprise, Eben Emael's garrison of over 1,000 soldiers soon meekly surrendered.

The most spectacular German airborne assault was the invasion of the Mediterranean island of Crete in May 1941 – the first exclusively airborne invasion in history. Some 5,000 paratroopers were dropped on the island, occupied by almost 30,000 British and Commonwealth troops. They seized and held the airfield at Maleme, where Luftwaffe

transport aircraft were then able to land with large-scale reinforcements and heavy equipment. However, even the operation at Crete showed the drawbacks of airborne assault. Many of the first wave of paratroops were killed either as they drifted defencelessly downwards or immediately on landing as they disentangled themselves from their parachutes. More determined action by the British forces could well have wiped them out. Success in Crete depended on aircraft also providing effective close air support and flying in fresh men and supplies to reinforce the initial attack.

Global conflict

By the end of 1941 the war had widened from a European conflict into a genuinely global struggle, with not only the Soviet Union but also the United States and Japan (already at war with China since 1937) entering the fray. There was a distinct difference in approaches to warfare on the two sides in the conflict. Germany, Italy, and

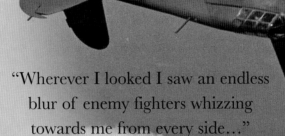

REICHSMARSCHAL GÖRING
Hitler invented the special title of Marshal of the Reich for his Great War ace Hermann Göring (front row, second from right) in the summer of 1940. But the morphine-addicted pilot later fell from favour as the Luftwaffe lost air superiority.

Japan all had regimes that encouraged a warrior spirit and praised war as a beneficent furnace in which men would be tempered to hardest steel. Their airmen were expected to triumph as the embodiment of the martial virtues – physical courage, ruthless aggression, patriotic self-sacrifice. Attitudes in the United States, Britain, and even the Soviet Union were more practical and pragmatic. The Soviets were second to none in the sacrifices they demanded of their people, but like the other Allied leaders Stalin knew the war would be won more by economic organization than by martial spirit. Victory in the air required great courage and skill from aircrews, untiring support from ground crews, the work of engineers constructing airfields, and the inventiveness of scientists and aircraft designers. But in the end the air war was won by industrial output.

Levels of productivity

The gulf in productivity between the major combatants was ultimately overwhelming. Japan and Italy simply did not have the industrial capacity or sophistication to keep up as the demands of aerial warfare intensified. The Germans in principle had both the factories and the expertise, but poor use was made of these assets. Disorganization, failure to commit resources, and bad decision-making meant that Germany only moved into top gear in 1943–44, by which time it was being asked to perform miracles under intensive air bombardment and with severe shortages of essential materials.

On the other side, Britain's air industry performed remarkably well – its ability to recover from the losses of front-line aircraft was one of the keys to the country's survival in the Battle of Britain. The achievements of the Soviet wartime aircraft industry would have been outstanding under any circumstances, but were doubly so given the conditions that prevailed after the German invasion of June 1941. Despite the primitive conditions and a lack of skilled personnel, raw materials, and machine tools, the Soviets turned out increasingly effective aircraft in remarkable quantities.

No one, however, could match the United States in mass production. The story of the US air industry in World War II is one of expansion at breakneck speed conducted with, on the whole, astonishing efficiency. The Douglas company offers a good example of the scale of output achieved. Where it had built fewer than 1,000 DC-3s up to 1941, during the war it made some 10,000 C-47s – the military transport

Junkers Ju 52/3m

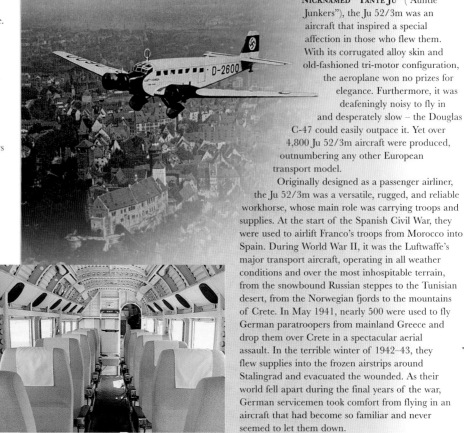

NICKNAMED "TANTE JU" ("Auntie Junkers"), the Ju 52/3m was an aircraft that inspired a special affection in those who flew them. With its corrugated alloy skin and old-fashioned tri-motor configuration, the aeroplane won no prizes for elegance. Furthermore, it was deafeningly noisy to fly in and desperately slow – the Douglas C-47 could easily outpace it. Yet over 4,800 Ju 52/3m aircraft were produced, outnumbering any other European transport model.

Originally designed as a passenger airliner, the Ju 52/3m was a versatile, rugged, and reliable workhorse, whose main role was carrying troops and supplies. At the start of the Spanish Civil War, they were used to airlift Franco's troops from Morocco into Spain. During World War II, it was the Luftwaffe's major transport aircraft, operating in all weather conditions and over the most inhospitable terrain, from the snowbound Russian steppes to the Tunisian desert, from the Norwegian fjords to the mountains of Crete. In May 1941, nearly 500 were used to fly German paratroopers from mainland Greece and drop them over Crete in a spectacular aerial assault. In the terrible winter of 1942–43, they flew supplies into the frozen airstrips around Stalingrad and evacuated the wounded. As their world fell apart during the final years of the war, German servicemen took comfort from flying in an aircraft that had become so familiar and never seemed to let them down.

NARROW CORRIDOR
The Ju 52/3m's long, narrow cabin, with a single row of seats on each side, had a maximum passenger capacity of 18. The original single-engined Ju 52s, which first flew in 1930, were used as civil transport craft.

Redecorated fin of modified Ju 52/3m shows its continuing popularity today

Modern radio antenna mast

Air intake for cabin ventilation

Corrugated alloy skin

Non-retractable tailwheel

A-702 HB-HOT

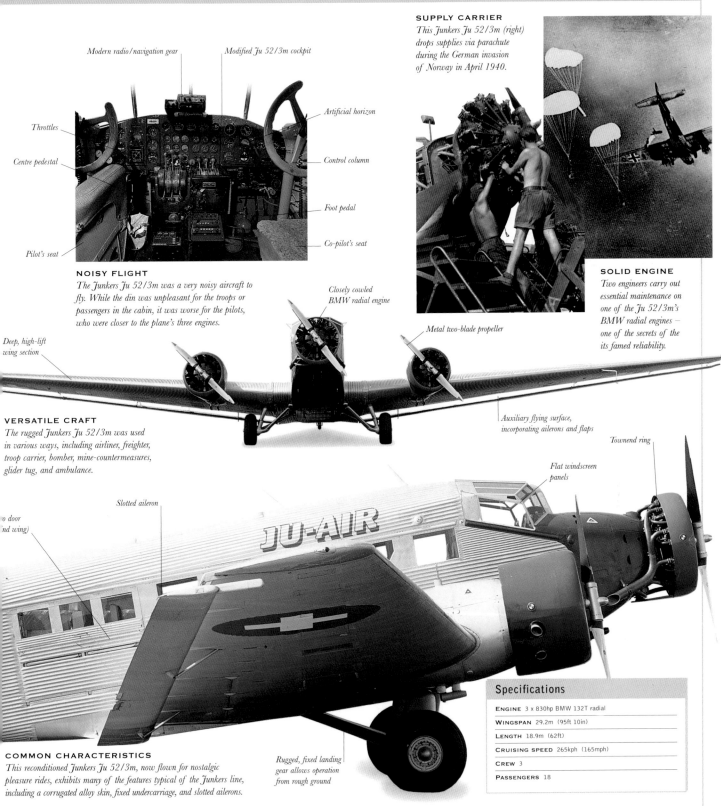

Modern radio/navigation gear

Modified Ju 52/3m cockpit

Throttles

Artificial horizon

Centre pedestal

Control column

Foot pedal

Pilot's seat

Co-pilot's seat

SUPPLY CARRIER
This Junkers Ju 52/3m (right) drops supplies via parachute during the German invasion of Norway in April 1940.

NOISY FLIGHT
The Junkers Ju 52/3m was a very noisy aircraft to fly. While the din was unpleasant for the troops or passengers in the cabin, it was worse for the pilots, who were closer to the plane's three engines.

Closely cowled BMW radial engine

Metal two-blade propeller

SOLID ENGINE
Two engineers carry out essential maintenance on one of the Ju 52/3m's BMW radial engines – one of the secrets of the its famed reliability.

Deep, high-lift wing section

VERSATILE CRAFT
The rugged Junkers Ju 52/3m was used in various ways, including airliner, freighter, troop carrier, bomber, mine-countermeasures, glider tug, and ambulance.

Auxiliary flying surface, incorporating ailerons and flaps

Townend ring

Flat windscreen panels

Slotted aileron

o door
nd wing)

JU-AIR

COMMON CHARACTERISTICS
This reconditioned Junkers Ju 52/3m, now flown for nostalgic pleasure rides, exhibits many of the features typical of the Junkers line, including a corrugated alloy skin, fixed undercarriage, and slotted ailerons.

Rugged, fixed landing gear allows operation from rough ground

Specifications

ENGINE	3 x 830hp BMW 132T radial
WINGSPAN	29.2m (95ft 10in)
LENGTH	18.9m (62ft)
CRUISING SPEED	265kph (165mph)
CREW	3
PASSENGERS	18

version of the aircraft. In all, Douglas produced almost 30,000 aircraft between 1941 and 1945. As well as expanding the output of aircraft manufacturers, the Americans turned automobile factories to aircraft production.

The result of rapid expansion might have been a sharp drop in quality, since large numbers of previously untrained workers had to be taken on. But the design of aircraft was intelligently modified to facilitate mass production and reduce the need for special skills in the workforce. The achievement of America's wartime aircraft industry did not come automatically from the United States being the world's leading industrial power. It was a feat of organization and applied intelligence that earned the success it deserved.

Although in retrospect the entry of the United States into the war in December 1941 doomed both Germany and Japan to eventual defeat, it took a long time to bring productivity and manpower to bear on the battlefield. The Americans and their Allies had to follow an arduous learning curve before their use of air power could match that of the Germans. In general, the Allies succeeded in sharply improving the performance of their aircraft and the training and experience of their pilots, while their enemies lost experienced pilots they could not adequately replace and often had to continue fighting in much the same aircraft with which they had started the war. In the spirit of "anything you can do, we can do better", the Allies learned to match and surpass the Germans in such areas as close air support and interdiction, and mount even bolder airborne offensives.

BUILDING FOR VICTORY
Over 18,000 four-engined Consolidated B-24 Liberator bombers rolled out of American factories during the war – an average of more than 10 a day. The ability to make aircraft in such unprecedented numbers was highly advantageous in terms of air superiority over enemies incapable of gearing up on the same scale.

BURMESE AIRDROP

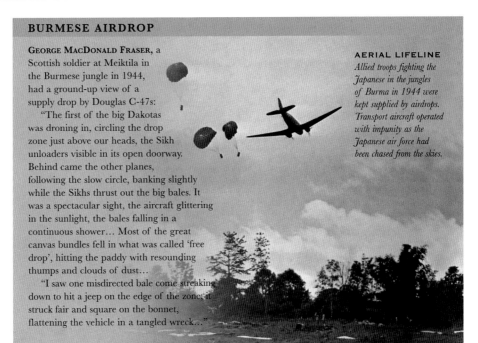

GEORGE MacDONALD FRASER, a Scottish soldier at Meiktila in the Burmese jungle in 1944, had a ground-up view of a supply drop by Douglas C-47s:

"The first of the big Dakotas was droning in, circling the drop zone just above our heads, the Sikh unloaders visible in its open doorway. Behind came the other planes, following the slow circle, banking slightly while the Sikhs thrust out the big bales. It was a spectacular sight, the aircraft glittering in the sunlight, the bales falling in a continuous shower... Most of the great canvas bundles fell in what was called 'free drop', hitting the paddy with resounding thumps and clouds of dust...

"I saw one misdirected bale come streaking down to hit a jeep on the edge of the zone; it struck fair and square on the bonnet, flattening the vehicle in a tangled wreck..."

AERIAL LIFELINE
Allied troops fighting the Japanese in the jungles of Burma in 1944 were kept supplied by airdrops. Transport aircraft operated with impunity as the Japanese air force had been chased from the skies.

War in Asia

One aspect of the widening war was the need for airmen and ground crews to operate in the world's most inhospitable terrains and climates. It was in the Sahara desert that the Allies learned the effective use of air power against ground forces during their battles against the Afrika Korps in 1942–43. Here, flying conditions were usually good, but keeping aircraft engines free of clogging sand was another matter and taxed the ingenuity of technicians.

The Burma-China theatre was more demanding still, becoming the site for some of the epics of World War II air combat. There, operating first out of Rangoon and then southern China, the Curtiss P-40s of Claire Chennault's truculently independent American Volunteer Group – better known as the Flying Tigers – inflicted heavy losses on numerically superior Japanese air forces in 1941–42. There also the remarkable guerrilla operations of General Orde Wingate's Chindit long-range penetration force were carried out in the Japanese-occupied Burmese jungle. Set up under Colonel Philip Cochran, the American Air Commando Group supported Wingate's imaginative forays with an array of aircraft, from gliders and C-47 transports to North American P-51 Mustang fighters and even a few early Sikorsky helicopters.

In their boldest operation, in March 1944, 67 of Cochran's Dakotas towed gliders carrying Chindit soldiers and American engineers, plus bulldozers and other heavy equipment, into Burma, releasing

them to fly down to jungle clearings. There they built landing strips so that transport aircraft could return the following night, landing troop reinforcements, artillery, jeeps, and more than a thousand mules. It did not work smoothly – half of the first wave of gliders was lost – but it did allow Wingate to, in his own words, "insert himself in the guts of the enemy".

FLYING SHARK
This distinctive sharkmouth decoration was adopted by 112 Squadron of Britain's Royal Air Force (RAF) in North Africa in the second half of 1941.

The American and British transport aircraft based in northern India not only had to supply Allied troops, but also, after the Japanese cut off the "Burma Road", became the only means of supplying the Chinese Nationalists and supporting US forces holed up in southwest China. Between 1942 and 1945, every single vehicle, weapon, round of ammunition, or drum of fuel was delivered "over the hump" from Dinjan in Assam across the Himalayas to Kunming. This was generally recognized as the most demanding route flown by transport aircraft anywhere during the war – a journey of 800km (500 miles) that took the C-47s, C-54s, and other transport planes over mountain ridges more than 5,000m (16,000ft) high.

If they hugged the mountain valleys, pilots ran into violent turbulence; if they put on their oxygen masks and flew high, they risked severe icing. During the monsoon season, dense cloud engulfed mountain ridges and valleys alike; pilots mostly flew on instruments, without sight of the ground, for the entire length of the journey. Flying the hump cost hundreds of aircrew their lives – the route was strewn with the wreckage – yet some 660,000 tonnes of supplies were delivered by the war's end. Nor was the experience necessarily a grim one for airmen. One RAF pilot recorded the exhilaration of flying a Dakota over the Himalayan peaks on a moonlit night, with his radio tuned to a fine programme of classical music from the BBC.

FLYING THE HUMP
A C-47 flies the India-to-China supply route over the Himalayas. Operating a transport service across some of the world's most unforgiving terrain was an unprecedented challenge. At the cost of many lives Allied airmen proved that it could be done, keeping the Chinese Nationalist forces supplied with food, fuel, and munitions.

Russian winter

The most inhospitable environment in which any fliers had to operate was Russia in winter. The Luftwaffe, like the German army in general, was poorly prepared for the challenge of temperatures as low as -50°C (-58°F). Hard snow did not make a bad landing surface, but keeping aircraft operational was a nightmare. Fuel tanks and engine lubricant froze, hydraulic pumps broke down, rubber tyres went brittle and cracked, flight instruments and radios refused to function. Often as few as one in four aircraft was fit to fly.

Yet under these conditions the Luftwaffe achieved some notable feats of air supply. At Demyansk in the winter of 1941–42, an army of 100,000 soldiers that had been encircled by the Soviets was kept supplied for three months by a fleet of Ju 52s and bombers, pressed into service as transports. But the following winter Hitler's demand that the Luftwaffe repeat this achievement at Stalingrad was simply beyond its capability.

The 250,000 men of General Paulus' Sixth Army, trapped in Stalingrad by encircling Soviet forces in November 1942, needed an airlift of at least 300 tonnes of supplies a day to survive. At best the Luftwaffe managed a third of this total, and on many days they could deliver nothing at all.

Stalingrad airlift

For German airmen the Stalingrad airlift developed into an epic of personal heroism, collective suffering, and organizational chaos. Despite the bitter cold, both ground and aircrews lived in makeshift unheated shelters, alongside airstrips that came under repeated Soviet air attack. The hastily assembled transport fleet consisted chiefly of Ju 52s and Heinkel 111 bombers, but also included a motley collection of training and communications aircraft, and even 18 four-engined Focke-Wulf Condors. Working in the open in the snow and ice, ground crews struggled around the clock to make these aircraft flyable – on occasion mechanics were frozen fast to an engine as they tried to service it. Pilots routinely took off and landed in almost zero visibility. During the flight to Stalingrad's Pitomnek airstrip and back, the transport aircraft were harassed by Soviet fighters and flak. They sometimes landed under artillery fire, dodging wrecked aircraft and craters. Harrowing incidents abounded: the transport packed with wounded soldiers that crashed immediately after take-off, apparently because the wounded had slid to the back of the plane as it rose; or the last-minute evacuation of Tazinskaya airstrip as it was overrun by Soviet tanks, in which a third of a fleet of 180 Ju 52s were lost, many crashing into their own colleagues as escape descended into chaos.

WINTER WAR
Soviet ground-crew members load up a bomber under the trying conditions of the Russian winter. The aircraft is a British-built Handley Page Hampden twin-engined bomber in Soviet markings.

PARACHUTES AND GLIDERS

ONE OF THE MOST IMPRESSIVE sights in the war was the departure for a major airborne operation, the air thick with hundreds of transport aircraft carrying parachutists and towing gliders. Unfortunately, this vision of order and power had a tendency to degenerate into something approaching chaos over the drop zone. Pilot Pierre Clostermann described the Allied airborne assault on the east bank of the Rhine in March 1945 as "an apocalyptic spectacle. Thousands of white parachutes dropped through an inferno of heavy, medium, and light flak, while Dakotas crashed in flames and gliders rammed high-tension cables in showers of blue sparks."

German successes early in the war convinced the Americans and British of the value of airborne operations. But airborne assaults were technically difficult to carry off. A man floating down on a parachute was desperately vulnerable if surprise had not been achieved. Despite rigorous training, paratroops were often widely scattered by the time they reached the ground and separated from their equipment. When the enemy reacted, the paratroops either needed to join up rapidly with friendly forces advancing on the ground or be backed up by close air support and aerial resupply.

Gliders could carry in equipment such as jeeps and light artillery as well as troops. A glider had a pilot and co-pilot sitting side by side, communicating with the "tug" pilot by a telephone wire in the tow rope. They did not have an easy job. If there was turbulence, or the glider inadvertently strayed into the tug's slipstream, the tow rope could break, leaving the glider to an uncertain fate. Those that reached the target zone had to land at upwards of 110kph (70mph) on whatever clear strip of land they could locate. Inevitably landing accidents were common.

Allied large-scale airborne operations – for example, during the D-Day landings and in the ill-fated Operation Market Garden in September 1944 – rarely went according to plan, but they sometimes proved useful.

"Thousands of white parachutes dropped through an inferno of flak, while Dakotas crashed in flames."

PILOT PIERRE CLOSTERMANN
AIRBORNE ASSAULT ON THE RHINE, 1945

OPERATION MARKET GARDEN
On 17 September 1944 some 10,000 Allied airborne troops were sent into German-occupied Holland by parachute and glider to take key bridges over the rivers Maas, Waal, and Rhine. It fell short of its full objectives because of the failure to capture the heavily defended bridge at Arnhem.

PARADROPPABLE MOTORCYCLE
During World War II, containers holding weapons, ammunition, radios, and even motorcycles such as this one were parachuted in to supply troops. Often even larger vehicles were dropped, including jeeps, which needed up to four parachutes to land undamaged.

GETTING AIRBORNE
British paratroops stoop to enter a Hotspur glider during a training exercise in 1942. Airborne troops, who quickly earned a reputation as elite units, added an extra dimension to warfare, allowing generals to insert forces behind enemy lines, seizing key points such as bridges or airfields.

In all, between November 1942 and Paulus'
surrender at the end of January 1943, the
Luftwaffe lost 490 machines, including 266 Ju 52s
and 165 He 111s.

The Eastern Front

From 1943 onwards, the Luftwaffe suffered an
ultimately unsustainable rate of attrition on the
eastern front and during the Allied strategic
bombing offensive over Germany. Because
German aircraft production belatedly expanded
under the direction of new Nazi industrial chief
Albert Speer, German airmen were able to
continue inflicting heavy losses on their enemies,
but it was a losing battle against mounting odds.

The Germans on the whole sent their less
experienced pilots and more obsolescent aircraft
to the Eastern Front, reserving the latest and best
for the defence of Germany. Yet Luftwaffe pilots in
Russia recorded astonishing kill rates: top ace Erich
Hartmann claimed 352 victories, and six other
German pilots are credited with over 200 each.
However, Soviet fighters were increasingly a match
for their enemy – the Yakovlev Yak-9 could hold its
own against Focke-Wulfs and Messerschmitts and
denied the Luftwaffe superiority over the battlefield.

"Tank-busting" aircraft on both sides – the
Soviet Il-2 Shturmovik and Lavochkin La-5, and
the Luftwaffe's Henschel Hs 129s and
Ju 87 Stukas fitted with armour-piercing
cannon – played a major part in the great tank
battle of Kursk in July 1943 and in the other
armoured clashes that followed as the Soviets
rolled the German armies back. A single Stuka
pilot, Hans-Ulrich Rudel, was reckoned to have
destroyed more than 500 Soviet tanks. But the
German panzers also suffered grievously under
air attack, and the Soviets were better able to
replace lost armour. It was the same story in
the struggle for the air. In all, the Luftwaffe
claimed to have shot down 44,000 Soviet
aircraft, yet German aeroplanes were still
outnumbered on the Eastern Front
throughout the later stages of the war.

The D-Day landings

In the west, air power was central to the
success of the D-Day invasion of
Normandy in June 1944 and
the Allied drive across Europe
that followed. At the start of the war
the American and British air forces
were ill-prepared for providing direct
support to ground forces. Even if they accepted
the goal of achieving air superiority over the
battlefield, they were little interested in providing
air support. Senior air commanders, who resented
the idea of being at the service of the army,
had done nothing to develop techniques for

Yakovlev Yak-3

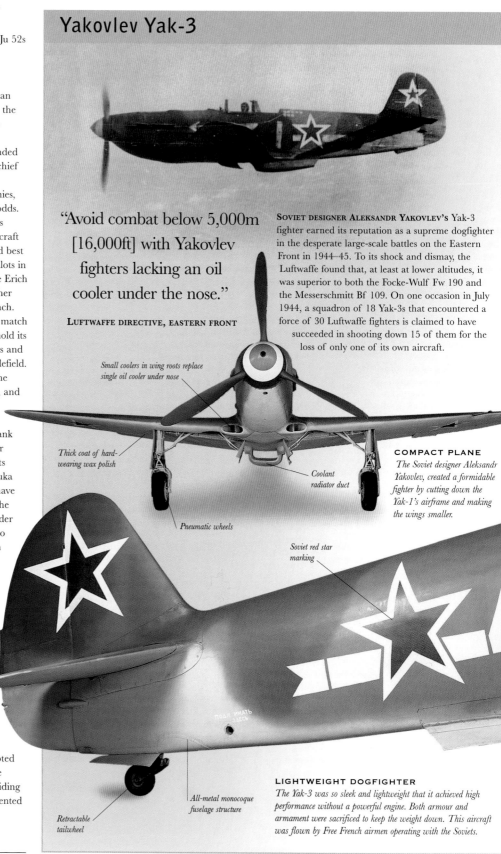

"Avoid combat below 5,000m
[16,000ft] with Yakovlev
fighters lacking an oil
cooler under the nose."

LUFTWAFFE DIRECTIVE, EASTERN FRONT

SOVIET DESIGNER ALEKSANDR YAKOVLEV'S Yak-3
fighter earned its reputation as a supreme dogfighter
in the desperate large-scale battles on the Eastern
Front in 1944–45. To its shock and dismay, the
Luftwaffe found that, at least at lower altitudes, it
was superior to both the Focke-Wulf Fw 190 and
the Messerschmitt Bf 109. On one occasion in July
1944, a squadron of 18 Yak-3s that encountered a
force of 30 Luftwaffe fighters is claimed to have
succeeded in shooting down 15 of them for the
loss of only one of its own aircraft.

Small coolers in wing roots replace
single oil cooler under nose

Thick coat of hard-
wearing wax polish

Coolant
radiator duct

Pneumatic wheels

COMPACT PLANE
The Soviet designer Aleksandr
Yakovlev, created a formidable
fighter by cutting down the
Yak-1's airframe and making
the wings smaller.

Soviet red star
marking

All-metal monocoque
fuselage structure

Retractable
tailwheel

LIGHTWEIGHT DOGFIGHTER
The Yak-3 was so sleek and lightweight that it achieved high
performance without a powerful engine. Both armour and
armament were sacrificed to keep the weight down. This aircraft
was flown by Free French airmen operating with the Soviets.

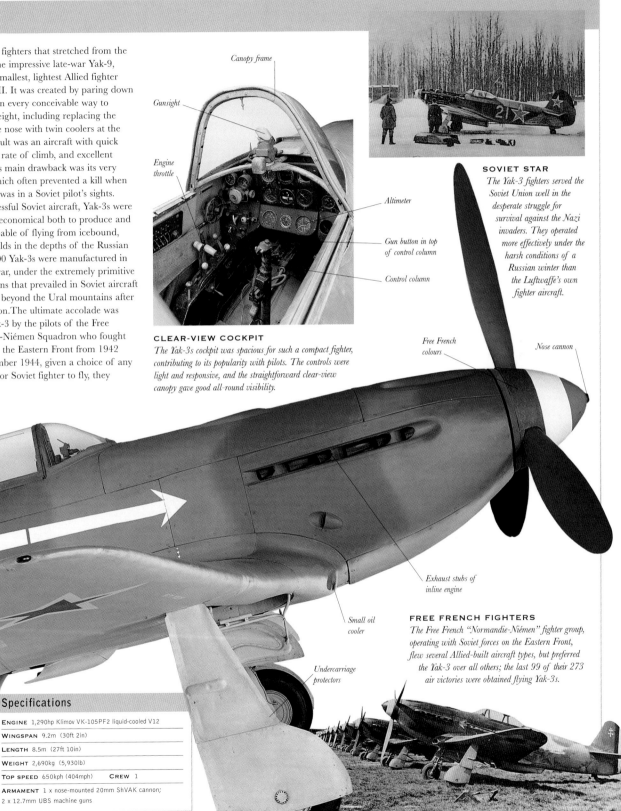

One of a family of fighters that stretched from the Yak-1 of 1939 to the impressive late-war Yak-9, the Yak-3 was the smallest, lightest Allied fighter flown during WW II. It was created by paring down the original Yak-1 in every conceivable way to reduce drag and weight, including replacing the oil cooler under the nose with twin coolers at the wing roots. The result was an aircraft with quick acceleration, a fine rate of climb, and excellent manoeuvrability. Its main drawback was its very light armament, which often prevented a kill when a Luftwaffe fighter was in a Soviet pilot's sights.

Like many successful Soviet aircraft, Yak-3s were no-frills machines, economical both to produce and to operate, and capable of flying from icebound, snow-covered airfields in the depths of the Russian winter. Almost 4,900 Yak-3s were manufactured in the course of the war, under the extremely primitive and trying conditions that prevailed in Soviet aircraft factories, relocated beyond the Ural mountains after the German invasion. The ultimate accolade was awarded to the Yak-3 by the pilots of the Free French Normandie-Niémen Squadron who fought with the Soviets on the Eastern Front from 1942 onwards. In September 1944, given a choice of any American, British, or Soviet fighter to fly, they picked the Yak-3.

Canopy frame

Gunsight

Engine throttle

Altimeter

Gun button in top of control column

Control column

SOVIET STAR
The Yak-3 fighters served the Soviet Union well in the desperate struggle for survival against the Nazi invaders. They operated more effectively under the harsh conditions of a Russian winter than the Luftwaffe's own fighter aircraft.

Free French colours

Nose cannon

CLEAR-VIEW COCKPIT
The Yak-3s cockpit was spacious for such a compact fighter, contributing to its popularity with pilots. The controls were light and responsive, and the straightforward clear-view canopy gave good all-round visibility.

Single-seater clear-view canopy

Exhaust stubs of inline engine

Small oil cooler

FREE FRENCH FIGHTERS
The Free French "Normandie-Niémen" fighter group, operating with Soviet forces on the Eastern Front, flew several Allied-built aircraft types, but preferred the Yak-3 over all others; the last 99 of their 273 air victories were obtained flying Yak-3s.

Undercarriage protectors

Coolant radiator duct

Specifications

ENGINE	1,290hp Klimov VK-105PF2 liquid-cooled V12
WINGSPAN	9.2m (30ft 2in)
LENGTH	8.5m (27ft 10in)
WEIGHT	2,690kg (5,930lb)
TOP SPEED	650kph (404mph) **CREW** 1
ARMAMENT	1 x nose-mounted 20mm ShVAK cannon; 2 x 12.7mm UBS machine guns

co-operation between the forces. Such techniques eventually evolved in action, first in the North African desert and then in Italy after the Allied invasion of 1943. These experiences left the Allies reasonably well prepared for the large-scale use of aircraft in support of the armies in Normandy.

Without air superiority the D-Day landings could not even have been attempted. As it was, Allied fighters largely insulated their ground forces and shipping from air attack, while the first wave of the invasion included the flying in of

three airborne divisions by parachute and glider. Allied bombers and fighter-bombers had destroyed bridges and other communication links in northwest France so comprehensively that the region was virtually cut off. German attempts to move in reinforcements and supplies encountered constant harassment from marauding aircraft, until movement was hardly possible. On the front line, heavy bombers were used for the first time to "carpet-bomb" enemy positions in preparation for an offensive. By 1944 no artillery barrage could

match the awesome quantity of explosives delivered by a bomber squadron.

Close air support was not without its problems. Determined efforts were made to bring air power to bear on the right targets at the right time, but under battle conditions this was never easy. Forward air controllers, either on the ground or airborne in light aircraft over the front (another innovation of this period), would call in and direct air-strikes, while various indicators such as smoke or flares were used to help identify targets. But

POWERFUL TYPHOON
Introduced as a fighter in 1941, the RAF's Hawker Typhoon was initially a flop, suffering a spate of engine and structural failures. Later in the war it was switched to a ground-attack role with devastating effect, proving ruggedly resistant to enemy fire and packing a powerful payload of bombs and air-to-ground rockets.

NORMANDY LANDINGS
American troops wade ashore on D-Day, 6 June 1944. This vast amphibious operation would have been impossible to attempt without air superiority, and extensive bombing of bridges, roads, and railways in northern France before the invasion made it impossible for the Germans to rush reinforcements to Normandy.

ROCKET ATTACK
This is the view from an RAF Typhoon as it fires a rocket at German vehicles on a road in Normandy in 1944. The Typhoons became renowned for the tank-busting power of their rocket attacks, though with unguided weapons the odds were always against a direct hit.

response times were often too slow and accuracy lacking. Time and again, Allied aircraft hit their own troops instead of the enemy – making soldiers inclined to fire on any aeroplane, whether "friendly" or not. Airmen were sometimes made to spend time with troops in the front line to gain an appreciation of the ground forces' perspective, but it made little difference.

But this is not to deny the impact ground-attack aircraft had on the battles in western Europe in 1944–45. Fighter-bombers such as the Hawker Typhoon and P-47 Thunderbolt earned a fearsome reputation for their striking power with machine guns, cannon, bombs, and rockets. Napalm was also part of the ground-attack armoury. As on the eastern front, even heavily armoured tanks proved vulnerable to air attack. The final German counter-attack in the Ardennes at Christmas 1944 was only possible because bad weather prevented Allied aircraft from operating.

As Allied fighters and fighter-bombers ranged over enemy-held areas – preying on trains, attacking airfields, shooting up convoys of trucks – they rarely suffered substantial losses at the hands of the heavily outnumbered German aircraft. But flak took a severe toll. Flying in at low altitude to strafe or bomb a target defended by AA guns required nerve and luck. Apart from

the chances of being hit by enemy fire, there was always a risk of flying straight into the target, or into a building or pylon. If the target exploded – because it was an ammunition truck, for example – the pilot might find himself careering through flying slabs of road or pieces of chassis. Most airmen consistently preferred the active challenge of air-to-air combat to the sense of passive vulnerability they felt in the face of ground fire.

Overwhelming air power

By the end of the war, both the German and Japanese air defences had been totally overwhelmed by Allied air power. The scale of Allied air operations was quite phenomenal. In the American and British strategic air offensive in Europe, thousand-bomber raids became commonplace. The first day of the Arnhem airborne offensive in September 1944 involved more then 4,000 Allied transport aircraft, fighters, bombers, and gliders. In March 1945, during its final drive on Berlin, the Soviet air force carried out over 17,500 flights in a single day. Throughout the whole period of conflict, the United States had built almost

BOLT FROM THE BLUE
American ground crew service a P-47 Thunderbolt in England in 1943. Under the fuselage is a drop tank, providing extra fuel for increased range. The engine cowling was painted white to avoid confusion with the rather similar German Focke-Wulf Fw 190, which put Thunderbolts at risk from friendly fire.

300,000 military aircraft, and pilots were trained by the hundreds of thousands. The war left a sorrowful legacy of destruction, much of it caused by the deployment of air power. But another legacy of the war was flight on an unprecedented scale.

VIEW TO A KILL

FRENCH PILOT PIERRE CLOSTERMANN flew both with the Free French Air Force and the RAF during World War II. Clostermann described leading an attack by four Tempest fighter-bombers on a train in the winter of 1944 – a kind of operation he said was one of "those inhuman, immoral jobs we had to do because... war is war":

"The four Tempests slid down to 3,000ft [900m] in the frozen air and their polished wings caught the first gleams of a dingy dawn. We obliqued towards the train and instinctively four gloved hands, benumbed by the cold, were already pushing the prop lever to fine pitch. We could now make out the locomotive and the flak truck in front of it and the interminable mixed train dragging painfully behind.

"Without dropping our auxiliary tanks, we went into a shallow dive at full throttle... 350... 380... 420... 450mph [550... 600... 675... 725kph]. The blood throbbed in my parched throat – still that old fear of flak. Only about a mile or two [1.5–3km] now. I began to set my aim for about 20 yards [18m] in front of the locomotive.

"Now! I leant forward, tensed. Only 800 yards [730m]. The first burst of tracer – the staccato flashes of the quadruple 20-mm flak mounting – the locomotive's wheels skidding with all brakes jammed on... I was skimming over the snow-covered furrowed fields. Rooks flew off in swarms.

My cannon roared – the engine driver jumped out of his cabin and rolled into the ditch. My shells exploded on the embankment and perforated the black shape which loomed in my sights.

"Then the funnel vomited a hot blast of flame and cinders, enveloped in the steam escaping from the punctured pipes. A slight backward pressure on the stick to clear the telegraph wires, a quick dive through the smoke, then, once again, the sky in my windshield, covered with oily soot... A glance backwards. The locomotive had disappeared, shrouded in soot and spurting steam. People were scrambling out of doors and tearing down the embankment like agitated ants."

FRENCH FIGHTER ACE
French fighter pilot Pierre Clostermann settles into the cockpit of his Hawker Tempest, marked with a cross for each of his kills. Clostermann regarded the Tempest as the best Allied fighter aircraft of the war.

"Strafing of trains in the grey dawn... those inhuman, immoral jobs we had to do... because war is war."

PIERRE CLOSTERMANN
THE BIG SHOW

WWII FIGHTERS AND FIGHTER-BOMBERS

WWII FIGHTER AIRCRAFT were asked to perform in a variety of roles. Fighters had to battle for air superiority with their opposite numbers on the enemy side; act as interceptors against enemy bombers; fulfil a ground attack role in support of armies; act as bomber escorts; and operate as night fighters. Although many aircraft worked well in several roles, none could excel at them all. In general the best air-superiority fighters were light single-seaters, fast in a dive and tight in a turn, such as the Spitfire and Messerschmitt Bf 109. The interceptor role suited aircraft that provided a stable gun platform with heavy firepower, such as the Hurricane. Heavier fighters that could take punishment and carry a substantial armament – such as the Hawker Typhoon and P-47 Thunderbolt – excelled at ground attack. The best night fighters were two- or three-seaters, because the pilot needed someone to operate complex radio and radar equipment. Escort fighters required the range to accompany bombers to their targets plus the fighting ability to see off enemy interceptors. The North American Mustang was peerless in this role.

SUPER SPITFIRE
The Supermarine Spitfire first entered service with the RAF in 1938 and remained in production during WWII. See pages 210–11.

Bell P-39D Airacobra

The American Airacobra is unusual in having the engine behind the pilot. This allowed room for the 20mm or 37mm cannon to fire through the propeller shaft. Although the P-39's performance was poor at high altitude, it was very effective at ground attack, making it a valuable fighter on the Eastern Front.

ENGINE	1 x 1,150hp Allison V-1710-35 liquid-cooled V-12 cylinder	
WINGSPAN	10.4m (34ft)	LENGTH 9.2m (30ft 2in)
TOP SPEED	592kph (368mph)	CREW 1
ARMAMENT	1 x 37mm M4 cannon firing through propeller hub, 2 x .5in nose-mounted and 4 x .3in wing-mounted Browning machine guns	

Bristol Blenheim IV

The Bristol Blenheim bomber entered RAF service in 1938. However, aircraft were developing so fast that it was already outclassed at the start of WWII. Despite this, Blenheim VIs were used to strike at targets in German-occupied Europe during 1940. The fighter version was the first aircraft to be fitted with airborne radar, and it formed the core of the night fighting force during 1940–41.

ENGINE	2 x 920hp Bristol Mercury XV 9-cylinder air-cooled radial	
WINGSPAN	17.7m (56ft 4in)	LENGTH 13m (42ft 7in)
TOP SPEED	428kph (266mph)	CREW 3
ARMAMENT	4 x .303in Browning machine guns (fighter version carried four extra guns under fuselage); 454kg (1,000lb) bombload	

Browning machine gun

Aerial mast

Curtiss P-40E Warhawk

The US's P-40 was already in production in 1939, for supply to Britain and France, but in combat it was outclassed by almost every other fighter. However, the RAF and USAAF learned to exploit its strengths, using it for close air-support of ground troops.

ENGINE	1,150hp Allison V-1710-39 liquid-cooled V-12 cylinder	
WINGSPAN	11.4m (37ft 4in)	LENGTH 9.5m (31ft 2in)
TOP SPEED	539kph (335mph)	CREW 1
ARMAMENT	6 x .5in wing-mounted machine guns; 1 x 227kg (500lb) and 2 x 45kg (100lb) bombs	

Focke-Wulf Fw 190A

In service from 1941 onwards, the streamlined, radial-engined Fw 190 was fast, strong, and heavily armed, with good all-round vision and excellent ground handling. It was superior in every way to the Messerschmitt Bf 109, which, however, it never replaced. In fact, at several stages during WWII the Fw 190 was better than the existing Allied fighters and was certainly the best fighter which Germany produced. It was easily adapted for "hit and run" bombing, ground attack, torpedo attack, tactical reconnaissance, and night fighting.

ENGINE	1,700hp BMW 801 Dg. air-cooled 18-cylinder two-row radial	
WINGSPAN	10.5m (34ft 5in)	LENGTH 8.8m (29ft)
TOP SPEED	653kph (408mph)	CREW 1
ARMAMENT	2 x 13mm machine guns, 4 x 20mm cannon in wings; 1 x 500kg (1,100lb) bomb	

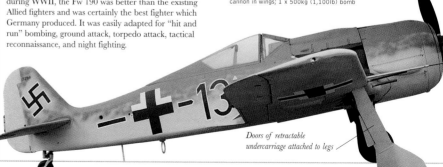

Doors of retractable undercarriage attached to legs

Hawker Hurricane IIB

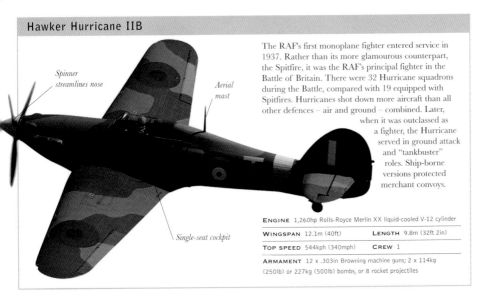

Spinner streamlines nose

Aerial mast

Single-seat cockpit

The RAF's first monoplane fighter entered service in 1937. Rather than its more glamourous counterpart, the Spitfire, it was the RAF's principal fighter in the Battle of Britain. There were 32 Hurricane squadrons during the Battle, compared with 19 equipped with Spitfires. Hurricanes shot down more aircraft than all other defences – air and ground – combined. Later, when it was outclassed as a fighter, the Hurricane served in ground attack and "tankbuster" roles. Ship-borne versions protected merchant convoys.

ENGINE	1,260hp Rolls-Royce Merlin XX liquid-cooled V-12 cylinder	
WINGSPAN	12.1m (40ft)	LENGTH 9.8m (32ft 2in)
TOP SPEED	544kph (340mph)	CREW 1
ARMAMENT	12 x .303in Browning machine guns; 2 x 114kg (250lb) or 227kg (500lb) bombs, or 8 rocket projectiles	

Hawker Typhoon IB

Designed to have twice the power of the previous generation of British fighters, the Typhoon was rushed into production in 1941 and suffered a series of structural and engine failures. Worse was the fact that its performance at high altitude was poor. However, it was extremely fast at low level and, with its heavy armament, proved to be a devastating ground attack aircraft, playing a valuable "tank-busting" role during the 1944 Normandy landings.

ENGINE	2,180hp Napier Sabre II liquid-cooled H-24 cylinder	
WINGSPAN	12.7m (41ft 7in)	LENGTH 9.7m (31ft 11in)
TOP SPEED	664kph (412mph)	CREW 1
ARMAMENT	4 x 20mm Hispano cannon; 8 rocket projectiles or 2 x 227kg (500lb) bombs under wings	

Hawker Tempest V

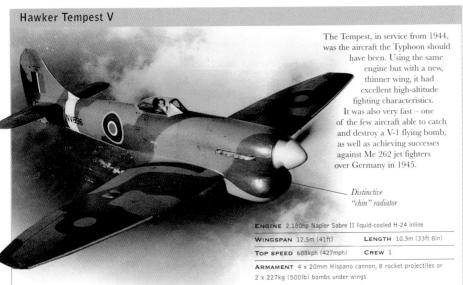

The Tempest, in service from 1944, was the aircraft the Typhoon should have been. Using the same engine but with a new, thinner wing, it had excellent high-altitude fighting characteristics. It was also very fast – one of the few aircraft able to catch and destroy a V-1 flying bomb, as well as achieving successes against Me 262 jet fighters over Germany in 1945.

Distinctive "chin" radiator

ENGINE	2,180hp Napier Sabre II liquid-cooled H-24 inline	
WINGSPAN	12.5m (41ft)	LENGTH 10.3m (33ft 8in)
TOP SPEED	688kph (427mph)	CREW 1
ARMAMENT	4 x 20mm Hispano cannon, 8 rocket projectiles or 2 x 227kg (500lb) bombs under wings	

Henschel Hs 129B-1/R2

The Hs 129 was a ground attack aircraft with an armoured cockpit and very heavy armament, designed to destroy tanks. The first model was grossly underpowered, and, as no other German engines were available, it was fitted with engines from Vichy France after 1940. These were unreliable and vulnerable to battle damage. The aircraft operated with some success on the Eastern Front, after being fitted with 37mm, and, even the huge 75mm anti-tank guns, in an effort to counter the thickly armoured Russian tanks.

ENGINE	2 x 700hp Gnome-Rhone 14M 4/5 14-cylinder radials	
WINGSPAN	14.2m (46ft 7in)	LENGTH 9.8m (32ft)
TOP SPEED	407kph (253mph)	CREW 1
ARMAMENT	2 x 7.9mm MG17 machine guns, 2 x 20mm MG151 cannon, 1 x 30mm MK101 cannon	

Ilyushin Il-2 M3 Shturmovik

The Shturmovik or armoured attack aircraft was a unique Soviet design and played a decisive role on the Eastern Front, with over 37,000 built. The armour-plated fuselage enclosed the engine, pilot, and fuel, allowing it to attack German tanks through a hail of small arms fire.

ENGINE	1,750hp Mikulin AM-38F liquid-cooled V-12 cylinder	
WINGSPAN	14.6m (47ft 11in)	LENGTH 11.6m (38ft 11in)
TOP SPEED	404kph (251mph)	CREW 2
ARMAMENT	2 x 23mm cannon, 2 x 7.62mm and 1 x 12.7mm machine guns; 8 rockets or 600kg (1,323lb) bombload	

Junkers Ju 87D-5

The German technique of armoured warfare involved close support of the tanks by dive-bombers ("Stukas"). This contributed to the success of *Blitzkrieg* in Poland, France, and, initially, in Russia, and the Ju 87 was regarded as a wonder weapon. But it was slow and vulnerable on the way to its target and modern fighters could easily destroy it. Ju 87s were hastily withdrawn from the Battle of Britain, after whole formations were shot down. As the Allies gained air superiority, the Stuka became increasingly ineffective.

ENGINE	1,400hp Junkers 211J liquid-cooled V-12 cylinder	
WINGSPAN	15m (49ft 3in)	LENGTH 22.5m (73ft 9in)
TOP SPEED	411kph (255mph)	CREW 2
ARMAMENT	1 x 7.9mm MG 81Z twin machine gun, 2 x 20mm MG 151/20 wing-mounted cannon; up to 1,800kg (4,000lb) bombload	

Lavochkin La-5 FN

Most Soviet combat aircraft of the early war years were built entirely of wood, as light aircraft alloys were scarce. The La-5, with this weight penalty, was not quite a match for the speed of the Messerschmitt Bf 109G, but it was still extremely manoeuvrable. After the German invasion of the USSR in 1941, the Russians put enormous effort into expanding aircraft production. It was obvious that not only

sheer numbers, but also improved designs, were required to combat the skilful and well-equipped Luftwaffe. Yakovlev fighters were initially the most successful, while the LA-5, by contrast, was too slow. In 1942, The FN variant was re-engined, transforming its performance and making it able to outperform even the Focke-Wulf Fw 190.

Wooden fuselage

ENGINE	1,700hp Shvetsov M-82FN air-cooled 14-cylinder radial	
WINGSPAN 9.8m (32ft 2in)		LENGTH 8.5m (27ft 11in)
TOP SPEED 650kph (403mph)		CREW 1
ARMAMENT 2 x 20mm ShVAK cannon above engine		

Lockheed P-38J Lightning

The Lockheed Lightning, designed for long-range missions, was the first fighter aircraft fitted with turbochargers and a tricycle undercarriage. It was complex and expensive to produce, and initially most went to the Pacific theatre where long flights over water made the twin-engine arrangement desirable. Their most famous action there was in 1943, when 16 aircraft, operating 885km (550 miles) from their base, shot down the Japanese Commander-in-Chief, Admiral Yamamoto. Many Lightnings reached Europe in 1944 and were used by the tactical air forces for ground attack, and photographic reconnaissance.

1,425hp Allison engine with turbocharger

Twin-fin stabilisers

Twin-boom fuselage

USAF markings

ENGINE	2 x 1,425hp Allison V-1710 V-12 cylinders with turbochargers	
WINGSPAN 15.9m (52ft)		LENGTH 11.5m (37ft 10in)
TOP SPEED 666kph (414mph)		CREW 1
ARMAMENT 1 x 20mm Hispano cannon and 4 x .5-in Browning machine guns; 1,452kg (3,200lb) bombload or 10 rocket projectiles		

Macchi M.C.202 Folgore

The Macchi M.C.200 Saetta was the main Italian fighter in 1940. It was a good dogfighter and able to perform well against the Hurricane, but it was underpowered. This disadvantage became crucial when faced with improving Allied types, and from 1941 the aircraft was fitted with a German engine – at first imported, and later licence-built – to become the M.C.202 Folgore ("Thunderbolt"). Still not quite up to Allied or German standards, the Folgore served in North Africa, Sicily, and Russia until the Italian surrender in 1943.

ENGINE	1,200hp Daimler-Benz DB 601A liquid-cooled V-12 cylinder	
WINGSPAN 10.6m (34ft 9in)		LENGTH 8.9m (29ft)
TOP SPEED 594kph (369mph)		CREW 1
ARMAMENT 2 x 12.7mm and 2 x 7.7mm Breda-SAFAT MGs		

Messerschmitt Bf 110 C-5

Designed as a heavy, long-range fighter to protect strategic bombers, the Bf 110 was first put to the test in the Battle of Britain. It proved to be no match even for the Hurricane, which could outmanoeuvre it easily, and had to be protected, in turn, by Bf 109s. In other theatres it was very effective as a ground-attack aircraft, as long as there was no fighter opposition. It came into its own as a night fighter – fitted with airborne radar – from 1942 onwards, inflicting significant losses on RAF Bomber Command raids over Germany.

ENGINE	2 x 1,100hp Daimler-Benz DB 601A-1 liquid-cooled V-12 cylinder	
WINGSPAN 16.2m (53ft 5in)		LENGTH 12.1m (39ft 9in)
TOP SPEED 541kph (336mph)		CREW 2
ARMAMENT 4 x 7.9mm MG 17 machine guns, 1 x 7.9mm MG 15 machine gun in rear cockpit		

Mikoyan MiG-3

One of the few Soviet fighters available at the time of the German invasion in 1941, the MiG-3 performed creditably against the Luftwaffe onslaught. However, its overall weight restricted the armament that could be carried, and its performance below high altitude was poor.

ENGINE	1,350hp Mikulin AM-35A liquid-cooled V-12 cylinder	
WINGSPAN 10.3m (33ft 9in)		LENGTH 8.2m (26ft 9in)
TOP SPEED 640kph (398mph)		CREW 1
ARMAMENT 1 x 12.7mm and 2 x 7.62mm machine guns in nose		

Nakajima Ki-84-Ia Hayate ("Frank")

Japanese fighter aircraft achieved their fearsome reputation by concentrating on manoeuvrability and speed over heavy armament. The introduction of the Ki-84 in 1944 brought a fast, heavily armed, and rugged machine that could outperform both the Hellcat and the Mustang. Fortunately for the Allies, heavy bombing reduced the quantity of aircraft produced and the quality of those that did reach the front line.

ENGINE	1,900hp Nakajima Ha-45 air-cooled 18-cylinder radial		
WINGSPAN	11.2m (36ft 10in)	LENGTH	9.9m (32ft 6in)
TOP SPEED	631kph (392mph)	CREW	1
ARMAMENT	2 x 12.7mm machine guns, 2 x 20mm cannon		

Northrop P-61A Black Widow

The first aircraft ever designed explicitly as a radar-equipped night fighter, the P-61 was ordered by the US Army following the Battle of Britain, when the RAF achieved the first successful radar interceptions. Huge and complex, it took until May 1944 to enter service – two months later it achieved its first kill. Surprisingly agile for an aircraft the size of a medium bomber, due to the innovative control systems devised by Northrop, it packed a devastating punch. Its large bombload meant that it was often used on "intruder" missions.

Twin-boom fuselage

2,000hp Double Wasp radials

ENGINE	2 x 2,250hp P&W R-2800-65 Double Wasp radials		
WINGSPAN	20.1m (66ft 1in)	LENGTH	14.9m (48ft 11in)
TOP SPEED	589kph (366mph)	CREW	3
ARMAMENT	4 x 20mm M2 cannon; 2,900kg (6,400lb) bombload		

North American P-51D Mustang

The Mustang, designed by North American Aviation to a British specification, was WWII's outstanding long-range fighter. Its first flight was in October 1940, and initially it was fitted with an Allison engine. However, the Mustang's performance was transformed after it was equipped with a Packard Merlin engine. This, combined with its low-drag wing and fuselage allowed it to fly faster and further than the Spitfire. In October 1943, the US strategic bombing campaign, using unescorted B-17s and B-24s, was suspended after high losses, but in February 1944, it was resumed with Mustangs escorting the bombing raids. Fitted with drop tanks, the P-51 could fly as far as Berlin, and, even at this range, its performance was superior to most German fighters.

Large teardrop canopy affords good visibility

3 x .5in wing-mounted machine guns

ENGINE	1,490hp Packard V-1650-7 Merlin liquid-cooled V-12 cylinder		
WINGSPAN	11.9m (37ft)	LENGTH	9.9m (32ft 3in)
TOP SPEED	703kph (437mph)	CREW	1
ARMAMENT	6 x .5in Browning machine guns; 2 x 454kg (1,000lb) bombs		

Republic P-47 Thunderbolt

Introduced in 1943, the P-47 was then the largest and heaviest single-seat fighter ever built. Originally used as an escort fighter for the US 8th Air Force's strategic bombing offensive, the Thunderbolt's tough construction and heavy firepower allowed it to devastate German armour, troops, transport, and airfields in close support of the American armies during 1944–5.

ENGINE	2,300hp Pratt & Whitney R-2800 radial with turbocharger		
WINGSPAN	12.4m (40ft 9in)	LENGTH	11m (36ft 1in)
TOP SPEED	690kph (429mph)	CREW	1
ARMAMENT	8 x .5in machine guns; 907kg (2,000lb) bombload		

Yakovlev Yak-9

The Yak-1 series of fighters, to which the Yak-9 belongs, eventually numbered nearly 37,000 examples, almost as many as the Il-2 Shturmovik. When the two-seat trainer developed for the Yak-1 was found to handle better than the parent aircraft, it was built as a fighter in its own right: the Yak-7. This in turn became the Yak-9 after mid-1942, when light alloy wing spars replaced wood, giving room for increased payload. Many variants were produced, including the 9D (long range) and 9DD (very long range). The Yak-9, in considerable numbers, made a decisive contribution to the victory at Stalingrad at the end of 1942.

ENGINE	1,260hp Klimov VK-105PF liquid-cooled V-12 cylinder		
WINGSPAN	10m (32ft 9in)	LENGTH	8.5m (28ft)
TOP SPEED	599kph (372mph)	CREW	1
ARMAMENT	1 x 20mm ShVAK cannon firing through propeller hub, 2 x 12.7mm BS machine guns above engine		

THE BATTLE FOR BRITAIN

SMALL IN SCALE COMPARED WITH AIR OPERATIONS LATER IN THE WAR, THE BATTLE OF BRITAIN AND THE BLITZ WERE CRUCIAL MOMENTS IN WORLD HISTORY

"Never in the field of human conflict was so much owed by so many to so few."

WINSTON CHURCHILL
BRITISH PRIME MINISTER, REFERRING TO RAF FIGHTER COMMAND, SEPTEMBER 1940

THE BATTLE OF BRITAIN was raised to heroic and legendary status even before it began. On 18 June 1940 prime minister Winston Churchill told the House of Commons: "I expect that the Battle of Britain is about to begin. Upon this battle depends the survival of Christian civilization... The whole fury and might of the enemy must very soon be turned on us... Let us therefore brace ourselves to our duties and so bear ourselves that, if the British Empire and the Commonwealth last for a thousand years, men will still say, 'this was their finest hour'."

Such grandiloquence invites deflation, and there have been plenty of mockers dedicated to demolishing the "myth of the finest hour". Yet more than 60 years after the event, the drama of the "Spitfire summer" shows no sign of losing its grip on the popular imagination. It still stands as the first major battle fought entirely in the air, just as the Blitz that followed was the first sustained campaign of strategic bombing.

The Battle of Britain was an aerial contest for which the British had prepared and the Germans had not. Since the mid-1930s the defence of Britain against an attack by the Luftwaffe had been a central focus of British military planning. For the Germans, the campaign was an improvised response to finding themselves, with surprising suddenness, in control of western Europe. Hitler was nervous about invading Britain, but something had to be done to make the British accept that they were beaten. An air offensive seemed to have every advantage. It might in itself force the British to negotiate a surrender, especially if backed up by the threat of an invasion; and if it went particularly well, the invasion might be possible for real.

WARTIME LEADER
In June 1940, Britain's prime minister Winston Churchill gave the airborne defence of his country its famous name, proclaiming: "The Battle of Britain is about to begin."

Since early July the RAF and the Luftwaffe had been clashing over the English Channel, as British ports and merchant convoys came under air attack from German aircraft. The main German onslaught on England, in response to Hitler's directive to the Luftwaffe – "to overcome the British air force with all means at its disposal and in the shortest possible time" – began on 13 August. Fleets of bombers, escorted by fighters, carried out daylight raids with special concentration on destroying airfields, aircraft factories, and radar installations. The determined resistance of RAF Spitfires and Hurricanes in the first days of the offensive led the Luftwaffe to focus increasingly on raiding Fighter Command airbases and on wearing down the fighter force in the air. Then on 7 September, at a time when Fighter Command was under maximum pressure, the focus of Luftwaffe operations shifted again, turning to mass bombing raids on London. By the end of October the Luftwaffe had given up its dream of air superiority, settling for night-time bombing raids on London and other British cities – the Blitz.

LEGENDARY FIGHTER
The Supermarine Spitfire entered aviation legend with its performance in the Battle of Britain in 1940. Although not the most numerous aircraft in Fighter Command, it was in many ways the best. Without it, RAF pilots would not have been able to take on the Messerschmitt Bf 109 so successfully.

SYMBOL OF DEFIANCE
St Paul's Cathedral stands unscathed amid the smoke of burning buildings after a Luftwaffe raid on the City of London. The British people took what comfort they could from such symbols of defiance during the dark months of the Blitz in 1940–41.

Clocks in operations rooms had colour-coded segments. Arrows placed on map tables used the same colour coding, showing when aircraft positions were last updated.

OPERATIONS ROOM
Members of the British Women's Auxiliary Air Force push wooden blocks and arrows representing enemy bombers around a map as information from Observer Corps Centres comes through on their headsets. Above them, controllers watch the progress of the enemy.

SCRAMBLE FOR THE SKIES
This photograph (taken at Duxford, England, before WWII) is of a demonstration given for the press by the RAF. The purpose of the exercise was to show the speed at which the airmen could reach their aircraft when called upon.

While the German commanders were confused in their objectives, the British were focused and organized to a single end. Fighter Command chief Hugh Dowding recognized that for the RAF, surviving as a coherent defensive force was enough to constitute victory. By avoiding committing too large a part of his resources to the combat prematurely, and making his fighters concentrate on knocking out German bombers, Dowding conducted a ruthless and calculating campaign of attrition.

British defences

The British air-defence system was the most sophisticated in the world, a triumph of organization and applied technology. In its front line were the radar stations that identified Luftwaffe aircraft approaching Britain's shores. Radar operators provided a dense flow of raw data that was processed at a centralized "filter room" and forwarded to the people who controlled operations. In the operations rooms members of the Women's Auxiliary Air Force converted the information into 3-D graphic form by pushing wooden blocks around on a map with croupier's rakes, watched from a balcony above by the controllers. The controllers at the headquarters of the four fighter groups – each responsible for the defence of an area of the country – decided when, where, and at what strength aircraft should be sent up to meet the Luftwaffe. Controllers at sector level, with typically three or four squadrons under them, were responsible for directing the fighters on to their targets by radioed instructions.

This system was fallible. The problem of distinguishing between friendly and enemy aircraft was tackled by fitting RAF fighters flying in a radar zone with IFF (Identification Friend or Foe) devices, which gave them a distinctive radar signature, but there were often not enough IFF sets to go round, and they did not always work. Also, the Bf 110s of the Luftwaffe's Erprobungsgruppe 210 discovered that by flying close to ground level they could creep under the radar unobserved. But the

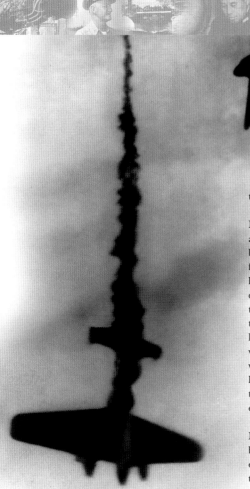

SUCCESSFUL INTERCEPTION
*A Messerschmitt Bf 110 is shot down by a
Hurricane over southern England. The heavy
German twin-engined fighter was no match for
RAF Spitfires or Hurricanes in a dogfight.*

guns, which was on top of the control column. The guns were as immobile as the pilot, and in combat the pilot aimed the plane at the target. He had enough ammunition to fire for about 13 seconds on each sortie.

The Luftwaffe fighter-escort pilots, in their Messerschmitt Bf 109s or 110s, liked to function as marauding hunters with their bomber force as bait. Lurking at high altitude in the loose "finger-four" formation – two pairs, each consisting of a lead pilot and his wingman – they would dive down to "bounce" the RAF fighters advancing in threes in tight V formations, especially picking off the vulnerable aircraft at the back of the V. This kind of combat could be over in seconds. The victim of a Messerschmitt diving out of the sun would probably never see the aircraft that shot him down. Exploiting the accumulated speed of the dive, the Messerschmitts could make their escapes before any response was possible.

If the RAF fighters could engage the Messerschmitts in a dogfight they stood a much better chance. On the whole, the Spitfires and even the slower Hurricanes could outmanoeuvre the Bf 109s, and they were certainly more agile than the Bf 110s. In combat, where the key manoeuvre was a tight turn to get on the enemy's tail, a good RAF pilot would probably come out on top.

Dowding's orders were that the fighters should concentrate on shooting down bombers. A division of labour developed, with the Spitfires holding off the Luftwaffe fighter cover while the Hurricanes went for the fleets of Heinkels, Junkers, and Dorniers. Bombers were not easy targets. The He 111s – the most numerous Luftwaffe bomber – were by this time partly armoured, and the Ju 88s were sturdy and fast. Their skilful gunners downed substantial numbers of RAF fighters. The most effective way of attacking a bomber formation was to fly straight at it from the front, an especially unnerving experience for bomber crews clustered in the nose of their aircraft behind plexiglass. However, few fighter pilots had the nerve to risk collision, and most attacked bombers from behind or, occasionally, the flank. It was relatively easy to score hits on a bomber, but bringing one down was more difficult. Many made it back to their airfields riddled with bullet holes, often carrying one or two seriously injured or dead crew members.

The Messerschmitts were not really suited to the role of bomber escort. The Bf 109 had inadequate range, allowing the briefest spell in the combat zone. The Bf 110 had the required range, but was not a good enough dogfighter. When mounting bomber losses forced the Messerschmitts to take their escort duties more seriously, making some fly in close support, instead of 3km (10,000ft) above, this was extremely unpopular with Messerschmitt pilots. RAF fighter pilots had their own discontents. Some chafed at the discipline imposed by ground controllers, fuming when their hunting instincts were frustrated in the interest of some wider tactical scheme. The technological advances in place by 1940 meant that the romantic image of the pilot as lone hunter, or even of an independent hunting pack, was already obsolete.

In the hot seat

Nevertheless, fighter pilots and their skills were ultimately the key factor determining victory or defeat. Thanks to the sterling efforts of British aircraft factories, there was never a serious chance

---

system was also efficient and robust, with most communications depending on standard telephone lines that were impossible to jam and easy to repair.

The call to scramble

The radar-based early-warning system relieved Fighter Command of the need to mount continuous combat air patrols, for which it simply did not have the resources. Nevertheless, response times were very tight. Fighter pilots always ran to their aircraft when the call to "scramble" came through, because every second meant a chance to gain more height before encountering the enemy. It typically took five minutes for a squadron to get airborne, which was rarely quick enough to avoid beginning the fight at an altitude disadvantage.

Once airborne, the pilot in a Spitfire or Hurricane – or for that matter in a Messerschmitt Bf 109 – was effectively welded to his machine. Strapped tightly into the small metal cockpit, with its plexiglass hood, a bulletproof windscreen in front of him, and, in most cases, an armour plate behind his back, he could barely move. He sat with his right hand on the control stick, his left on the throttle, and his two feet on the rudder bar. His right thumb rested on the firing button for his

---

Supermarine Spitfire

"SOME MEN FALL IN LOVE WITH YACHTS", said RAF fighter pilot Bob Stanford-Tuck, "or some with women... or motor cars, but I think every Spitfire pilot fell in love with it as soon as he sat in that nice tight cosy office [RAF slang for cockpit] with everything to hand". Responsive to a touch of the fingertips on the joystick, or the feet on the rudder pedals, the Spitfire was a joy for a good pilot to fly.

Prototyped in 1936, the Spitfire went into production in 1938. Its quality depended on a marriage between an imaginative airframe design and a superb new engine, the Rolls Royce Merlin. Incorporating all the most advanced features of aircraft of its day – adjustable-pitch propeller, all-metal monocoque construction, retractable undercarriage, enclosed cockpit – it achieved uniqueness through its elliptical wing. This ingeniously solved the problem of housing eight machine guns and a retracted undercarriage, while providing enough strength to withstand the stress of high-speed manoeuvres.

The Spitfire had a few drawbacks. Its novel wing initially posed problems for mass production; the pilot could not see in front of the aircraft when taxiing; and early versions of the engine (those in service prior to 1941) had a habit of cutting out going into a dive, since negative gravity cut off the fuel supply – a problem the Messerschmitt's fuel-injected engine did not have.

But the overall quality of Mitchell's design was proven by the Spitfire's ability to hold its own up to the end of the war.

"The Spitfire had style and was obviously a killer."

"SAILOR" MALAN
CO 74 SQUADRON AND BATTLE OF BRITAIN ACE

GROUND SUPPORT
The ground crew, here members of the WAAF (Women's Auxiliary Air Force), prepare to close the side hatch. The pilot entered and exited the craft via this hatch.

Coolant header tank

Exhaust stub

Upper fuel tank ahead of cockpit

Rear view mirror

Streamlined spinner covers pitch-change mechanism

Laminated wood propeller blade

CITY

CONTINUOUS EVOLUTION
Over 20,000 Spitfires had been built by the end of WWII in a score of variants. Later marks showed that the power, weight, and firepower of the fighter could all be doubled without altering the basic design. This Mark V, a version first introduced in 1941, has been personalised for a Canadian pilot and shows nine kills.

Steel tube engine bearer

Wingtips detached to improve low-altitude manoeuvrability

Pitot tube registers airspeed in cockpit

Main undercarriage leg

Radiator flap, operated from cockpit, controls cooling

Specifications (F.Mk.V)

ENGINE 1,470hp Rolls-Royce Merlin liquid-cooled V12	
WINGSPAN 9.8m (32ft 2in)	
LENGTH 9.1m (29ft 11in)	
TOP SPEED 575kph (357mph)	**CREW** 1
ARMAMENT 2 x 20mm Hispano cannon, 4 x .303in Browning machine guns	

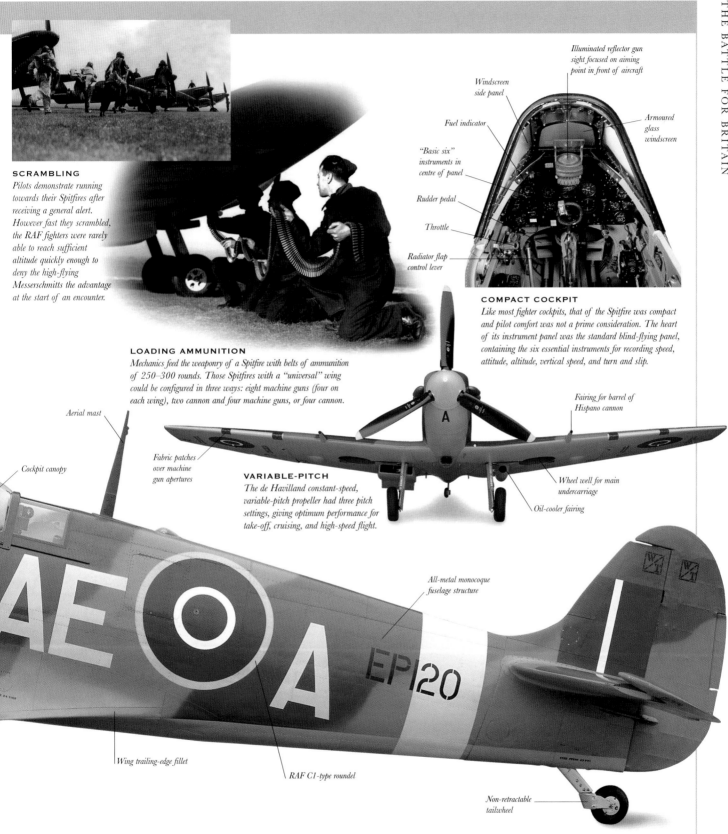

SCRAMBLING
Pilots demonstrate running towards their Spitfires after receiving a general alert. However fast they scrambled, the RAF fighters were rarely able to reach sufficient altitude quickly enough to deny the high-flying Messerschmitts the advantage at the start of an encounter.

Illuminated reflector gun sight focused on aiming point in front of aircraft

Windscreen side panel

Fuel indicator

"Basic six" instruments in centre of panel

Rudder pedal

Throttle

Radiator flap control lever

Armoured glass windscreen

COMPACT COCKPIT
Like most fighter cockpits, that of the Spitfire was compact and pilot comfort was not a prime consideration. The heart of its instrument panel was the standard blind-flying panel, containing the six essential instruments for recording speed, attitude, altitude, vertical speed, and turn and slip.

LOADING AMMUNITION
Mechanics feed the weaponry of a Spitfire with belts of ammunition of 250–300 rounds. Those Spitfires with a "universal" wing could be configured in three ways: eight machine guns (four on each wing), two cannon and four machine guns, or four cannon.

Fairing for barrel of Hispano cannon

Aerial mast

Fabric patches over machine gun apertures

VARIABLE-PITCH
The de Havilland constant-speed, variable-pitch propeller had three pitch settings, giving optimum performance for take-off, cruising, and high-speed flight.

Wheel well for main undercarriage

Oil-cooler fairing

Cockpit canopy

All-metal monocoque fuselage structure

EP120

Wing trailing-edge fillet

RAF C1-type roundel

Non-retractable tailwheel

that the RAF would run out of aeroplanes. But the rate of attrition among experienced fliers meant that men with virtually no experience of flying single-seat fighters were soon being drafted in. On their first few excursions, novices could only hope to survive through pure luck.

Britain was fortunate in having its Commonwealth and occupied Europe to draw on for personnel. Fighter Command was an international force, with Canadians, New Zealanders, South Africans, Australians, many Poles and Czechs, some Belgians and Free French, and even a handful of Americans serving alongside the British. Each contingent had its own distinctive character – the Poles were described by Dowding as "very dashing and totally undisciplined" – but they all shared youthful courage and enjoyed the status their profession gave them with women.

Among British pilots, class distinction was rife. Pilot officers who had come into the RAF through the peacetime Auxiliary Air Force and University Air Squadrons – run as exclusive clubs for young men from well-off families – were sharply distinguished from the sergeant pilots who had

DOUGHTY INTERCEPTORS
RAF Hawker Hurricanes fly in close formation. During the Battle of Britain, Hurricanes were the bomber-killers of choice – intercepting Dorniers and Heinkels – while Spitfires engaged the Bf 109s that shadowed the German bombers at high altitude.

ADOLF "SAILOR" MALAN

ADOLF "SAILOR" MALAN (1905–63), one of the RAF's top fighter aces, was brought up on a farm in South Africa. Used to firing a gun since childhood – which may account for his excellent shooting in the cockpit – Malan joined the RAF in 1935. Leader of 74 Squadron from August 1940, he schooled his men in the need for dedicated teamwork and constant practice. He applied himself to killing Germans in a professional and methodical spirit, but had no time for "score chasing". A tactical experimenter, he broke away from the rigid tight-V formation officially required by the RAF, and circulated his own "Ten Rules of Air Fighting".

"SAILOR" AND MASCOT
Malan was called "Sailor" because he had once been a merchant seaman. Aged 30 at the time of the Battle of Britain, he was an old man by fighter-pilot standards.

graduated from the non-exclusive Volunteer Reserve. Officers and sergeants flew side by side, but they were positively discouraged from socializing.

In the air, an experience shared by many RAF pilots in the the Battle of Britain was that of being shot down. Pilots in the busiest sectors could expect to be forced to bale out at least once a month. With men more valuable than machines, it made little sense to perform heroics nursing a damaged aeroplane to safety – though some pilots still did. If having parachutes was the sharpest difference between the experience of World War II fliers and their World War I predecessors, the most striking resemblance was the terrible fear of fire. Pilots wrapped themselves round with clothing to cover any bare flesh, hoping to extend by a second or two the time they might have to escape the cockpit without disfiguring burns.

"The few"

As in World War I, a disproportionate number of kills in the Battle of Britain were recorded by a few gifted individuals. The required combination of flying skills, sharp eyesight, fast reaction times, and killer instinct was rare. The top Battle of Britain ace in the RAF, credited with 17 kills, was a Czech pilot, Josef Frantisek, flying with a Polish squadron. The highest score for a British RAF pilot was recorded by Sergeant Ginger Lacey – one in the eye for the class system. In the Luftwaffe, the two most trumpeted aces, Adolf Galland and Werner Mölders, vied with Major Helmut Wick in a highly publicized contest for top score. Although some pilots in Britain achieved fame – for example, the legless Douglas Bader – the RAF discouraged competition over kills and individual hero-worship. Even in Germany, the aces did not have the same public status as in World War I. The need for air heroes to cement public support for the war was not felt as keenly.

Fighter Command pilots were definitely, as Churchill dubbed them, "the few" to which so much was owed. The number killed in the Battle of Britain was 544 – about one in five of those RAF pilots who took part. Luftwaffe airmen died in much greater numbers – around 2,700 of them. The difference was mostly due to the toll taken of bomber aircrews. Overall, the RAF is reckoned to have shot down around 1,900 Luftwaffe aircraft for the loss of just over a thousand of their own.

There were days in late August and early September when Fighter Command was severely stretched. Luftwaffe raids on airfields had ground crew and civilian support services battling heroically to fill craters and restore communications. Repeatedly scrambled to meet harrying attacks, RAF pilots suffered periods of

"If you are new to the game and if you are required to fly within a few feet of your neighbour's wingtip, it is a dicey experience."

ROALD DAHL
ON FORMATION FLYING IN A HURRICANE

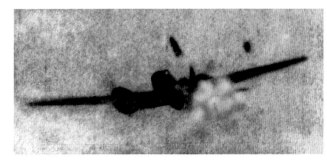

demoralization and exhaustion. But to win, the Luftwaffe had to break Fighter Command's resistance and, despite stretching it severely, the Germans never achieved that aim.

The turning-point in the battle came on 15 September, which is celebrated as Battle of Britain Day in the United Kingdom. Indeed, this was a day of heavy fighting, with Luftwaffe bombers and fighters arriving in two waves, the largest comprising almost 500 aircraft. It was also spectacular for Londoners – a good deal of the combat happened right above their heads. But RAF claims to have shot down 185 enemy aircraft were grossly inflated; the actual figure was probably 56, about twice the RAF's own losses. And the battle in no sense stopped – roughly the same number of German aircraft were downed on 27 September.

But the continued ability of the RAF

GAS MASKS
Civilian gas masks differed from the military models (left), but their purpose was the same. However, gas bombs were not used by either side, chiefly for fear of retaliation.

LUCKY ESCAPE
This still taken by a gun camera shows what is probably the pilot of a Heinkel He 111 jumping out of his cockpit seconds before a direct hit. Depsite their vulnerability, He 111s formed the core of the German bomber force in the Blitz.

to put up a fighter defence in strength, along with the onset of autumn weather, meant that German invasion plans were definitively called off, and by the end of October the Luftwaffe was devoting all its resources to the night bombing of Britain's cities.

The Blitz

From the first major daylight bombing of London – on 7 September 1940 – through to May 1941, Britain's cities were subjected to sustained aerial bombardment on a scale never before attempted. This was Douhet's concept of air war at last put to the test – an attempt to break civilian morale by using fleets of bombers to devastate enemy cities and industrial centres. Not only London was battered, but also ports such as Plymouth,

Portsmouth, Cardiff, Liverpool, Glasgow, and Belfast, and inland industrial cities such as Birmingham and Coventry. It was an unprecedented experiment, with the British people as guinea pigs. Would their morale hold or would social order fall apart in some catastrophic manner under the strain of bombardment? The result is now known, but at the time it could not be comfortably predicted.

The scale of the bombing during the Blitz was later dwarfed by Allied bombing operations against Germany and Japan – in the heaviest raids of 1940–41 the Luftwaffe never dropped

KEEPING WATCH
Bofors anti-aircraft gunners keep watch over a seaside town on the southern coast of England, as a friendly aircraft flies past. Sneak raids were an insoluble problem along the Channel coast.

Ground speed/ drift scale

Aperture

Turn and drift knob

Disk speed drum

Data table

JUNKERS JU 88 BOMB-SIGHT

German bomb-aiming systems were far superior to those of the RAF in the early stages of the war. German bombers were often fitted with two bomb-sights – one for level bombing, and one for dive or glide bombing.

more than 500 tonnes of bombs, whereas as early as 1943 the RAF was dropping over 2,000 tonnes on a single night over Germany. But the Luftwaffe raids were no powder-puff punch. Some 40,000 civilians were killed in the Blitz and local devastation – for example, in London's East End and at Coventry – could be of awesome intensity.

The impact of the Luftwaffe's campaign was undoubtedly limited by the lack of a heavy bomber. The Heinkel He 111 had a maximum bombload of around 2,000kg (4,500lb), well under half that of a four-engined RAF bomber such as the Halifax, and even the Junkers Ju 88 had a maximum load of only 3,000kg (6,600lb). But in other ways the Luftwaffe was much better prepared for a strategic bombing campaign than the RAF.

FIREPROOF

A British poster publicizes the Fire Guard, an organization set up in 1941 to counter the effect of incendiary bombs. Many urban areas were destroyed by fire in the Blitz.

GUIDED DESTRUCTION OF COVENTRY

ON THE EVENING OF 14 November 1940 a dozen Heinkel He 111s took off from Vannes, Brittany, to spearhead the bombing of the industrial city of Coventry in the British Midlands. They were part of Kampfgruppe 100, the Luftwaffe's elite pathfinders used to mark targets for the bomber swarms. The Heinkel crews had the X-gerät target-finding system (see page 216).

To find the city itself was child's play for the experienced Heinkel pilots. Tuned in to a *Knickebein* radio beam directed from a transmitter in Britanny to Coventry, the pilot adjusted his course each time the steady hum of the "equisignal" broke up into morse dots or dashes, telling him he was wandering to one side or the other of the beam. Behind the pathfinders, similar beams guided almost 500 other bombers towards the city.

Laid across the *Knickebein* radio beam ahead of Kampfgruppe 100 were three "crossbeams" transmitted from further east. The first alerted the bomber's crew with an "advance signal", telling them they were approaching Coventry.

> "The usual cheers that greeted a direct hit stuck in our throats. The crew gazed down on the sea of flames in silence."
>
> **LUFTWAFFE PILOT**
> REMEMBERING THE BOMBING OF COVENTRY

The second beam crossed their course exactly 30km (19 miles) from their pre-established target. It was the signal for the navigator to activate the X-gerät device, a primitive bomb-aiming computer. The third beam was laid 15km (9½ miles) from the target, giving the signal for pressing another key, setting the X-gerät to automatically release the bombload. As long as the pilot held a steady course, speed, and altitude for the last 15km, the X-gerät would release the bombs directly on the target.

Aware of the Luftwaffe's use of radio beams, the British were frantically attempting to jam or confuse them with electronic countermeasures, but on this occasion they failed. The Heinkels dropped their incendiaries and saw flames flickering across the centre of Coventry, marking the target for the fleets of bombers advancing behind them. More than 500 tonnes of high explosives and incendiaries were dropped on the city in the course of the night. More than 500 people were killed, and over a thousand were injured.

MORNING AFTER

The morning after the bombing raid on Coventry in November 1940, local people go about their business. Over 500 civilians died in the raid, a small number compared with the mass deaths in German and Japanese cities later in the war but a severe shock at the time.

Remote-controlled bombing

In particular, the Germans had given considerable thought to finding and hitting a target, a problem that Bomber Command had strangely neglected to take seriously. The Luftwaffe's experience in Spain had underlined the importance of night-flying and had led to the adaptation of civil radio-navigation techniques. The result was that, unlike the airships and Gothas of World War I, German night bombers did not always have to feel their way blindly through the darkness over blacked-out Britain.

All German bombers were equipped to follow a *Knickebein* ("dog-leg") radio beam to reach their targets. The elite Kampfgruppe 100, sent in first to mark targets with incendiaries, had even more sophisticated target-finding equipment, the X-gerät and Y-gerät. The Y-gerät system, introduced in December 1940, constituted fully remote-controlled bombing. The aircraft flew along a usual beam, but re-radiated the beam back to its source, so the ground station could precisely track its progress. When the bomber arrived over its target zone, it went on to automatic pilot, and as soon as it reached the map co-ordinates of its target, the ground station transmitted a signal that released the bomb. This technique was reckoned to be able to put a bomb within a radius of 90m (100 yards), at a distance of 400km (250 miles).

Luftwaffe superiority

In practice, the Luftwaffe's radio-navigation systems did not always run smoothly. The British rapidly developed countermeasures, jamming or distorting the beams in a secret electronic war. The bombers still always preferred to attack on a clear night, when they could orient themselves by the stars and by features on the ground – the

Thames estuary proved as useful a marker for German navigators in 1940 as it had in 1916. And, for all the technical wizardry, most bomb-aiming remained very approximate.

Whatever might be the case by day, by night the Luftwaffe had air superiority over Britain. Their fleets of 100 to 500 bombers roamed over Britain virtually unscathed. Anti-aircraft batteries made a lot of noise and reassured people on the ground that something was being done, but in practice they stood little chance of hitting their targets. RAF night-interceptor squadrons operated with increasing success, especially using two-seater Bristol Beaufighters or Boulton Paul Defiants, which either flew blind, directed on to their targets by ground controllers, or used air-interception radar sets. But either way, their chances of a kill were modest. A German bomber crew, tightly packed inside the perspex nose of a Junkers, Heinkel, or Dornier over the shadowy flaming confusion of a British city, could be reasonably confident of returning home intact.

BUTTERFLY BOMB
This small anti-personnel bomb is an example of the thousands dropped by German bombers over England. It has two folding wings that revolved, slowing the rate of descent and arming the fuse.

Victory in survival

Some of the most ferocious nights of the Blitz came in April and May 1941, but this was deceptive. The Germans were by then thoroughly engaged in preparations for the invasion of the Soviet Union. As spring turned to summer, the bombers shifted eastwards and the British people, wary and watchful, slowly realized that they had come through.

London, the most frequently battered target, had been subjected to 57 nights of aerial bombardment. This had imposed a sometimes near-intolerable strain on the civilian population and on the emergency services. Yet there had been no general breakdown of society or popular pressure on the government to surrender. In the Blitz, as in the Battle of Britain, survival had amounted to a kind of victory.

SHOOTING AT V1S
British anti-aircraft guns open fire at night against V1 flying bombs. The V1s were visible in the dark because of their fiery engine exhaust. Many were shot down, exploding either when hit or on impact with the ground.

SECRET WEAPONS

IN THE SUMMER OF 1944 London was subjected to a second blitz, this time by unpiloted V1 flying bombs. The Allies had been aware for some time that the Germans were developing "secret weapons", and devastating bombing raids both on the experimental centre at Peenemunde and on launch sites under construction in France delayed the V1's deployment. But by June 1944 the Germans had switched to smaller launchers that were not detected by Allied reconnaissance aircraft. Shortly after D-Day, Hitler ordered a flying-bomb offensive against London.

In summer 1944 about 100 V1s a day crossed the Channel, by day and night, and in all weather conditions. Although they were unarmed and unescorted, the flying bombs posed a novel challenge to air defences. A combination of early-warning radar, AA guns,

V1 AUTOPILOT
This autopilot from a V1 flying bomb fed signals to the bomb's elevators and rudder to control altitude and direction. The terminal dive was initiated when a preset distance had been flown.

and fighters was deployed, as against any intruders, but Spitfires, Typhoons, and Mustangs found it hard to cope with such a fast, small target. Only the new high-performance Hawker Tempests could easily catch the flying bombs; jet-powered Gloster Meteors were also sent up as interceptors, the first operational use of jets by the Allies. Shooting down a V1 was a hazardous action, since if it exploded it could easily destroy its attacker. Tempest pilots developed a technique of flying alongside a V1, lifting their wing under the flying bomb's wing, and tipping it over so that it spiralled down out of control.

After V1 launch sites in northern France were overrun by Allied troops in August 1944, the Germans began air-launching them from Heinkel He 111s flying in at low level over the North Sea. About half of the 8,000 V1s launched against Britain were shot down by aircraft or ground fire.

Hitler's other secret weapon, the V2 ballistic missile, was not interceptible, although Allied air forces made heroic efforts to destroy heavily defended V2 launch and production sites. Fortunately, the Germans did not have an atomic warhead to put on the end of it. The V weapons killed almost 9,000 British people in 1944–45.

SPOILS OF WAR
After the war's end, an American soldier studies a V2 in the enormous, underground rocket-assembly plant at Nordhausen, Germany. Slave labourers worked under appalling conditions to produce the rockets; many thousands died of ill-treatment.

Warhead

Controls compartment

Alcohol tank

Liquid-oxygen tank

Rocket engine

Stabilizing fin

V2 ROCKET
The V2 rocket, here stripped of its outer casing, was the first ballistic missile used in warfare. Travelling at up to five times the speed of sound, it exploded before its victims could hear it coming. Some 3,200 V2s were launched in the war.

FLYING-BOMB LAUNCH
A German V1 flying bomb captured by the Allies at the end of the war – and bearing US insignia – is test-fired. Propelled by a primitive jet engine and flying at over 640kph (400mph), the V1 was packed with a tonne of explosives. When it hit the ground, the destructive effect was impressive. Yet V1s killed only 5,475 people in Britain – less than one for each V1 launched.

AIR WAR AT SEA

AIRCRAFT WERE AS CRUCIAL IN NAVAL WARFARE AS IN THE WAR ON LAND, ESPECIALLY IN THE PACIFIC WHERE CARRIER FORCES BATTLED AT LONG RANGE

"The destiny of our homeland hinges on the decisive battle in the Southern Seas, where I shall fall like a blossom from a radiant cherry tree."

JAPANESE FLYING OFFICER
WRITING ON THE EVE OF A KAMIKAZE MISSION

O**N THE EVENING OF** 7 May 1942, a formation of aircraft with their landing lights on approached the US Navy carriers Yorktown and Lexington as they steamed through the Coral Sea in the South Pacific. When one of the carriers' destroyer escorts opened fire on the planes, a sharp message was radioed to its captain telling him to stop shooting at friendly aircraft. The destroyer's skipper snapped back that he knew Japanese planes when he saw them. Sure enough, they were Japanese carrier aircraft that had become confused in the failing light and mistaken the American ships for their own. As every gun in the US fleet opened up, the Japanese pilots switched off their lights and scattered into the darkness. For the next half hour, American radio operators could hear their Japanese opposite numbers talking their aircraft down somewhere out in the dark ocean.

This kind of incident was to be expected as two navies invented a new way of making war. The Japanese and American forces had been feeling blindly for one another for days, with aerial reconnaissance inhibited by much low cloud. The Japanese carriers *Zuikaku* and *Shokaku* were veterans of the attack on the US naval base at Pearl Harbor (see page 222), but the American carrier aircraft had had their first experience of war only earlier that same day, when Devastator torpedo-bombers and Dauntless dive-bombers sank the Japanese light carrier *Shoho* – an occasion celebrated by pilot Bob Dixon, who led the attack, with the pithy message: "Scratch one flat-top".

PILOTED BOMB
Its nose packed with explosives, the Japanese Ohka22 piloted bomb was carried into action by a bomber aircraft. After its release, the kamikaze pilot ignited rockets and guided the bomb to its target.

PACIFIC ADMIRALS
Admiral Chester Nimitz (right) held command in the Pacific area from 1941. Vice-Admiral Marc Mitscher notably led Carrier Task Force 58 in the 1944 Battle of the Philippine Sea.

At first light on 8 May, the Americans sent off reconnaissance patrols and soon located their enemy. The two carrier forces were 160km (100 miles) apart. At 8.30am, about 90 US Navy aircraft took off from the carriers. As they headed for their target, the Japanese carrier aircraft headed off towards the US ships. The American fliers found the Japanese carriers in a tropical rainstorm. The *Zuikaku* slipped away into the mist, but the *Shokaku* was repeatedly hit by the American dive-bombers and was left still afloat but ablaze. Unfortunately for the *Yorktown* and *Lexington*, the sky over their patch of

ocean was clear as Lieutenant-Commander Kuichi Takahashi led his Aichi D3A (Val) dive-bombers and Nakajima B5N (Kate) torpedo-bombers into the attack. As they flew into an intense barrage of anti-aircraft fire, Takahashi's aircraft was blown to pieces and so were a number of others. But too many of the aeroplanes got through.

It was momentarily a curiously personal close-range encounter as the torpedo-bombers came in at just above flight-deck height: a Japanese pilot remembered seeing "American sailors staring at my plane as it rushed by". Both carriers took punishment, especially the "Lady Lex". Racked by fires and internal explosions, she had to be abandoned. The *Yorktown* struggled back to Hawaii for repair. The Battle of the Coral Sea was not a

COMING IN TO LAND
A Curtiss SB2C Helldiver, the US Navy's prime dive-bomber in the last two years of the war, prepares to land on a carrier deck. For a pilot to find his way back to his carrier after a mission was by no means straightforward, especially when short of fuel.

decisive encounter, but it was a turning-point in the history of warfare. For the first time, two naval forces had fought using carrier aircraft alone, far beyond the range of even the most powerful warship's guns.

Seaplanes and flying boats

Carrier aircraft were far from being the only aerial presence at sea during World War II. Warships were generally equipped with catapult-launched seaplanes for reconnaissance – aircraft such as the slow and ungainly Supermarine Walrus, much loved by its British crews, or the German Arado Ar 196. And large flying boats such as the American Consolidated Catalina and Martin Mariner and the British Short Sunderland (a military version of the Empire flying boats that cruised to Cape Town and Sydney before the war) also patrolled the oceans, searching for enemy submarines and ships, as well as fulfilling an invaluable air-sea rescue role. Although slow by the standards of most World War II aircraft, the Catalinas and Sunderlands had invaluable range – the Catalina could stay in the air for 24 hours. These giants were not so gentle either, packing considerable defensive firepower, along with their bombs and depth-charges. The Sunderland became so noted for fending off harrying attacks

A shipborne Walrus flying boat is brought out of its hangar, wings folded for compact storage. The Walrus would be launched by catapult from the deck and land alongside the ship on returning from its mission.

by Luftwaffe fighters over the Bay of Biscay that the Germans nicknamed it "the Porcupine".

Convoy protection

One of the most important uses of aircraft in World War II was for the defence of merchant convoys. German U-boats and aircraft took a heavy toll on merchant shipping sailing to and from Britain. In the winter of 1940, Fw 200 Condors based in occupied Norway and France launched long-range bombing raids on Atlantic convoys. Where shipping sailed closer to land – for example, the Arctic convoys from Britain to the Soviet ports of Murmansk and Arkhangelsk passing Norway – shorter range Ju 88s and

He 111s also struck with devastating effect.

The convoys needed air-cover to stop these attacks. One desperate measure in the early days was to catapult a Hawker Hurricane from a merchant ship – a one-off mission, since the pilot's only recourse when he ran out of fuel was to ditch in the sea and hope to be picked up. Later, escort carriers provided a less wasteful solution, although there were never enough of them. Long-range flying boats, like the Catalinas and Sunderlands, patiently quartered the ocean in search of submarines. At first they were mostly limited to deterring U-boats from surfacing, but by 1942–43, the Allied aircraft were fitted with improved radar and depth-charges, and they became more aggressive.

Allied commanders were slow to devote adequate air resources to the Battle of the Atlantic, but the allocation of long-range Consolidated B-24 Liberator bombers to the ocean in 1943 marked a decisive turning-point. From then until the end of the war Allied aircraft imposed such heavy losses on the German U-boat fleet that it ceased to pose a major threat to merchant shipping.

Warship defences

No one could accuse traditional naval commanders of ignoring air power in their war preparations. They had taken great trouble to arm their warships against air attack – the density

AIRCRAFT AGAINST SUBMARINES
Aircraft proved an effective answer to German U-boats in the Battle of the Atlantic. Left, the crew of an RAF Coastal Command Sunderland flying boat keep watch on patrol over the Atlantic. Below, an American aircraft escorts a convoy of merchant ships bringing vital supplies across the ocean from the United States to beleaguered Britain.

WWII ANTI-SUBMARINE AIRCRAFT

AIRCRAFT PLAYED A DECISIVE ROLE IN ANTI-SUBMARINE WARFARE during the Battle of the Atlantic from 1940–43. The protection of merchant ships bound for Britain from attack by German U-boats was crucial to the country's survival. Although anti-submarine duties were also undertaken by shipborne aircraft, the major burden was carried by long-range aircraft flying from coastal bases. Under the circumstances, Britain gave surprisingly low priority to RAF Coastal Command, which was responsible for Atlantic anti-submarine patrols. Advocates of strategic bombing wanted long-range aircraft allocated to the bombing offensive against Germany, resenting the diversion of resources to help keep shipping lanes open. One of the major requirements in anti-submarine warfare aircraft was range. Until late 1942 the Allies had no aircraft patrolling the mid-Atlantic, and this gap in convoy air defence was ably exploited by German U-boats. The gap was closed by the introduction of a maritime version of the B-24 Liberator, the Consolidated PB4Y-1, which had a range of around 4,500km (2,800 miles). New equipment made aircraft effective U-boat killers by the end of 1942. They contributed greatly to the effective defeat of the U-boat menace in 1943.

LONG-RANGE CATALINA
One of the most familiar aircraft of its time, the durable, dependable "Cat" had a distinguished service record in many theatres of war with both the American and Allied air forces.

Consolidated PBY Catalina

The PBY Catalina was the US Navy's anti-submarine patrol aircraft from 1936. RAF Coastal Command also ordered large numbers to complement its own Sunderland. More Catalinas were built (over 4,000) than any other flying boat in history. Though slower and less well-armed than the Sunderland, it was tough and adaptable.

ENGINE	2 x 1,200 hp P&W R-1830 Twin Wasp air-cooled radials	
WINGSPAN	31.7m (104ft)	LENGTH 19.5m (63ft 10in)
TOP SPEED	314kph (196mph)	CREW 7
ARMAMENT	5 x .5in machine guns; 1,614kg (4,000lb) of torpedoes, depth charges, or bombs	

Consolidated PB4Y-1 Liberator

In November 1941, the RAF was the first to use the Liberator for very long-range anti-submarine patrols. A US Navy PB4Y-1 Liberator sunk its first U-boat in November 1942. The Navy had almost 1,000 PB4Y aircraft including a special remote-controlled version designed to attack the V 1 missile sites in France. In August 1944, Lt. Joseph Kennedy, brother of the future US President, was killed in one such operation.

ENGINE	4 x 1,200hp P&W R-1830-65 Twin Wasp 14-cylinder radials	
WINGSPAN	33.3m (110ft)	LENGTH 20.5m (67ft 3in)
TOP SPEED	449kph (279mph)	CREW 9–10
ARMAMENT	8 x .5in machine guns; 3,628kg (12,800lb) bombload	

Fairey Swordfish Mk.III

The Swordfish, nicknamed the "Stringbag", was so successful as the Royal Navy strike aircraft that, despite its antiquated appearance, it outclassed its intended replacement and served throughout the War. In November 1940, 21 Swordfish torpedo carriers and bombers from the carrier HMS *Illustrious*, sank the Italian fleet at Taranto, for the loss of only one aircraft. From 1943, the Swordfish Mk. III was fitted with radar and operated from convoy escort carriers in the North Sea and Atlantic, sinking many U-boats.

ENGINE	750hp Bristol Pegasus 30 air-cooled 9-cylinder radial	
WINGSPAN	13.9m (45ft 6in)	LENGTH 10.9m (35ft 8in)
TOP SPEED	224kph (138mph)	CREW 2–3
ARMAMENT	2 x .303in Vickers machine guns; 1 x torpedo or 681kg (1,500lb) mines, bombs or depth charges, 8 x 27kg (60lb) rockets	

Lockheed (A-28/A-29) Hudson I

ENGINE	2 x 1,200hp Wright R-1820 Cyclone 9-cylinder radials	
WINGSPAN	20m (65ft 6in)	LENGTH 13.5m (44ft 4in)
TOP SPEED	396kph (246mph)	CREW 5
ARMAMENT	5 x machine guns; 340kg (750lb) bombload	

Developed in 1938 as a maritime reconnaissance bomber from the Model 14 Super Electra airliner, the portly Hudson was the first American-built aircraft in RAF service during WWII. Other firsts included the first RAF aircraft to destroy an enemy aircraft and the first to sink a U-boat with rockets. Nearly 3,000 Hudsons served both Allied air forces to the end of the war.

Martin PBM Mariner

Designed in 1937, four years after the more famous Catalina, the Model 162 Mariner flying boat, with its gull wing and distinctive canted tail fins, finally entered service in 1941. Over 1,300 were built in a number of variants, the principal wartime version being the PBM.

ENGINE	2 x 1,700hp Wright R-2600-12 Cyclone 14-cylinder radials	
WINGSPAN	36m (118ft)	LENGTH 24.4m (80ft)
TOP SPEED	319kph (198mph)	CREW 7–8
ARMAMENT	7 x machine guns; 1,814kg (4,000lb) bombload	

Short S.25 Sunderland

The Sunderland was developed by the Short brothers from their famous commercial "Empire" class flying boats for RAF Coastal Command service. Equipped with radar, its job was to patrol the Atlantic for up to 13 hours looking for German U-boats. Submarines caught on the surface at night were often sunk before they had time to dive.

ENGINE	4 x 1,200hp P&W Twin Wasp air-cooled 14-cylinder radials	
WINGSPAN	34.4m (112ft 10in)	LENGTH 26m (85ft 4in)
TOP SPEED	343kph (213mph)	CREW 10
ARMAMENT	12 x .303in Browning machine guns, 2 x .5in machine guns in beam positions; 2,250kg (4,960lb) bombload or depth charges	

ATTACKING PEARL HARBOR

TIME FOR ACTION
Japanese sailors look on as a Mitsubishi "Zero" takes off from their flight-deck. Seaman Iki Kuramoti, on the carrier Akagi, said of the attack on Pearl Harbor: "An air attack on Hawaii! A dream come true!"

"THIS IS NO DRILL"
This message, telling the US fleet that Pearl Harbor was under attack, was broadcast minutes after the raid began. It was no news to personnel already fighting for their lives.

AT DAWN ON 7 DECEMBER 1941, Japanese pilots for the first wave of the attack on Pearl Harbor climbed into their cockpits, each carrying a rations pack for the flight – rice and plums, chocolate, and pep pills. Propellers spun and engines roared into life as the leader of the attack, Commander Mitsuo Fuchida, donned the traditional *hachimaki* headband. Despite a choppy sea, 183 aircraft took off without incident and formed up to set course for the Hawaiian island of Oahu, observing strict radio silence. For much of the flight there was low cloud. Fuchida corrected his bearing as he closed on the target by taking a fix on a music programme broadcast by a Honolulu radio station – which also gave an update on local weather conditions. Then the cloud broke and the pilots were looking down on the lush green island. As Pearl Harbor came into view, Fuchida saw "the whole US Pacific fleet in a formation I would not have dared to dream of in my most optimistic dreams". A total of 90 ships lay at anchor or in dry dock, including eight battleships. In the excitement of the moment, Fuchida radioed the carrier force with the signal for

> ## "The moment has arrived.
> ## The rise or fall of our empire
> ## is at stake…"
>
> **ADMIRAL YAMAMOTO**
> MESSAGE TO THE FLEET BEFORE PEARL HARBOR

victory – "Tora, tora, tora" – as the first strike went in, the torpedo-bombers skimming in low over the water and dive-bombers sweeping down from 3,500m (12,000ft). The air was thick with the smoke of explosions by the time the horizontal bombers, led by Fuchida, advanced in single file through anti-aircraft fire, while Mitsubishi A6M Reisen ("Zero") fighters swooped down to strafe the military airfields below. Almost 300 American planes were damaged or destroyed on the ground, and the Japanese had the air almost to themselves. In the midst of the mayhem, a flight of B-17 bombers arriving from the mainland United States were badly shot up.

The Japanese pilots were oblivious to the human drama unfolding on the ships and ground below, concerned only to carry out their tasks successfully. Many returned to their carriers in personal shame amid the general euphoria, convinced they had missed their targets, letting down their colleagues and their emperor. But Fuchida stayed over Pearl Harbor during the 170-plane second wave of the attack and was able to report to Admiral Nagumo on the damage caused. The admiral commented, "We may then conclude that anticipated results have been achieved." That just about summed it up.

BLAZING WARSHIPS
A rescue launch attempts to pick up survivors from US warships surrounded by burning oil. Eighteen ships were sunk or severely damaged at Pearl Harbor, including five battleships. (This is a colourized version of a black-and-white print.)

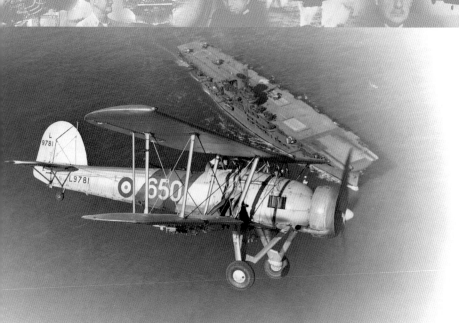

of anti-aircraft fire a battleship or cruiser could throw up was daunting. But the vulnerability of surface ships to air attack, in the absence of adequate air-cover, still came as a shock. One of the worst days of the war for the Royal Navy was 10 December 1941, when the battleship *Prince of Wales* and the battle-cruiser *Repulse* were attacked by Japanese Mitsubishi G4M ("Betty") medium bombers and Nakajima Kate torpedo-bombers off the coast of Malaysia. The *Prince of Wales* alone had 175 anti-aircraft guns capable of firing 60,000 shells a minute, and the capital ships and their destroyer escorts were in open water, able to manoeuvre at speed. Yet in little over two hours, the *Prince of Wales* and the *Repulse* had been sunk, for the loss of only three Japanese aircraft.

Even Britain, which had lost its pre-eminence in carrier development, was able to score major victories by using aircraft against surface ships. The German battleship *Bismarck*, racing through the stormy North Atlantic for the safety of Brest harbour in May 1941, would have eluded the pursuit of the Royal Navy had it not been spotted by a Catalina from RAF Coastal Command, and then damaged by Fairey Swordfish-delivered torpedoes from the carrier *Ark Royal*. That these slow-moving, open-cockpit biplanes were sent to attack the world's most high-tech battleship seems extraordinary in itself, let alone that they should have given it a crippling wound.

The Swordfish's other memorable success was the raid on the Italian fleet at Taranto in November 1940. Attacking a heavily defended shallow harbour by night, 21 of these seemingly obsolescent torpedo-bombers from the carrier *Illustrious* sank three battleships and a destroyer. A result out of all proportion to the force applied, the Taranto raid was studied with interest by naval experts around the world – including the Japanese.

STRIKING SWORDFISH
The strike on the Italian fleet at Taranto by Swordfish biplanes from the Royal Navy carrier Illustrious *in November 1940 partly inspired the subsequent Japanese attack on Pearl Harbor.*

Tora! Tora! Tora!

The surprise Japanese attack on Pearl Harbor on 7 December 1941 is one of the most celebrated, or infamous, uses of aircraft in the history of aviation. It was devised by senior Japanese commanders in a spirit of desperation, since Japan's determination to control China and southeast Asia had put it on a collision course with the United States, a country that they could not realistically hope to defeat. Admiral Isoroku Yamamoto, the naval commander-in-chief, hoped that if he could take out the American Pacific fleet at the same time as invading southeast Asia, Japan might at least buy time to organize a defence of its conquests. Admiral Nagumo, entrusted with commanding the surprise attack, was opposed to it and doubted that it would work.

Although Americans were understandably outraged at an attack timed to coincide with a declaration of war rather than follow it, the Pearl Harbor raid has to be recognized as a technically masterful naval air operation, in both its preparation and its execution. Japanese technical experts developed a torpedo that worked in the shallow water of the American harbour – normal torpedoes would have stuck in the seabed – and a bomb to pierce the battleships' armoured decks, made by adding fins to an artillery shell. Their pilots, an elite group who had survived a harsh process of elimination, rehearsed the attack meticulously. And a fleet of 31 ships, including six aircraft carriers, was assembled and sailed undetected across 1,600km (1,000 miles) of ocean

JAPANESE AIRCRAFT USED IN THE RAID ON PEARL HARBOR

Aichi D3A ("Val")

The main dive-bomber type of the Imperial Japanese Navy at the time of Pearl Harbor, 126 "Vals" (as they were nicknamed by the Allies) took part in that attack. Aichi D3As sank more Allied naval ships than any other type of enemy aircraft during the war.

ENGINE 1,080hp Mitsubishi Kinsei 44 air-cooled 14-cylinder radial	
WINGSPAN 14.4m (47ft 1in)	**LENGTH** 10.2m (33ft 5in)
TOP SPEED 460kph (272mph)	**CREW** 2
ARMAMENT 2 x 7.7mm machine guns, 1 x 7.7mm machine gun; 1 x 250kg (551lb) bomb, 2 x 30kg (66lb) bombs under wings	

Mitsubishi A6M5 Reisen ("Zero")

Japanese rising sun emblem

The most famous of all Japanese combat aircraft, when the "Zero" first appeared in 1940, it outclassed every Allied fighter in the Pacific. This lightly built aircraft displayed outstanding manoeuvrability and had unparalleled range for the time.

ENGINE 1,300hp Nakajima NK1C Sakae 21 14-cylinder radial	
WINGSPAN 11m (36ft 1in)	**LENGTH** 9.1m (29ft 11in)
TOP SPEED 557kph (346mph)	**CREW** 1
ARMAMENT 2 x 20mm cannon in wings, 2 x 7.7mm machine guns in fuselage; wing racks carry 2 x 60kg (132lb) bombs	

Nakajima B5N2 ("Kate")

Whereas the British Navy's torpedo-bomber in 1940 was the antiquated Swordfish biplane, the Japanese developed this sleek all-metal monoplane design, bristling with modern features. After playing a major role at Pearl Harbor, it went on to help in the sinking of the aircraft carriers *Yorktown*, *Lexington*, and *Hornet*.

ENGINE 1,115hp Nakajima Sakae 21 air-cooled 14-cylinder radial	
WINGSPAN 15.5m (50ft 11in)	**LENGTH** 10.3m (33ft 9in)
TOP SPEED 378kph (235mph)	**CREW** 3
ARMAMENT 2 x 7.7mm machine guns, 1 x 7.7mm machine gun in rear cockpit; 800kg (1,764lb) torpedo under centreline	

BOMBERS ON DECK

Sixteeen North American B-25 Mitchell bombers are parked nose to tail on the deck of USS Hornet, on their way across the Pacific for the April 1942 Doolittle bombing raid on Tokyo. The B-25s were too large to be stowed below decks.

to within striking distance of Hawaii, refuelling from tankers in heavy seas.

While the Japanese were greatly aided by the peace-numbed laxness of American defences, their pilots carried through the two-wave attack with skill and determination. They had the best naval aircraft in the world at that time, in the Mitsubishi A6M Reisen (Zero) fighter and the Nakajima Kate attack aircraft, used as both a torpedo-bomber and a conventional horizontal bomber. The other type used in the Pearl Harbor operation, the Aichi Val dive-bomber, was broadly similar in its strengths and weaknesses to the German Stuka.

The practical impact of the Japanese action was limited by the fortuitous absence of US carriers from Pearl Harbor on that day and by their failure to destroy oil tanks, which allowed a faster American recovery than might otherwise

AMERICAN VENGEANCE

The "sneak" attack on Pearl Harbor created a desire for instant revenge that was partially satisfied by the air-raid on Tokyo in 1942.

have been the case. Still, 18 ships were sunk or seriously damaged and some 164 aircraft destroyed on the ground, for the loss of only 29 Japanese aircraft out of a strike force of 353 planes.

Pearl Harbor convinced most remaining doubters of the power of carrier-borne aircraft as the decisive strike-force in naval warfare. Also, the destruction or temporary disablement of so much of America's surface fleet left the carriers as the US Navy's key surviving warships in the Pacific. The carriers' intended role had originally been a subordinate one. Now they would steal centre stage, with the chief function of the surface warships becoming to provide a protective screen for the carriers.

The Tokyo raid

The Pearl Harbor raid and the setbacks that followed in its wake left Americans thirsting for vengeance and in need of a lift to morale. Looking for a spectacular way of hitting back at the Japanese, the Americans hatched the plan of a carrier-launched bombing raid on Tokyo. Because no carrier aircraft had the range to strike Japan from far enough out, the USAAF was called on to provide North American B-25 Mitchell bombers. No one had ever thought that one of these aircraft could be flown off a carrier flight-deck, but now they would have to.

Volunteer USAAF aircrews led by the renowned Lieutenant-Colonel James Doolittle were put through an intensive course of training in short take-offs and low-level, long-distance flight before sailing from San Francisco on the

> "The raid had a great many things going for it, but… the biggest thing was morale for the American people."
>
> REAR ADMIRAL HENRY L. MILLER
> ON THE DOOLITTLE RAID

carrier *Hornet* in April 1942. Sixteen bombers were tethered on the flight-deck because they were too big to be stowed below, leaving no room for the *Hornet*'s own aircraft to take off or land. The carrier *Enterprise* joined up to provide air-cover. The plan was for the bombers to take off late on 18 April, when the *Hornet* would be about 640km (400 miles) from Tokyo. They would raid the city under cover of darkness and fly on to land at Chuchow airfield in China, which was held by friendly forces. But early on the morning of the 18th, the task-force was spotted by Japanese patrol boats. It was decided to launch the bombers immediately, although Tokyo was now 1,050km (650 miles) away and the raiders would arrive in daylight.

Doolittle was the first to take off. With the aircraft laden with bombs and extra fuel, and the carrier pitching in a heavy sea, conditions were hardly ideal for the first sea-launching of a B-25. Fortunately there was a 30-knot wind, which, with the 20-knot progress of the carrier, gave a

TAKING OFF FOR TOKYO
Doolittle takes off from USS Hornet *on 18 April 1942. Each B-25 had to lift off just as the pitching deck swung up towards the crest of a wave.*

THE THACH WEAVE

EVEN THE BEST US Navy pilots were disturbed by having to fly against the Mitsubishi Zero fighter, which could outperform the Grumman F4F Wildcat. Fortunately, the navy had a number of exceptional pilots who developed tactics to counter the Zero. Jimmy Thach, the commander of Fighting Squadron 3, had the idea of abandoning the three-aircraft V formation – a leader and two wingmen – that was standard for US Navy fighters and previously had been for the RAF. Instead, he had his pilots adopt a four-plane combat unit made up of two two-plane sections, the same solution that the Luftwaffe Messerschmitt pilots had come up with.

Using the four-aircraft formation, Thach developed a manoeuvre specifically designed to negate the Zero's superiority in performance, which would theoretically give it the upper hand in any combat situation. Known as the "Thach Weave", the manoeuvre went like this. The two pairs of Wildcats flew with just enough air between them for a tight turn. They watched each other's tails. If a Zero came down behind one of the left-hand pair of aircraft, for example, the right-hand pair would wait until the Japanese aircraft was almost in shooting range and then turn sharply towards their

FLYING FOURS
US Navy Grumman F4F Wildcats fly in the four-plane fighting formation pioneered by Jimmy Thach. Each combat unit of four aircraft is made up of two two-plane sections. A section consists of a leader and his wingman.

colleagues. Seeing the turn, the targeted Wildcat would know he had a Zero on his tail and dive down and to his right. The right-hand pair of Wildcats would get a side shot at the Zero if he pulled out or a head-on shot if he turned to keep his target in his sights. Thach had the chance to implement this tactic at the Battle of Midway.

windspeed of 50 knots to aid lift-off. Doolittle made it look easy and the other 15 somehow got off behind him with only one minor incident.

The flight to Tokyo took four hours and achieved total surprise. The B-25s did not have a sufficient bombload to cause much damage, but their sudden appearance was a severe shock to the Japanese. The aftermath to the raid did not go as planned. Most of the bombers ran out of

fuel and all were lost, as crews either baled out or crash-landed. Of the 80 aircrew, 73 survived the raid, including Doolittle who returned to make a further distinguished contribution to World War II as a commander in the European theatre. Of little tactical significance, the Doolittle raid shocked the Japanese high command into rushing their Pacific-expansion plans.

The Battle of Midway

The instructions to the US Pacific Fleet in 1942 were to "hold what you've got and hit them when you can". There were only six full-size American carriers when the war started and their numbers were soon reduced by enemy action. The Japanese enjoyed superiority both in the quantity and overall quality of their naval aircraft and pilots. The Battle of the Coral Sea was a setback for Japan – a lesson that the Americans were still in the ring and fighting. But Japanese naval commanders remained convinced that if they could draw the US Pacific Fleet into battle, they could destroy the carriers and control the ocean. This was the scenario they envisaged when they invaded Midway Island in June 1942.

Because American cryptographers had cracked Japan's naval codes, the US was forewarned of the Midway operation and so undistracted by a simultaneous Japanese move against the Aleutian Islands. Unable to assemble a naval force comparable to the 200 ships Yamamoto sent into action, Admiral Chester Nimitz, newly appointed commander of the Pacific Fleet, had to depend on his airmen to stop the Japanese invasion. Thanks to heroic efforts by dockyard workers at Pearl Harbor, the *Yorktown* – originally estimated to need 90 days to recover from its battering in the Coral Sea – was repaired in three days, and joined the carriers *Enterprise* and *Hornet* in Nimitz's fleet. Nimitz also had at his disposal Boeing B-17 bombers stationed on Midway. The Japanese, unaware of the strength or position of US naval forces, sent four carriers to win air superiority in preparation for the invasion – other carriers had still not been refitted after the Coral Sea or were dispersed elsewhere.

The Japanese and US forces clashed off Midway on 4 June 1942. The result was an overwhelming American victory, with all four Japanese carriers sunk for the loss of the *Yorktown*. But the Battle of Midway did not give America instant air or naval superiority in the Pacific. Although Japanese naval air power had suffered a severe setback, American navy fliers were still outnumbered. The loss of the *Lexington* at the Coral Sea and the *Yorktown* at Midway was followed by the sinking of the *Wasp* and *Hornet* in the fighting around Guadalcanal in the second half of 1942. During this period, the Americans had only one carrier operational in the Pacific at any given time. Some airmen found themselves without a deck to fly off and operated alongside the Marines from the precarious airstrip at Henderson Field on Guadalcanal.

The Essex-class carriers

During 1942–43, the most powerful carrier fleet in the world was taking shape in American shipyards. Back in 1940, the US Navy had been authorized to build a new generation of heavy carriers, which emerged as the Essex class. The first of these came into operation in 1943. The design of these modern ships was based on lessons learnt from previous experience of carrier warfare, and great attention was paid to the practicalities of flying and servicing aeroplanes in a severely limited space. Much thought was also given to fire-fighting and damage control, with the result that Essex-class ships had a far better chance of surviving an enemy air attack than earlier carriers. Alongside

BATTLE OF MIDWAY

TORPEDO-BOMBERS
Douglas TBD Devastators (top) made up the bulk of the American torpedo-bomber force at the Battle of Midway, although half a dozen newer Grumman TBF Avengers (above) also took part. All suffered heavy losses – of the TBDs seen here on the deck of the Enterprise *before the battle, only four survived.*

BATTLE WAS JOINED AT MIDWAY on 4 June 1942. The Japanese carrier aircraft opened the action with an early morning raid on the airfields on Midway Island. Japanese commanders were preparing a follow-up raid before they first became aware that there might be American carriers within striking range. By that time *Enterprise* and *Hornet* had launched their aircraft to attack the Japanese carriers. Unfortunately, the American pilots had trouble locating their target and squadrons became split up. The 15 Devastator torpedo-bombers from the *Hornet* found the carriers first, but had lost touch with their fighter cover. Slow and highly vulnerable as they flew in at low altitude to deliver their attack, the Devastators were pounced on by Zero fighters. Not a single one survived.

The aircraft from the *Yorktown* had been launched well after those from the other two carriers, but now arrived in proper formation, with fighters, dive-bombers, and torpedo-bombers prepared for a co-ordinated attack. The Wildcat fighters, led by Jimmy Thach, were hopelessly outnumbered by the Zeros of

BATTLE PANORAMA
This image, from a dioramic representation of the events, shows the torpedo squadron from USS Yorktown *attacking the Japanese carriers* Soryu *and* Akagi *on the morning of 4 June. Having to hold a steady course low over the sea to deliver their torpedoes, the bombers ran the gauntlet both of anti-aircraft fire from warships and predatory fighters.*

DAUNTLESS DESTROYERS
Artist R.G. Smith's impression of the destruction of the Japanese carrier Akagi: *Douglas Dauntless dive-bombers turn away after releasing their bombs on to the carrier.*

"A number of black objects suddenly floated eerily from their wings. Bombs! Down they came straight towards me."

JAPANESE COMMANDER MITSUO FUCHIDA
ON THE ATTACK OF THE *AKAGI* BY US DIVE-BOMBERS

the Japanese combat air patrol. Although they made a brave job of occupying the Japanese fighters, they could not prevent the torpedo-bombers again being decimated. But concentrating on the low-flying aircraft, the Zeros missed the American dive-bombers – Douglas Dauntlesses not only from the *Yorktown*, but also from the *Enterprise* – that fortuitously arrived high above the enemy ships at that critical moment. As the dive-bombers turned into their steep attack, Jimmy Thach remembered: "I saw this glint in the sun, and it just looked like a beautiful silver waterfall, these dive-bombers coming down… I'd never seen such superb dive-bombing. It looked to me like almost every bomb hit…" In about five minutes three Japanese carriers – the *Akagi*, *Kaga*, and *Soryu* – were reduced to burning hulks, blackening the sky with columns of black smoke.

Despite the shock that the Japanese had received, they still had one carrier intact, the *Hiryu*. It flew off a wave of dive-bombers followed by a wave of torpedo-bombers, to deliver a counterstrike against the *Yorktown*. Since there were only half a dozen Zeros to provide cover for each wave of bombers, they were savaged by the fighters of *Yorktown*'s combat air patrol, as well as by the ship's anti-aircraft fire.

But the Japanese pilots pressed their attack regardless of losses. Remembering a Japanese with his aircraft on fire still holding steady to deliver his torpedo, Thach said: "As far as determination was concerned, you could hardly tell any difference between the Japanese pilots and the American pilots. Nothing would stop them…" Damaged by the dive-bombers and then crippled by air-launched torpedoes, the *Yorktown* was finally finished off by a torpedo from a Japanese submarine. The day ended with the destruction of the *Hiryu* by dive-bombers from the *Enterprise*.

The invasion of Midway was abandoned. The Japanese had lost four carriers and about 330 aircraft to America's one carrier and roughly 150 aircraft. Japan had also lost a large proportion of its most experienced and skilful pilots. This was a crushing defeat and is rightly regarded as the turning-point in the Pacific War.

Vought F4U Corsair

"It got to be a very fine plane once the bugs were out of it."

HERBERT D. RILEY
ADMIRAL, US NAVY

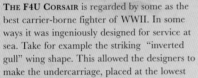

INTO ACTION
Fighter pilots in the Solomon Islands rush from a "ready room" to their waiting aeroplanes. Corsairs were flown by the US Marines (shown here), the US Navy, and Britain's Fleet Air Arm. They remained in service until the 1950s.

THE F4U CORSAIR is regarded by some as the best carrier-borne fighter of WWII. In some ways it was ingeniously designed for service at sea. Take for example the striking "inverted gull" wing shape. This allowed the designers to make the undercarriage, placed at the lowest point of the wing, short and sturdy – ideal for carrier landing, while still keeping the very large propeller clear of the deck. Yet in other ways

the Corsair was unsuited for carrier service. When the first Corsairs were delivered to the US Navy in October 1942, pilots found that with the long engine stretching in front of them, they had to lean out of the side of the cockpit to see where they were going. This proved especially problematic when attempting to land on a carrier deck, which became an extremely hazardous procedure. Consequently, the Corsair was first deployed operationally in 1943 with shore-based squadrons (mostly US Marines) in the Pacific. The aircraft was not cleared for

carrier service until April 1944, by which time the pilot's seat and cockpit canopy had been raised to improve visibility.

Once this and a number of other adjustments had been made and pilots had learned how to cope with the machine's peculiarities, the Corsair proved itself a truly outstanding aircraft. It was successful both as an air-superiority fighter and as a strike aircraft carrying either bombs or rockets. In combat with Japanese fighters such as the "Zero", kill ratios of around 11 enemy planes shot down to every Corsair lost were achieved.

ROCKETS ABOARD
These servicemen at Okinawa are loading the underwing of a Corsair with 5in rockets, June 1945.

High-visibility propeller tip

Distinctive "inverted gull" wing shape

Long nose

Sliding canopy

Large diameter propeller

Undercarriage (retracts backwards)

WHISTLING DEATH
The supercharger intercoolers on the F4U created a whistling sound in flight. The Japanese dubbed the craft "Whistling Death".

Landing-gear doors

Hydraulically operated flap

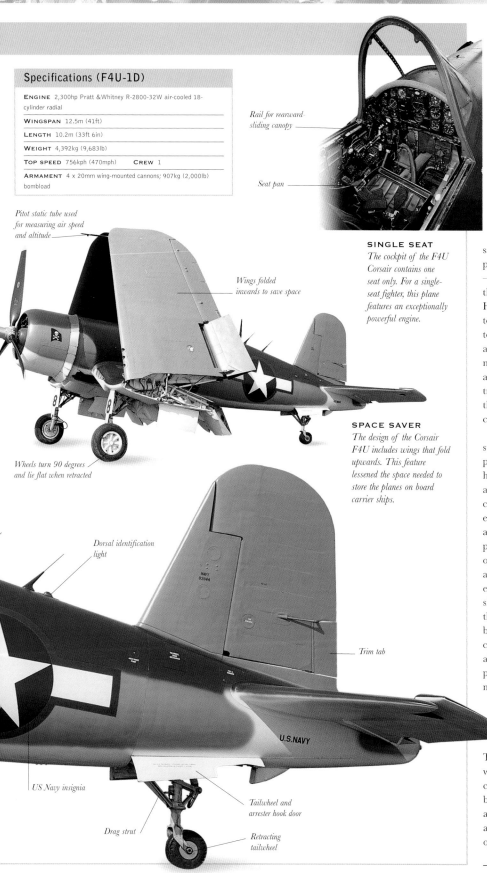

Specifications (F4U-1D)

ENGINE 2,300hp Pratt &Whitney R-2800-32W air-cooled 18-cylinder radial	
WINGSPAN 12.5m (41ft)	
LENGTH 10.2m (33ft 6in)	
WEIGHT 4,392kg (9,683lb)	
TOP SPEED 756kph (470mph)	**CREW** 1
ARMAMENT 4 x 20mm wing-mounted cannons; 907kg (2,000lb) bombload	

Rail for rearward-sliding canopy

Seat pan

SINGLE SEAT
The cockpit of the F4U Corsair contains one seat only. For a single-seat fighter, this plane features an exceptionally powerful engine.

Pitot static tube used for measuring air speed and altitude

Wings folded inwards to save space

SPACE SAVER
The design of the Corsair F4U includes wings that fold upwards. This feature lessened the space needed to store the planes on board carrier ships.

Wheels turn 90 degrees and lie flat when retracted

Dorsal identification light

Trim tab

US Navy insignia

Drag strut

Tailwheel and arrester hook door

Retracting tailwheel

these heavy carriers, new light carriers, and escort carriers, were coming into service. By the autumn of 1943, the Americans had 19 carriers of all kinds in the Pacific, and the production drive was continuing at pace.

There would be no point in having carriers without the aircraft to fly off them and the trained pilots to fly the aircraft. The size of the US naval air programme was prodigious – the number of new aeroplanes needed was initially set at 27,500. To cope with pilot training on an unprecedented scale, the navy pioneered the use of flight simulators, primitive by today's standards but good enough to allow trainees to develop the skills for deck landing, take-off, and instrument-flying far more quickly and safely than had been possible before. The mass production of a new generation of naval aircraft – the Grumman TBF Avenger torpedo-bomber, the Vought F4U Corsair and Grumman F6F Hellcat fighters, and the Curtiss SB2C Helldiver to replace the Dauntless dive-bomber – required tough, practical decisions. It was no use producing an aircraft with optimal performance if it could not be cranked out in sufficient numbers with the available factories and machine tools. It is a tribute to the American genius for organization that aeroplanes and pilots were ready for the new carriers as they rolled off the production lines.

Not all of the new aircraft were an instant success. The Helldiver was disliked by many pilots, still attached to their old Dauntlesses that had performed so well at Midway. Helldivers were a challenge for both their pilots and maintenance crews – they were complex and often faulty, especially when they had been produced by automobile companies roped in to aircraft production for the war effort. The first version of the Corsair had a glaring defect as a carrier aircraft, in that pilots could not see over the engine as they attempted to land on deck. A simple solution was eventually found by raising the pilot's seat and cockpit canopy by 16cm (6in), but the Corsair was not authorized to fly off carriers until April 1944. Yet together, these new aircraft marked a giant stride forwards in performance that was unmatched by comparable major advances on the Japanese side.

The Pacific offensive

By 1944 the United States was ready for a Pacific offensive spearheaded by carriers. The fleet was organized into carrier task-groups, with two or three carriers sailing at the centre of concentric circles composed of first cruisers and battleships providing a screen of anti-aircraft fire, and then destroyers, primarily responsible for anti-submarine defence. A number of task-groups operating together, known as a fast carrier force,

CARRIER POWER

*A carrier task-group, led by USS Essex, heads into action in
the Philippine Sea in 1944. The scale of US naval air
operations grew to overwhelming proportions, while Japanese
carrier aviation shrank to a shadow of its former power.*

NEW AVENGER
Compact and robust, the Grumman TBF Avenger became America's standard torpedo-bomber from 1943. The Japanese could match neither the quality nor the quantity of the aircraft deployed by the Americans in the later years of the war.

constituted an impressively potent weapon of attack.

From the start of operations in 1944, it was evident that a gulf had opened up between the Americans and Japanese in terms of technology, flying skills, and tactical organization. Early in February, the heavily defended Japanese naval base at Truk was subjected to a series of raids by aircraft from nine American carriers, including a night attack by Avengers using airborne radar to identify their targets. As well as suffering heavy loss of shipping, the Japanese lost ten aircraft for every one American aeroplane shot down – a sign of things to come.

The Japanese were without Admiral Yamamoto, who had himself become a victim of American air power in April 1943 after the decoding of a radio intercept revealed his itinerary for a tour of inspection of Japanese forces in the southwest Pacific. Lockheed Lightning P-38 long-range fighters led by Major John Mitchell were dispatched from Henderson Field on Guadalcanal to take the admiral. Their task was to fly over more than 640km (400 miles) of ocean and intercept a small flight of Japanese aircraft – two Mitsubishi Betty bombers carrying Yamamoto and his staff, and an escort of six Mitsubishi Zeros – without the help of onboard radar or sophisticated navigational instruments. Mitchell found his way in traditional style by dead-reckoning, using a compass, a watch, and his airspeed indicator. Thanks to his skill, and to Japanese punctuality, the P-38s found their target over the coast of Bougainville. Admiral Yamamoto's aircraft was shot down and crashed in the jungle. The admiral was thus spared from witnessing the destruction of his navy.

JAPANESE MASTERMIND
Admiral Yamamoto, the man who planned the Japanese attack on Pearl Harbor, was shot down by the Americans in 1943.

Japanese disaster
The Battle of the Philippine Sea in June 1944 was planned by Japanese commanders as a masterstroke against the US fleet, which was supposed to be trapped between a powerful carrier force and airbases on the Marianas. Japanese aircraft could in theory shuttle back and forth

> "Our pilots… were improving every day, and the Japanese fleet was not capable of countering our fleet again."
>
> VICE-ADMIRAL WILLIAM I. MARTIN
> DISCUSSING THE LAST YEAR OF THE WAR

between the carriers and the island airstrips, hitting the Americans in between. What actually happened is now known as the "Marianas Turkey Shoot".

Between 8.30am and 3.00pm on 19 June, four waves of Japanese aircraft were sent in to attack the US fleet. Fed early warning of the enemy's approach by radar operators, shipborne Combat Information Centers scrambled their fighters and vectored them on to the incoming targets. With no element of surprise, the largely inexperienced Japanese pilots were pounced on by the Hellcats. Any that slipped through were cut down by naval gunners. Meanwhile American bombers attacked the island airfields, and their fighters preyed on Japanese aircraft trying to land there. In all, the Japanese lost over 300 aircraft in the day's fighting, while two Japanese carriers were sunk by American submarines.

Admiral Marc Mitscher, commander of Fast Carrier Task-force 58, felt that the victory would be incomplete without a counterstrike against the Japanese fleet. The enemy was located late in the afternoon of 20 June. Mitscher decided to strike immediately, although it would take his aircraft to the extreme limit of their range and they would not be back before nightfall. Over 200 carrier aircraft set off, the usual mix of torpedo-bombers, dive-bombers, and fighters. Watching their fuel levels dropping on the way out, pilots were keenly aware of the problem they were going to have getting back.

It was evening by the time the Japanese fleet was found and attacked. Lieutenant Don Lewis, flying a Dauntless, recalled the end of his dive from 4,500m (15,000ft) towards a Japanese carrier: "The last time I glanced at my altimeter it registered 3,000ft [900m]. Stopped below, the big carrier looked even larger. It was completely eneveloped in a sort of smoke haze. It was hard to stay in my dive this long. Under some conditions, a person can live a lifetime in a few seconds. It was time. I couldn't go any lower. Now! I pulled my bomb release, felt the bomb go away, started my pullout. My eyes watered, my ears hurt, and my altimeter indicated 1,500ft [450m]… I had already closed my dive flaps and had 280 knots, but I couldn't seem to go fast enough… Everywhere I looked there seemed to be ships with every gun blazing. The sky was just a mass of black and white puffs, and in the midst of it, planes already hit, burning and crashing into the water below…"

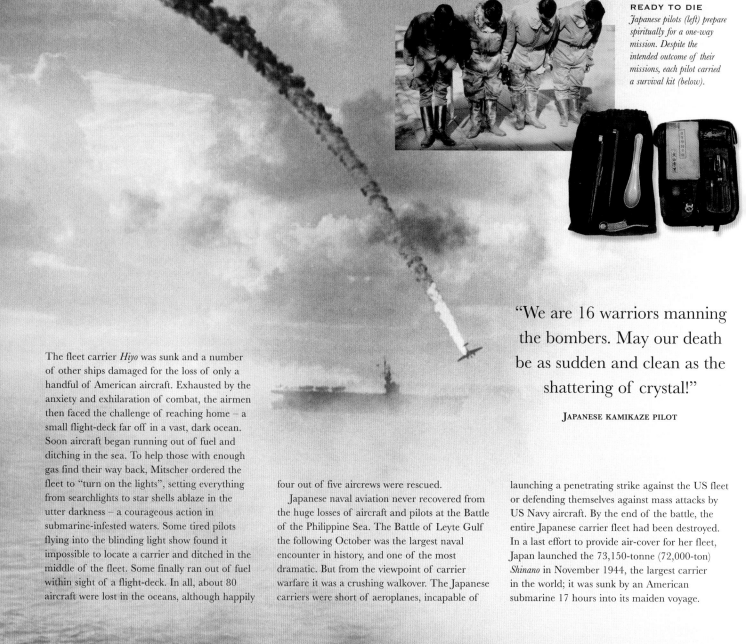

> "We are 16 warriors manning the bombers. May our death be as sudden and clean as the shattering of crystal!"
>
> JAPANESE KAMIKAZE PILOT

The fleet carrier *Hiyo* was sunk and a number of other ships damaged for the loss of only a handful of American aircraft. Exhausted by the anxiety and exhilaration of combat, the airmen then faced the challenge of reaching home – a small flight-deck far off in a vast, dark ocean. Soon aircraft began running out of fuel and ditching in the sea. To help those with enough gas find their way back, Mitscher ordered the fleet to "turn on the lights", setting everything from searchlights to star shells ablaze in the utter darkness – a courageous action in submarine-infested waters. Some tired pilots flying into the blinding light show found it impossible to locate a carrier and ditched in the middle of the fleet. Some finally ran out of fuel within sight of a flight-deck. In all, about 80 aircraft were lost in the oceans, although happily

four out of five aircrews were rescued.

Japanese naval aviation never recovered from the huge losses of aircraft and pilots at the Battle of the Philippine Sea. The Battle of Leyte Gulf the following October was the largest naval encounter in history, and one of the most dramatic. But from the viewpoint of carrier warfare it was a crushing walkover. The Japanese carriers were short of aeroplanes, incapable of

launching a penetrating strike against the US fleet or defending themselves against mass attacks by US Navy aircraft. By the end of the battle, the entire Japanese carrier fleet had been destroyed. In a last effort to provide air-cover for her fleet, Japan launched the 73,150-tonne (72,000-ton) *Shinano* in November 1944, the largest carrier in the world; it was sunk by an American submarine 17 hours into its maiden voyage.

FUTILE SACRIFICE

A Japanese fighter aircraft on a suicide mission is shot down by US Navy anti-aircraft guns. The majority of kamikaze missions ended like this – without inflicting damage on Allied warships.

Kamikaze tactics

It was during the Battle of Leyte Gulf that Japanese naval airmen first adopted kamikaze tactics. On 19 October 1944, Admiral Takijira Ohnishi, commander of the First Air Fleet, suggested to commanders based at Mabalacat airfield in the Philippines that "the only way of assuring that our meagre strength will be effective to a maximum degree" would be "to organize suicide attack units… with each plane to crash-dive into an enemy carrier." Twenty-six pilots enthusiastically volunteered to form the first "special attack unit". They were dubbed *kamikaze* ("divine wind") after a typhoon that had miraculously saved Japan from Mongolian invasion in the 13th century.

Nothing was spared in the effort to bolster morale in men effectively condemned to death. Ohnishi assured them that they were "already gods, without earthly desires". A ritual was improvised just before take-off: the kamikaze pilots drank a glass of water or sake, sang a traditional martial song, and donned the *hachimaki* headband once worn by the samurai. Thus encouraged, they went off to die for the emperor. On 25 October, a kamikaze pilot crashed a Mitsubishi Zero through the flight-deck of the escort carrier *St Lo*, dowsing the hangars in burning gasoline that ignited stored ammunition. Ripped apart by a violent explosion, the *St Lo* sank within an hour. It was a notable success for Japan amid abject failure. Over the following months, kamikaze tactics were adopted throughout the now land-based Japanese naval air force and the army air force.

There was a clear military logic to turning their aircraft into manned, guided missiles. Technologically inferior to the Americans and forced to throw poorly trained pilots into battle, the Japanese could see no other way of reaching and hitting their targets. Japan's airmen had been flying off in their hundreds to die for the emperor without inflicting the slightest damage on the US fleet. Now they would still die, but not in vain. Kamikaze pilots were presented as an elite who proved through their sacrifice the superiority of the Japanese warrior spirit even in defeat.

The reality was different. As soon as kamikaze attacks became a general tactic, it was obvious that suicide missions would be an absurdly quick way of using up the limited number of experienced pilots. Inevitably, the suicide planes were entrusted to second-raters, dispensable and in more plentiful supply. The experienced pilots flew escort, using their skills to fend off the American fighters. So even in the early days, when suicide attacks were carried out by small groups of aircraft, the kamikaze pilot was hardly a member of an elite.

FIGHTING THE KAMIKAZES

CLOSE ENCOUNTER
A Japanese Zero fighter tries to crash on to the deck of the USS Missouri. *The final moments of a kamikaze attack brought a pilot virtually face to face with his enemies.*

"**THE JAPANESE AIRCRAFT DIVED THROUGH** a rain of steel. It had been hit in several places and seemed to be trailing a banner of flame and smoke, but it came on, clearly visible, hardly moving, the line of its wings as straight as a sword. The deck was deserted; every man, with the exception of the gunners, was lying flat on his face. Flaming and roaring, the fireball passed in front of the 'island'… The entire vessel was shaken, some forty yards [35m] of the flight-deck folded up like a banana skin…" This was the terrifying reality of being on the receiving end of a kamikaze attack, in this case a hit on the carrier *Enterprise*, described by George Blond.

The damage a kamikaze inflicted was often extreme because of the combination of the aircraft's impact, the explosion of its bombload, and burning aviation fuel. To meet this menace, in the words of US pilot Jimmy Thach, "we needed to have more than a good air defence; we had to have a completely airtight defence". The number of fighter aircraft on carriers was increased, at the expense of bombers, and combat air patrols were kept in the air over the outer line of destroyer pickets, ready to intercept intruders. A careful watch was kept on returning aircraft, because kamikaze pilots would try to sneak through the defences with them. A system of "blanket air patrols" was instituted, to keep American aircraft over Japanese airbases almost round the clock, making it impossible for them to launch sorties.

LOW-LEVEL APPROACH
A Nakajima B6N Tenzan (Jill) torpedo-bomber flies in to attack a US carrier through heavy anti-aircraft fire – the splashes in the water are made by American naval gunfire.

But some Japanese aircraft always got through, and then it was up to the navy gunners to bring them down. The Japanese pilots practised all the tricks of any bombing attack on warships – for example, a pair coming in one high one low, to divide the anti-aircraft fire, or attacking simultaneously from different directions. The problem for the American gunners was, of course, that unless the Japanese aeroplane was completely destroyed or rendered uncontrollable, even if it was thoroughly shot up, the kamikaze pilot would still be able to complete his mission.

By April 1945, Japanese commanders were herding idealistic young men to slaughter in droves. As the Americans began their invasion of Okinawa, a force of over 2,000 aircraft dedicated to suicide attacks was assembled on the southern Japanese island of Kyushu under the command of Admiral Matome Ugaki. They were launched in mass attacks on the US fleet, hundreds of planes at a time attempting to overwhelm the American defences. The pilots for these *kikusui* ("floating chrysanthemum") raids were often recently drafted students who barely knew how to fly. Kanoya, the main naval air-force base on Kyushu, was under constant threat from B-29 bomber raids. The pilots were housed in half-ruined buildings, bedding down on the bare floor. In these uncomfortable and insecure surroundings, they awaited their first and last mission, most convinced that their death would be honourable and worthwhile.

Kamikaze lagacy

Kamikaze attacks undeniably had both a psychological and physical impact on the US fleet. The bewilderment and sheer terror experienced by American sailors when they first encountered suicide bombing cannot be quantified, although it never undermined their disciplined response. The physical damage inflicted is reckoned at 34 vessels sunk and 288 damaged – a considerable battering for the US Navy and, in the later stages, its British allies. After the war, the US Strategic Bombing Survey concluded that if the attacks had been carried out "in greater power and concentration they might have been able to cause us to withdraw…". But Japan did not have the resources to sustain mass suicide bombing for long. Whereas on 6 and 7 April 1945, at the height of the kamikaze frenzy, more than 300 planes a day attacked the US fleet, by June the Japanese were hard pressed to find 50 aircraft for a raid. Some 2,000 Japanese aircraft and pilots were lost in suicide attacks, far more than could be replaced.

In the end, pitting the samurai spirit of heroic self-sacrifice against overwhelming industrial might was bound to fail. The Americans organized better for production and for combat. Commanders who valued the lives of their men – and airmen who valued their own lives – fought more effectively than those who glorified death in battle.

When the Japanese emperor broadcast his country's surrender on 15 August 1945, kamikaze commander Admiral Ugaki took off with ten other pilots on a final suicide mission. On his aircraft radio he announced: "I am going to make an attack on Okinawa where my men have fallen like cherry blossoms. There I will crash into and destroy the hated enemy in the true spirit of *bushido*…". The admiral and his pilots were never seen again.

WWII US CARRIER AIRCRAFT

THE COMPLEMENT OF aircraft on a WWII US Navy carrier was largely made up of three types: dive-bombers, torpedo-bombers, and fighters, whose role was to defend the fleet and escort the bombers to their targets. When the Pacific War began in 1941, the US Navy had recently re-equipped with monoplanes such as the Douglas Dauntless, Grumman Wildcat (rushed into British service as the Martlet), and Brewster Buffalo. Within two years these were being replaced by a new generation of aircraft, including the Curtiss Helldiver, Grumman Hellcat and Avenger, and Vought Corsair. Although not all immediately popular with pilots, they gave the Americans a clear qualitative advantage over the Japanese.

CORSAIRS READY TO GO
Operating aircraft off the crowded deck of a carrier – pictured here after World War II – required discipline and organization. See pages 228–29.

Brewster F2A-3 Buffalo

ENGINE	1,200hp Wright R-1820-40 Cyclone 9-cylinder radial	
WINGSPAN	10.7m (35ft)	LENGTH 8m (26ft 4in)
TOP SPEED	517kph (321mph)	CREW 1
ARMAMENT	4 x .5in machine guns	

The first monoplane fighter to serve on US Navy aircraft carriers, the Buffalo went into service in 1939. Most were released for export (British markings shown), notably to Finland, where they fought on the Axis side throughout the war. The shortcomings of the design for combat in the Pacific were soon recognized and it was rapidly replaced by the Wildcat.

Curtiss SB2C-3 Helldiver

The replacement for the Dauntless was in theory better in every way – it was faster, had a longer range, and could carry more bombs, which were held in an internal bay. But it was late being delivered and when it was, it did not work. There were so many problems that serious thought was given to cancelling the programme, but this was politically impossible. Eventually over 7,000 were delivered (and made to work) and it became, by weight of numbers, the most widely used Allied dive-bomber of the war. That failed to stop pilots saying that SB2C stood for "Son of a Bitch, 2nd Class".

ENGINE	1,900hp Wright R-2600-20 Cyclone 14-cylinder radial	
WINGSPAN	15.1m (49ft 8in)	LENGTH 11.2m (36ft 8in)
TOP SPEED	472kph (293mph)	CREW 2
ARMAMENT	2 x 20mm cannon, 2 x .3in machine guns; 907kg (2,000lb) bombload or torpedoes, 2 x 227kg (500lb) bombs under wings	

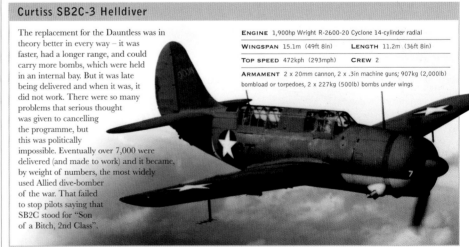

Douglas SBD-3 Dauntless

Two-man crew

Radar antenna mast

The US Navy's main dive-bomber at the start of the war, the Dauntless, based on the Northrop BT-1, was the best available and performed excellent service as a bomber and reconnaissance aircraft. The SBD-3 first entered service in March 1941, and in May and June 1942, during the battles of the Coral Sea and Midway, Dauntless squadrons sank five Japanese aircraft carriers, changing the course of the war in the Pacific. The aircraft carried a much heavier bombload than its Japanese counterpart, and its vulnerability to fighters was compensated for by its ability to absorb damage – it had the lowest loss rate of any US Pacific carrier type. Nearly 6,000 Dauntlesses were built, sinking more Japanese shipping than any other Allied weapon.

ENGINE	1,000hp Wright R-1820-52 Cyclone 9-cylinder radial	
WINGSPAN	13.9m (45ft 6in)	LENGTH 10.1m (33ft 1in)
TOP SPEED	402kph (250mph)	CREW 2
ARMAMENT	2 x .5in machine guns in nose, 2 x .3in machine guns in rear cockpit; 545kg (1,200lb) bombload	

Grumman TBM Avenger

Grumman named their new torpedo-bomber for the US Navy "Avenger" (TBM-1 with British marking shown), on the day that the Japanese attack on Pearl Harbor brought the US into WWII. It proved to be just that, playing a major part in the sinking of over 60 ships of the Imperial Japanese Navy. Two features made the Avenger outstanding. It was the first single-engined American aircraft to incorporate a power-operated gun turret, and the first to carry the heavy 577mm (22in) torpedo. In total, 9,836 Avengers were produced.

ENGINE	1,900hp Wright R-2600-20 Cyclone 14-cylinder radial	
WINGSPAN	16.5m (54ft 2in)	LENGTH 12.5m (40ft 11in)
TOP SPEED	444kph (276mph)	CREW 3
ARMAMENT	4 x .5in machine guns; 907kg (2,000lb) bombload, or 1 x 577mm (22in) torpedo and 8 x 12.7cm (5in) rockets	

Grumman F6F-5 Hellcat

When Grumman designed a replacement for the Wildcat, they consulted the pilots and produced the F6F, to the same formula but with more of everything. The engine was nearly twice as powerful, more ammunition and fuel were carried, and it was covered in over 91kg (200lb) of armour plate. In service from January 1943, the Hellcat's most famous action took place on 19 and 20 June 1944, during the Battle of the Philippine Sea (the infamous "Marianas Turkey Shoot"), when the Japanese lost nearly 350 aircraft to the Americans' 20. This effectively ended Japanese naval air power.

ENGINE	2,000hp Pratt and Whitney R2800-10W 18-cylinder radial	
WINGSPAN	13.1m (42ft 10in)	LENGTH 10.2m (33ft 7in)
TOP SPEED	612kph (380mph)	CREW 1
ARMAMENT	6 x .5in machine guns; 907kg (2,000lb) bombload, 6 x 12.7cm (5in) rockets	

Radio mast

Turbocharged Double Wasp radial engine with improved cowling

Grumman F4F Wildcat

At the outbreak of war, the Japanese "Zero" was faster, more manoeuvrable, and better armed than the Wildcat. US pilots, in many heroic defensive actions, found their aeroplane's one advantage – durability due to armour plate and self-sealing fuel tanks – could be used to defeat the Zero. Wildcats were involved in all major actions in the Pacific – including the desperate fighting on Guadalcanal – until the end of 1943, when they were replaced by the Hellcat and Corsair. They also operated from small US and Royal Navy convoy escort carriers in the Atlantic, in anti-submarine teams with torpedo-bombers.

Radio mast

USN marking

1,350hp Wright Cyclone radial engine

ENGINE	1,200hp P&W R1830-86 Twin Wasp air-cooled radial	
WINGSPAN	11.6m (38ft)	LENGTH 8.8m (28ft 11in)
TOP SPEED	512kph (318mph)	CREW 1
ARMAMENT	6 x .5in machine guns in wings	

DEATH FROM THE AIR

ALLIED BOMBING REDUCED ENEMY CITIES TO RUINS,
BUT AT A COST THAT SHOCKED THOSE WHO BELIEVED
THAT "THE BOMBER WOULD ALWAYS GET THROUGH"

> "There are a lot of people who say that bombing cannot win the war. My reply to that is that it has never been tried… and we shall see."
>
> AIR CHIEF MARSHAL
> SIR ARTHUR HARRIS
> HEAD OF RAF BOMBER COMMAND, 1942

THERE ARE FEW more apocalyptic images of total war than the mass bomber formations of World War II in action – at times more than a thousand aircraft filling the sky, some stretch of earth beneath them battered into an inferno of dust, smoke, and fire. Yet if the bomber raid offered a spectacle of power and impersonal destruction on a grand scale, the men who crewed the bombers were very far from invulnerable. For much of the war, Allied bomber crews had a higher chance of dying than the people they were bombing.

USAAF 15th Air Force bombardier Howard Jackson recalled his feelings before flying in a raid over Germany in World War II: "The terror starts on the night before the mission. This should not be confused with fear. Fear is when you have to ask a girl to dance who might say no… Terror is anxiety, dreams, rationalization of excuses not to fly, headaches, loose bowels, shaking and silence." Such feelings were almost universal, and perfectly rational. No one in the American and British armed forces had a more dangerous job than bomber crews. The USAAF Eighth Air Force – the "Mighty Eighth" – which carried out a daylight bombing offensive against Germany from 1942 to 1945, lost 26,000 men – one in eight of those who flew into action – as well as another 40,000 either wounded or shot down and made prisoners of war.

RAF Bomber Command lost a staggering 56,000 of its British and Commonwealth crew members – more than half of those who took part in its night-time strategic bombing campaign.

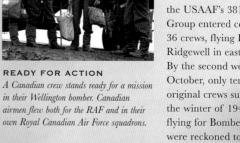

READY FOR ACTION
A Canadian crew stands ready for a mission in their Wellington bomber. Canadian airmen flew both for the RAF and in their own Royal Canadian Air Force squadrons.

At times the youthful bomber aircrews – typically men aged between 19 and 22 – faced a tour of duty with little more chance of survival than a World War I infantryman sent "over the top" at the Somme or Verdun. In late June 1943, the USAAF's 381st Bomb Group entered combat with 36 crews, flying B-17s out of Ridgewell in eastern England. By the second week in October, only ten of the original crews survived. In the winter of 1943–44, airmen flying for Bomber Command were reckoned to have a one in five chance of surviving a 30-mission tour of duty, and those were better odds than they had sometimes faced earlier in the war.

Strategic bombing

The bomber crews were asked to take this punishment because senior commanders believed that strategic bombing could make a vital contribution to winning the war, or even win it outright. Men such as Sir Arthur Harris, in charge of Bomber Command, and General Carl Spaatz of the USAAF were convinced that if they were only given the resources, they could end the war without the need for land battles that would cost hundreds of thousands of soldiers' lives.

The RAF's bombing campaign got off to a slow start. Before the war, RAF commanders had argued for the "deterrent" effect of a strategic bomber force – the threat of having his cities destroyed from the air would stay the aggressor's hand. But when Hitler invaded Poland, no attempt was made to force a German withdrawal by bombing cities. Britain's political leaders were anxious to avoid provoking German reprisals and keen to curry favour with President Roosevelt,

JUNKERS NIGHT FIGHTER
Designed as a fast bomber, the versatile Junkers Ju 88 also gave excellent service as a night fighter defending Germany against Allied bombing raids. The aerials protruding from the nose are part of the Lichtenstein airborne radar system.

BOMBING THE OILFIELDS
Consolidated B-24 Liberators of the USAAF 15th Air Force bomb the Romanian Ploesti oilfields in May 1944. More than 3,000 American airmen lost their lives in raids on Ploesti, but the eventual destruction of the oil facilities was a crucial blow against the Nazi German war machine.

who had called on all combatants to refrain from using aircraft as a weapon of terror. Until May 1940, RAF bombers were restricted to attacks on naval targets and to dropping propaganda leaflets.

In any case Bomber Command was woefully ill-equipped to carry out a strategic bombing campaign. Its twin-engined Bristol Blenheim, Armstrong Whitworth Whitley, Handley Page Hampden, and Vickers Wellington bombers were inadequate in bombload, speed, and armament. Their crews were mostly short on training and had to operate with poor bomb-sights and navigational equipment limited to a compass and a map. Yet it was presumed that they would

WALLIS' WELLINGTONS
Designed by Barnes Wallis, the twin-engined Vickers Wellington was the RAF's most advanced bomber at the start of the war. Hopelessly vulnerable in daylight raids, it proved a reasonably effective night bomber, taking part in raids on Germany until withdrawn from front-line service in 1943.

NAVIGATIONAL EQUIPMENT
Bomber Command's twin-engined bombers had to rely, for the most part, on maps and compasses to navigate to their targets. The map (far left) is Sheet N53, Berlin, of a British wartime series intended for both aircrews and ground troops. The astro-compass (left) used the Sun, the Moon, and other heavenly bodies to help aircrews plot accurate courses.

simply fly to their targets in daylight without fighter escorts and drop their bombs on designated spots. This fantasy cost hundreds of young men their lives. Losses rose to as high as 50 per cent on a single mission. And even if the bombers reached their targets, they could not bomb accurately enough to hit them.

From the spring of 1940, the unsustainable losses in daylight operations led to a switch to night bombing, for which Bomber Command was even more ill-prepared. Britain's political leaders were still reluctant to envisage attacks on civilians, authorizing raids against targets such as synthetic oil plants and railroad yards. But an official survey carried out in 1941 estimated that, groping blindly through the darkness over enemy territory, only one bomber in three managed to drop its bombload within 8km (5 miles) of its target.

Civilian targets

Logically, Britain should probably have abandoned strategic bombing altogether, at least until it could develop a fighter with sufficient range to act as a bomber

SIR ARTHUR HARRIS

THE REPUTATION OF Air Chief Marshall Sir Arthur Harris (1892–1984), head of Bomber Command from 1942 to the end of the war in Europe, has become the subject of impassioned public controversy, yet among the men he led, respect for his leadership was almost universal.

Harris had been a pilot in the Royal Flying Corps during World War I. He understood and fought for things his airmen needed – for better aircraft and equipment, for proper pay and leave, and for official recognition of their efforts – even if he had no hesitation in sacrificing their lives by the thousand in pursuit of victory. Harris never wavered in his advocacy of the bombing of cities as a war-winning strategy. His stubborn refusal to disown the February 1945 bombing of Dresden when it had become politic to do so was a piece of plain honesty that cost him dearly. He was shunned by the British establishment at the end of the war and was subsequently vilified for having carried through a policy that, in the war years, had enjoyed widespread support.

MAN ON A MISSION
Harris was a blunt and aggressive man, often to the point of rudeness. The force of his personality was felt through Bomber Command, stiffening aircrews' resolve to keep going in the face of heavy losses.

escort. But at that stage in the war it constituted the only way the British had to carry out offensive action against Germany. Accepting that technical limitations made night attacks on any target smaller than a city futile, in February 1942 Bomber Command was directed to focus on "the morale of the enemy civil population and, in particular, of the industrial workers". The "area bombing" of cities was what the bombers were capable of, so that was what they would do.

RAF bomber crews on the whole responded to the switch to bombing of civilians with dutiful indifference or outright enthusiasm. One airman recalled: "When we were briefed it was made clear that for the first time we were not attacking any military targets, but were bombing a town

indiscriminately. A great shout of excited agreement greeted the news as most aircrews had come from towns that had suffered heavily from German air attacks." Perhaps more typical was the attitude described by Avro Lancaster pilot Jack Currie: "I think we felt that we were in one hell of a battle for survival, and that we had to do, without too many qualms, the duties for which we were selected and equipped."

The adoption of area bombing coincided with the arrival of a new commander, Arthur Harris, who had total faith in the war-winning potential of the bombing of enemy civilians. He soon made his mark by organizing three headline-grabbing "thousand-bomber raids" on Germany in May and June 1942. Since Bomber Command

EVENING TAKE-OFF
As the light fades, RAF Short Stirlings line up to take off for a night raid on Germany. Stirlings could not reach the altitude achieved by Handley Page Halifaxes and Avro Lancasters and were often hit by bombs dropped by higher flying colleagues.

CITY DESTRUCTION
Hamburg was one of the most heavily damaged German cities. Until very late in the war, however, Allied bombers only rarely managed to inflict devastation on this scale.

Avro 683 Lancaster

> "We saw them coming like relief coming to a hard-pressed army; they were unconquerable; the days of heavy losses were over."
>
> **DON CHARLWOOD**
> RAF LANCASTER NAVIGATOR, 1942

THE LANCASTER WAS THE RAF'S best heavy bomber of WWII, developed from the unsuccessful twin-engined Avro Manchester. The four-engined Lancaster saw its first action in April 1942 and was greeted with widespread enthusiasm by its bomber crews. It might have been available at the start of the war had it not been for problems finding a suitable engine. The adoption of the powerful Rolls-Royce Merlin created a bomber with a superior bombload-to-range ratio than that of the American B-17 or B-24 bombers.

Lancasters had fewer guns than American bombers, partly because manpower shortages made the British baulk at a ten-man crew. The Lancaster usually flew with a seven-man crew: the pilot, flight engineer, radio operator, navigator, bombardier, and two gunners. Although the pilot was the "skipper", he was usually a sergeant, while other members of the crew – for example the navigator – might be officers. The gunners in the rear and mid-upper turrets had the worst positions in the aircraft, sitting through long flights in cold isolation, breathing oxygen through their masks, and only linked to their colleagues by their earphones. The fate of the crew depended, to a great extent, on their alertness when Luftwaffe night fighters homed in.

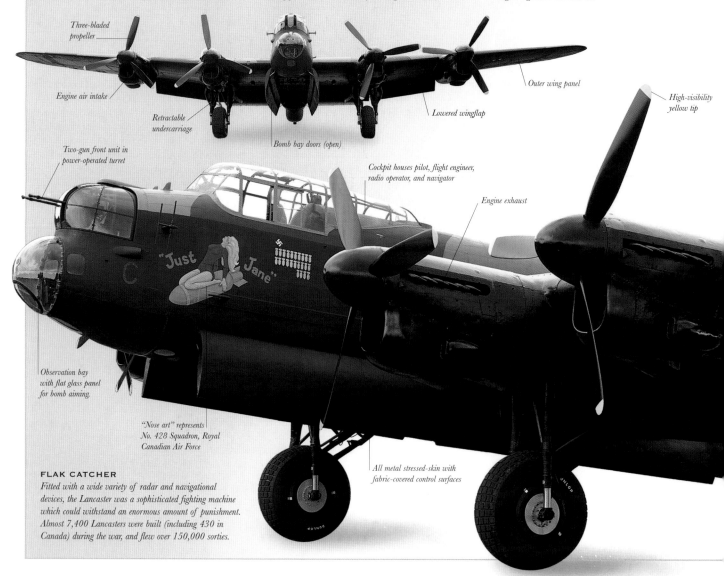

Three-bladed propeller

Engine air intake

Retractable undercarriage

Two-gun front unit in power-operated turret

Bomb bay doors (open)

Lowered wingflap

Outer wing panel

High-visibility yellow tip

Cockpit houses pilot, flight engineer, radio operator, and navigator

Engine exhaust

"Just Jane"

Observation bay with flat glass panel for bomb aiming.

"Nose art" represents No. 428 Squadron, Royal Canadian Air Force

All metal stressed-skin with fabric-covered control surfaces

FLAK CATCHER
Fitted with a wide variety of radar and navigational devices, the Lancaster was a sophisticated fighting machine which could withstand an enormous amount of punishment. Almost 7,400 Lancasters were built (including 430 in Canada) during the war, and flew over 150,000 sorties.

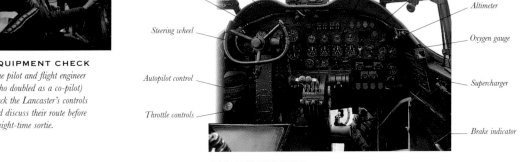

EQUIPMENT CHECK
The pilot and flight engineer (who doubled as a co-pilot) check the Lancaster's controls and discuss their route before a night-time sortie.

Rate of climb indicator
Compass
Boost counters for each of the four engines
RPM counters for each of the four engines
Rate of turn
Altimeter
Steering wheel
Oxygen gauge
Autopilot control
Supercharger
Throttle controls
Brake indicator

TUNING IN
A radio operator finds his wavelength and checks his equipment prior to take-off. Communications were conducted in morse code – note the morse key beside his right hand. The operator also handled the aircraft's defensive radar, alerting the gunners to incoming enemy fighters.

Specifications (Mk. X)

ENGINE	4 x 1,390hp Packard Merlin 28-piston
WINGSPAN	31m (102ft)
LENGTH	21m (68ft 11in)
HEIGHT	5.9m (19ft 6in)
WEIGHT	30,855kg (68,000lb)
TOP SPEED	434kph (270mph) CREW 7
ARMAMENT	8 x .303in machine guns; 6,350kg (14,000lb) bombload or 1 x 9,084kg (22,000lb) Grand Slam bomb

COCKPIT INTERIOR
The Lancaster was fully equipped for night flying, with an impressive range of flight instruments, including two sets of four dials for its high-powered Merlin engines. The flight engineer doubled as co-pilot although he was rarely well trained enough to land the bomber safely. If the pilot was killed the crew usually bailed out.

Landing light
Streamlined engine nacelle
Aileron
Rudder
Four-gun tail unit in power-operated turret
Rudder trimtab
Radar "blister"
Retractable tailwheel
NX611

PRONE POSITION
The Lancaster (shown here being loaded with a 1,800kg/4,000lb bomb) was one of the most effective long-range bombers of WWII.

IMPRESSIVE BOMBLOAD
RAF armourers check 115-kg (250-lb) bombs before they are loaded into a Stirling. Despite such careful scrutiny, about one in seven bombs dropped on Germany failed to explode.

had fewer than 500 operational front-line bombers, this involved scraping the bottom of several barrels to make up numbers, including using crews that were still in training. But at least one of the raids, on Cologne, caused considerable devastation, and a pattern was set for staging concentrated attacks by large-scale forces against a single city.

Bomber crews

At the start of the war RAF bomber crews were often poorly trained. Apart from the pilots, they consisted mostly of young men who had joined the air force to learn a trade and then signed up to fly for the extra pay. As training improved, the specialist skills of the navigator, radio operator, bomb-aimer, and gunners

complemented those of the pilot. At the end of training, men were not usually allotted to a crew, but teamed up through a process of self-selection – they were simply put together and told to sort themselves out into crews. Once formed, a crew traditionally stuck together on the ground as in the air. Lancaster pilot Jack Currie observed that for "a gunner to be in company with the pilot of another crew more than once or twice would be thought unnatural or disloyal". On a mission, all the members of the crew depended upon one another for survival; they lived or died together. One British flier observed that "friendships thus forged had a depth and unique quality that never existed with friendships before, and for me never after".

RADARS

THE NIGHT FIGHTING over Germany during World War II accelerated the development of compact airborne radar sets. The Luftwaffe, for example, fitted Bf 110 and Ju 88 night fighters with Lichtenstein interception radar, which greatly improved the aircraft's effectiveness. Manned by the radio operator, it had a range of 3km (2 miles), and was used when the aircraft were closing in on targets. British bombers were equipped with H2S radar sets (below) to help them find their targets. Various radar-jamming devices were also deployed by both sides. Radar detectors enabled the Germans to locate enemy bombers by their H2S emissions, while similar equipment warned bomber crews of the approach of radar-guided night fighters.

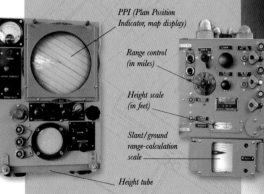

PPI (Plan Position Indicator, map display)
Range control (in miles)
Height scale (in feet)
Slant/ground range-calculation scale
Height tube

H2S RADAR SYSTEM
The RAF's H2S airborne radar helped night bombers identify targets on the ground. It was also used by the USAAF on daylight raids when there was heavy cloud.

Night-bombing operations involved many hazards, even apart from enemy flak and fighters. A heavily laden Lancaster or Halifax was not easy to get into the air, and difficult night landings were legion – even if the bomber was undamaged, it was still not simple to land at night when airfields could not be lit because of possible enemy intruders. The weather was frequently a source of anxiety. Bombers would ice up dangerously, and many pilots' worst nightmare stories were of being caught in thunderclouds, which could send a Lancaster whirling out of control like a leaf in a gale. Flying in a bomber

stream was in itself a considerable feat of airmanship, with the risk of a mid-air collision always present. Over the target, there was often the risk of being hit by bombs falling from another bomber at a higher altitude. Yet pilots also recalled moments of relief and even joy flying these mighty aircraft through the night – skimming along the top of a thick cloud layer under the stars, or relaxing with a cigarette on the return flight across the North Sea, the autopilot on and the oxygen switched off, a breakfast of bacon and eggs in prospect.

Navigational aids

The development of Bomber Command's night offensive against German cities and industrial areas through 1942 and 1943 was aided by major advances in equipment and operational tactics. Firstly, they had much better aircraft in the four-engined Halifax and Lancaster bombers, capable of carrying an impressive bombload into the heart of Germany. Then there were

navigational aids that allowed them to match the night-bombing accuracy that the Luftwaffe had achieved during the Blitz.

The first of these was Gee: a grid of radio beams projected over Germany from sites in Britain were picked up by aircraft receivers, allowing navigators to plot their progress accurately. This was enough to ensure that the average bomber crew could find a German city in the dark. Next came Oboe: one beam guided a bomber to its target, while a series of crossbeams told the crew when they were drawing near to their goal and when to release their bombs. Finally, in 1943, there was an airborne look-down radar, H2S, which improved chances of identifying targets on the ground even at night and in heavy cloud (see page 242).

Oboe and H2S were mostly used by elite squadrons of the Pathfinder Force (PFF). Flying Wellingtons or Stirlings (and, later, de Havilland Mosquitos), their job was to accurately mark targets with incendiaries for the less skilful pilots

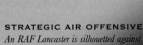

THE BATTLE OF HAMBURG

IN THE CAT-AND-MOUSE GAME that was night war over Germany, a new move giving a temporary advantage could have devastating results. This is what happened over the port city of Hamburg in late July 1943. The RAF had a new trick called "window", the scattering of strips of aluminium foil out of the bomber's flare chute to confuse German radar. Without radar, the anti-aircraft guns, the night fighters, and their controllers were blinded and let the bombers through. During the short time it took the German defenders to adapt to the new situation, a series of raids by the RAF laid Hamburg to waste.

On the night of 27–28 July, 735 bombers dropped 2,326 tonnes of explosives and incendiaries in little over an hour, starting a firestorm that killed over 40,000 people. The devastation of Hamburg represented the ideal to which Bomber Command aspired – an act of destruction on such a scale that, if repeated night after night, might have made the Germans think seriously about ending the war. But the bombers could not achieve this. They were not to repeat such destruction until Dresden in February 1945.

STRATEGIC AIR OFFENSIVE
An RAF Lancaster is silhouetted against flames, smoke, and flak during a Bomber Command attack on Hamburg in 1943.

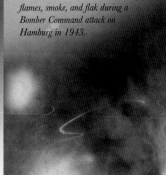

DAY-FIGHTER FAILURE
The two-seater Messerschmitt Bf 110 was a failure as an escort fighter in the Battle of Britain, but a resounding success as a night interceptor, playing a leading role in the defence of Germany.

and bomb-aimers to home in on. The main bomber force followed in behind the pathfinders in a "bomber stream" that hopefully would overwhelm the German defences by sheer numbers. By the end of 1943, bombers were fitted with VHF radio for air-to-air communication, so that a pathfinder "master bomber", circling the target during the attack, could direct the operations.

German defences

But none of these technical improvements could give the RAF air superiority in the night skies over Germany. A chain of radar stations denied the

bombers any chance of surprise. Night fighters stood ready to scramble from their airfields and intercept the attackers from whichever direction they came. Potential targets were surrounded by a dense barrier of searchlights and defended by radar-directed anti-aircraft guns. It was no wonder that Bomber Command losses remained high.

In the early stages of the war German night fighters, guided towards the bombers by ground controllers, had to rely on their eyesight to close for the kill. The introduction of the Lichtenstein airborne interceptor radar (see page 242) greatly improved their effectiveness, even if the strange assembly of direction-finding antennae fixed to the fighter's nose – dubbed the "barbed wire fence" by German pilots – produced drag that slowed the aircraft down. Later still, the Germans were equipped with infra-red detectors that could pick up on a Lancaster's engine exhausts.

Until mid-1943 the German night fighters were committed to a rigidly planned sector defence, staying within "boxes" under strict ground control. During the crisis of the Battle of Hamburg, however, more flexible tactics began to be adopted.

AIRCRAFT RECOGNITION
Luftwaffe recruits use models to learn to identify German and Allied aircraft. Airmen on both sides suffered from "friendly fire" when misidentified by anti-aircraft gunners or fellow fliers.

Pilot Hajo Herrmann pioneered what became known as *Wilde Sau* ("Wild Boar") tactics, in which single-seat day fighters roamed freely over the target area at high altitude, swooping down on bombers they spotted silhouetted against ground fires. This was so successful that specialist *Wilde Sau* squadrons were established. After *Wilde Sau* came *Zahme Sau* ("Tame Boar"), in which night fighters with onboard radar shadowed or mixed with the bomber stream, picking off targets at will.

Single-seat fighters like the Bf 109, used by *Wilde Sau* squadrons, were not equipped for night fighting. Most night operations were entrusted to two-seater Bf 110s or converted bombers such as the Ju 88. These aircraft had the crew to operate the radio and radar equipment that made night flying reasonably safe and effective, and were quite fast and nimble enough to take on unescorted bombers.

The deadly battle of wits in the night war brought both complex technical advances and inspired improvisations. For example, to exploit the fact that RAF bombers had no gun under the fuselage, some German Bf 110s were fitted with two upward-firing cannon, known rather bizarrely as *Schräge Musik* – a slang term for jazz. The interceptor positioned itself in level flight underneath a bomber and, using a reflex sight for aiming, fired directly upwards, ideally trying to hit the fuel tanks. The RAF never really found an

THE DAMBUSTERS

THE RAID ON THE RUHR DAMS on the night of 16–17 May 1943 was perhaps Bomber Command's finest hour. It started as an idea in the head of a scientist, Dr Barnes Wallis, chief designer of the Wellington bomber. Wallis worked out that a well-placed 2,700-kg (6,000-lb) bomb could breach the massive Möhne and Eder dams. He devised a "bouncing bomb" that would skip across water, hit the side of a dam, and detonate after sinking.

An elite bomber force, 617 Squadron, was created specifically for the mission. Led by 25-year-old Guy Gibson, its aircrews included Britons, Australians, Canadians, New Zealanders, and even a couple of Americans. They would have to drop their bouncing bombs at an exact altitude and distance from the dam wall at night and under fire. To achieve the precise flying height of 60ft (18.3m), the Lancasters were fitted with a pair of spotlights angled to meet on the surface of the water when the aircraft was at the right altitude. The bombers were also fitted with VHF radio so that Gibson could control operations, a novelty later to come into general use.

A total of 19 Lancasters set out for the Ruhr dams just after 9.00pm on 16 May. Five were shot down or forced to turn back before reaching their targets. At the heavily defended Möhne dam Gibson made the first bomb run, skimming in low through heavy flak, his gunners furiously returning fire. Gibson's bomb exploded against the dam but it was not breached. The next Lancaster was hit by flak and crashed into a hillside. Three more bomb runs were made, this time with Gibson and another Lancaster flying wingtip-to-wingtip with the bomb-carrying aircraft to distract the German gunners and concentrate fire on them. After a third hit, the dam at last broke, releasing a racing flood of water down into the valley below.

Later in the night the Eder dam was also breached and two other dams were damaged. The cost was heavy – eight of the 19 Lancasters were lost – and, although the raid was good for British morale, the material effect was limited.

LANCASTER CREW
Wing commander Guy Gibson (left) poses with members of the crew of his Lancaster, including bomb-aimer "Spam" Spafford on his left and navigator Terry Taerum, far right.

DAM BREACHED
A representation of the raid (left) by war artist Frank Wooten, and an aerial reconnaissance photograph (below) shows the breach in the Möhne dam and water flooding into the valley below.

BOMB BOFFIN
The idea for the bouncing bomb sprang from the brain of Dr Barnes Wallis, who later developed the concept of a swing-wing fighter.

BOUNCING BOMB
The Lancasters were adapted to carry Wallis' bouncing bombs under the fuselage, rather than in the bomb bay. An electric motor put backspin on the bomb before it was released. This helped it to hop over the anti-torpedo barrier in front of the dam and sink straight down after hitting the dam, so it would detonate right against the concrete wall.

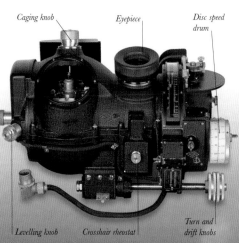

Caging knob   *Eyepiece*   *Disc speed drum*

Levelling knob   *Crosshair rheostat*   *Turn and drift knobs*

NORDEN BOMB-SIGHT
The bombardier fed the Norden bomb-sight with data on airspeed, wind direction, and other relevant factors, and waited for two crosshairs to fix on the target several miles ahead. The Norden then told him how to fly the aircraft and when to drop the bombs.

answer to *Schräge Musik*, but they did develop a whole range of countermeasures to jam or confuse German radars and radios, as well as fitting bombers with short-range radars to detect incoming fighters.

The effectiveness of German air defences could be measured by Bomber Command losses. Between August 1943 and March 1944, a period when the main target of the RAF night offensive was Berlin, on average about one in 20 bombers was lost on each raid. And there was, from the RAF's point of view, no sign of improvement. Right at the end of March 1944, in one particularly disastrous raid on the city of Nuremberg, 95 out of 795 bombers sent out did not return.

American involvement

During 1942, the USAAF joined in the European bombing campaign. Despite the discouraging British experience earlier in the war, American commanders were convinced that their unescorted bombers could penetrate the German defences in daylight, carrying out precision raids on key

strategic targets. In theory, the Boeing B-17s and Consolidated B-24s would fly too high to be hit by flak and the interlocking gunfire of their mass formations would face off enemy fighters. Attacking targets in Germany, the bombers would "shoot their way in and shoot their way out again".

The air and ground crews of the US Eighth Air Force arriving at airbases in the east of England had more to get used to than just tea and warm beer. The reality of war in Europe proved radically different from training in Texas or Arizona. First, there was the weather. One airman said of Britain: "I love that country. The people are fighters and made of the right stuff, but the climate was not my cup of tea and hell to fly in." The advantage of day bombing over night bombing resided in visibility. But in Europe cloud cover was so common and so unpredictable that this advantage often did not exist.

The key to USAAF expectations of hitting precision targets was the Norden bomb-sight (see left), a sophisticated proto-computer that could "drop a bomb in a pickle barrel". These

bomb-sights were considered so vital a secret device that between missions they were removed from aircraft and kept under armed guard. Bombardiers had strict orders to destroy them before bailing out if shot down over enemy territory. But the Norden bomb-sight was so sophisticated that only a highly trained bombardier could operate it successfully. The USAAF soon adopted a system in which only the lead bomber of the formation had the bomb-sight; the other bombers in the formation dropped their bombs when he did. But even the best of bombardiers needed to be able to see the target to use the bomb-sight. Cloud, mist, and smoke could cause havoc with bombing accuracy.

Through 1943 the USAAF daylight offensive gathered momentum, launched from bases not only in England but also in North Africa and, later, in Italy. But the bombers took heavy punishment. For instance, on 17 August 1943, B-17s from bases in England carried out simultaneous raids on the ball-bearing factory at Schweinfurt and the aircraft factory at Regensburg. Of 376 aircraft that set out on the dual mission, 60 were shot

CURTIS LEMAY

GENERAL CURTIS LEMAY (1906–90) was known as the toughest officer in the USAAF – his men were reputed to look forward to a spell in an enemy prisoner-of-war camp as a soft option compared with working for "Old Iron Ass".

It was a toughness that brought results. LeMay had been born on the wrong side of the tracks, and forged a path into the air force and up through its ranks by hard work and willpower. He took the B-17s of 305th Bomb Group to England in the fall of 1942 and proved his outfit the most effective in the "Mighty Eighth", with the tightest bomb pattern and lowest losses. He transferred to the Pacific theatre in 1944 and personally led missions against Japan to take

stock of the task. He devised the strategy of low-level night incendiary raids that reduced Japanese cities to ashes in spring 1945. After the war LeMay commanded the US nuclear bomber force. In the 1960s he was involved in right-wing politics, famously threatening the North Vietnamese with being "bombed back to the Stone Age".

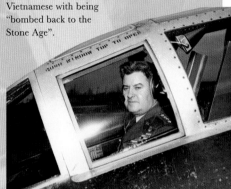

CURTIS LEMAY
As head of US Strategic Air Command in the early years of the Cold War, LeMay's hawkish personality contributed to the credibility of nuclear deterrence – here was a man who was definitely ready, if required, to unleash nuclear war.

FIERY DEATH
A B-24 Liberator erupts in flames in a raid over Austria in 1944. The USAAF photographer who took the picture said: "I felt guilty, helplessly snapping a death picture while the men were burning inside."

Boeing B-17 Flying Fortress

THE WARTIME MODELS of the Boeing B-17 – the B-17C, D, E, F, and G – were extraordinary fighting machines. Also known as the "Flying Fortress", the B-17 was bristling with machine guns, could fly at an altitude of over 9,000m (30,000ft), and – when in mass formation – was capable of delivering a staggering tonnage of explosives. In clear weather, the sophisticated Norden bombsight allowed the B-17 to strike a relatively small target, otherwise the USAF came to rely on blind bombing techniques.

For the B-17's crew of ten, conditions were cramped and uncomfortable. The aircraft was not

"The B-17 was a very sturdy, easy-to-fly airplane that would take lots of damage and get you home..."

DICK ATKINS
USAAF PILOT

Viewing panel for aiming bombs

Hamilton Standard propeller

Plexiglas nose

Navigation light

Outer wing panel

De-icing strip

TURBO ENGINES
The B-17G had four turbo-supercharged engines, each with a 3.5m (11ft 6in) diameter propeller. This enabled it not only to carry a heavy bombload but also to cruise at high altitude.

Remote-controlled chin turret

Fabric-covered rudder

Specifications (B-17G)

ENGINES 4 x 1,200hp Wright R-1820 air-cooled radial with General Electric B-22 turbo-supercharger	
WINGSPAN 31.6m (103ft 9in)	
LENGTH 22.7m (74ft 4in)	
WEIGHT 16,391kg (36,135lb)	
TOP SPEED 486kph (302mph) **CREW** 10	
ARMAMENT 13 x M-2 Browning .5in machine guns; 2,724kg (6,000lb) bombload	

Hand-held waist gun

124485

Rudder trim tab

Tail gun turret

Tailwheel

Twin Browning machine guns

pressurized, and the effects of altitude sickness were highly unpleasant. The crew also had to sit for hours in freezing cold – the tail and waist gunners sometimes suffered frostbite. German fighters quickly discovered that front on was the best way to attack a B-17. The glass nose offered no protection, and until the "chin turret" was introduced in the B-17G, it had only ball and socket guns in the plexiglass nose. Fortunately the B-17 was robust: many B-17s were able to return to base despite losing large chunks of wing, fuselage, or tail.

Since the B-17 was employed in large fleets, mass production of the craft was essential. For every B-17 the Germans shot down, American factories produced more than two. As a result, there were more B-17s available in the last months of the war than at any time previously.

ESSENTIAL PREPARATIONS
This B-17 is undergoing essential pre-flight preparations at a US bomber/fighter base in England. The B-17 was able to transport a bombload of up to 2,724kg (6,000lb) a distance of approximately 3,200km (2,000 miles).

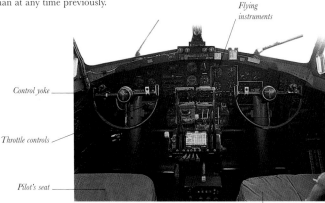

Flying instruments

B-17 COCKPIT
The pilot and co-pilot of a B-17 were afforded excellent front and side visibility from the cockpit. The most important flying instruments are situated between the two control yokes, so that both pilots can see them.

CREW CONDITIONS
Conditions were cramped in the B-17. This waist gunner (above) – wearing full flying gear, including an oxygen mask and flak jacket – is unable to stand upright. And the only way to reach the rear gun turret (right) was to crawl on all fours.

Control yoke

Throttle controls

Pilot's seat

Co-pilot's seat

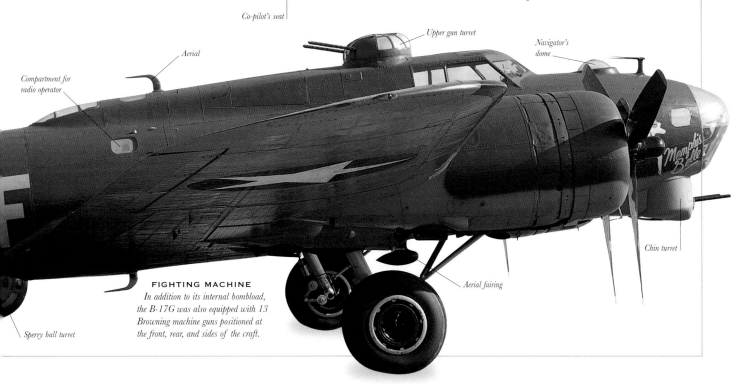

Upper gun turret

Navigator's dome

Aerial

Compartment for radio operator

Chin turret

Aerial fairing

Sperry ball turret

FIGHTING MACHINE
In addition to its internal bombload, the B-17G was also equipped with 13 Browning machine guns positioned at the front, rear, and sides of the craft.

LOADING THE FORTRESS
American ground crew load the bombs on to a Boeing B-17E Flying Fortress. The B-17's normal bombload of around 2,700kg (6,000lb) was about half that of an RAF Lancaster, making the B-17, in British terms, a medium rather than heavy bomber.

down and another 11 so badly damaged on their return that they were written off to be cannibalized for spare parts. It is a tribute to the courage and commitment of American aircrews that they held steady in the face of such losses.

Bombing routine

The routine of a US Eighth Air Force bomber operation started early. The men would be roused at around 2.00am by an operations officer snapping on the light in their hut and calling off the crews that were to fly on the day's mission. They stumbled out into the darkness

of a raw East Anglian night to walk to a briefing session, where the announcement of the target for the day, almost inevitably deep inside Germany, was greeted with a traditional groan and cursing. By 4.00am the aircrews were eating breakfast, if they had the stomach for it, while ground crews swarmed over the B-17s readying and loading them for the mission. An hour later the airmen would be in the dispersal tent by their aircraft, waiting for the red flare from the control tower that meant "start engines". Although each member of the crew meticulously carried out the practical preparations for his role on the mission, most also carried a lucky coin or a love letter about their person – faced with the lottery of death, superstition flourished.

The "Forts" took off at 30-second intervals, lifting off sluggishly, loaded with around 50 bombs, 9,500 litres (2,100 UK gallons) of fuel, and heavy crates of ammunition for the guns. Then the ball-turret gunner climbed into his position and at 3,500m (10,000ft) the crew went on to oxygen, while the aircraft continued climbing, taking up its place in the formation. Setting off for Germany, the bombers were an awesome sight. A pilot recalled his "exhilaration and pride" at the spectacle: "The great battle formations were something to see! As far as the eye could see there were B-17s, some of them olive-drab Fs, others the new silver Gs." The flight was cold and uncomfortable, especially for the ball-turret gunner suspended in space with his knees pulled up almost to his chest.

Facing the flak

As soon as the day bombers were beyond the range of fighter escort, they came under attack from swarms of German fighters, mostly Bf 109s and Focke-Wulf Fw 190s. Discomfort and cold were immediately forgotten as the terror and excitement took over. The Luftwaffe pilots assigned to defence of the fatherland were the cream of the service and quite undeterred by the bombers' bristling guns. They often attacked head-on, trying to break up the bomber formation and exploiting the lack of armour and armament on the B-17's nose. Or they would launch beam attacks, raking the bombers with machine-gun and cannon fire as they cut across them. Gunners blazed away as the German aircraft buzzed around the formation. Soon parachutes would be peppering the sky as airmen jumped from their burning B-17s; some Fortresses exploded in mid-air, giving the crews no chance of escape.

Approaching the target, inevitably through heavy flak, the bombardier and his Norden bomb-sight took over control of the B-17 from the pilot. Holding straight and steady on the run into the target was essential to bombing accuracy, but

FLYING FORTRESS RAID

AT THE START OF THE WAR, US commanders believed that formations of B-17 Flying Fortresses would be immune to fighter attack, because of their speed, high altitude, and the weight of fire from their many guns. But German fighters were not easily deterred. They discovered that the best way to attack a B-17 was from straight ahead. The glass nose provided no protection and, until the chin gun was introduced in the B-17G, there was no forward-firing armament. Fighters attacking head-on sometimes shot a tail-gunner in the back with fire passing down the inside of the fuselage. Coming under fighter attack brought an intense experience of terror mixed with

"None of us ever thought of going back."

LIEUTENANT ROBERT MORRILL
B-17 PILOT

adrenaline-fuelled excitement. Lieutenant Robert Morrill, a B-17 pilot in the raid on Regensburg, 17 August 1943, recalled: "Time after time my entire ship shook as every gun fired. The air was filled as the formation fired thousands upon thousands of tracers. I glanced behind me and found the top turret gunner standing in a heap of shell cases that covered the entire floor. In the cockpit my hands were glued to the wheel and throttles. I don't believe I could have let go if I had tried. In spite of the cold, sweat was running from my hair under the helmet and down across the oxygen mask, falling on to my jacket and freezing there. But none of us ever thought of turning back."

Fortunately the B-17 had a remarkable ability to survive punishment. Robust and reliable, many B-17s made it back to base even after receiving hits that took off large chunks of wing, fuselage, or tail. This was the key to its popularity with aircrews. As one American officer put it: "The B-17 was a very sturdy, easy-to-fly aeroplane that would take lots of damage and get you home."

FLAK JACKET
To protect them from anti-aircraft fire, US aircrew wore reinforced flak jackets, which were introduced in 1942.

KEEPING OUT THE COLD
Like all the crew, the B-17 waist gunners wore heated flying suits to protect them from the cold at high altitudes. The temperature inside the aircraft could drop to -45°C (-50°F).

OFFICER PILOTS
The pilot and co-pilot of a USAAF bomber were always officers, and usually college graduates. The B-17 afforded them an excellent view of the sky and ground.

a gift to German gunners on the ground. The relief was palpable when the crew felt the bomber lift upwards as the weight of the bombs dropped away – and then there was the small question of getting home. Many of the B-17s would by then be carrying significant damage; some would have dead or wounded aircrew on board. The hours would seem long before, in mid-afternoon, the bombers returned to their airfields, counted in by anxious ground crews.

Fighter support

The only way that the effectiveness of day bombing, and the survival rate of bomber crews, could be improved was through fighter escorts to protect them from attack. But the P-38 Lightnings and P-47 Thunderbolts used as escorts in 1943,

with a range of about 725km (450 miles), there and back, could only accompany the B-17s and B-24s as far as the German border. Obviously, the Germans fighters waited until the escorts had turned away and then attacked the bombers at will. The Americans desperately needed an aircraft with a range nearer to 1,600km (1,000 miles), yet capable of holding its own in air combat with the Messerschmitts and Focke-Wulfs.

The need for a long-distance escort fighter had not been foreseen before the war and, given the long lead-time involved in developing and putting into production a new aircraft, it was, on the face of it, unlikely that the need could be met. A fortuitous solution came from an unlikely quarter. The North American P-51 Mustang was a fighter aircraft that had failed. Produced in the

THE RED TAILS

REPUTEDLY THE ONLY American escort fighter group never to lose a bomber to enemy fighters was the 332nd, operating out of Italy. The 332nd Fighter Group also had another distinction: all its pilots were black. The African-American aviators only won the right to fly in combat through a long, hard battle against prejudice. Even after the USAAF reluctantly agreed to train them, at Tuskegee in Alabama, senior commanders stubbornly resisted sending them into combat, convinced that blacks were only suitable for use as ancillary staff. Protest and political pressure finally saw the first all-black fighter unit sent to North Africa in 1943.

The four-squadron 332nd, commanded by Lieutenant Colonel Benjamin O. Davis, assumed escort duties in mid-1944, flying on missions to well-defended targets such as Berlin and the Ploesti oilfields. Because of the colour of the paint on their

99TH SQUADRON
The 99th Pursuit Squadron (whose insignia is shown right) was awarded two Presidential Unit Citations before joining the 332nd.

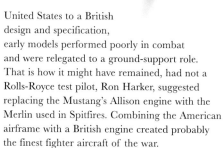

Mustangs, bomber pilots called them the Red Tails. The Tuskegee airmen more aggressively dubbed themselves the "Spookwaffe". In the strictly segregated US armed forces, the black pilots faced humiliating restrictions – they had to use separate R&R facilities and messes. But bomber crews soon learned that they were safer with the Red Tails than with any other escorts.

ALL-BLACK SQUADRON
Pilots of the 332nd Fighter Group pose beside one of their Mustangs in Italy in 1944. The Tuskegee airmen were credited with destroying 261 enemy aircraft, and black pilots earned 95 Distinguished Flying Crosses during the war.

United States to a British design and specification, early models performed poorly in combat and were relegated to a ground-support role. That is how it might have remained, had not a Rolls-Royce test pilot, Ron Harker, suggested replacing the Mustang's Allison engine with the Merlin used in Spitfires. Combining the American airframe with a British engine created probably the finest fighter aircraft of the war.

Equipped with drop tanks, the Mustang could fly to Berlin and back. And with a top speed of over 700kph (440mph), plus an impressive rate of climb and operational ceiling, it outclassed the Bf 109s and Fw 190s in combat performance. Introduced into the fighter-escort role from December 1943, the Mustang soon shifted the balance of the air war in favour of the American intruders, reducing bomber losses and sharply increasing the number of Luftwaffe aircraft shot down. By the spring of 1944 the USAAF was sending its bombers into Germany with up to 1,000 fighters in support.

Allied success

During 1944, the Allies won air superiority over Germany – the precondition for truly effective strategic bombing. Although the diversion of bombers to prepare for and support the D-Day

MIGHTY MUSTANG

*The North American P-51 Mustang was the most
renowned American fighter of World War II. Its range
and high performance enabled it to give Allied bombers
an effective escort to targets deep inside Germany.*

**PREPARED
FOR TAKE-OFF**

*A ground crewman gives the signal
for a Mustang to take off. The use
of a Merlin liquid-cooled engine,
like that in the Spitfire, gave
the Mustang a very different
look from most US fighters,
which had air-cooled
radial engines.*

landings temporarily reduced the pressure on
Germany in the early summer, the subsequent
liberation of France rendered the Germans'
defensive position ever more desperate. Even so,
it took a long time and a lot of bombs to seriously
affect Germany's ability to continue the war.
Damage to transport links was countered by
swift reconstruction; damage to industrial sites
was limited by the relocation of key factories
underground. But the targeting of fuel supplies,
both through raids on the Ploesti oilfields
of Romania and on factories in Germany
manufacturing fuel out of coal, did eventually
begin to cripple the German war machine.

Even before the fuel began to run out, the
once-proud Luftwaffe was a shadow of its former
self. Despite the destruction of aircraft factories

in Allied raids in the spring of 1944, German
aircraft output actually reached its peak in
that year, with 40,000 planes
produced. But losses of pilots
could not be made good as readily
as losses of aircraft. As the war
of attrition in the air took its
toll, the Luftwaffe began
throwing its pilots into
combat with inadequate
training. In general,
neither German aircraft
nor the men flying them
were any longer a match
for their enemies.
Outnumbered and
outfought, with their

WWII GERMAN JET AND ROCKET AIRCRAFT

ALTHOUGH THE GERMANS ACHIEVED the first jet flight, with a Heinkel He 178, a week before the outbreak of WWII, they experienced many difficulties and delays in developing practical military jets. Experience led to essential design advances such as replacing the standard tail wheel on piston-engined aircraft with a nose wheel, to stop the jet efflux hitting the runway at take-off. But producing a sufficiently reliable jet engine that gave enough thrust proved time-consuming. Junkers eventually made their Jumo 004 an effective turbojet engine and supplied it to Heinkel, Messerschmitt (whose Me 262 emerged as the most prominent jet aircraft of the war), and Arado. Jets remained essentially experimental aircraft, difficult to fly and unreliable, but were not without impact late in the war.

LATE ENTRY
Introduced in 1944, the jet-propelled Me 262A-1 fighter hit top speeds over 160kph (100mph) faster than conventional Allied fighters.

Arado Ar 234B Blitz

The world's first jet bomber (234A type shown) was originally designed as a long-range reconnaissance aircraft. When it went into service in August 1944, it finally provided the Germans with photographic intelligence which years of Allied air superiority had prevented. With the weight of the externally carried bomb slowing it down, the Arado carried out relatively few operations during the final defence of Germany.

ENGINE	2 x 900kg (1,980lb) thrust Junkers Jumo 004B turbojet		
WINGSPAN	14.2m (46ft 3in)	**LENGTH**	12.7m (41ft 5in)
TOP SPEED	742kph (461mph)	**CREW**	1
ARMAMENT	2 x 20mm MG151 cannon; 1 x 500kg (1,100lb) bomb		

Bachem Ba 349B-1 Natter

An emergency interceptor design developed in late 1944, the Bachem airframe was cheaply built of wood and disposable after one flight. The rocket-powered jet, launched vertically from a ramp, was aimed at the target by the pilot, who fired the rocket missiles and then baled out. Although test flights were made, the Natter was never used in action.

Wooden airframe

ENGINE	2,000kg (4,410lb) thrust Walter 509C liquid fuel rocket, plus 4 x 300kg (660lb) thrust Schmidding solid fuel rocket boosters		
WINGSPAN	c.4m (c.13ft 2in)	**LENGTH**	c.6m (c.19ft 9in)
TOP SPEED	c.998kph (c.620mph)	**CREW**	1
ARMAMENT	24 x 73mm Hs 217 Föhn rocket missiles		

Heinkel He 162A-2

An emergency programme for a lightweight mass-produced jet fighter was started in September 1944. Amazingly, the first 46 were delivered in February 1945, but by Germany's surrender, only a couple of dozen were operational. Effective in skilled hands – of which there was a shortage at that time– it recorded no "kills" in the few actions in which it participated.

Lightweight fuselage   *920kg thrust BMW turbojet engine*

ENGINE	920kg (2,028lb) thrust BMW 003E-2 turbojet		
WINGSPAN	7.2m (23ft 7in)	**LENGTH**	9m (29ft 8in)
TOP SPEED	840kph (522mph)	**CREW**	1
ARMAMENT	2 x 20mm Mauser MG151 cannon		

Heinkel He 178

The He 178 made the world's first jet-powered flight on 27 August 1939, nearly two years before the Whittle-engined Gloster E.28/39. Its designer, Dr. Hans Joachim Pabst von Ohain, began work on turbojets around the same time as Frank Whittle, but working for Heinkel (Whittle was a serving RAF officer) he was able to get a demonstration aircraft built sooner.

ENGINE	380kg (838lb) thrust Heinkel HeS 3b turbojet		
WINGSPAN	7.2m (23ft 7in)	**LENGTH**	7.5m (24ft 6in)
TOP SPEED	c.600kph (c.373mph)	**CREW**	1
ARMAMENT	None		

Messerschmitt Me 163 Komet

Development of this revolutionary rocket-powered interceptor started in 1938 under the title "Project X". On take-off, the pilot jettisoned a small trolley, and the aircraft climbed with phenomenal speed to engage Allied bombers with its heavy cannon armament.

ENGINE	1,701kg (3,750lb) thrust Walter liquid fuel rocket		
WINGSPAN	9.3m (30ft 7in)	**LENGTH**	5.7m (18ft 8in)
TOP SPEED	960kph (596mph)	**CREW**	1
ARMAMENT	2 x 30mm Rheinmetal-Borsig MK108 cannon		

Messerschmitt Me 262A-1a Schwalbe

While the excellent sweptwing Messerschmitt Me 262 – the first operational jet fighter in the world – was ready by 1941, it did not enter service until July 1944, due to continual engine problems. By then, they were heavily outnumbered by Allied aircraft.

ENGINE	2 x 900kg (1,980lb) thrust Junkers Jumo 004B turbojet		
WINGSPAN	12.5m (40ft 11in)	**LENGTH**	10.6m (34ft 9in)
TOP SPEED	870kph (540mph)	**CREW**	1
ARMAMENT	4 x 30mm Rheinmetal-Borsig MK108 cannon		

Survivors search through the ruins of the city of Dresden, destroyed by Allied bombers in February 1945. An estimated 60,000 people died in the raids on Dresden, which became a focus for critics of the morality of strategic bombing.

airfields coming under attack and their fuel reserves shrinking, by 1945 the Luftwaffe was more or less a spent force.

German jet fighter

German fighter ace and air-defence chief Adolf Galland always held that the outcome could have been different if Germany had correctly played its trump card: the Messerschmitt Me 262 jet fighter. When he first flew an Me 262 in May 1943, he described the experience as like being "pushed by angels". Around 160kph (100mph) faster than any propeller-driven fighter, the jet was the defensive weapon Galland had been looking for, capable of penetrating any fighter cover that the bombers might receive.

The Me 262 project had been subject to frustrating delays – the airframe had been ready in 1941, but it had taken a long time to settle on a suitable engine and tackle problems with take-off and landing. Now, Galland hoped that it would be mass-produced and deployed as an impenetrable shield around Germany. Hitler did not agree. Obsessed with his search for an offensive "secret weapon" to win the war, the Führer insisted that the Me 262 be developed as a bomber. Against his wishes, a small number of the jet fighters were produced and an experimental combat unit was set up at Lechfeld in southern Germany to try them out. The Me 262 had many problems. It was difficult to fly and downright dangerous to land – the landing speed was high, requiring a long runway and putting excessive strain on the tyres, which were liable to burst. Engine "flame-outs" were disturbingly common. And although the outpaced Allied fighter pilots could not pursue the Me 262s, they found that they could pick them off by waiting over their airfields and pouncing as the jets returned to land.

Despite its faults, the Me 262 did take its toll of Allied bombers. In the last months of the war, the jets proved they could survive in the air when Bf 109s and Fw 190s were being shot down in their hundreds by the numerically superior Allied

fighters. In the final months of the war, out of favour with the Nazi hierarchy, Galland reverted to a role as squadron commander, and, grouping some of the few surviving Luftwaffe aces around him, formed Jagdverband 44 (JV44) to fly jets in a final self-consciously futile gesture of defiance. Nothing, however, could disguise the fact that German cities were by then wide open to bombing by day or night. The destruction of Dresden in February 1945 and the reduction of Berlin to gaunt, smouldering ruins constituted a belated fulfilment of the apocalyptic vision of the advocates of strategic bombing.

New bomber

Japan was spared from bombing until late in the war because of its sheer distance from its enemies. After the one-off carrier-borne Doolittle raid of April 1942 (see page 225), targets in Japan were not struck again until the summer of 1944, when America's new B-29 bombers came into operation. The B-29 Superfortress marked an impressive advance in military aviation. It had a range of around 6,400km (4,000 miles) – double that of a B-17 – and a top speed of 560kph (350mph). The crew worked in conditions that made earlier bombers seem primitive, with heated, pressurized cabins to avoid the unpleasant effects of high-altitude flight and remote-controlled guns operated by gunners using

computerized sights. The downside was that the USAAF rushed the new bomber into service with inadequate testing. From June 1944, B-29s based in India, operating via Nationalist-controlled southwest China, were sent to attack Japan. Mounting these raids from Asia posed formidable logistical problems, and also led to substantial losses through various kinds of equipment failure and engine fires, as well as enemy action.

Far East offensive

In late November 1944, the B-29s began a sustained offensive from bases on the Marianas Islands in the Pacific. For three months they pounded Japan from high altitude in daylight, using the same "precision-bombing" tactics as had been employed over Germany, to limited effect. Flying at 9,000m (30,000ft) in the Siberian jetstream, encountering winds of 320kph (200mph), navigators frequently failed to find their targets and, with cloud cover most days, bombardiers could not hit them if they did.

In March 1945 General Curtis LeMay instigated new tactics. The B-29s were sent in by night at low altitude. Stripped of their armament to save weight, they carried a maximum load of mostly incendiary bombs. On the night of 9 March, 279 B-29s started a firestorm that destroyed almost a quarter of Tokyo and may have killed 80,000 people. Bombers arrived back at their island bases blackened with soot. Night after night, other cities suffered a similar fate. Soon most Japanese people had fled to the countryside, industrial production had plummeted, and transport links were cut. Even before the dropping of the atomic bombs on Hiroshima and Nagasaki in August 1945, strategic bombing had crippled Japan's ability to continue the war.

WALTER NOWOTNY

THE MAN CHOSEN TO LEAD the Luftwaffe's first fully operational jet-fighter squadron was an Austrian, Major Walter Nowotny (1920–44). Aged 23, Nowotny was already a renowned fighter ace, a veteran of the Russian front with 225 credited victories, and a proud wearer of the Knight's Cross with Oak Leaves, Swords, and Diamonds. In October 1944 the Me 262s of Kommando Nowotny were based at Achmer and Heseper airfields near the border with Holland, on the main flight path for US bombers attacking Germany. For a month Nowotny led his pilots into battle in their often-faulty, accident-prone jets, before meeting an almost inevitable fate. On 8 November, during a visit to the Achmer airfield by fighter chief Adolf Galland, Nowotny took off for the last time to engage the bombers and their fighter escorts. He shot down a Consolidated

LUFTWAFFE ACE
Major Walter Nowotny was credited with 258 victories in aerial combat, mostly on the Eastern Front, before his death flying a Messerschmitt Me 262.

B-24 and possibly a Mustang before one of his engines burst into flames. His jet dived into the ground alongside the airfield. RAF pilot Pierre Clostermann recalled that when news of Nowotny's death came through, he was remembered "almost with affection" as a brave enemy, belonging to the fraternity of fighter pilots who knew "no ideologies, no hatred, and no frontiers".

Postwar analysis

In the postwar period, Allied strategic bombing retrospectively came under criticism on practical and moral grounds. Practically, it was said to have wasted resources that could have been better used to other military purposes. Morally, it was attacked for deliberately or inadvertently causing the deaths of hundreds of thousands of civilians. Certainly, it had taken a long time to achieve air superiority and to field the numbers of bombers and develop the bombs needed to devastate the enemy's heartland. Yet in the end it had done, and few Germans or Japanese would be inclined to minimize the bombing's impact. As for morality, in wartime it soon comes to seem natural to harm your enemy as much as you can and in any way that you can, and this truth governed the use of bombers from Warsaw and Rotterdam to Hiroshima and Nagasaki.

> "Thank God the war is over and I don't have to get shot at any more. I can go home."
>
> **THEODORE VAN KIRK**
> ENOLA GAY NAVIGATOR

Morality of war

The bomber commanders were hard men. A story is told of Arthur Harris that once during the war he was stopped by a British policeman for speeding in his car: "You'll kill someone if you go on driving like that," says the constable. "Young man," Harris replies, "I kill hundreds of people every night." When LeMay was asked after the war about the morality of what the bombers had done, he was typically forthright: "To worry about the morality of what we were doing? Nuts! A soldier has to fight. We fought. If we accomplished the job in any given battle without losing too many of our own folks that was a pretty good day." Airmen have expressed similar views. Melvin Larsen, who flew in B-17s, stated that bombing never gave him concern because: "I knew that each time we dropped our bombs... helped to bring the ending of the war that much closer." The airmen had a job to do, and they did it, at great risk to their lives.

DAYLIGHT RAIDERS
B-29 Superfortresses release their powerful bombloads on a high-level daylight mission. The American bombers achieved their maximum effectiveness in conventional raids on Japan when they were switched to low-altitude night raids with incendiaries.

THE BOMBING OF HIROSHIMA

ON 6 AUGUST 1945, AT 2.45AM local time, a B-29 bomber took off from Tinian Island in the Marianas carrying the world's first operational atomic bomb. In command was Colonel Paul W. Tibbets, head of 509th Composite Group, set up the previous year to deliver America's secret weapon. Tibbets had his mother's maiden name, Enola Gay, painted on the nose.

With the 4.4-tonne *Little Boy* bomb on board, the *Enola Gay* was overweight and gave a few anxious moments on take-off, but after that the six-hour flight went without notable incident. On the way the *Enola Gay* rendezvoused with two other B-29s that were to accompany it to the target and observe the big event.

The crew did their jobs – navigation, preparing the bomb – and snacked on coffee and ham sandwiches. Weather-reconnaissance aircraft radioed that the weather over the first-choice target, the city of Hiroshima, was mostly clear.

At 8.40am the *Enola Gay* approached the city at over 9,000m (30,000ft). Co-pilot Robert Lewis, who was scribbling a commentary on the mission, wrote: "There will be a short intermission while we bomb our target." Bombardier Major Thomas Ferebee took over control of the aircraft through the Norden bomb-sight and released *Little Boy* to airburst over the Aioi Bridge. The bomber jumped as the weight dropped away, then filled with an unbearably bright light. The first shock wave struck the plane with such force that Tibbets thought they had been hit by flak.

As the mushroom cloud rose and the ground below boiled, Tibbet announced: "Fellows, you have just dropped the first atomic bomb in history." The crew had trained hard for that moment and were relieved that it had worked. Navigator Theodore Van Kirk thought: "Thank God the war is over and I don't have to get shot at any more. I can go home."

LITTLE BOY
The atomic bomb dropped on Hiroshima was 3m (10ft) long and weighed 4,400kg (9,700lb). It exploded with the force of 12,500 tonnes of TNT.

ENOLA GAY CREW
The men who dropped the atomic bomb on Hiroshima: the leader of the mission, Colonel Tibbets, is third from the right in the back row, flanked by bombardier Major Ferebee on his right and co-pilot Captain Lewis on his left.

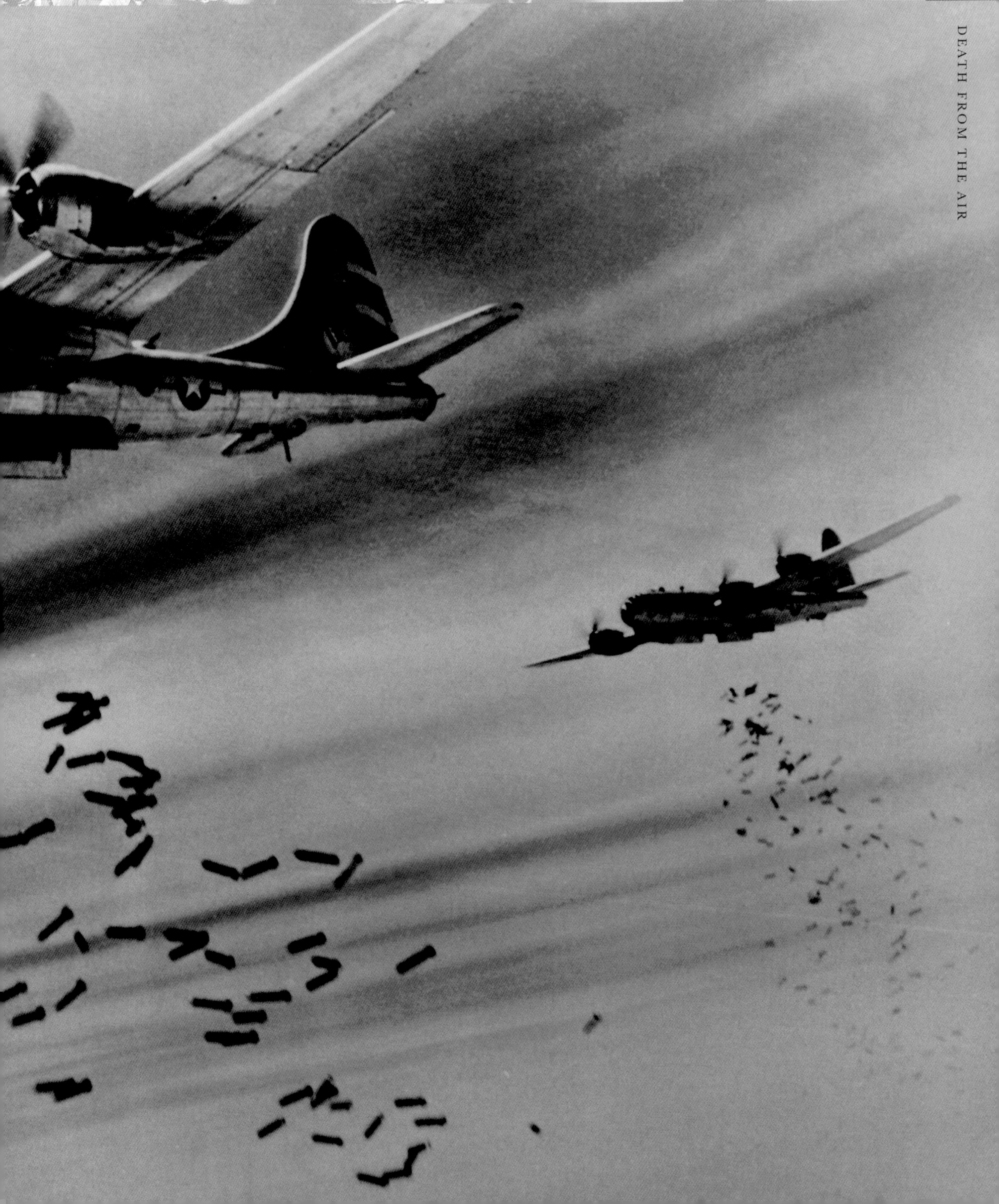

WWII BOMBERS

WWII BOMBER AIRCRAFT were powerful machines, capable of delivering a far heavier punch than their WWI predecessors. Vulnerable to ground fire and enemy fighters, bombers relied on substantial firepower along with fighter escorts or night cover for their survival. The United States and Great Britain were committed to using bombers as a strategic weapon and invested heavily in long-range heavy bombers that could be used for mass-formation raids by day or by night. Their aim was to cripple the opposing war effort. However, the impact of the Allied bombing effort on German morale was limited, but the attacks on industry – particularly oil and communications – and cities, from April 1944 onwards, eventually paid off, despite high losses. Japan suffered even more heavily under American air bombardment during 1944 and 1945.

AVRO LANCASTERS
The Avro Lancaster was the most effective British heavy bomber, able to carry a bomb load of 6,350kg (14,000lb). It featured most notably in the German dam raids of 1943. See pages 240–41.

Boeing B-29 Superfortress

Double Cyclone engine with two superchargers

Remotely controlled, power-operated gun turret

Directly controlled tail turret

Narrow, long-span wings give high lift

Introduced in 1944, the Boeing B-29 was the largest and technically most advanced bomber of the war. The B-29 was the ultimate long-range heavy bomber, responsible for dropping the atomic bombs on Hiroshima and Nagasaki that helped end the war in the Pacific. The B-29's main weakness was found in its engines being prone to fire leading to engine failure and flying accidents rather than enemy fire.

ENGINE	4 x 2,200hp Wright R-3350-23 Double Cyclone radial	
WINGSPAN	43.1m (141ft 3in)	LENGTH 30.2m (99ft)
TOP SPEED	603kph (357mph)	CREW 10–14
ARMAMENT	12 x .5in machine guns; 1 x 20mm cannon; 9,000kg (20,000lb) bombload	

Consolidated B-24J Liberator

This versatile aircraft was used as a bomber, transport, tanker, maritime patrol, reconnaissance, and anti-submarine aircraft. Built in larger numbers than any other US aircraft in history, its long range made it particularly useful in the Pacific theatre.

ENGINE	4 x 1,200hp Pratt & Whitney R-1830 Twin Wasp radial	
WINGSPAN	43.1m (141ft 3in)	LENGTH 30.2m (99ft)
TOP SPEED	507kph (300mph)	CREW 12
ARMAMENT	10 x .5in machine guns; 3,990kg (8,800lb) bombload	

de Havilland D.H.98 Mosquito B.16

The de Havilland Mosquito was a highly versatile fighter-bomber and one of the oustanding aeroplanes of WWII. The Mosquito was adapted to a wide variety of roles, including minelaying, ground attack, shipping strike, reconnaissance, and pathfinding.

Wooden construction

ENGINE	2 x 1,290hp Rolls-Royce Merlin 73 inline	
WINGSPAN	16.5m (54ft 2in)	LENGTH 26.6m (87ft 3in)
TOP SPEED	689kph (408mph)	CREW 2
ARMAMENT	1,812kg (4,000lb) bombload	

Handley Page Halifax II

Front gun-turret

Exhaust pipes fitted with flame dampers to conceal them from night fighters

Easy to fly, faster than the Wellington, and possessing a good range and bomb load, the Handley Page Halifax was the RAF's second four-engined bomber. Its only downside was its lack of defensive firepower, bringing heavy losses in raids. In 1942, the earlier variants were taken out of bomber command and given a new lease of life as minelayers and torpedo-bombers for coastal command.

ENGINE	4 x 1,390hp Rolls-Royce Merlin XXII water-cooled inline	
WINGSPAN	31.8m (104ft 2in)	LENGTH 21.4m (70ft 1in)
TOP SPEED	477kph (282mph)	CREW 7
ARMAMENT	9 x .303in Browning machine guns; 2,630kg (5,800lb) bombload	

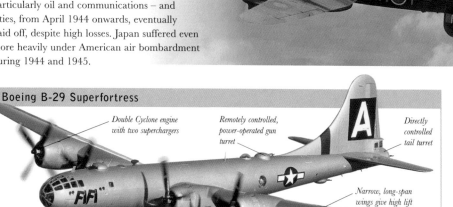

Junkers Ju 88R-1

First conceived in 1935 in response to official requirements for a high-speed medium "Schnellbomber", the Junkers Ju 88 became the main German bomber of the war. Just under 15,000 (more than all the other German bombers combined) were produced. Originally designed for level and dive bombing, the Ju 88 was also successfully adapted as a fighter, reconnaissance, and training plane. Its versatility meant that the Ju 88 was one of the oustanding planes of the war and could be called on for minelaying, torpedo bombing, pilotless missile attack, night-fighting, long-range escort, ground and sea attack, tank busting, and pathfinding duties.

ENGINE	2 x 1,600hp BMW 801MA air-cooled radial	
WINGSPAN	19.8m (65ft 8in)	LENGTH 15.6m (51ft)
TOP SPEED	470kph (292mph)	CREW 2
ARMAMENT	3 x 7.9mmMG17, 3 x 20mm MG FF/M	

Low/mid-set wing
Night-fighter variant radar
Air-cooled BMW 801MA radial engine
Ventral gondola housing 2 x 20mm cannon

Mitsubishi Ki-21 ("Sally")

The Ki-21 Type 97 heavy bomber was the Japanese Army Air Force's standard bomber at the time of Pearl Harbor. It played an important role throughout the war, although it had become obsolescent long before 1945. First flown in 1937, it went through various modifications, including the addition of a dorsal gun turret. After 1944, it was replaced by the Ki-67 ("Peggy").

ENGINE	2 x 1,490hp Mitsubishi Ha. 101 air-cooled radial	
WINGSPAN	24.9m (72ft 10in)	LENGTH 19.7m (52ft 6in)
TOP SPEED	460kph (297mph)	CREW 7
ARMAMENT	1 x 12.7mm and 5 x 7.7mm machine guns; 750kg (1,654lb) bombload	

North American B-25H Mitchell

A medium bomber, the North American B-25 Mitchell made its maiden flight in August 1940, and was in service by the time of Pearl Harbor. During the war, 870 B-25s were supplied to Russia. Later models carried a 75mm cannon along with provision for torpedoes.

ENGINE	2 x 1,850hp Wright R-2600-29 Cyclone radial	
WINGSPAN	20.6m (67ft 7in)	LENGTH 16.1m (52ft 11in)
TOP SPEED	438kph (275mph)	CREW 6
ARMAMENT	13 x .5in machine guns; 1,800kg (4,000lb) bombload	

Savoia-Marchetti SM.79-II Sparviero ("Hawk")

ENGINE	3 x 1,000hp Piaggio P.XI RC 40 radial	
WINGSPAN	21.2m (69ft 7in)	LENGTH 16.2m (53ft 2in)
TOP SPEED	456kph (270mph)	CREW 4
ARMAMENT	4 x 12.7mm Breda-SAFAT; 1 x 7.7mm Lewis machine gun; 1,240kg (2,750lb) bombload or 2 x torpedoes	

Despite its unorthodox "hunchback" appearance, the SM.79 was a highly efficient medium bomber, rated by many as the best land-based torpedo bomber of the war. It first appeared in 1934 as a commercial transport, before being tested as a bomber during the Spanish Civil War, on the side of Franco's nationalists. Making up over half of Italy's bomber force, the SM.79 was widely used around the Mediterranean area for anti-shipping, reconnaissance, and conventional bombing duties. It sank numerous British ships, including the *Malaya* and *Argos* and was also used to support Italy's North African campaign.

Short S.29 Stirling I

The Short Stirling was the first four-engined bomber to enter RAF service. Unfortunately its short wingspan, limiting its payload and altitude, proved troublesome and it was soon superseded by later heavy bombers such as the Handley Page Halifax and the Avro Lancaster. By 1943, it was being used as a transport and glider tug.

ENGINE	4 x 1,590hp Bristol Hercules XI radial	
WINGSPAN	30.2m (99ft 1in)	LENGTH 26.6m (87ft 3in)
TOP SPEED	440kph (260mph)	CREW 7–8
ARMAMENT	8 x .303in Browning machine guns; 6,350kg (14,000lb) bombload	

Vickers Wellington X

Direction-finding loop
1,675hp Bristol Hercules VI radial engine
Power-operated tail turret
Fabric-covered geodetic (latticework) structure provides protection

The twin-engined Vickers Wellington was a medium bomber renowned for the large amount of damage it could withstand, thanks to its geodetic structure, designed by the "bouncing bomb" inventor, Barnes Wallis. The Wellington was Britain's most effective night bomber until the arrival of the four-engined "heavies" and played a vital role during the dark days of 1941 and 1942. It dropped the first 1,880kg (4,000lb) "blockbuster" bomb in the Emden raid in 1941. During the mid-war years, it was effectively adapted to other roles such as maritime patrol, transport, and training.

ENGINE	2 x 1,675hp Bristol Hercules VI radial	
WINGSPAN	26.2m (86ft 2in)	LENGTH 19.7m (64ft 7in)
TOP SPEED	431kph (255mph)	CREW 6
ARMAMENT	6 x .303in Browning machine guns; 2,722kg (6,000lb) bombload	

COLD WAR, HOT WAR

5

THE PRIME DRIVING FORCE BEHIND technological developments in aviation in the 40 years after the end of World War II was the Cold War confrontation between the Western allies and the communist bloc. The imperative of national defence in a nuclear age made competition for the edge in performance intense and unabating. The result was remarkable progress in aircraft design, jet engines, avionics and weaponry. Meanwhile, shooting wars were fought in many parts of the world, in which air forces tried out the latest technology in action. In the post-Cold War period from the end of the 1980s, air power remained prominent as a means of enforcing the will and protecting the interests of states advanced enough to possess the latest technology.

NORTH AMERICAN X-15
Moments after being dropped from its Boeing B-52 mothership, a rocket-powered X-15 research plane begins a flight into the upper atmosphere. The X-15 flew faster than any other airplane, reaching over six times the speed of sound.

BEYOND THE SOUND BARRIER

IN THE POSTWAR PERIOD, TEST PILOTS IN JET AND ROCKET AIRCRAFT TOOK OUTRAGEOUS RISKS PUSHING HIGH-SPEED FLIGHT TO THE LIMITS

"Climbing faster than you can even think… You've never known such a feeling of speed while pointing up in the sky… God what a ride!"

TEST PILOT CHUCK YEAGER
ON FLYING THE BELL X-1

EARLY BRITISH JET
The British de Havilland D.H.112 Venom jet fighter-bomber debuted in 1949. De Havilland was among the leading experimenters in jet aviation, producing the Vampire in 1944 and the ill-fated D.H.108 in 1946.

ALTHOUGH JET AIRCRAFT played only a marginal role in World War II, by the end of the war it was clear that jet propulsion would hold the key to air supremacy in any future conflict. Piston-engined, propeller-driven aircraft had been pushed to the limit of their potential. Jets could fly faster – in November 1945 a British jet fighter, the Gloster Meteor, set a new official world speed record of 975.46kph (606.25mph) – and operate at higher altitudes.

Jet flight might have developed in a relatively measured, leisurely manner but for the

onset of the Cold War through the second half of the 1940s. The deepening mutual suspicions of the Soviet Union on one side, and the United States and the countries of Western Europe on the other, gave urgency to the quest for progress in military aviation technology. World War II had shown what a country with command of the air could do to its enemies. Even in countries officially at peace, there was a readiness to take

calculated risks with pilots' lives in the pursuit of a level of performance that might ultimately give an air force the edge over its potential foes. Jet development was to be rapid, secretive, and costly.

New heroes

In the race to develop high-performance jets, test pilots became the new heroes of aviation. With no hot war to fight in (most of the time), and with the decline of air racing and of long-distance flying stunts (who would any longer be impressed by solo flights across the Atlantic?), it was

ROCKET EXPERIMENT
The third Bell X-1 rocket plane – which flew only once, in 1951 – is positioned to be "mated" with the Boeing B-50 mother ship that will carry it into the air for an experimental flight at Edwards Air Force Base, California, USA. Standing on jacks, the bomber was lowered over the X-1, which was then shackled beneath the bomb bay. The B-50 released the X-1 at 6,000m (20,000ft).

flight testing that attracted the young airmen with the best natural skills and the nerve to put their lives on the line. From the late 1940s through the 1950s, they would fly faster than any humans had ever travelled before, in aircraft that could never be predicted to perform safely. The wind tunnels and other forms of simulation available at the time were simply not good enough to allow designers to iron out the flaws in their aircraft or to model the aerodynamic problems of high-speed flight. Experiments had to be conducted for real, with a pilot at the controls.

JET FIGHTER AT SEA
A Lockheed F-80 Shooting Star is serviced on board a US Navy carrier in 1946. The navy was holding trials to see whether a jet aircraft could operate from a flight deck.

T-33 jet trainer, in which generations of American fighter pilots were to learn the basics of their trade. The Soviet Union also used British engine technology in its jet-powered MiG-15s.

Airframe design

For crucial progress in airframe design, the Americans and Soviets learned from the Germans. The Me 262 had been designed with a partially swept-back wing. Captured documents showed that German aerodynamics experts had intended to sweep it back more fully, since their data indicated that this would reduce drag at very high speeds. They had not done so because a swept wing caused instability at low speeds, and they had no answer to this problem.

The key players

The development of military jets was a high-cost, high-technology business that was only going to be open to a very few players. After the Germans – clear leaders in jet-aviation technology – were put out of contention by defeat in the war, Britain found itself temporarily out in front in 1945, with its Gloster Meteor and de Havilland Vampire fighters. But the British did not have the resources to lead the way for long, and by the end of the decade they had been matched even in Europe, by the French with their Dassault Ouragan fighter and by Sweden's Saab 29 Tunnan. Inevitably, though, it was the United States and the Soviet Union that devoted the greatest resources to military jet development and were soon reaping the rewards.

The United States was surprisingly tardy in turning to jet propulsion. The Americans in effect let the Germans and British do the ground-breaking research and then built on the results. The first US jet fighter, the Bell P-59 Airacomet, was constructed around British designer Frank Whittle's jet engine, licensed to General Electric to build. It first flew in 1942, but performed disappointingly and was never sent into combat. The Lockheed P-80 Shooting Star, designed by Kelly Johnson's Skunk Works in 1943, also originally had a British engine, but only came into its own when refitted with the all-American Allison J33. As the F-80, the Shooting Star was America's first operational jet fighter – although too late for service in World War II – and gave birth to the

The German documents were made available to North American, which in 1945 was working on a straight-wing jet fighter designated as the XP-86. It led them to radically overhaul their design, adopting a fully swept-back wing and coping with low-speed instability by adding leading-edge slats. The result was the famous F-86 Sabre, which entered production in 1948. Its Soviet contemporary, the MiG-15, was also a swept-wing fighter – not so much a case of thinking alike as of poaching from the same source.

The F-86 and MiG-15 were designed to push to the edge of supersonic flight. Whether they would be able to pass Mach 1 – the speed of sound – was, at the time of their conception, unknown.

CLARENCE "KELLY" JOHNSON

INNOVATIVE DESIGNER
Kelly Johnson stands over a model of the Lockheed P-80 Shooting Star, the product of the original Skunk Works. Johnson's aircraft design work was characterized by a taste for radical innovation and imaginative solutions.

AIRCRAFT DESIGNER Clarence "Kelly" Johnson (1910–90) joined the Lockheed company in 1933, armed with a degree in aeronautical engineering from the University of Michigan. He became Lockheed's chief engineer while still in his twenties, designing the P-38 fighter and the Constellation airliner. In 1943 Johnson led the project to build the P-80 jet fighter, taking it from initial design to first flight in seven months. He achieved this by setting up a small, tightly integrated team working in total secrecy with the minimum of bureaucratic interference. Because it had to put up with the smell from a nearby plastics factory, the team was dubbed the "Skunk Works". The Skunk Works concept – defined by Johnson as "a few good people solving problems far in advance… by applying the simplest and most straightforward methods possible" – was subsequently applied to a string of Johnson-designed Lockheed aircraft, including the F-104 Starfighter, and the U-2 and SR-71 high-altitude reconnaissance aircraft.

ROCKET-LAUNCHED THUNDERJET
A Republic F-84 Thunderjet heads into the sky during an experiment in the use of zero-length launching. First flown in 1946, the Thunderjet was the successor to the same company's propeller-driven P-47 Thunderbolt and was produced by the same design team, headed by Alexander Kartveli. The zero-length launch of the F-84 was designed to provide a quick-reaction, tactical nuclear response.

HAZARDOUS EXPERIMENT

In experiments with zero-length launching at Edwards Air Force Base, California, both F-84 Thunderjets and F-100 Super Sabres were launched from missile platforms by a jettisonable rocket pod under the tail. Note the ambulance standing ready during this hazardous trial of a new technology. This series of zero-length launches was aimed at allowing fighters to get airborne even if enemy action had totally destroyed their runways.

BREAKING THE SOUND BARRIER

STOOP TO ENTER

Chuck Yeager demonstrates how to enter the cockpit of the Bell X-1 Glamorous Glennis. For experimental flights, the pilot left the ground as a passenger on the B-29 or B-50 mother ship and climbed down to the X-1 cockpit as the bomber gained altitude.

MOTHER SHIP READY

On 14 October 1947, the day that Yeager broke the sound barrier, the B-29 mother ship stands ready to take off, with the Bell X-1 mounted half inside its bomb bay. Experimental flights always took place early in the morning before the desert air heated up.

MOMENT OF TRUTH

The Bell X-1 is released from its mother ship at above 6,000m (20,000ft). The X-1 pilot then put down the nose to gather speed in a gliding dive before igniting one or more rocket chambers. Yeager described this as "the moment of truth: if you are gonna be blown up, this is likely to be when".

ON 14 OCTOBER 1947 a sound never before heard on earth echoed over the dry lake beds of the Mojave desert, California: the sonic boom from an aircraft reaching Mach 1. The aircraft was the Bell X-1 and the man at the controls was a fighter pilot from West Virginia, Chuck Yeager.

Yeager had nearly not made it into the air that morning. Two days earlier, he had ridden a horse into a gate and cracked two ribs. But there was no way the West Virginian pilot was going to pull out of a flight planned to test the feared sound barrier. Yeager's injury meant that climbing down the ladder into the X-1 cockpit, as it hung in the bomb bay of its B-29 mother plane at 2,000m (7,000ft), was agony. And it meant adding to Yeager's normal flight equipment a piece of broomstick, to use as a

lever to help him lock the cockpit door from the inside – an inspired piece of improvisation made up just before take-off.

When the B-29 had carried the X-1 up to 6,000m (20,000ft), the pilot asked Yeager whether he was ready to go. "Hell, yes," came the West Virginian drawl, "Let's get it over with." As was by then routine – it was Yeager's ninth powered X-1 flight, gradually pushing up the Mach numbers – the B-29 went into a dive and released the rocket plane like a gliding bomb. Yeager graphically described the experience of being air-launched in the X-1: "The bomb shackle release jolts you up from your seat, and as you sail out of the dark bomb bay the sun explodes in brightness."

Yeager put the nose down to gather more speed and avoid stalling, then levelled out and ignited all four rocket chambers in quick succession. Slammed back in his seat, he pointed the nose up and accelerated into the upper atmosphere.

At .88 Mach, the X-1 began to shake. This was an experience Yeager had run into before, the encounter with shock waves that would be followed

"I thought I was seeing things! We were flying supersonic! And it was as smooth as a baby's bottom: Grandma could be sitting up there sipping lemonade."

CHUCK YEAGER
ON EXCEEDING THE SPEED OF SOUND

HEADING FOR THE BARRIER
Photographed from a chaser plane, the rocket-powered Bell X-1 accelerates towards supersonic speed. The experimental flight pictured here took place in 1951, after Glamorous Glennis had been repainted white instead of its original orange.

There was widespread speculation that a "sound barrier" might block the path to any further rise in airspeed. Since no one had ever flown faster than sound, this was at least plausible. Pilots who had reached high subsonic speeds flying late-World War II piston-engined aircraft in steep dives reported violent turbulence. Another disturbing experience approaching the speed of sound was sudden loss of control of the aircraft – the controls seemed to freeze, as if all the cables to the control surfaces had been cut.

The sound barrier

High-speed flight has since become such a commonplace, it is difficult to grasp the anxieties that surrounded rising velocities in the late 1940s. No human being had ever travelled at anything approaching the speed of these jet aircraft. It was natural to ask whether there was a limit to what the human frame could stand, or a limit to the speed at which an aircraft could fly.

Speculation about the sound barrier was heightened by unexplained crashes. Chief MiG test pilot Alexei Grinchik was killed in May 1946, when the MiG-9 prototype he was flying went out of control. This concerned Soviet researchers but was unknown to the outside world. However, when Geoffrey de Havilland, son of the founder of the eponymous British aircraft company and its chief test pilot, was killed in a crash in September 1946 it caused a sensation. De Havilland was flying the D.H.108, an experimental, tailless swept-wing aircraft with the fuselage of a Vampire, when the aircraft broke up in flight. The popular press marked him down immediately as a victim of the invisible barrier in the sky.

The "right stuff"

The mystique of the sound barrier gave a special edge to experiments with the air-launched Bell X-1 rocket aircraft that began in 1946 at Muroc (later Edwards Air Force Base) in the Californian desert. The X-1 was designed purely as a test vehicle to allow NACA scientists to monitor the effects of high-speed flight. The 18 test pilots chosen for the programme included Bell's Chalmers "Slick" Goodlin and the USAAF's Captain Charles "Chuck" Yeager. It was the military flier who had the temperament to "push the outside of the envelope".

There were plenty of scares as, in flight after flight, Yeager pushed the X-1 towards the sound barrier. At .86 Mach, violent buffeting set in, like "driving on bad shock absorbers over uneven

by loss of control over the elevator – the very phenomenon that had led many people to believe that aircraft would never fly beyond the speed of sound. But Yeager was armed with a technique for avoiding loss of control, by shifting from controlling pitch with the elevator to manipulating the movable stabilizer on the tail.

Yeager rocketed on up to 13,000m (42,000ft), still gathering speed. "I noticed that the faster I got, the smoother the ride," he wrote. Then came the great moment. "Suddenly the Mach needle began to fluctuate. It went up to .965 Mach – then tipped right off the scale. I thought I was seeing things! We were flying supersonic! And it was as smooth as a baby's bottom: Grandma could be sitting up there sipping lemonade."

When Yeager glided down to land at Muroc, there was no hero's welcome waiting. The secrecy surrounding the X-1 programme extended even to a ban on too flamboyant a private celebration. A few beers were drunk and a few steaks were eaten, and that was all. But the crucial point had been made. As Yeager later wrote: "The real barrier wasn't in the sky, but in our knowledge and experience of supersonic flight."

MACHOMETER
The X-1's instrument panel featured a machometer – Mach 1 being the speed of sound. The one used during the first supersonic flight measured Mach numbers only up to Mach 1. This machometer was installed for subsequent flights.

Bell X-1

> "The four chambers blow a 30-foot [9-m] lick of flame. Christ, the impact nearly knocks you back into last week!"
>
> **CHUCK YEAGER**
> ON FLYING THE X-1 ON FULL THRUST

"**ANYONE WITH BRAIN CELLS** would have to wonder what in hell he was doing in such a situation – strapped inside a live bomb that's about to be dropped out of a bomb bay." That was Chuck Yeager's description of sitting inside the Bell X-1 rocket plane, waiting for release from its B-29 mother ship. A "live bomb" was not a bad description for the X-1, which was designed as a manned projectile with wings. Intended for ground launch, the X-1 was air-launched to maximize its flying time – two and a half minutes on full power. It would have used up all its fuel climbing towards operational altitude.

The X-1 was a strictly experimental aircraft, fitted out to collect data. Design began in 1944 when the USAAF and NACA jointly agreed on the need to investigate the problems of high-speed flight. The Bell design team created an airframe maximized both for speed and high strength – the chief concern was that the aircraft stand up to what were expected to be exceptional

stresses or buffeting. As an extra safety feature Bell engineers made the stabilizer – the horizontal tailplane – movable so that it could be used for pitch control if the elevators failed.

The X-1 was fuelled by alcohol and liquid oxygen (LOX), which had to be kept at -182.7°C (-297°F). The LOX was stored directly behind the pilot, making this, according to Yeager, "the coldest aeroplane ever flown". Piloting the X-1 was "like trying to work and concentrate inside a frozen food locker".

The pilot had the choice of igniting the four rocket chambers one at a time or in combination; all four together gave maximum thrust. Most of the flights started at around 6,000m (20,000ft) and went up to around 14,000m (45,000ft). When the day's experiment was completed, the pilot switched off the engine, jettisoned any remaining fuel, and came down in an unpowered glide. The three original X-1s finally reached Mach 1.46.

FRONT VIEW

Airspeed dial

Altitude indicator

Stabiliser position

Fuel supply

Mach number gauge

Rocket chamber pressure indicator

CRAMPED COCKPIT
The cockpit of the Bell X-1 was extremely cramped and only allowed the pilot a limited view. Pulling or pushing the control wheel moved the elevator to give pitch control up to .94 Mach; above this speed the movable stabilizer had to be operated.

Specifications

ENGINE	Reaction Motors XLR11-RM3 rocket engine
WINGSPAN	8.5m (28ft)
LENGTH	9.4m (30ft 10in)
HEIGHT	3.3m (13 ft 5 in)
WEIGHT	5,557kg (12,250lb)
TOP SPEED	1,556kph (967mph)

Pressurized cockpit

Fuselage – and even the windscreen – is shaped like a .50 caliber bullet

Horizontal stabiliser

U.S. AIR FORCE
6062

GLAMOROUS GLENNIS
Chuck Yeager named his sound-barrier-breaking Bell X-1 Glamorous Glennis in tribute to his wife. The aircraft was painted bright orange because it was thought this would help cameras and chase aircraft tracking the flights. It was later found that white was a better colour for the purpose.

Thin yet exceptionally strong wing sections improved flying control

FASTER MODELS
The Bell X-1A and X-1B were improved models of the experimental rocket aircraft that were produced between 1949 and 1952. They could carry more fuel than the original X-1, and the redesigned cockpit canopy gave the pilot a better view. Yeager reached Mach 2.44 in the X-1A in December 1953.

paving stones". At .94 Mach, the pitch control went dead as the shock wave immobilized his elevators – a solution was found in controlling pitch through tilting the stabilizer, the horizontal part of the tail assembly. At .96 Mach, the project came close to disaster in an incident unrelated to speed, when Yeager's cockpit windshield iced over at 13,000m (43,000ft), completely blocking his view; he had to make a blind landing with wholly inadequate instruments, talked down by the pilot of the chase aircraft shadowing his flight.

Finally broken

At last, on 14 October 1947, the famous sound barrier was passed without incident – so easily that Yeager recorded a feeling of let-down. "There should've been a bump on the road," he wrote, "something to let you know you had just punched a hole through that sonic barrier." But the only way he knew that he had reached Mach 1 was because his cockpit instruments, and those of the NACA engineers monitoring the flight, said so.

Breaking the sound barrier was only a start. Fliers at Edwards Air Force Base went on pushing speed and altitude records to the limit. In the early 1950s the Douglas D-558-2 Skyrocket began to set the pace, flown by US Navy pilot Bill Bridgeman and NACA test pilot Scott Crossfield. With a small swept wing and almost perfect streamlined shape, the Skyrocket reached Mach 1.88 – 1,992kph (1,238mph) at 20,100m (66,000ft) – in August 1951, and passed the new milestone of Mach 2 in November 1953. Fired by the competitive spirit so crucial to the "right stuff", Yeager quickly struck back with the Bell X-1A, recording a speed of Mach 2.44 the following December.

Perilous pursuit

These rocket-plane flights, pushing back the limits of aviation technology, continued to be extremely hazardous. The Skyrocket suffered from "supersonic yaw", which would suddenly send the plane skidding on an oblique course. The Bell X-1s were plagued by mystery explosions, which were eventually traced to a minor but fatal flaw in gaskets in the rocket system. But this was only one of, in Yeager's words, "a dozen different ways that the X-1 can kill you". Yeager himself survived total loss of

control in the X-1A during his record-setting Mach 2.44 flight of December 1953 – pushed too hard, the aircraft tumbled 15,250m (50,000ft) before he was able to wrestle it back under control. Another test pilot, Milburn Apt, was less fortunate under similar circumstances. In September 1956 Apt took the Bell X-2 to Mach 3.2, becoming the first person to fly at over 2,000mph (3,218kph). But setting the record killed him. The X-2 plummeted from high altitude out of control and, slipping in and out of consciousness, Apt was unable to bail out before hitting the ground.

CHUCK YEAGER

CHARLES ELWOOD "CHUCK" YEAGER was born in 1923 in the backwoods town of Hamlin on the Mud River, West Virginia. In 1941, fresh out of high school, he joined the air force and trained to be a fighter pilot. Sent to Europe in 1943, Yeager was shot down over Nazi-occupied France but escaped back to England via neutral Spain. He returned to combat during the invasion of Normandy. Flying Mustangs, he claimed 13 and a half kills, five of them in one day, including among his victims a Messerschmitt Me 262 jet fighter.

After the war Yeager trained as a test pilot at Wright Field in Dayton, Ohio. Selected to fly in the X-1 program at Muroc, he became the first man to break the sound barrier in October 1947. When his achievement was belatedly made public, he handled the fame and honours with the same quiet laid-back confidence that he had brought to piloting a rocket plane. He set a further speed record in 1953, reaching Mach

2.44 in the X-1A, but was also content to take the support role, often flying "chase" in a fighter while another pilot flew the test aircraft.

Yeager was unusal among test pilots in not having a college education. This lack of formal qualification prevented him being chosen as one of the first astronauts. Yeager led a fighter wing during the Vietnam War and retired from the air force as a general in 1975.

HELMETED AIRMAN
Shown here in a standard pilot's helmet, Yeager improvised his first helmet using a leather football helmet. He cut holes in it to accommodate his earphones and oxygen supply.

JACQUELINE AURIOL

FRENCH AVIATOR Jacqueline Auriol (1917–2000) was one of the few women to trespass into the macho territory of the jet test pilot. When she took up flying in the late 1940s, Auriol was already a minor celebrity, as the glamorous daughter-in-law of the president of the French Republic. She was gaining a reputation as a competition flier when, in 1949, a hydroplane in which she was travelling crashed in the river Seine. Auriol suffered serious facial injuries and underwent 22 operations to restore her features.

Undeterred, Auriol continued flying and qualified to pilot jets through a military training course in the United States. In May 1951, she flew a British Vampire at 818kph (511mph), a new record for a woman. From then until 1964, she and American pilot Jacqueline Cochran vied for the title of "fastest woman in the world". Both Jacquelines broke the sound barrier in 1953, Cochran first in an F-86 Sabre, then Auriol on board a Dassault Mystère II. By 1959 the informal competition was being conducted at speeds above Mach 2. Auriol set her last record in June 1963, when she flew a Dassault Mirage III R at 2,039kph (1,274mph). Cochran topped this in 1964, emerging as the overall winner.

SUPERSONIC JACQUELINE
Jacqueline Auriol, France's leading woman pilot of the postwar era, relaxes in her flying suit in front of a Dassault Mirage III R, the aircraft in which she reached a record-breaking speed of 2,039kph (1,274mph) in 1963.

Supersonic jets

Jet fighters were soon catching up with the rocket aircraft for speed and altitude. Piloted by George Welch, the prototype of the F-86 Sabre passed Mach 1 in a dive as early as April 1948. The F-100 Super Sabre, first flight-tested at Edwards Air Force Base in 1953, was the first jet fighter designed to go supersonic in level flight. By 1958, the F-104 Starfighter was able to set a jet speed record of 2,259kph (1,404mph) and fly to an altitude of over 30,500m (100,000ft). In fact, by the end of the 1950s, Mach 2 in level flight could be regarded as a standard requirement for state-of-the-art interceptors – not only the American aeroplanes, but aircraft such as the Soviet MiG-21, the French Dassault Mirage III, and the British Lightning.

Turbojet engines were given the potential for short bursts of very high power through the use of afterburners – spraying fuel into the hot exhaust gases to give an extra kick. And dazzlingly innovative airframes were designed to minimize drag and maximize performance. The Mirage adopted a delta-wing that gave a large wing surface for a small aircraft; the MiG-21 was a "tailed" delta-wing aircraft. At the other extreme, the F-104 – habitually described as "the missile

EJECTION SEATS

THE HIGH SPEEDS ACHIEVED through jet propulsion made a traditional parachute escape from a crippled aircraft impossible. In the forefront of the development of jets, the Germans inevitably also pioneered ejection seats, fitting them in the Heinkel He 280 prototype jet fighter from 1941. At the end of World War II the Americans lifted an ejection seat from a captured German jet and used it as the basis for developing their own.

Pilots soon found that the "bang seats" were almost as dangerous as they were necessary. At the tug of a handle, the system had to first jettison the canopy, then hurl the pilot out of the aircraft in half a second, firing him clear fast enough to miss the tail. Such rapid acceleration was bound to put enormous strain on the airman's body. And the aviator's arms and legs tended to flail around during ejection, leading to some nasty amputations. No wonder that pilots in the 1950s generally preferred to try nursing a damaged jet to a landing field, even at the risk of

ending up in a fireball or a crater in the ground. Today, pilots still only activate ejection seats as a last resort. But the British Martin-Baker seats and Russian Zevzda models have become ever safer. Strapping the pilot's arms and legs to the seat has reduced injury. Seats have also become more aerodynamically stable, and will self-right if the pilot ejects with the aircraft upside down.

IN THE HOT SEAT
A crewman ejects from the rear cockpit of a fighter as part of an ejector seat test. The percussive violence of the experience is obviously something to be avoided if possible.

DUMMY SEAT
An assistant holds up a canister containing the charge that fires an ejection seat beside a spring of the size needed to provide the same energy.

DELTA-WING SOLUTION
A Dassault Mirage III serving with the Swiss Air Force shows off its paces. The tailless delta configuration, which became virtually a Dassault trademark, was one design solution for supersonic jet fighters, combining the high-speed performance advantages of a swept-back wing with a large area for maximum lift.

with a man in it" – had a tiny wing that was razor-thin and only 2.3m (7½ft) from base to tip.

Unprecedented power and innovative design did not spell an easy life for pilots. It was estimated that in the 1950s one in four American fighter pilots would end his career by dying in a flying accident. The F-104 was a notoriously unforgiving aircraft. When it was delivered in large numbers to West Germany, attempting to rebuild an air force disbanded after the war, it killed more than 100 pilots in a decade.

By the end of the 1950s, some test pilots in the United States and the Soviet Union were preparing for a new role as astronauts or cosmonauts. Meanwhile, the X-15, star of the Edwards Air Force Base research programme from 1959 to 1968, rocketed into space – although not into orbit – reaching Mach 6.7 and an altitude of about 108km (67 miles). The heroic age of the jet pilot was transforming into the space age.

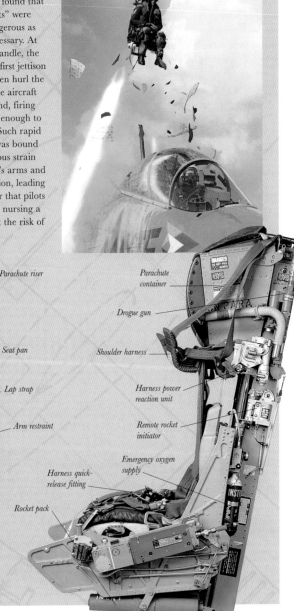

Head pad
Parachute riser
Back pad
Seat pan
Firing handle
Lap strap
Arm restraint
Personal survival pack
Leg restraint

Parachute container
Drogue gun
Shoulder harness
Harness power reaction unit
Remote rocket initiator
Emergency oxygen supply
Harness quick-release fitting
Rocket pack

EJECTOR SEAT
When activated, the Martin-Baker Type 10A ejector seat shown here, fitted in the Panavia Tornado, accelerates aviators to a speed of 160kph (100mph) in 0.4 seconds.

EARLY MILITARY JETS

ALTHOUGH THE GERMANS LED the way in the development of military jets during WWII (see pages 254–55), the British were not far behind. After successful experiments with the Gloster E.28/39, they built the first of the Gloster Meteor series, some of which flew in 1945. Initially using British jet engines, the Americans produced first the Airacomet and then the Shooting Star. Nearly all of these fighters had two engines, because a single-engined aircraft could not achieve an acceptable level of performance. Postwar jet fighters were characterized by the adoption of swept wings – an idea copied from the Germans – which reduced drag significantly at high subsonic speeds. Aircraft designers also took advantage of rapid advances in jet engine technology which made the thrust from a single engine more than adequate for such lightweight machines.

MIG INSPECTION
A North Korean defector's MiG-15 (the F-86 Sabre's main opponent in the Korean War) is inspected by USAF personnel. See pages 280–81.

Bell P-59B Airacomet

The British shared the secret of jet flight with the Americans in June 1941 by sending them a Whittle engine prototype. Bell designed a twin-engined aircraft and in October 1942, the General Electric powered P-59 became the first American jet.

ENGINE 2 x 907kg (2,000lb) thrust GE J31-GE-3 turbojet		
WINGSPAN 13.9m (45ft 6in)	**LENGTH** 11.6m (38ft 2in)	
TOP SPEED 658kph (409mph)	**CREW** 1	
ARMAMENT 1 x 37mm M4 cannon, 3 x .5in machine guns		

Dassault Mystère IVA

After WWII, Dassault played a leading part in the French aircraft industry's revival. The prototype Mystère flew in September 1952 and the French Air Force began receiving the IVA in 1955. The following year, during the Suez Crisis, Mystère IVAs flown by French and Israeli Air Force pilots, fought against Egyptian Air Force MiG-15s and MiG-17s.

ENGINE 2,855kg (6,280lb) thrust Hispano-Suiza Tay 250A turbojet	
WINGSPAN 11.1m (36ft 6in)	**LENGTH** 12.9m (42ft 2in)
TOP SPEED 1,114kph (696mph)	**CREW** 1
ARMAMENT 2 x 30mm DEFA 551 cannon; 2 x 454kg (1,000lb) bombs or 12 rockets	

de Havilland D.H.100 Vampire (F.1)

The second British jet fighter to enter service, the Vampire arrived too late to see action during WWII. As the engine and airframe were designed by the same company, much effort was put into matching the two together. This led to the twin-boom layout and short jet-pipe which increased the efficiency of the relatively low-powered jet. The Vampire was fast and manoeuvrable and equipped the first RAF aerobatic display team.

ENGINE 1,420kg (3,100lb) thrust de Havilland Goblin turbojet	
WINGSPAN 12.2m (40ft)	**LENGTH** 9.4m (30ft 9in)
TOP SPEED 869kph (540mph)	**CREW** 1
ARMAMENT 4 x 20mm Hispano cannon	

English Electric Canberra B.2

The first British jet bomber, the Canberra (B.1 shown) continued the concept already proved by the wartime Mosquito, of an unarmed medium bomber with such high performance that it could avoid fighter opposition. It entered RAF service in 1951 and saw action during the Suez Campaign of 1956. It was also built under licence in the United States and served with the USAF as the Martin B-57. The Canberra was also successful in other roles, including night intruder and photo-reconnaissance.

Upward-angled tailplane

2,948kg thrust Rolls-Royce Avon RA.3 turbojet engine

ENGINE 2 x 2,948kg (6,500lb) thrust Rolls Royce Avon 101 turbojet		
WINGSPAN 19.5m (64ft)	**LENGTH** 20m (65ft 6in)	
TOP SPEED 917kph (570mph)	**CREW** 3	
ARMAMENT 2,722kg (6,000lb) bombload		

Gloster E.28/39

The project to build the first British jet aircraft became a reality in 1940 when the Air Ministry issued a contract for an aircraft using the engine designed by Flight Lieutenant Frank Whittle. Whittle had lobbied for years to have his revolutionary idea taken up, and the E.28/39 made its first flight on 15 May 1941.

ENGINE 390kg (868lb) thrust Power Jets W.1 turbojet	
WINGSPAN 8.8m (29ft)	**LENGTH** 7.6m (25ft 3in)
TOP SPEED 749kph (466mph)	**CREW** 1
ARMAMENT None	

Gloster Meteor F.8

The Meteor was the first British jet fighter. Meteor Is first went into action in July 1944, and were used against V-1 flying bombs and in the ground attack role, during the last year of the war. The Meteor F.8 entered RAF service in 1950, replacing the earlier F.4 in Fighter Command. Royal Australian Air Force Meteor F.8s were the only British-built jet fighters to operate during the Korean War and took part in the largest air battle of that conflict, in September 1951.

ENGINE	2 x 1,633kg (3,600lb) thrust Rolls-Royce turbojet	
WINGSPAN	11.3m (37ft 1in)	LENGTH 13.5m (44ft 5in)
TOP SPEED	962kph (598mph)	CREW 1
ARMAMENT	4 x Hispano 20mm cannon	

Grumman F9F-2 Panther

Grumman, traditional suppliers of fighters to the US Navy, made their first jet design in typical sturdy fashion. The first aircraft flew in 1947 and deliveries to squadrons began in 1949. Along with McDonnell Banshees, Panthers were the first US jet aircraft in action in Korea, mainly in the ground attack role.

ENGINE	2,270kg (5,000lb) thrust Pratt and Whitney J42-2 turbojet	
WINGSPAN	11.6m (38ft)	LENGTH 11.4m (37ft 3in)
TOP SPEED	926kph (575mph)	CREW 1
ARMAMENT	4 x 20mm M3 cannon; 907kg (2,000lb) bombload	

Ilyushin Il-28 "Beagle"

One of the first Soviet jet bombers, the Il-28, went into service in 1950. Using the same Rolls-Royce derived engine as the MiG-15, it was unknown in the West for several years, but around 10,000 examples were produced. They equipped all Warsaw Pact light bomber units up until 1970, and included reconnaissance and torpedo-bomber types.

ENGINE	2 x 2,700kg (5,952lb) thrust Klimov VK-1 turbojet	
WINGSPAN	21.5m (70ft 4in)	LENGTH 17.7m (57ft 10in)
TOP SPEED	900kph (560mph)	CREW 3
ARMAMENT	2 x 23mm cannon in nose, 2 x 23mm cannon in tail turret; 3,000kg (6,500lb) bombload	

Lockheed P-80A Shooting Star

The first operational American jet fighter, the P-80 arrived in Europe too late to see action during WWII. Although it was designed around the de Havilland Goblin engine, production models were fitted with an American power unit. By the time of the Korean War (1950–53), the (redesignated) F-80 was the USAF's front-line fighter.

ENGINE	2,087kg (4,600lb) thrust Allison J33-9 turbojet	
WINGSPAN	11.9m (38ft 11in)	LENGTH 10.5m (34ft 6in)
TOP SPEED	898kph (558mph)	CREW 1
ARMAMENT	6 x .5in machine guns	

McDonnell F2H-2 Banshee

The US Navy ordered its first jet fighter in 1944 from a new firm, McDonnell, as Grumman were too busy building piston-engined planes. This was the FH-1 Phantom, which, when delivered in 1945, was too low-powered to be anything but a trainer. With improved engines the design was developed into the F2H Banshee, deliveries of which began in March 1949, beating the Grumman Panther into service by a few months. They were soon in action in the Korean War, mainly as fighter-bombers.

ENGINE	2 x 1,474kg (3,250lb) thrust Westinghouse turbojet	
WINGSPAN	13.7m (44ft 10in)	LENGTH 12.2m (40ft 2in)
TOP SPEED	856kph (532mph)	CREW 1
ARMAMENT	4 x 20mm M2 cannon; 1,361kg (3,000lb) bombload	

Saab 29 "Tunnan"

Universally known by its Swedish nickname "Tunnan", after its barrel shape, the Saab 29 was the first swept-wing fighter built in Western Europe after WWII. Entering service in 1951, the type only retired in 1976. It set several world speed records during the 1950s, and was the only Swedish aircraft ever to engage in combat. This was during the Congo Crisis of 1961 when five UN J29Bs destroyed the Katangan air force.

ENGINE	2,270kg (5,000lb) thrust Svenska Flygmotor RM2 turbojet	
WINGSPAN	11m (36ft 1in)	LENGTH 10.1m (33ft 2in)
TOP SPEED	1,060kph (659mph)	CREW 1
ARMAMENT	4 x 20mm Hispano cannon; 990kg (2,200lb) bombload under wings	

North American F-86A Sabre

North American Aviation's second jet, begun in 1944, incorporated the German-influenced swept-wing design, to allow for higher speeds. In 1948, an F-86 exceeded the speed of sound in a shallow dive, though not in level flight. In November 1950, just a few months into the Korean War, the USAF was shocked to discover that their jets could not match the Soviet MiG-15 for speed. Sabre squadrons, not yet fully operational, were rushed to Korea. While the Sabre was slightly inferior to the MiG, the more highly skilled US pilots soon established air superiority.

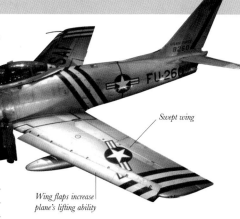

Swept wing

Wing flaps increase plane's lifting ability

ENGINE	2,725kg (6,000lb) thrust GE J47-GE-13 turbojet	
WINGSPAN	11.3m (37ft 1in)	LENGTH 11.2m (36ft 7in)
TOP SPEED	1,086kph (675mph)	CREW 1
ARMAMENT	6 x .5in machine guns; 2 x 454kg (1,000lb) bombs	

Yakovlev Yak-15

In February 1945, Soviet aircraft designers were instructed to produce jet fighters, but as no Soviet engines were available, they used captured German ones instead. Yakovlev produced this remarkable design by adapting the Yak-3 and replacing its piston engine with a jet one. An all-metal tail wheel was also fitted to resist the jet blast.

ENGINE	900kg (1,984lb) thrust RD-10 turbojet	
WINGSPAN	9.2m (30ft 2in)	LENGTH 8.7m (28ft 6in)
TOP SPEED	805kph (500mph)	CREW 1
ARMAMENT	2 x 23mm cannon above engine	

COLD WAR WARRIORS

IN THE DECADES AFTER THE END OF WORLD WAR II, NUCLEAR-ARMED AIR FORCES TRAINED TIRELESSLY TO FIGHT A THIRD WORLD WAR THAT NEVER CAME

"The swept wings gave an impression of arrow swiftness; the shining body, of brightness and cleanness; the eight great engines, of power and pure functional efficiency… [In the bomb bay] were stored two thermonuclear bombs."

PETER GEORGE
Dr Strangelove or: How I Learned to Stop Worrying and Love the Bomb

AT THE HEIGHT OF THE COLD WAR, on any day or night, about 600 American nuclear-armed bombers stood fully fuelled on alert, dispersed at airbases across the United States and in allied countries. Near the bombers their crews waited, studying weather briefings and mission plans, playing cards or watching movies, theoretically prepared at any moment for the call to be airborne in 15 minutes, before their bases could be vaporized by a Soviet surprise attack. For extra insurance, between a dozen and 70 nuclear bombers were on permanent airborne alert over the Atlantic, flying exhausting 24-hour shifts that covered 16,000km (10,000 miles), ready and waiting to be directed towards the Soviet Union. If the call came, the B-52s or B-47s would go in low, ducking under Soviet radar, shuddering and bucking over the ground contours to deliver their weapons of mass destruction. This was the world of Dr Strangelove and of Mutually Assured Destruction – in which the only rational way to keep the peace was to maintain the real threat of instant annihilation. And airmen were the intended agents of this Armageddon.

The dropping of the atom bombs on Hiroshima and Nagasaki in 1945 had opened a new era in aerial warfare. By the 1950s, both sides in the Cold

LOCKHEED U-2R
The Lockheed U-2 high-altitude reconnaissance aircraft was used as a spyplane to overfly the USSR during the Cold War. Its altitude was supposed to put the U-2 out of reach of Soviet interceptors and missiles, but two were shot down in the 1960s.

War had "the bomb". The chief function of air forces to the east and west of the Iron Curtain was to anticipate and intercept an enemy nuclear attack and penetrate enemy air defences to deliver a nuclear strike of their own. In 1947 the independent US Air Force (USAF) was created to replace the USAAF. Within it, in recognition of the importance of the nuclear role, the Strategic Air Command (SAC) was established to handle the nuclear bomber force.

In the days before the introduction of intercontinental ballistic missiles (ICBMs) and reconnaissance satellites – that is, through to the early 1960s – the nuclear confrontation was almost exclusively a business for aircraft and airmen. High-altitude spyplanes kept watch on enemy military preparations, providing photos of potential targets and air-defence installations. High-performance jet interceptors stood ready to scramble against incoming bombers. They would be directed by a chain of early-warning radar stations spread out across the north of Canada and Greenland down into Britain, feeding information to control rooms in underground bunkers. And above all there were the strategic bombers. Those theorists of aerial warfare in the 1920s and 1930s who had believed bombers were invincible and could win a war quickly on their own seemed to have been proved wrong in World War II. But the nuclear age, in an unforeseen manner, made them seem prescient. Now a bomber force really could, in principle, deliver a shock attack that would end a war in days. At least a few bombers would always get through, and armed with such destructive power, that was enough.

ROCKET-ASSISTED BOMBER

A Boeing B-47 jet bomber makes a rocket-assisted take-off. The aircraft needed the extra boost to lift off the ground with a full load of bombs and fuel. Short on range, it was turned into a credible intercontinental nuclear bomber by in-flight refuelling.

Yet while the focus of military spending and air-force planning was on nuclear offence and defence, from the start of the Cold War quite other demands were made on military aircraft and their crews as a variety of local crises and hot wars flared without escalating to nuclear conflict. The first of these was the Berlin Airlift of 1948–49, which brought hardworking

REMEMBRANCE STAMP
A West German commemorative stamp celebrates the "air bridge" that kept West Berlin supplied with food and fuel through the winter of 1948–49. Only a few years earlier, Allied aircraft had been bombing the city.

American and British transport aircraft briefly into the limelight usually hogged by the fighters and bombers.

The Berlin Airlift

The greatest supply operation in aviation history, the origins of the Berlin Airlift lay in the intendedly temporary division of defeated Germany and its ruined capital, Berlin, between the victorious allies. This left the Western powers in control of three sectors of Berlin deep within the Soviet-controlled area that later became East Germany. In the summer of 1948 the Soviets decided to take over the whole of Berlin. They blocked the overland corridor that linked the city to western Germany. With this lifeline cut off, the population of West Berlin – some 2.5 million people – would starve or freeze to death unless they accepted

communist rule and the Western allies withdrew. It was calculated that the Berliners would need 4,500 tonnes of food and fuel a day to survive the winter. When the airlift began, the USAF had 102 C-47s stationed in Europe, each capable of carrying a load of 3 tonnes. To anyone with simple arithmetic skills, it seemed obvious that an airlift could only be a token gesture. The American Military Governor of Germany, General Lucius Clay, said that even trying it would make people think he was "the craziest man in the world". Yet the airlift quickly gathered pace. Four-engined C-54 Skymasters, capable of carrying a 10-tonne load, were flown in from as far off as Hawaii and Alaska and swelled the supply armada. So did RAF Sunderland flying boats, landing on Berlin's Havel lakes, and Avro Yorks, transports developed

> ## "The sound of the engines is music to our ears."
>
> **AN ANONYMOUS BERLINER**
> WRITING AT THE TIME
> OF THE BERLIN AIRLIFT

from Lancaster bombers. New airstrips were built at Tempelhof and Gatow to take the mounting flow of traffic. By September, in good weather the airlift was comfortably exceeding its minimum target. But as winter drew in, pilots found themselves flying through at best persistent low cloud and rain, and at worst fog, ice, and snow.

Under the intense pressure of the occasion, with the future of a city turning on the success or failure of the airlift, the normally unheroic business of air-traffic control, instrument flying, and logistical organization took on heroic status. There was little enemy fire to fear – the Soviets made no serious attempt to intervene – but maintaining the flights at intervals of between three and six minutes by day and night imposed heavy risks. In all, 54 Allied airmen lost their lives in the airlift, and limiting losses to that level required the utmost discipline from all those involved – from the airmen and air-traffic controllers to the ground crews who kept the transports faultlessly operational.

By February 1949 the Combined Airlift Task Force was shifting 8,000

BLOCKADE LIFTED
A cheer is raised at the news in May 1949 that the Soviet blockade of Berlin has been lifted. The airlift continued until the supply situation had returned to normal in the following September.

tonnes of supplies a day. It was obvious to the Soviets that their ploy had failed, and they ended the land blockade in May. The airlift continued until September, by which time the transports had made around 277,000 flights.

The Berlin Airlift had only been over for ten months when the next Cold War crisis erupted in Asia, as Communist North Korea invaded South Korea and

a US-led coalition went in under the UN flag to take on the aggressors. The Chinese soon joined the war in support of North Korea and Soviet "volunteers" flew planes for the North Korean air force. The Korean conflict was hot war and no doubt about it, yet it shared with the Berlin Airlift the characteristic feature of aerial conflict in the Cold War period – the imposition of rules to prevent a full-scale war between the major powers. In Germany, the Soviets did not use their fighters or anti-aircraft capability against Allied transport aircraft; in Korea, the Americans not only did not use nuclear weapons but banned their airmen from striking against airbases inside China, from which enemy fighters operated.

The Korean War

The first stage of American involvement in the war was a classic demonstration of the importance of air power on the modern battlefield. The North Korean advance on the ground was first stopped and then pushed back, partly through the devastating impact of attacks by bombers and fighter-bombers flying close air support and

POPULAR CARGO-CARRIER
Children standing on rubble in West Berlin wave to a Douglas C-47 in the early stages of the Berlin Airlift. The C-47s could not carry a sufficient weight of cargo to supply the city and were soon replaced as the key component of the airlift by the four-engined Douglas C-54 Skymasters.

GIANT GLOBEMASTER
USAF personnel and civilians unload flour from the cargo hold of a Douglas C-74 Globemaster at Gatow airfield, Berlin. The C-74 was the largest transport aircraft used in the Berlin Airlift.

DIVING SABRE

A diving North American F-86 Sabre jet fighter fires rockets at a target range at Nellis Air Force Base, Nevada, around 1953. Sabres were not equipped with rockets or missiles when taking on MiGs during the Korean war. As in World War II the air battles were fought with guns, although combat took place at the higher speeds made possible by jet engines.

FRANCIS "GABBY" GABRESKI

FRANCIS "GABBY" GABRESKI (1919–2002) was one of only seven pilots to achieve ace status both in World War II and the Korean War. Born in a small town in Pennsylvania, the son of Polish immigrants, Gabreski fell in love with flying as a boy in 1932, when he was taken to see James Doolittle win the Thompson Trophy Race in a GeeBee Sportster. In 1940 he applied to join the Army Air Corps but, awkward and nervous, was almost washed out of flight training, narrowly qualifying in a last-chance "elimination flight".

Sent to fight in Europe, he flew with a Polish squadron of the RAF before becoming a member of the USAF's 56th Fighter Group. Flying P-47 Thunderbolts, he recorded 28 kills before being shot down and made a prisoner of war in the summer of 1944.

Gabreski was in his thirties when the Korean War gave him a second bite at air combat.

Leading 51st Fighter-Interceptor Wing based at Suwon, he was credited with shooting down six and a half MiGs (one shared with another pilot). Like many pilots who had fought in World War II, Gabreski disdained new-fangled technology such as radar-controlled gunsights. An inveterate gum-chewer, he would pick some gum out of his mouth as he entered combat and stick it to his Sabre's windshield to line up his shot. His natural ability never let him down.

FAMED ACE
Gabreski's ace status made him famous – he was invited to meet the president – but he always remained at heart just a small-town boy in love with flying.

had such importance. Returning from the mission on which he recorded his fifth kill, pilot Frederick "Boots" Blesse claims that he said a prayer: "Lord, if you have to take me while I'm over here, don't do it today. Let me get back and tell someone I finally got number five." There was competition for scores not only among individual pilots but between the two Sabre-equipped formations, the 4th and 51st fighter-interceptor wings. After a mission on which he shot down two MiGs, Blesse was embraced by his commander, Colonel Harrison Thyng of 4th wing: "Damn, Boots," said Colonel Thyng, "it's about time somebody in this wing was the leading ace."

Many of the American pilots were World War II veterans – the average age of pilots in Korea was around 30, which would have been old for an airman in 1941–45. The war's "top gun", Captain Joseph McConnell, conformed to the classic Korean War profile – 30 years old, World War II veteran (although as a navigator on B-24s). Aces such as James Jabara and "Gabby" Gabreski had been fighter pilots in the earlier conflict. They were as keen to repeat or improve on earlier achievements as the younger pilots were to make their mark for the first time. They had a maximum of 100 missions to show what they were worth before being transferred back to the United States.

interdiction missions. The aircraft were mostly of World War II vintage – Mustangs used in a ground-attack role, US Navy Corsairs operating from carriers, B-26 and B-29 bombers – along with F-80 Shooting Star jets almost as old. In the absence of any serious opposition, these well-tried aeroplanes ruled the air. But when China entered the conflict, it brought with it the state-of-the-art MiG-15, and Korea became the first war of the jet age.

Jet combat

The first shooting down of one jet aircraft by another took place in November 1950, when American pilot Lieutenant Russell J. Brown, flying an F-80, downed a Chinese MiG-15. Operating from safe bases around Antung in Chinese Manchuria, the jets began to intercept US bomber missions, threatening to deny the Americans the air superiority they had previously enjoyed. In December 1950, the United States sent in the F-84 Thunderjet as a ground-attack aircraft, and its latest fighter, the F-86 Sabre, to take on the MiGs. The Sabres were deployed on offensive sweeps, engaging the communist jets over the Yalu River in northwest Korea, and effectively preventing enemy aircraft from interfering

with American air operations further south.

There were strange echoes of World War I in the Korean conflict. As on the Western Front in 1914–18, the most important function of air power was to support ground troops engaged in a grim and desperate war of attrition. But once again, it was the fighter pilots and their contest for air superiority that grabbed the headlines. The duel between MiG-15s and Sabres in "MiG Alley" over the Yalu was awarded legendary status as a classic of aerial combat even while it happened.

The MiG pilots were anonymous, generally operating in formations more than 50 strong. The far less numerous Sabre pilots were a self-conscious elite, aggressive, eager for action, and intensely competitive. Not since World War I had the pursuit of ace status – the famous five kills –

READY FOR ACTION
US Marines disembark from Piasecki HRP-1 tandem-rotor transport helicopters during an early air assault operation. The HRP-1, which first flew in 1947, was known as the "Flying Banana" because of the distinctive shape of its fuselage.

MiG Alley

The fighting in MiG Alley was a gripping study in contrasting tactics and styles of warfare, allowing a direct comparison not only between two closely matched jet fighters but also between pilots trained in very different traditions.

The great strength of the MiG-15s was their performance at high altitude. Cruising at 14,500m (48,000ft) and at just below Mach 1, the MiG was effectively unreachable by the Sabres. If they stayed at this kind of altitude, as they often did, the Sabres were denied the possibility of combat. When the MiG pilots were more aggressively inclined, a number of them would peel away from their mass formation and dive on the Sabres patrolling far below. They would try to get in a shot at the Americans before pulling back up, using their superior rate of climb to escape pursuit. These tactics were known to the American pilots as "Yo-Yo" or, when the MiGs came out of the sun and climbed back towards it again, as "Zoom and Sun".

The Sabre pilots had an aircraft that was faster than the MiG in level flight and distinctly more effective at lower altitudes. If they could engage the MiGs in a dogfight they had a high chance of success. One effective tactic was dubbed "Jet Stream". Sixteen Sabres would enter MiG Alley in flights of four at a few minutes' interval. If the MiGs could be tempted to dive on one of the flights, the others would converge to counter-attack.

The Sabres operated at a significant disadvantage in that they were always far from their bases and even with auxiliary fuel tanks – jettisoned before entering combat – they never had more than 20 minutes in the battle zone. The MiGs mostly remained within a few minutes' flying time of their bases over the Chinese border. The Sabres were also constantly outnumbered, usually by at least three or four to one. Yet they had by far the better of the fighting. Although estimates vary, it may be that 792 MiGs were destroyed in air combat by Sabres, compared with 78 Sabres shot down by MiGs – a kill ratio of 10 to 1.

The crucial factor was pilot quality. Although the communists deployed some excellent Soviet pilots, these experienced fliers were very much in the minority. Most communist pilots were good at following instructions, but showed little initiative or aggression and often made basic errors in dogfights. The American pilots displayed an outstanding hunger for battle – many had plotted and schemed for months or years to arrange a transfer from some safe posting in the United States or Europe to the Korean front line. They had the "right stuff" in plentiful supply, and this proved decisive.

Mikoyan–Guryevich MiG-15

> "It was a beautiful sports car of a fighter... It looked like a first-class airplane."
>
> LIEUTENANT-COLONEL BRUCE HINTON
> FIRST SABRE PILOT TO SHOOT DOWN A MiG-15

THE MiG-15 WAS A PRODUCT of the Soviet design bureau headed by Artyem Mikoyan and Mikhail Guryevich. The aircraft first flew in December 1947, two months after the maiden flight of the US F-86 Sabre. These two aeroplanes were destined to be the key players in the battle for air supremacy during the Korean War (1950–3). The Sabre and the MiG-15 were similar in their swept-wing configuration but different in purpose. The Sabre was an air-superiority fighter; the MiG-15 was primarily intended as an interceptor. The MiG-15 was originally produced to protect the Soviet Union from the threat of fleets of American bombers flying into Soviet airspace at high altitude. The designers therefore created an aircraft with a service ceiling of about 15,500m (51,000ft) and a rate of climb of 2,750m (9,000ft)

per minute. And they armed it with powerful cannons – preferable to machine guns for striking a bomber but less effective in a dogfight.

US pilots in Korea were upset to find that the MiG-15 had an advantage over their Sabres of about 900m (3,000ft) per minute in a climb. Furthermore, the MiG-15s could operate at altitudes that the Sabres simply could not reach. When a North Korean pilot defected with a MiG-15 in 1953, no less a person than Chuck Yeager (the first man to break the sound barrier) was flown out to Japan to look it over. He found nothing revolutionary – just a tough, agile, well-designed aircraft with a suitably powerful engine. Korean War ace Captain James Jabara also felt that the MiG-15 was nothing special, stating: "The F-86 is the best jet fighter in the world and the MiG is the second best".

Radio antenna mast

Bubble canopy

1120

23mm cannon

Retractable nosewheel

BG:

Underwing fuel tank

Specifications

ENGINE	Klimov VK/1FA turbojet
WINGSPAN	9.6m (31ft 6in)
LENGTH	11.3m (36ft 11in)
WEIGHT	4,182kg (9,220lb)
TOP SPEED	1,074kph (667mph) CREW 1
ARMAMENT	2 x 23mm cannons and 1 x 37mm cannon; plus additional bombs or unguided rockets

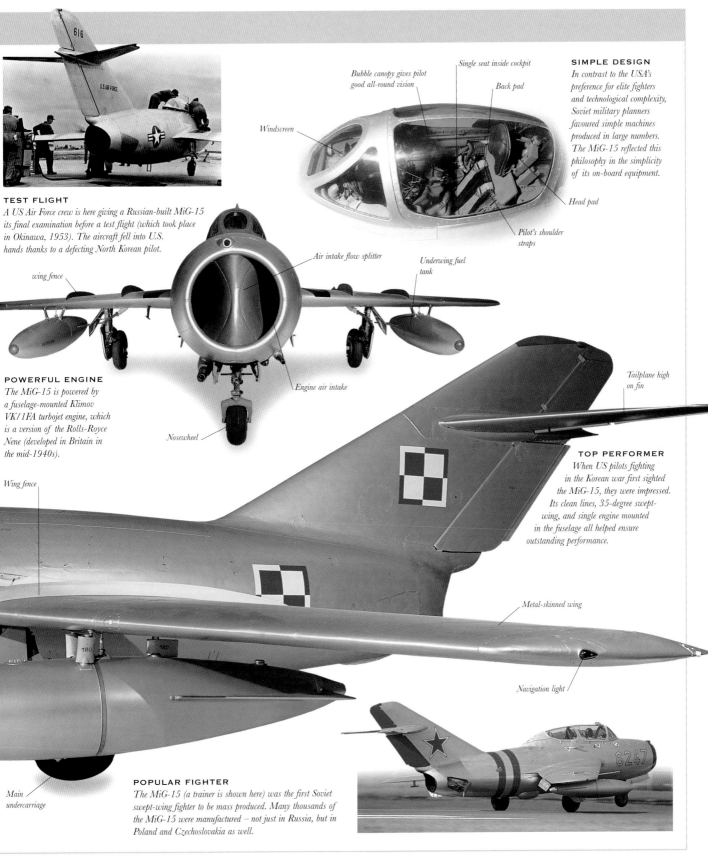

TEST FLIGHT
A US Air Force crew is here giving a Russian-built MiG-15 its final examination before a test flight (which took place in Okinawa, 1953). The aircraft fell into U.S. hands thanks to a defecting North Korean pilot.

SIMPLE DESIGN
In contrast to the USA's preference for elite fighters and technological complexity, Soviet military planners favoured simple machines produced in large numbers. The MiG-15 reflected this philosophy in the simplicity of its on-board equipment.

Bubble canopy gives pilot good all-round vision

Single seat inside cockpit

Back pad

Windscreen

Head pad

Pilot's shoulder straps

wing fence

Air intake flow splitter

Underwing fuel tank

POWERFUL ENGINE
The MiG-15 is powered by a fuselage-mounted Klimov VK/1FA turbojet engine, which is a version of the Rolls-Royce Nene (developed in Britain in the mid-1940s).

Engine air intake

Nosewheel

Wing fence

Tailplane high on fin

TOP PERFORMER
When US pilots fighting in the Korean war first sighted the MiG-15, they were impressed. Its clean lines, 35-degree swept-wing, and single engine mounted in the fuselage all helped ensure outstanding performance.

Metal-skinned wing

Navigation light

Main undercarriage

POPULAR FIGHTER
The MiG-15 (a trainer is shown here) was the first Soviet swept-wing fighter to be mass produced. Many thousands of the MiG-15 were manufactured – not just in Russia, but in Poland and Czechoslovakia as well.

Military helicopters

Combat between jet fighters was not the only innovation of the Korean War. Another of its novelties was the first extensive military use of helicopters. They were not used offensively, but still had an immediate impact, primarily as a means of evacuating wounded troops from the battlefield. Sikorsky H-5s, Bell H-13s, and Hiller H-23s carried the wounded in panniers attached to the helicopter fuselage. Receiving speedy medical attention at Mobile Army Surgical Hospitals (MASH) radically improved a wounded

soldier's chances of survival. Many men had helicopters to thank for saving their lives. Many downed airmen also owed the helicopter a debt of gratitude, for it was US Navy HO3Ss and USAF H-5s and H-19s that carried out combat rescue missions when aircraft were shot down behind communist lines or ditched in the ocean.

These errands of mercy did not exhaust the helicopters' usefulness. H-19 transport helicopters ferried troops and cargo, and helicopters were also employed as airborne command posts and for aerial observation of the battlefield.

At the same time that the Americans were discovering some of the uses of the helicopter in conventional warfare, the British were flying a version of the Sikorsky H-5, which they called the Dragonfly, in counter-insurgency operations against communist guerrillas in Malaya – a foretaste of the prominent role that helicopters would later play in Vietnam.

The Korean War ended in a cease-fire in 1953, leaving Korea divided as it had been before the fighting started. The evidence from this "limited war" suggested that, at least when prevented by

EVOLUTION OF THE HELICOPTER

AIRCRAFT DESIGNER IGOR SIKORSKY said that "the idea of a vehicle that could lift itself vertically from the ground and hover motionless in the air was probably born at the same time that man first dreamed of flying." It was certainly an option explored, by Sikorsky among others, as early as the opening decade of the 20th century. Frenchman Paul Cornu is often credited with getting the first helicopter briefly off the ground in 1907, but it would be a long time before the goal of controlled, manned, sustained rotor flight was attained.

Of the many problems confronting helicopter designers, the most intractable was control. For a start, the torque generated by the rotating blade – the helicopter's equivalent of the airplane's wing – would automatically make the helicopter spin in the opposite direction to the blade. In 1912, Russian experimenter Boris Yuriev showed that the torque could be overcome by mounting a smaller vertical propeller on the tail, but his work was largely ignored.

Even if the helicopter did not spin like a top, it was hard to figure out how it was to be made to rise and fall, move backward or forward, change direction, or hover. Groundbreaking progress on this point was made by Argentinian engineer Raul Pateras de Pescara, who worked out how to vary the pitch of each rotor blade so that the helicopter would tilt in different directions. This "cyclic control" would allow a pilot to fly the aircraft forward or backward, left or right.

Further progress with rotors came from Spanish engineer Juan de la Cierva's 1920s invention, the Autogiro. Like an airplane, the Autogiro was driven by a propeller, but lift was provided by rotors, which were spun around by the air as the aircraft moved forward. To make this hybrid vehicle work, De la Cierva had to exhaustively explore the properties of spinning rotors. The example of his machines gave a new impetus to helicopter research.

By the second half of the 1930s, French, German and American experimenters were all vying to produce a practical helicopter. Frenchman Louis Breguet is generally accepted to have won the race in 1935, but Heinrich Focke significantly improved on his design with the Fa 61 the following year. In the United States, in 1939 Sikorsky returned to the experiments he had abandoned almost 30 years earlier, producing the VS-300. Factory production of helicopters started during World War II.

Sikorsky emerged as the individual publicly identified with the success of this new form of aviation. His ultimate dream was that the helicopter

CORNU'S HELICOPTER
Frenchman Paul Cornu claimed to have risen from the ground for a few seconds in his primitive twin-rotor helicopter in November 1907. Modern studies suggest that it could not have sustained its own weight and that of a pilot.

would become the successor to the automobile, parked on every front lawn in America. It did not turn out like that, but given the many and varied uses rotating-wing craft have found, this hardly counts as a disappointment.

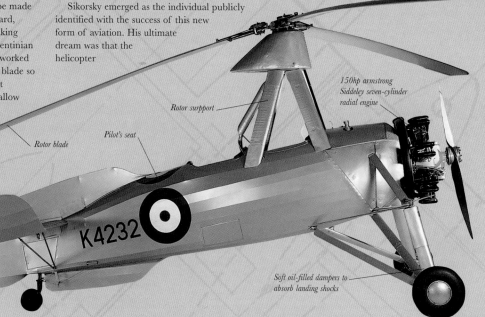

Rotor blade

Pilot's seat

Rotor surpport

150hp armstrong Siddeley seven-cylinder radial engine

Soft oil-filled dampers to absorb landing shocks

CIERVA C-30 AUTOGIRO
Juan de la Cierva's Autogiros were the first practical rotary-wing aircraft. Since the rotor was unpowered it provided lift but not propulsion. The Autogiro shown here is a C-30 built under license by Avro for the RAF as a Rota 1.

K4232

EARLY HELICOPTERS

GERMAN PLANEMAKER HEINRICH FOCKE, a founder of the Focke-Wulf company, led the world in helicopter development in the late 1930s. Focke-designed helicopters were ordered into production for military use during WWII, but few entered service. Meanwhile, Igor Sikorsky's experimental 1939 single-rotor VS-300 led to the R-4 Hoverfly, which went into production for the US Army in 1942. Unlike its German equivalents, the R-4 had a chance to prove its usefulness in combat. The Bell company also began helicopter development during the war, and in 1946 the highly successful Bell 47 became the first helicopter licensed for civilian use. By the early 1950s, some 30 types of helicopter were flying, including: variants on the classic Sikorsky design with a single horizontal rotor and vertical tail rotor; machines with two counter-rotating horizontal rotors; and helicopters with small jets at the tips of the rotor blades, which obviated the need for a tail rotor.

BELVEDERE ANCESTOR
The prototype Type 173 first flew in 1952 and developed into the RAF's successful Type 192 Belevedere.

Cierva C.8 Mk.IV Autogiro

Rotary-winged flight was achieved long before the first practical helicopter appeared. In 1920, Juan de la Cierva began experimenting with this form of aircraft which he called an "Autogiro". It had an unpowered rotor, so it could not take off vertically or fly backwards, but it came close to a vertical landing. On 18 September 1928, a C.8L made the first crossing of the English Channel by a rotating wing aircraft.

ENGINE	200hp Armstrong-Siddeley Lynx IVC 7-cylinder radial	
ROTOR SPAN	12.1m (39ft 8in)	LENGTH 8.7m (28ft 6in)
TOP SPEED	161kph (100mph)	CREW 1
PASSENGERS	1	

Fairey Jet Gyrodyne

The original Gyrodyne first flew in December 1947, combining autogyro and helicopter features. In 1948 it set a new world rotorcraft speed record of 200kph (124mph). The Jet Gyrodyne appeared in January 1954, powered by compressed air from the engine fed to jet units at the tips of the two-blade rotor.

ENGINE	525hp Alvis Leonides 9-cylinder radial	
ROTOR SPAN	18.3m (60ft)	LENGTH 7.6m (25ft)
TOP SPEED	225kph (140mph)	CREW 2
PASSENGERS	1	

Focke-Achgelis Fa 61

The Fa 61 was the first fully controllable helicopter and flew in June 1936. Although influenced by the Cierva autogiro, the new machine used twin rotors which gave remarkable agility. After establishing a number of world rotorcraft records, the aircraft was flown indoors in the Berlin Deutschlandhalle in February 1938 by the brilliant pilot, Hanna Reitsch, who demonstrated its ease of control and manoeuvrability to the crowd.

ENGINE	160hp Bramo Sh.14a 7-cylinder radial	
ROTOR SPAN	7m (23ft)	LENGTH 7.3m (24ft)
TOP SPEED	100kph (62mph)	CREW 1
PASSENGERS	None	

Piasecki YH-16A Transporter

Together with Sikorsky, Frank Piasecki was a pioneer of American helicopter design. From the mid-1940s, his company built a series of twin-rotor machines from which the current Boeing Chinook is a direct descendant. An improved version of the 1953 prototype, with shaft turbines replacing the piston engines, first flew in 1955. A year later the project was cancelled.

ENGINE	2 x 1800shp Allison YT-38-10 turboshaft	
ROTOR SPAN	25m (82ft)	LENGTH 23.8m (77ft 7in)
TOP SPEED	235kph (146mph)	CREW 3
PASSENGERS	40 troops	

Sikorsky VS-300

A young Igor Sikorsky built an unsuccessful helicopter in 1909, before turning his attention to giant aeroplanes and flying boats. Thirty years later, he returned to his first love, which would make his name synonymous with helicopters. His VS-300, which made a tethered flight on 13 September 1939, was the first successful American-built helicopter.

ENGINE	75hp 4-cylinder Lycoming (initial tethered flight)	
ROTOR SPAN	8.5m (28ft)	LENGTH 20.1m (66ft)
TOP SPEED	103kph (64mph)	CREW 1
PASSENGERS	None	

Sikorsky R-4 Hoverfly

The VS-300 led to the world's first production helicopter, the R-4 Hoverfly. Ordered by the US Army Air Corps, trials of the prototype in 1942 led to an unprecedented order for 100 R-4Bs. Used during WWII by the Allied forces, the Hoverfly proved that the helicopter was a practical machine.

ENGINE	180hp Warner R-550 Super Scarab radial	
ROTOR SPAN	11.6m (38ft)	LENGTH 10.7m (35ft 3in)
TOP SPEED	121kph (75mph)	CREW 2
PASSENGERS	None	

US nuclear bombers

political rules from carrying out full-scale strategic bombing, air forces could have a major impact on the progress of a conflict, but not a decisive one.

The heavy losses suffered by the forces of the United States and its allies in Korea, and the failure to achieve a decisive victory there, led America to reassess its defence policy. Nuclear deterrence seemed to offer a more effective, and more cost-effective, way of blocking communist expansion than the threat of conventional warfare. Thus in the 1950s the central fact of military aviation was that aircraft constituted the major (and, at first, the only) delivery

system for the nuclear weapons that the United States had come to depend on for its defence.

Nuclear scientists on both sides of the Iron Curtain had no difficulty creating bigger and bigger bangs. The first hydrogen bomb, exploded in 1952, was 500 times more powerful than the device that had destroyed Hiroshima. But even the United States, with all its resources, took time to create an intercontinental bomber force for the nuclear age. In 1947 the Strategic Air

NUCLEAR GIANT
A group of mechanics stand on the horizontal stabilizer of the massive Convair B-36 nuclear bomber, one of the largest aircraft ever built. The rudder was as tall as a five-storey building.

> "I believe we can get the B-36 over a target and not have the enemy know it is there until the bombs hit."
>
> **GENERAL CURTISS LeMAY**
> HEAD OF THE US STRATEGIC AIR COMMAND

Command's potential nuclear strike force consisted of just ten B-29 bombers. And it was widely accepted that the B-29 was not up to the job, in range, altitude, or speed.

The process of finding a replacement bomber for the nuclear role was plagued by indecision and politics, centred around disputes as to which kind of bomber would work and whether what would work was worth the price. Boeing initially came up with the B-47, a seminal jet design with swept wings and the engines in pods on

NUCLEAR GIANT
The Consolidated B-36 was powered by an array of pusher propellers and turbojets. It was so large that the crew of 15 moved from the front to the back of the bomber by pulling themselves along a tunnel on a trolley, and slept in bunks when off duty.

YB-49, as America's strategic nuclear bomber. But after much discussion, the air force instead opted for the Consolidated B-36, an aircraft that was remarkable above all for its sheer size – one pilot said it was "like flying an apartment house". The B-36's wingspan of 70m (230ft) was more than double that of a World War II B-17 – in fact considerably bigger than today's Boeing 747. A physical monument of the transition to the jet age, it had six pusher propellers plus four turbojets, an array of power that made the ground shake as it flew overhead. Deployed alongside the B-47, the B-36 provided a stop-gap solution until the Boeing B-52 took over as the hub of America's nuclear bomber force in the second half of the 1950s.

Soviet counterparts

To the dismay of Americans, the Soviet Union seemed able, time and again, to match the progress of Western technology or even surpass it. In retrospect, it is obvious that the American perception of Soviet strength was frequently exaggerated, with misleading intelligence reports fuelling ill-considered paranoia. Never was this more true than during the "bomber gap" crisis of the mid-1950s, when many Americans came to believe that the Soviet Union was well ahead of the United States in the creation of a strategic nuclear bomber force.

In fact, the Soviets found it even more difficult than the United States to create a credible intercontinental nuclear bomber. When the Soviet Union exploded its first atomic bomb in 1949, it could only have been used against the United States on a suicide mission, since the Soviets' Tupolev Tu-4 bomber would have had the range for a one-way journey only. It was not until the Tu-95 entered service in 1956 that the Soviets had a bomber of true intercontinental range – and this was a turboprop aircraft.

DETERRENT FORCE
A still from a 1949 documentary, Target: Peace, *shows a crewman's view of a fleet of B-36s in operation. After it was withdrawn from service, the B-36 was nicknamed "Peacemaker" to stress that its purpose was to prevent war through deterrence.*

Interceptors

Still, the threat of aircraft destroying a city with a single bomb understandably gave both sides in the Cold War an urgent desire to develop interceptor aircraft. With the speed of bomber aircraft increasing constantly, it was reckoned that from the time an enemy incursion was identified by radar, an interceptor would have little more than ten minutes to reach and destroy its target.

The specialist interceptors designed in the 1950s were pilot-operated missile-platforms maximized for speed in a straight line and rate of climb, with electronic systems on board linked directly to ground controllers. The F-106, dubbed the "Ultimate Interceptor", flew at over Mach 2, could climb to almost 12,000m (40,000ft) in a minute, and, having no guns, depended entirely on its air-to-air missiles.

Aircraft with this level of performance and complexity of electronic systems were massively expensive. An F-106 cost more than ten times as much as an F-86 Sabre to build, and its operating costs were similarly astronomical. For countries trying to keep up with the United States and the Soviet Union, the economic demands were daunting. Nonetheless, the French, keen to assert

struts under the wing. It was considered fast enough to penetrate Soviet air defences, cruising at around 890kph (560mph), but it lacked intercontinental range – therefore it would either have to be based on the territory of America's allies nearer the Soviet border, or rely heavily on in-flight refuelling.

For a true intercontinental bomber, the USAF looked at one point as if it would turn to America's most innovative designer, Jack Northrop. Since the early 1940s, Northrop had been working on the XB-35, a propeller-driven flying-wing bomber with neither fuselage nor tail. He proposed a jet-powered version, the

SOVIET BEAR
Introduced into service in 1956, the Tupolev Tu-95 Bear turboprop bomber gave the Soviet Union, for the first time, the capacity to deliver a nuclear strike against the United States.

Boeing B-52 Stratofortress

> "Moving from a B-47 bomber to the 'Buff' was like progressing from a sportster to a stretched limousine."
>
> **CAPTAIN GENE DEATRICK**
> 1950S TEST PILOT

THE BOEING B-52, KNOWN TO AIRMEN as the "Buff", has proved the most durable military aircraft in aviation history. First delivered to the US Strategic Air Command in June 1955, it is still used in frontline service today. The B-52 was designed specifically to drop nuclear bombs on the Soviet Union. But it turned out to be highly adaptable, able to change from high-altitude missions to low-level attack, to provide a platform for Cruise missiles, and to operate as a conventional bomber of awesome power in regional conflicts.

The B-52 was able to carry an impressive bombload in the bomb-bay or on pods under the wings. The "Big Belly" modification of the B-52D, for example, could carry 27,200kg (60,000lb) of bombs – almost five times the capacity of a World War II "heavy" such as the Lancaster.

The B-52 features a twin-deck forward fuselage. The pilot and co-pilot sit above, while

the navigator and radar navigator crouch in the "black hole" below. The navigators' ejection seats fire downwards – a worrying fact on low-level missions. The electronic warfare officer (EWO) sits facing backwards at the rear of the upper deck, with the gunner alongside, operating a tail gun by remote control. In early models of the B-52, the gunner sat out in a tail turret, buffeted and shaken a good deal. Coming in to the forward fuselage (in later models) made this job less isolated and uncomfortable.

Because it was originally thought that the B-52 would be attacked by missiles, it does not feature all-round guns (and partly accounts for the small crew of six). Missiles were instead dealt with by electronic countermeasures and chaff that blocked or distracted the missiles' homing systems.

TALL TAIL

Although over the years the B-52 has progressed through a series of models and modifications, one of the constant features has been the tall vertical tail. However, on the B-52G (shown below), the tail is almost 2.5m (8ft) shorter than those found on previous models.

CRAMPED CADILLAC

Despite the fact that the B-52 has been dubbed "the Cadillac of the skies", there is little room in the cockpit – or elsewhere. A person of average size is unable to stand upright anywhere in the craft, and only the pilots are able to see outside.

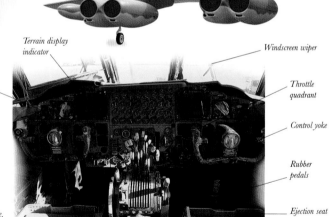

Weapons pylon attachment point
Terrain display indicator
Attitude indicator
Windscreen wiper
Throttle quadrant
Control yoke
Rubber pedals
Ejection seat

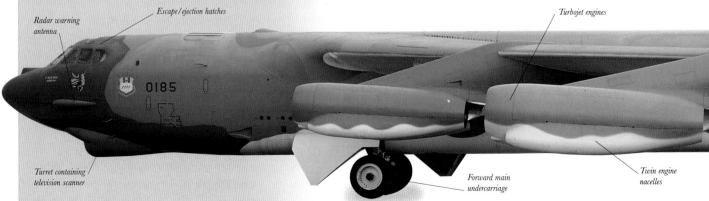

Radar warning antenna
Escape/ejection hatches
Turret containing television scanner
Forward main undercarriage
Turbojet engines
Twin engine nacelles

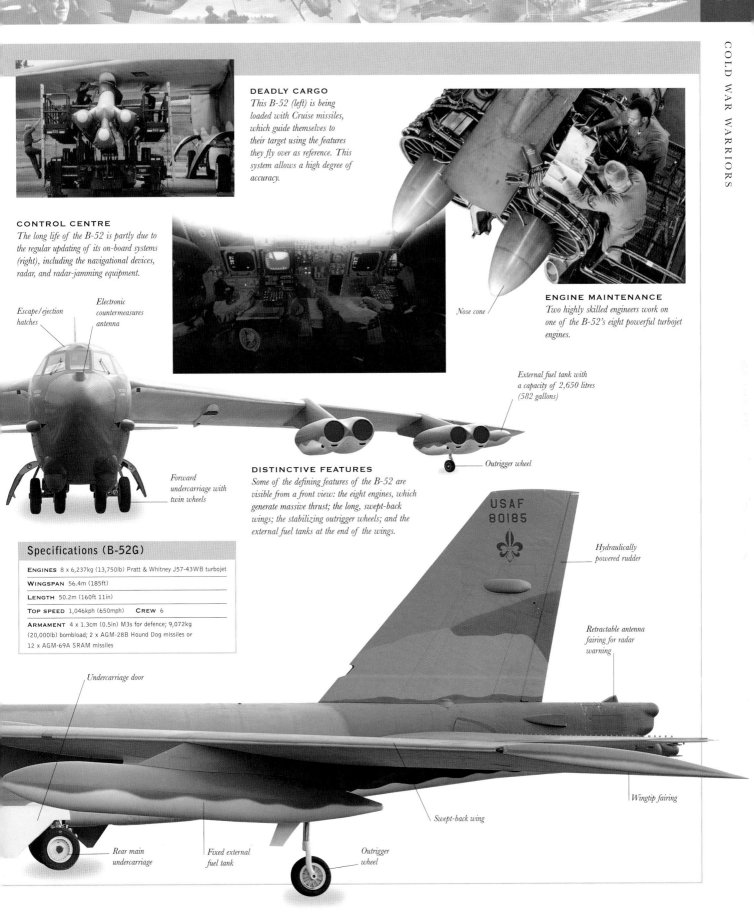

DEADLY CARGO
This B-52 (left) is being loaded with Cruise missiles, which guide themselves to their target using the features they fly over as reference. This system allows a high degree of accuracy.

CONTROL CENTRE
The long life of the B-52 is partly due to the regular updating of its on-board systems (right), including the navigational devices, radar, and radar-jamming equipment.

Escape/ejection hatches

Electronic countermeasures antenna

ENGINE MAINTENANCE
Two highly skilled engineers work on one of the B-52's eight powerful turbojet engines.

Nose cone

External fuel tank with a capacity of 2,650 litres (582 gallons)

Outrigger wheel

Forward undercarriage with twin wheels

DISTINCTIVE FEATURES
Some of the defining features of the B-52 are visible from a front view: the eight engines, which generate massive thrust; the long, swept-back wings; the stabilizing outrigger wheels; and the external fuel tanks at the end of the wings.

USAF 80185

Hydraulically powered rudder

Retractable antenna fairing for radar warning

Specifications (B-52G)
ENGINES 8 x 6,237kg (13,750lb) Pratt & Whitney J57-43WB turbojet
WINGSPAN 56.4m (185ft)
LENGTH 50.2m (160ft 11in)
TOP SPEED 1,046kph (650mph)   **CREW** 6
ARMAMENT 4 x 1.3cm (0.5in) M3s for defence; 9,072kg (20,000lb) bombload; 2 x AGM-28B Hound Dog missiles or 12 x AGM-69A SRAM missiles

Undercarriage door

Rear main undercarriage

Fixed external fuel tank

Outrigger wheel

Swept-back wing

Wingtip fairing

MILITARY MISSILE
The US Navy tests an early surface-to-air missile (SAM) in 1950. SAMs were to revolutionize air defences and place new demands on aircraft design.

their independence from the United States, produced high-performance fighter aircraft such as the Super Mystère and the Mirage III through the Dassault company in the 1950s, and went on to develop an independent nuclear strike force in the following decade. Britain was also committed to developing its own nuclear deterrent in the 1950s and produced effective aircraft for offence and defence: the three "V bombers" – Victor, Valiant, and Vulcan – and the Hawker Hunter and English Electric Lightning interceptors. But by the time the Lightning entered service in 1959, economic stringencies were forcing Britain to drop out of independent development of aircraft for strategic nuclear war.

Missile development

One of the reasons the British government gave for cancelling expensive fighter-interceptor projects was that air defence would soon be exclusively a business for ground-based missiles. The increasing effectiveness of surface-to-air missiles (SAMs) in the late 1950s led to some major rethinking all round on strategic air war. The nuclear bomber forces were geared up to

> "I was a pilot flying an aeroplane and... where I was flying made what I was doing spying."
>
> **FRANCIS GARY POWERS**
> INTERVIEWED AFTER HIS RETURN TO THE US

penetrate the enemy's defences through speed and high altitude. But once SAMs had proved they could shoot down high-flying aircraft, the bombers had to turn to very low-level attack, attempting to fly under enemy radar and missile screens.

Cold-war spyplanes

That Soviet SAMs could strike aircraft at high altitude was proven in 1960 when a U-2 spyplane was downed over Sverdlovsk. The U-2 was designed to operate at an altitude of more than 21,000m (70,000 ft), which was hoped to be out of reach of Soviet radar, ground-based missiles, and interceptors. Run by the CIA, U-2s overflew the Soviet Union repeatedly from 1956, bringing back detailed photographs of military installations. In May 1960, however, a U-2 piloted by Francis Gary Powers was sent into a fatal spin by a SAM exploding nearby. Powers escaped from the cockpit and parachuted to the ground, having failed to activate the U-2's self-destruct mechanism, and declined to use the cyanide pill provided to avoid capture. Soviet leader Nikita Khrushchev drew maximum diplomatic advantage from the incident, flouncing out of a

AIR-TO-AIR MISSILES

THE FIRST GENERATION of air-to-air guided missiles (AAMs) was developed in the 1940s. They were either radar controlled or heat-seekers. The radar-controlled missiles, such as the Falcon and Sparrow, were designed for intercepting high-altitude bombers. A radar on board the interceptor locked on to its target and the missile homed in on the reflections from the interceptor's radar. Keeping the target illuminated with the radar during the missile's flight meant that the interceptor had to stay both straight and level – meaning that the missile was useless for combat between fighters.

In contrast, the heat-seeking Sidewinder was a fire-and-forget missile. In classic dogfight style, the

RADAR-GUIDED SPARROW
A US Navy F-14A Tomcat fires an AIM-7 Sparrow radar-guided, air-to-air missile. The Sparrow, which is widely deployed by US and NATO forces, can destroy targets more than 16km (10 miles) away.

FALCON MISSILES
A pilot shows off radar-guided (left) and heat-seeking Falcon missiles, which came into service in the mid-1950s.

pilot manoeuvred his aircraft behind the enemy's tail and fired the missile towards the hot exhaust. An infrared detector in the missile guided it on to the heat source. Although effective, until the 1970s heat-seeking missiles had drawbacks: they were useless fired head-on and could easily pick up the wrong target if another jet exhaust presented itself.

Inevitably, guided missiles of all kinds bred countermeasures. Radar-guidance systems could be jammed or confused by chaff; heat-seekers could be distracted by firing flares or shaken off by swift manoeuvre.

AERIAL PHOTOGRAPHY

The Lockheed U-2's effectiveness depended on the development of cameras capable of reconnaissance photography from over 22,000m (70,000ft). This camera (right) is a Hycon B used in U-2 reconnaissance over Cuba in late 1962. The picture far right shows missile erectors and launch stands in Cuba during the missile crisis.

Programme and junction box

Optical system assembly

Oblique head assembly

Shutter

SPY IN THE SKY

Built to meet a joint CIA/USAF requirement, the top-secret Lockheed U-2 spyplane was a unique source of intelligence for the United States until the advent of reconnaissance satellites in the 1960s.

Satellites took over much of the burden of military reconnaissance, although the SR-71 Blackbird coming into service in 1966 proved that with sufficient speed and altitude it was possible for an aircraft to defy any missile defence system.

Although the Cuban Missile Crisis ushered in a period of relative détente in the Cold War, the nuclear arms race never stopped. From the mid-1960s, however, attention was distracted from the possibility of a nuclear war by the reality of conventional warfare. While airmen continued to train and stand on alert for the nuclear "big show" that never materialized, hot war flared in the Middle East and Southeast Asia.

summit conference with President Eisenhower and other Western leaders. Powers was put on trial for espionage and the Cold War entered its most unstable phase.

Soviet basing of missiles in Cuba precipitated the crisis that brought the world to the brink of nuclear war in 1962. Overflying Cuba, which had become an outpost of communism, U-2s from Edwards Air Force Base photographed preparations to install Soviet medium-range ballistic missiles on the island. As the American government demanded a withdrawal of Soviet missiles, the nuclear bomber force went on red alert, with 70 B-52s airborne at all times on 24-hour shifts. Before the crisis ended, another U-2 was shot down by a Soviet SAM, this time over Cuba. Its pilot was the only casualty in a crisis that could have resulted in the deaths of millions.

Changing arsenals

The Cuban Missile Crisis took place at a moment when major shifts were taking place in the nuclear arsenal. Intercontinental ballistic missiles (ICBMs) and submarine-launched missiles were beginning to supersede bombers as the main delivery systems for nuclear devices, although aircraft remained an important part of the panoply of strategic nuclear weaponry. The need to penetrate enemy defences at low level meant that fighter-bombers were increasingly prominent in a diversified strategic strike force alongside heavy bombers, and that the high-altitude interceptor was out of date.

SPYPLANE PILOT

Gary Powers was tried for espionage in the Soviet Union after being shot down in 1960. He was later "swapped" for a Soviet spy.

Lockheed SR-71 Blackbird

> "I took off late on a winter afternoon, heading east where it was already dark... you streaked from bright day and flew into utter black."
>
> COLONEL JIM WADKINS
> SR-71 PILOT

A PRODUCT OF LOCKHEED'S "Skunk Works" and chief designer Kelly Johnson, the SR-71 Blackbird may well qualify as the most remarkable aircraft ever built. Developed as a spy plane for the CIA under conditions of total secrecy in the late 1950s, it involved radical innovation in almost every feature, from the materials used for the airframe and engine, through to the hydraulic system and the fuel. The result was the fastest jet-powered, crewed aircraft ever, capable of flying at altitudes of up to 30,000m (100,000ft) at more than three times the speed of sound.

When test pilots first saw the Blackbird prototypes at the CIA's Groom Lake site in Nevada in the early 1960s, they were taken aback by their shape and size – the elongated, slender fuselage and the huge engine pods mounted on the wings. The engines are actually wider than the main body of the fuselage. The radar-absorbent black paint, which gives the plane its name, covers a skin of titanium alloy, a lightweight, heat-resistant material that posed several problems for Lockheed engineers. It was too hard and brittle for most existing machine tools to work, and it was extraordinarily sensitive to contamination. Yet Johnson's instinct in persisting

FILLING UP
Since oxygen is explosive when transported at the speeds attainable by the SR-71, the fuel tanks are purged with liquid nitrogen prior to filling (top). In-flight refuelling (above) is used to extend the SR-71's already impressive range.

with titanium alloy proved justified. It resists very high temperatures in Mach 3 flight, although the metal skin on the aeroplane's nose regularly wrinkles from the heat. The ground crew smooth it out after a flight using a blowtorch – SR-71 pilot Colonel Jim Wadkins described the process as "like ironing a shirt".

The SR-71 entered service in 1966 and successfully performed the global reconnaissance role for which it was designed. Lockheed had ambitious plans for other versions of the aircraft, especially a high-altitude fighter interceptor prototyped as the YF-12A. The company claimed that 93 of these aircraft armed with air-to-air missiles would be able to defend the entire United States from attack by Soviet bombers. But the United States government would not fund this ambitious and costly project.

SHOCK DIAMONDS
As the SR-71B takes off, "shock diamonds" appear in the exhaust gases. These form when shock waves are reflected as they exit the engine's exhaust nozzles. The diamonds glow brightly as the surplus fuel ignites.

U.S. AIR FORCE

Equipment bay, with aperture for panoramic camera

Rear cockpit for reconnaissance systems officer

Forward-retracting undercarriage

Fuselage houses interchangeable reconnaissance equipment packs

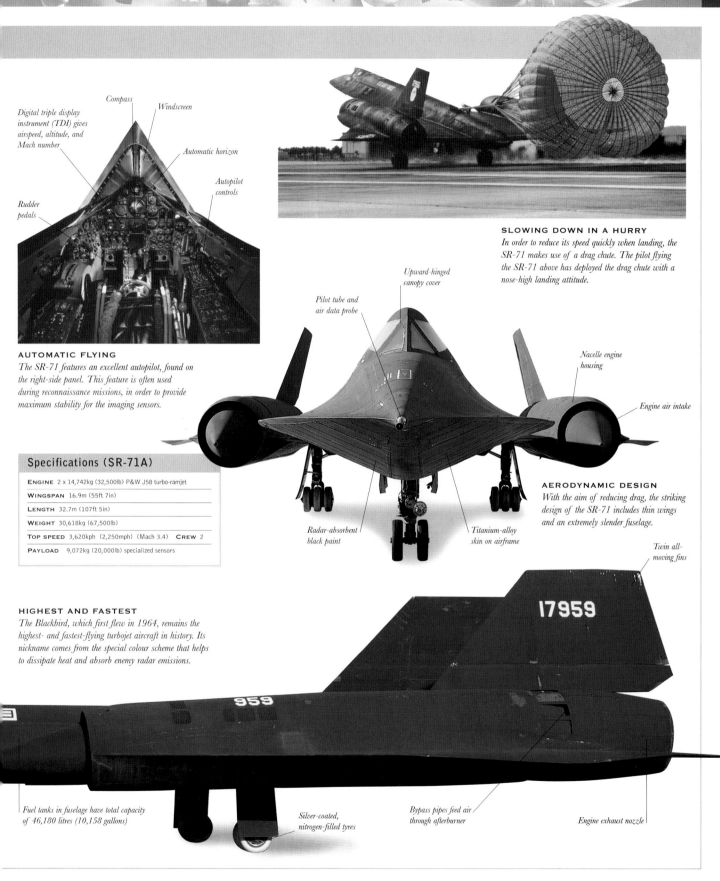

Digital triple display instrument (TDI) gives airspeed, altitude, and Mach number

Compass

Windscreen

Automatic horizon

Autopilot controls

Rudder pedals

SLOWING DOWN IN A HURRY
In order to reduce its speed quickly when landing, the SR-71 makes use of a drag chute. The pilot flying the SR-71 above has deployed the drag chute with a nose-high landing attitude.

Upward-hinged canopy cover

Pilot tube and air data probe

Nacelle engine housing

Engine air intake

AUTOMATIC FLYING
The SR-71 features an excellent autopilot, found on the right-side panel. This feature is often used during reconnaissance missions, in order to provide maximum stability for the imaging sensors.

Specifications (SR-71A)

ENGINE	2 x 14,742kg (32,500lb) P&W J58 turbo-ramjet
WINGSPAN	16.9m (55ft 7in)
LENGTH	32.7m (107ft 5in)
WEIGHT	30,618kg (67,500lb)
TOP SPEED	3,620kph (2,250mph) (Mach 3.4) CREW 2
PAYLOAD	9,072kg (20,000lb) specialized sensors

Radar-absorbent black paint

Titanium-alloy skin on airframe

AERODYNAMIC DESIGN
With the aim of reducing drag, the striking design of the SR-71 includes thin wings and an extremely slender fuselage.

Twin all-moving fins

HIGHEST AND FASTEST
The Blackbird, which first flew in 1964, remains the highest- and fastest-flying turbojet aircraft in history. Its nickname comes from the special colour scheme that helps to dissipate heat and absorb enemy radar emissions.

17959

959

Fuel tanks in fuselage have total capacity of 46,180 litres (10,158 gallons)

Silver-coated, nitrogen-filled tyres

Bypass pipes feed air through afterburner

Engine exhaust nozzle

PRE-1970s NUCLEAR BOMBERS

IN THE SECOND HALF OF THE 1940s, the first nuclear-bomber force in the United States consisted of aircraft conceived for conventional strategic bombing in World War II. Meanwhile, work started on devising a custom-built nuclear bomber. It was assumed that the best way to penetrate air defences would be to fly as high and as fast as possible. For both these accomplishments, jet engines offered the best solution, but with the serious drawback that they could not at the time achieve the required range. It did not help that early nuclear bombs were large and heavy. Various ingenious solutions were tried involving rocket-assisted take-off and combined jet and piston power. Eventually the range problem was resolved by improved jet engines and in-flight refuelling. From 1957 the installation of SAMs, plus improvements in the performance of interceptors, led to a radical rethink. The bombers were forced to adopt a low-altitude attack profile, attempting to creep under radar. At the same time nuclear devices were getting smaller. By the end of the 1960s, strike aircraft such as the fighter-designated F-111 were preferred for low-level penetration, with ICBMs as the core of nuclear deterrence.

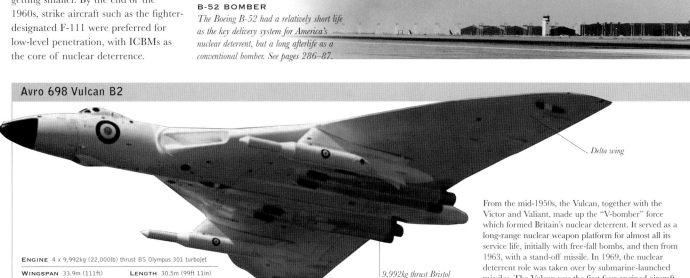

B-52 BOMBER
The Boeing B-52 had a relatively short life as the key delivery system for America's nuclear deterrent, but a long afterlife as a conventional bomber. See pages 286–87.

Avro 698 Vulcan B2

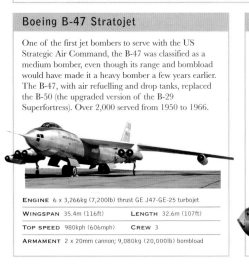

Delta wing

From the mid-1950s, the Vulcan, together with the Victor and Valiant, made up the "V-bomber" force which formed Britain's nuclear deterrent. It served as a long-range nuclear weapon platform for almost all its service life, initially with free-fall bombs, and then from 1963, with a stand-off missile. In 1969, the nuclear deterrent role was taken over by submarine-launched missiles. The Vulcan was the first four-engined aircraft with a delta wing – chosen because it offered a unique combination of good load-carrying capabilities, high subsonic speed at altitude, and long range.

9,992kg thrust Bristol Siddeley Olympus turbojet engine

ENGINE 4 x 9,992kg (22,000lb) thrust BS Olympus 301 turbojet	
WINGSPAN 33.9m (111ft)	**LENGTH** 30.5m (99ft 11in)
TOP SPEED 1,029kph (640mph)	**CREW** 5
ARMAMENT 21 x 454kg (1,000lb) bombs, nuclear bombs, or 1 x Blue Steel missile	

Boeing B-47 Stratojet

One of the first jet bombers to serve with the US Strategic Air Command, the B-47 was classified as a medium bomber, even though its range and bombload would have made it a heavy bomber a few years earlier. The B-47, with air refuelling and drop tanks, replaced the B-50 (the upgraded version of the B-29 Superfortress). Over 2,000 served from 1950 to 1966.

ENGINE 6 x 3,266kg (7,200lb) thrust GE J47-GE-25 turbojet	
WINGSPAN 35.4m (116ft)	**LENGTH** 32.6m (107ft)
TOP SPEED 980kph (606mph)	**CREW** 3
ARMAMENT 2 x 20mm cannon; 9,080kg (20,000lb) bombload	

Boeing B-50D

The most powerful and sophisticated bomber of WWII, the Boeing B-29 dropped thousands of tons of conventional bombs and two nuclear bombs on Japan. At the time of the surrender, there were thousands of B-29s on order which were all cancelled, except for a few of a new model, B-29D, which was redesignated B-50. This would carry America's nuclear deterrent until new longer range and faster aircraft came into service. This took much longer than anticipated, and the B-50 was in front-line service from 1948 to 1953.

ENGINE 4 x 3,500hp P&W R-4360-35 Wasp Major air-cooled radial	
WINGSPAN 43.1m (141ft 3in)	**LENGTH** 30.2m (99ft)
TOP SPEED 611kph (380mph)	**CREW** 8
ARMAMENT 12 x .5in machine guns in four remote turrets, 1 x 20mm cannon in manned turret; 9,080kg (20,000lb) bombload	

High, sweeping fin

3,500hp Pratt & Whitney Wasp Major radial engine

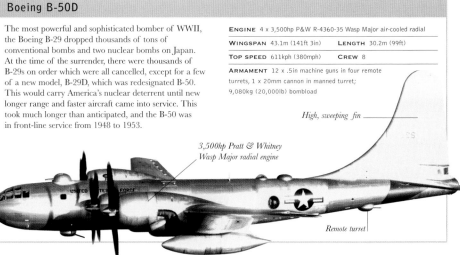

Remote turret

Convair B-36J

Ordered in 1941, when it seemed the USA might have to attack Germany and Japan from its own territory, development of the B-36 was slowed down or speeded up depending on how the war was going. In 1945, it gained a new lease of life as its 16,090km (10,000 mile) range was capable of taking a nuclear attack to the United States' new enemy, the Soviet Union.

ENGINE	6 x 3,800hp P&W R-4360-53 Wasp Major air-cooled radial	
WINGSPAN 70.2m (230ft)	LENGTH	49.4m (161ft 1in)
TOP SPEED 661kph (411mph)	CREW	15
ARMAMENT 16 x 20mm cannon in eight remote-controlled turrets; 32,710kg (72,000lb) bombload		

Convair B-58 Hustler

The United States began development of this supersonic nuclear bomber in 1946. The weapon and much of the fuel were carried in the under-fuselage pod, which would be dropped on the target, making the aircraft faster and more fuel-efficient on the return journey. The project was on the limit of available technology and, after many delays and several near-cancellations, only entered service in 1961. By then, anti-aircraft missiles had improved so much that the Hustler was vulnerable, and all aircraft were phased out in 1970.

ENGINE	4 x 6,815kgp (15,000lbst) thrust General Electric J79-GE-5B turbojet with afterburner	
WINGSPAN 17.3m (56ft 10in)	LENGTH	29.5m (96ft 9in)
TOP SPEED 2,122kph (1,319mph) Mach 2	CREW	3
ARMAMENT 1 x 20mm M-61 rotary cannon in tail; nuclear bomb and fuel carried in large underfuselage pod		

Myasishchev M-4 (Mya-4) "Bison"

The Soviet equivalent of the B-52 and entering service at about the same time, this strategic bomber was hardly known in the West for some years. Given the name "Bison" by NATO, its effectiveness, like that of the B-52, was eroded by improvements in fighter and missile defences. However, it was not developed like the American aircraft into a tactical conventional bomber.

ENGINE	4 x 8,700-kg (19,180-lb) thrust Mikulin AM-3D turbojets	
WINGSPAN 50.5m (165ft 7in)	LENGTH	47.2m (154ft 10in)
TOP SPEED 998kph (620mph)	CREW	8
ARMAMENT 7 x 23-mm NR-23 cannon in two remote-controlled and one manned turrets and in nose; 15,000kg (33,000lb) bombload		

Tupolev Tu-16 "Badger"

The Soviet Union's dependence on Rolls-Royce derived jet engines was ended by the development of the hugely powerful Mikulin AM-3. The Tu-16, known in the West as "Badger", was a long-range medium bomber using two of these engines. It entered service in 1954, about the same time as the American B-47, which needed six engines to achieve the same performance. Exported models were used in action by Indonesia and Egypt during the 1960s and 1970s.

ENGINE	2 x 9,500kg (20,945lb) thrust Mikulin AM-3M 301 turbojet	
WINGSPAN 32.9m (108ft)	LENGTH	36.3m (118ft 11in)
TOP SPEED 992kph (616mph)	CREW	6
ARMAMENT 7 x 23mm NR-23 cannon; 6,000kg (13,000lb) bombload, or 2 x air-to-surface missiles.		

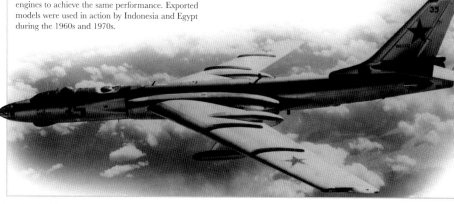

Tupolev Tu-95 "Bear"

In Soviet air force service from 1955, the "Bear" is the only strategic bomber ever to use turboprop engines. It is also the only propeller-driven aircraft with a swept wing, which helped give the aircraft both speed and efficiency. The prototype was still faster than any other propeller aircraft. Although rendered obsolete as a bomber by advances in fighter and missile defences, its huge range ensured a new role in reconnaissance and maritime patrol. A long-range anti-submarine variant (Tu-142) was developed for the Soviet Navy.

Soviet red star marking

Swept wing

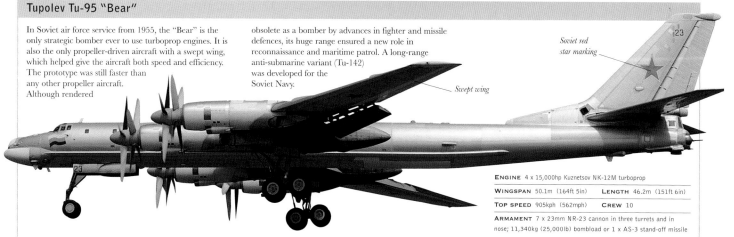

ENGINE	4 x 15,000hp Kuznetsov NK-12M turboprop	
WINGSPAN 50.1m (164ft 5in)	LENGTH	46.2m (151ft 6in)
TOP SPEED 905kph (562mph)	CREW	10
ARMAMENT 7 x 23mm NR-23 cannon in three turrets and in nose; 11,340kg (25,000lb) bombload or 1 x AS-3 stand-off missile		

INTERCEPTORS OF THE 1950s AND 1960s

STRATEGISTS CONTEMPLATING A POSSIBLE World War III assumed that it would start with air strikes by nuclear bombers. The prime purpose of fighter aircraft, in this respect, was to intercept enemy bombers before they could deliver their nuclear strike, with success depending above all, on speed of response. Interceptors were not expected to engage with enemy fighters, so manoeuvrability was not a pressing concern. They were designed for maximum acceleration and rate of climb. Rising to an altitude of around 12,200m (40,000ft) within a minute of take-off, they would be directed on to the bombers by ground controllers and shoot them down with radar-guided air-to-air missiles. An average mission was expected to last about 10 minutes. Many of the aircraft developed for this role in the 1950s ran into problems as the new jet technology was pushed to the limits. Some Western interceptors – such as the F-106 – were dangerous if not flown by the most skilful of pilots. As usual, the Soviet Union went for simpler, cheaper machines that were easier to produce and fly.

ULTIMATE INTERCEPTOR
A Convair F-106 refuels from a KC-135 Stratotanker, while another waits its turn. The F-106 was dubbed the "Ultimate Interceptor".

Convair F-102 Delta Dagger

Although Convair had built the world's first delta-winged aircraft in 1948, this supersonic all-weather interceptor caused them severe embarrassment when it failed to exceed Mach 1. However, a rapid emergency re-design programme, reshaping the fuselage to reduce drag, ensured that the F-102 entered service three years late in 1956.

ENGINE	7,802kg (17,200lb) thrust P&W J57-P-23A turbojet	
WINGSPAN	11.6m (38ft 1in)	LENGTH 20.8m (68ft 4in)
TOP SPEED	1,328kph (825mph) (Mach 1.25)	CREW 1
ARMAMENT	2 x nuclear or 6 x conventional Falcon guided air-to-air missiles in fuselage bay	

Convair F-106 Delta Dart

ENGINE	11,130kg (24,500lb) thrust P&W J75-P-17 turbojet	
WINGSPAN	11.7m (38ft 3in)	LENGTH 21.6m (70ft 8in)
TOP SPEED	2,035kph (1,265mph) (Mach 1.9)	CREW 1
ARMAMENT	1 x nuclear Genie unguided rocket, 4 x conventional Falcon guided air-to-air missiles in fuselage bay	

While the F-102 problems were being fixed, a second, more complete re-design was started. Originally known as F-102B, this became the F-106, as it was completely different, with a more powerful engine and the "wasp-waist" shape necessary for supersonic flight incorporated into the main design. More sophisticated fire control equipment was fitted, including an automatic link to ground detection systems. The F-106 entered service in 1959.

English Electric Lightning F.1A

The Lightning was the RAF's first supersonic fighter and the first aircraft to exceed the speed of sound in Great Britain. The Lightning could climb to over 18,000m (60,000ft) and was equipped with radar which enabled the pilot to "lock on" to the target. After 1974, Phantoms began to replace Lightnings.

ENGINE	2 x 6,545kg (14,430lb) thrust Rolls-Royce 210 turbojets	
WINGSPAN	10.6m (34ft 10in)	LENGTH 15.2m (50ft)
TOP SPEED	2,230kph (1,386mph) (Mach 2.1)	CREW 1
ARMAMENT	2 x 30mm Aden cannon in fuselage; 2 x Firestreak guided air-to-air missiles	

Gloster Javelin F(AW).9

The Javelin was the main RAF all-weather fighter from 1956–64. It was the world's first twin-jet, delta-wing fighter, designed to intercept bombers at high altitudes and fitted with all-weather electronic equipment. After many development problems, the aircraft was finally made reliable, but by then, the Javelin was not fast enough to catch increasingly fast bombers, and was replaced by the supersonic Lightning in the early 1960s.

ENGINE	2 x 5,579kg (12,300lb) thrust Bristol Siddeley turbojet	
WINGSPAN	15.9m (52ft)	LENGTH 17.2m (56ft 9in)
TOP SPEED	1,130kph (701mph)	CREW 2
ARMAMENT	4 x 30mm Aden cannon in wings; 4 x Firestreak guided air-to-air missiles	

Thick delta wing houses engine and fuel to increase aerodynamic efficiency

Radar carried in nose

Hawker Hunter

The RAF's main interceptor fighter from 1954 to 1960, the swept-wing Hunter replaced the Meteor and served until the introduction of the supersonic Lightning in the early 1960s. Over 1,100 were sold abroad, serving with some 20 air forces throughout the world.

ENGINE	4,542kg (10,000lb) thrust Rolls-Royce Avon turbojet	
WINGSPAN	10.2m (33ft 8in)	LENGTH 14m (45ft 11in)
TOP SPEED	1,150kph (715mph)	CREW 1
ARMAMENT	4 x 30mm Aden cannon in removable pack under fuselage; 454kg (1,000lb) bombs or rocket batteries under wings	

Lockheed F-104A Starfighter

Kelly Johnson of Lockheed believed US aircraft in the Korean War were too heavy, so he sold the idea of a lightweight fighter to the US Air Force and produced this extraordinary "missile with a man in it". While it was very fast – breaking the world speed record in 1958 – it had very short range and poor manoeuvrability due to its small wings.

ENGINE	6,713kg (14,800lb) thrust GE J79-GE-3 turbojet	
WINGSPAN	6.7m (21ft 11in)	LENGTH 16.7m (54ft 9in)
TOP SPEED	2,330kph (1,450mph) (Mach 2.2)	CREW 1
ARMAMENT	1 x M-61 20mm rotary cannon in fuselage; 2 x Sidewinder air-to-air missiles on wing tips	

McDonnell F-101B Voodoo

The Voodoo was originally intended as a supersonic escort for the B-36 bomber but, when it became clear that the range could not be achieved, it was developed into a tactical, nuclear-strike aircraft. From this came the F-101B all-weather interceptor, with the latest "collision course" radar fire-control system. This version entered service in 1959 and over 400 were delivered. Although fast and heavily armed, it was never easy to fly; about a fifth were lost in crashes. It was phased out by 1970.

ENGINE	2 x 6,800kg (14,990lb) thrust P&W J57-P-55 turbojet	
WINGSPAN	12.1m (39ft 8in)	LENGTH 20.6m (67ft 4in)
TOP SPEED	1,963kph (1,220mph) (Mach 1.85)	CREW 2
ARMAMENT	3 x Falcon guided air-to-air missiles in fuselage bay, 2 x Genie nuclear-tipped, unguided air-to-air rockets	

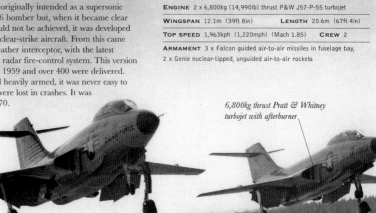

6,800kg thrust Pratt & Whitney turbojet with afterburner

Mikoyan-Gurevich MiG-17 "Fresco"

Radio antenna

Tailpipe of engine afterburner

Air-data boom, including pitot head

High-speed handling problems that limited the effectiveness of the MiG-15 were largely eliminated in the MiG-17, introduced in 1952. Although it was claimed that the prototype exceeded Mach 1 in level flight, production models could not quite achieve this. During the Vietnam War (1961–73) – although by then technically obsolete – the MiG-17 gave a good account of itself against heavily-laden US attack aircraft. Around 8,000 MiG-17s of all types were produced, with licensed production in Poland, Czechoslovakia, and China.

ENGINE	3,380kg (7,452lb) thrust Klimov VK-1F turbojet	
WINGSPAN	9.5m (31ft)	LENGTH 11.1m (36ft 3in)
TOP SPEED	1,145kph (711mph)	CREW 1
ARMAMENT	1 x 37mm and 2 x 23mm cannon under nose	

Mikoyan-Gurevich MiG-19S "Farmer"

The first Soviet production fighter to go supersonic, it was either just beaten by or just preceded the American F-100 as the world's first, depending on whose account you read. Whatever the truth, by late 1954 the Soviet air force was receiving MiG-19s just as the USAF was getting Super Sabres. Exported to many of the Soviet Union's allies and client states, the MiG-19 saw action in Vietnam and an improved Chinese-built version continued in production for many years.

ENGINE	2 x 3,040kg (6,700lb) thrust Mikulin AM-5 turbojet	
WINGSPAN	9m (29ft 6in)	LENGTH 12.5m (41ft 2in)
TOP SPEED	1,452kph (903mph) (Mach 1.4)	CREW 1
ARMAMENT	3 x 30mm cannon and missiles or rockets	

Saab J35A Draken

This versatile high performance fighter was a key part of the very advanced Swedish air defence system in the 1960s and 1970s. The Draken had a unique "double delta" wing shape, which gave both manoeuvrability and exceptional take-off and landing performance.

ENGINE	6,895kg (15,200lb) thrust Svenska Flygmotor RM6B	
WINGSPAN	9.4m (30ft 10in)	LENGTH 16m (52ft 4in)
TOP SPEED	1,915kph (1,190mph) (Mach 1.8)	CREW 1
ARMAMENT	2 x 30mm Aden cannon; 4 x Sidewinder missiles	

Yakovlev Yak-28P "Firebar"

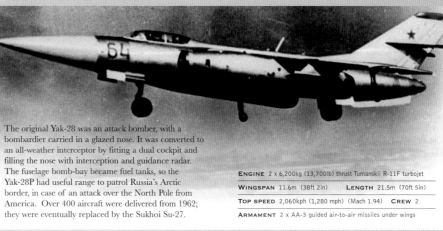

The original Yak-28 was an attack bomber, with a bombardier carried in a glazed nose. It was converted to an all-weather interceptor by fitting a dual cockpit and filling the nose with interception and guidance radar. The fuselage bomb-bay became fuel tanks, so the Yak-28P had useful range to patrol Russia's Arctic border, in case of an attack over the North Pole from America. Over 400 aircraft were delivered from 1962; they were eventually replaced by the Sukhoi Su-27.

ENGINE	2 x 6,200kg (13,700lb) thrust Tumanskii R-11F turbojet	
WINGSPAN	11.6m (38ft 2in)	LENGTH 21.5m (70ft 5in)
TOP SPEED	2,060kph (1,280 mph) (Mach 1.94)	CREW 2
ARMAMENT	2 x AA-3 guided air-to-air missiles under wings	

AIR POWER IN ACTION

WARS IN SOUTHEAST ASIA AND THE MIDDLE EAST REVEALED BOTH THE POTENTIAL AND LIMITATIONS OF AIR POWER IN THE JET AND MISSILE AGE

"The Vietnam War symbolized a new era of aerial technology, searing its way into public consciousness through the clatter of a helicopter or the bright yellow flame of a bomb explosion."

JOHN L. PIMLOTT
MILITARY HISTORIAN

FROM JANUARY TO MARCH 1968, more than 5,000 US Marines and South Vietnamese soldiers were besieged by communist forces in a combat base at Khe Sanh, South Vietnam. Fielded in their support was an array of air power such as only the world's most technologically advanced nation could deploy. Lockheed C-130 and Fairchild C-123 transports flew in supplies, by day and night under artillery and mortar fire, to the airstrip around which the base had been constructed. Three times a day "super gaggles" of transport helicopters, escorted by helicopter gunships and preceded by a wave of attack aircraft, ventured across hostile territory to isolated marine outposts around the base, taking in supplies and carrying out the wounded. Ground-attack aircraft battered the communist forces around the camp perimeter with high explosives and napalm – an average of 300 air-strikes a day came in, roughly one every five minutes. Further out, B-52s reduced swathes of terrain to moonscape, dropping their formidable bombload from 11km (7 miles) high.

The air operations often appeared desperate and chaotic. The weather, as so often in Vietnam, was dreadful for flying. The official marine history, describing operations around Khe Sanh, says: "Only those who have experienced the hazards of monsoon flying can fully appreciate the veritable madhouse that often exists when large numbers of aircraft are confined to restricted space beneath a low-hanging overcast sky." Mortar and artillery bombardment of the airstrip became so hot that transports would land, taxi, and take off again without stopping, or not land at all and simply roll out their load from open cargo doors.

And yet the air operations worked. Fourteen years earlier, a French force besieged under similar circumstances by Vietnamese forces had been overrun at Dien Bien Phu, a catastrophe that precipitated French withdrawal from their colonies in Southeast Asia. But despite dire predictions in the media, Khe Sanh held out and was eventually relieved. The difference was air power.

If air power could make a decisive difference to the outcome of a battle in Vietnam, why could it not in the end give the United States victory in the war? US air operations were astonishing in variety and quantity: for example, fleets of troop-transport helicopters providing mobility for the "air cavalry"; converted transports

FIGHTING PHANTOM
The McDonnell F-4 Phantom II was the leading western fighter of the 1960s. During the Vietnam War it served with the US Navy, Marines, and Air Force. Phantoms were adapted for a variety of roles, including reconnaissance and electronic warfare.

AIRMOBILE TROOPS

Infantrymen of the US 1st Cavalry Division leap from their Bell UH-1 helicopter to conduct a reconnaissance patrol in South Vietnam in 1967. Airmobile units such as the 1st Cavalry were central to American combat tactics in the Vietnam War, which relied on the mobility offered by helicopters as well as the fire-power that could be brought to bear from the air.

fitted out as gunships; reconnaissance aircraft fitted with the latest electronic and infrared detection devices; and the many hundreds of land- and carrier-based fighter and strike aircraft that took the war to North Vietnam. But despite the skill and bravery of US aircrews and the technological ingenuity shown in adapting to the specific demands of the conflict, it remained the wrong war for American air forces to be fighting, with the wrong kind of enemy and the wrong kind of rules.

The problem of Vietnam, as it presented itself to America's political leaders, was how to prevent a communist takeover in South Vietnam without escalating the conflict into a confrontation with China or the Soviet Union. It was decided that North Vietnam, whose communist leaders were regarded as in control of the guerrillas in the South, could not be invaded but could, within limits, be bombed. Bombing the North was extremely attractive politically because it would minimize American casualties and was relatively easy to control – the heat could be turned up or down by political decision-makers. But the desire to avoid escalation meant that the unbridled destruction of targets in and around North Vietnam's main cities was ruled out. Under these circumstances, strategic air power had no chance of deterring a determined enemy. From 1965, the United States was forced to commit massive ground forces to

South Vietnam, supported by tactical air power on an unprecedented scale.

Numbers of American airmen had already been involved in the conflict in South Vietnam since the early 1960s. For example, American helicopter pilots accompanied South Vietnamese troops into combat and American crews flew transport aircraft loaded with Agent Orange, defoliating large tracts of the countryside in order to deny shelter to the guerrilla forces. Once the US Army and Marines went in to take over the main combat role from the South Vietnamese, aircraft became central to their commanders' strategy for defeating the guerrillas. Their thinking centred on the twin concepts of fire-power and mobility, and aircraft could supply both. Helicopters would overcome the problem of operating against an elusive enemy in difficult country by moving men and material rapidly to engage guerrilla forces wherever they showed themselves. Fire from the sky would destroy that enemy as aircraft acted as mobile artillery.

The helicopter strength deployed in Vietnam was unprecedented. They performed on a larger scale all the tasks outlined in Korea. They ferried troops and equipment around

> "The sight of 60 helicopters flying in formation… at treetop level was one which none who witnessed will ever forget."
>
> LIEUTENANT GENERAL BERNARD W. ROGERS
> DESCRIBING OPERATION CEDAR FALLS, 1967

SIKORSKY SKYCRANE
A Sikorsky CH-54 Tarhe helps build a US Army fire-base in the South Vietnamese jungle. The lack of a conventional closed cargo bay let the CH-54 carry objects of almost any size or shape.

– including heavy artillery hung under Sikorsky Skycranes. They kept fire-bases deep in hostile country supplied and equipped. They evacuated casualties, carried out low-level reconnaissance, and acted as aerial command posts. But helicopters also took on an unprecedentedly active combat role, attacking the enemy on the ground with gun and rocket fire and allowing troops to ride into battle as an "aerial cavalry". Mass heliborne operations were an impressive spectacle. Lieutenant General Bernard W. Rogers described one that occurred during Operation Cedar Falls in 1967: "The sight of 60 helicopters flying in formation and zooming into Ben Suc at treetop level was one which none who witnessed will ever forget… In less than one and one-half minutes an entire infantry battalion, some 420 men, was on the ground…"

NAPALM AND PHOSPHORUS
During the Vietnam War the US air forces controversially made extensive use of napalm and phosphorous bombs against enemy guerrillas who were dug into concealed positions.

FLIGHT BOOK
A US Marine Corps helicopter pilot's flight-crew checklist notebook shows a hand-drawn map of the approach to an air facility, and heading and distance details for radio navigation.

Since North Vietnamese aircraft did not venture into the South, the only threat to US helicopters came from ground fire. When a helicopter landing zone was closely hemmed in by Viet Cong, this was a very considerable threat. Each Bell UH-1, the war's ubiquitous utility helicopter, had a gunner stationed at the open door, "flying shotgun" like the guard on a Wild West stagecoach. The helicopter crews' war was close up and personal, as journalist Frank Harvey wrote: "They didn't hurl impersonal thunderbolts from the heights in supersonic jets. They came muttering down to the paddies and hootch lines, fired at close range… They took hits through their plastic windshields and through their rotor blades." The helicopter's role extended still further to that of aerial artillery, especially with the arrival of AH-1 Cobras in 1967. A salvo of rockets from the helicopter gunship was reckoned equivalent to a barrage from 105-mm howitzers.

The helicopter gunship joined a broad array of aircraft that the army could call on for close air support. The dependence of ground troops on air power in Vietnam has often been criticized as excessive. Infantry patrols that encountered guerrillas might make little effort to take them on, hastily calling in air support to do the job for them. There are stories of jittery lieutenants summoning an air-strike to deal with a sniper. But the desire to avoid pitched infantry battles was understandable, and the aircraft were there to be

SEARCH AND DESTROY
A wave of combat helicopters of the 1st Air Cavalry Division flies over a remote landing zone in the jungle during a "search and destroy" mission in South Vietnam. An entire battalion could be flown to the point of battle. One Vietnam War pilot commented that formations of helicopters "always looked sloppy… because no two ships were ever at the same altitude".

called on. At the height of the fighting, the US Air Force, Navy, and Marines were conducting an average of 800 tactical air sorties a day. Aging piston-engined aircraft, especially A-1 Skyraiders, were notably successful in the ground-attack role, their slower speed helpful in picking out targets, but the bulk of close air-support missions were flown by high-performance jets such as F-4 Phantom IIs and F-100 Super Sabres.

Heavy bombing

The B-52s constituted the most powerful attack force used in South Vietnam. Until the very end of America's involvement in the conflict, these quintessential strategic bombers were barred from operating over North Vietnam, denying them their obvious role. They spent most of the war carpet-bombing areas of paddy-field and jungle, seeking to destroy communist supply routes and base camps as well as troop formations. Extraordinary technological ingenuity was devoted to finding targets for the B-52s. Aircraft scattered listening devices and other sensors across the hills and forests where the famed Ho Chi Minh Trail brought men and supplies from North Vietnam into the South. When troop or transport movements were detected, flights of B-52s would fly from their base on Guam in the Pacific, and release their bombloads, devastating several square kilometres of ground at a time.

The lack of proportion between the air power deployed and its results was nowhere greater than in the campaign against these supply routes. It is reckoned that a larger tonnage of bombs was dropped on the Trail than had been dropped on all fronts during World War II. Although estimates are speculative, it may have, on average, taken 100 tonnes of bombs to kill a single communist soldier.

Paradoxically, the fire-power that the Americans had at their disposal in South Vietnam was in fact excessive. When operating in remote unpopulated zones, the scale of the devastation US aircraft caused was no real matter for concern. But when fighting moved into populated rural areas, the "collateral damage" inflicted on local non-combatants was a grave embarrassment to the United States, weakening support for the war at home. The use of napalm, which had first been employed in World War II, was a particular focus of controversy and protest. To the US military it was simply a very effective weapon against troops dug in to tunnels or trenches. But a photograph of a young Vietnamese girl burnt by napalm became the single most famous image of the war.

Bell AH-1 Cobra

THE IDEA OF AN ATTACK HELICOPTER emerged in the early 1960s when armies realized how vulnerable their troop-carrying helicopters were to ground fire, especially during counter-insurgency operations. Since it was difficult to fly fixed-wing aircraft with helicopters during a mission, it was decided that the best helicopter escort would be another helicopter.

At the start of the Vietnam War, the Americans used the UH-1 Iroquois ("Huey") helicopter to suppress ground fire. This was simply a more heavily armed version of a troop-transporter. In a sense, the AH-1 Cobra represented only a limited step forward: about 85 per cent of its components were identical to those found in the UH-1. It was, however, the first rotating-wing aircraft specifically designed as a gunship, and, as such, was a milestone in the development of military helicopters. First flown in 1965, AH-1s arrived in Vietnam two years

later. As well as "escorting" duties, the AH-1s were also sent on search-and-destroy missions, working in tandem with OH-6 Cayuse scout helicopters in what was termed a "Pink Team". Shadowed by a Cobra at about 600m (2,000ft), the low-flying scout helicopters would seek out the Viet Cong. When the enemy opened fire on the OH-6, the Cobra would fly down and lay waste to the area from where the fire was coming. After the Vietnam War, the AH-1 was adapted for a new role as a tank-buster. It was fitted

with guided anti-armour missiles and an upgraded engine to give it more chance of survival on a conventional battlefield (where it would have to contend with ground fire, hostile aircraft, and missiles). Equipped with a multitude of advanced avionics – such as fire-control computers and infra-red receivers for night fighting – by the 1980s the Cobra had evolved into an extremely sophisticated fighting machine.

Turboshaft engine's exhaust

Tailskid

Elevator with inverted aerofoil section

Tailboom

UNITED STATES AR

SPITTING COBRA
The AH-1S can mount formidable attacks, using anti-tank missiles and rockets positioned on either side of the helicopter's stub wings.

NARROW TARGET
To ensure that the Cobra presented as small a target as possible to the enemy, its design featured a streamlined fuselage and a narrow profile.

Main rotor blades

Blade-pitch control rod

Rotor-head fairing

7.62mm Gatling gun

Armoured windscreen

TOW launcher

Landing skids

FRONT SEAT
The gunner's seat is at the front of the craft; the pilot sits behind (in an elevated seat). The gunner has several electronic aiming devices at his disposal.

Sighting system viewfinder

Artificial horizon

Airspeed indicator

Radio compass

Gunner's seat

Sight control / trigger

AWESOME FIREPOWER
The AH-1S boasts an impressive array of weaponry, found at the front and side of the craft. One eyewitness from the Vietnam War said that when an AH-1 opened up with all its armament, it felt like being "inside an exploding ammo factory".

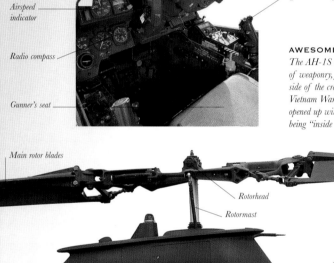

Main rotor blades

Rotorhead

Rotormast

Pilot's seat

Upward-hinged side door to cockpit

Turret housing XM-200 Rocket pods and Grenade Launcher

Specifications (AH-1G)

ENGINE	1,400shp Lycoming T53-L-13 turboshaft
ROTOR DIAMETER	13.4m (44ft)
LENGTH	13.5m (44ft 5in)
WEIGHT	4,275kg (9,500lb)
TOP SPEED	274kph (171mph)　　**CREW** 2
ARMAMENT	1 x 7.62mm M134 minigun; 1 x M129 40mm grenade launcher; 14 x 70mm M159 rockets

Rolling Thunder

From 1965 to 1968, an air campaign against North Vietnam ran in tandem with the fighting in the South. After a number of earlier "retaliatory strikes" against the North, a sustained air offensive, codenamed Rolling Thunder, was set in motion in March 1965. Although often described as a strategic bombing campaign, Rolling Thunder was really nothing of the kind. Wary of escalating the war or alienating world opinion, the US government placed a series of limits on the use of air power that precluded in advance any possibility of achieving the prime aim of strategic bombing – that is, the destruction of the enemy's industrial and military infrastructure. Whole areas of North Vietnam were declared "sanctuaries", out of bounds for bombing, including the chief cities, Hanoi and Haiphong, and a broad zone near the Chinese border. And even within areas where bombing was permitted, specific targets were often ruled out – for example, port facilities and, initially, air-defence systems.

The campaign was executed not by America's strategic bombers but by tactical strike aircraft – F-100s, F-4s, F-105s – focusing initially on targets such as bridges, roads, and supply dumps. This was in essence an interdiction campaign, attempting to impede the flow of men and supplies from the North into South Vietnam. Yet the US government persisted in hoping that it might achieve a strategic objective by inflicting enough damage to persuade the North Vietnamese to stop supporting the war in the South. The rules governing the campaign were

ANTI-AIRCRAFT DEFENCES
North Vietnamese SAMs and anti-aircraft guns combined to erect a formidable barrier against American air-raids. The anti-aircraft guns were deadly at low altitude, forcing the American planes to fly higher, where the SAMs were most effective.

STRIKE PACKAGE

FOR RAIDS ON NORTH Vietnam, the US air forces assembled "strike packages" in which support aircraft greatly outnumbered those tasked with hitting the target. Furthest from the action were KC-135 tankers, which refuelled the combat aircraft. An RC-121 radar-surveillance aircraft acted as an aerial command post, giving early warning of MiGs taking off. EB-66B Destroyers, packed with electronic countermeasures equipment, flew high above the strike force, escorted by F-4s. The EB-66Bs were intended to jam ground radars, "blinding" the SAMs and anti-aircraft guns. In case they failed, Wild Weasel F-105s or F-4s armed with anti-radiation missiles went ahead to shoot it out with the SAMs. At the heart of the package were 20 or 30 F-105 strike aircraft armed with bombs and Bullpup missiles, surrounded by F-4s to fight off any MiGs. A single mission could easily involve 100 aircraft.

THUD STRIKE
The Republic F-105 Thunderchief, known to pilots as the "Thud", was America's prime ground-attack aircraft in Vietnam. Fast, robust, and packed with complex electronic equipment, it was designed to penetrate the most sophisticated air defences carrying an impressive payload of missiles and bombs.

GUIDED BULLPUP
The Bullpup was America's standard air-to-ground guided missile in the early years of the Vietnam War. It was radio-guided on to its target by a controller in the launch aircraft using a small joystick. The Bullpup was superseded by fire-and-forget missiles that "locked on" to their targets.

periodically changed to build up pressure – a new type of target allowed or the area in which bombing was permitted extended. On occasion a "bombing pause" was declared, to give the communists time for reflection. None of this worked at all because it was based on the false idea that a bombed government and people would respond according to rational cost analysis, rather than with the gut emotion of defiance.

So Rolling Thunder was a strange concoction in which essentially tactical bombing was used for allegedly strategic purposes – bombing bridges to undermine a country's will to resist. The tight rules of engagement, which put pilots' lives at risk and blunted the effectiveness of air operations, were partly designed to limit the numbers of civilian casualties caused. And yet this did not prevent the United States being criticized worldwide for carrying out the bombing campaign, which is estimated to have killed around 50,000 North Vietnamese. Even US

HELICOPTERS IN VIETNAM

THE START OF FULL-SCALE US military involvement in Vietnam occurred at a time when a major increase in the number and variety of American military helicopters was taking place. The Chinook was already established as a transport workhorse, and the Sikorsky Skycrane had recently arrived to give a new heavy-lift capacity, while the first attack helicopter, the Bell Cobra, was undergoing flight tests. Going into the war, there was still scepticism in the US Army about the ability of rotary-wing craft to operate effectively in combat – they were widely seen as too vulnerable to enemy fire. But this soon evaporated as experience proved that helicopters could survive a lot of punishment, as well as dish it out themselves, if well-armed. The core of the helicopter fleet remained the all-purpose "Huey", but increasing specialization was visible, for example, by the deployment of the Cayuse as an effective observation platform.

COBRA GUNSHIP
Cobra gunships (AH-1T shown) first showed their effectiveness as close support aircraft in the Vietnam War. See pages 300–1.

Bell UH-1 Iroquois ("Huey")

Following the success of the helicopter in casualty evacuation in the Korean War, Bell designed this turbine-powered machine. Officially named Iroquois, it at once became universally known as the "Huey", after its original designation, HU-1. On any one day during the Vietnam war, there might be 2,000 Huey sorties.

ENGINE 1,100hp Lycoming T53-L-11 turboshaft	
ROTOR SPAN 13.4m (44ft)	LENGTH 12.7m (41ft 7in)
TOP SPEED 222kph (138mph)	CREW 2
PASSENGERS Three stretcher cases or up to seven seated	

Boeing-Vertol CH-47A Chinook

The first Chinook was an enlarged version of the Vertol 107, first flying in 1961 as the Vertol 114. The initial production variant was the CH-47A which entered service in 1961 and soon became the "Huey's" indispensable heavy-lift partner in Vietnam. By 1999, over 800 Chinooks had been built in numerous variants.

ENGINE 2 x 2,200shp Lycoming T55-1 turboshaft	
ROTOR SPAN 18m (59ft 1in)	LENGTH 15.5m (51ft)
TOP SPEED 270kph (168mph)	CREW 2–3
PASSENGERS 44 troops	

Hughes OH-6A Cayuse

In the early 1960s, Hughes was one of 12 companies competing for the US Army's Light Observation Helicopter (LOH) competition. Nicknamed the "Flying Egg", Model 369 was declared the winner in May 1965. In Vietnam, the Cayuse performed a variety of duties, including artillery spotting, escort, and reconnaissance. Part of its success derived from its advanced structural design, which balanced strength and rigidity with a low weight and streamlined fuselage.

ENGINE 317shp Allison T63-A turboshaft	
ROTOR SPAN 8m (26ft 4in)	LENGTH 7.1m (23ft 2in)
TOP SPEED 244kph (152mph)	CREW 2
PASSENGERS 4	

Sikorsky CH-53D Sea Stallion

First used in Vietnam in 1967, the CH-53 evolved out of the amphibious Sikorsky S-65, borrowing the Skycrane's rotor system to give the US Marines a much-needed heavy assault helicopter and heavy-lift capability. The USAF also modified the type into the HH-53, their principal rescue helicopter, aka the "Super Jolly".

ENGINE 2 x 3,925shp General Electric T64-GE-413 turboshafts	
ROTOR SPAN 22m (72ft 3in)	LENGTH 20.5m (67ft 2in)
TOP SPEED 315kph (196mph)	CREW 3
PASSENGERS 55 troops	

Sikorsky CH-54A Tarhe

The Sikorsky S-64A which flew in May 1962, was the first Skycrane. Successful field trials in Vietnam, under the military designation of YCH-54A Tarhe (an American-Indian word meaning "crane"), led to a large order and by 1965, the ungainly Skycrane was fully operational. It could transport damaged aircraft, artillery, and armoured vehicles, while the cargo pod could be a field hospital, barracks, or a command post.

ENGINE 2 x 4,500shp Pratt & Whitney T73 P-1 turboshafts	
ROTOR SPAN 22m (72ft)	LENGTH 21.4m (70ft 3in)
TOP SPEED 203kph (126mph)	CREW 3–4
PASSENGERS 87 troops or 9,072kg (20,000lb) load	

Sikorsky H-34 Choctaw

The H-34 Choctaw, which entered US army service in 1962, was a derivative of the US Navy's 1954 HSS-1 submarine hunter/killer. In 1962, the Choctaw was deployed to Vietnam, where it was used for staff transport and airborne search. However, it was the US Marines' who became the primary user of the type with their UH-34 Seahorse variant, because the Army had concerns over vulnerability to ground fire. The US Army's last Choctaws were retired in the early 1970s.

ENGINE 1,525hp Wright R-1820 radial	
ROTOR SPAN 17.1m (56ft)	LENGTH 14.3m (46ft 9in)
TOP SPEED 196kph (122mph)	CREW 2–3
PASSENGERS 18 troops	

Mikoyan–Guryevich MiG-21

THE MiG-21, known to NATO by the codename "Fishbed", is an aircraft that benefits from a design focused on limited, attainable objectives. It was conceived in the aftermath of the Korean War, when the Soviet military decided they needed a new-generation short-range interceptor and air-superiority fighter. The aircraft had to be fast – capable of flying at Mach 2 – and manoeuvrable, with all other features sacrificed to high performance. It also needed to be simple, reliable, easy to maintain, and cheap enough to be manufactured in large numbers.

The design produced by the Mikoyan and Gureyvich bureau was a "no frills", stripped-down, classic dogfighter and bomber-killer. In 1959, at the same time as the US was developing the F-4 Phantom – which was heavy enough to require two engines and needed an Electronic Warfare Officer to operate its array of electronic gadgetry – the MiG-21 emerged as a single-seat, single-engine, lightweight fighter, with a simple radar, two heat-seeking missiles, and a cannon. When F-4 pilots first encountered MiG-21s over North Vietnam, their craft's advanced electronics and extra engine power did not necessarily translate into combat victories. In fact the MiG was more nimble and tighter in a high-speed turn, and its gun gave a definite advantage over the US fighters, which initially did not feature a gun.

Over the years, the MiG-21 evolved away from its original lightness and simplicity. Later models had more sophisticated radar and an extra fuel tank to give the aircraft a greater range. The engine was modified to allow the aeroplane to carry more missiles as well as the extra fuel load. But the virtues of cheapness, reliability, and high performance remained. Over 13,000 MiG-21s were produced. They went into service with air forces around the world and saw action not only in Vietnam but also in other areas, such as the Middle East. Many were still operational at the start of the third millennium.

> "It was superb to fly, tough, simple and easy to build in large numbers..."
>
> IVAN RENDALL
> COMMENTING ON THE MiG-21
> IN *ROLLING THUNDER*

POPULAR CRAFT
Since it was first produced (in 1959), the MiG-21 has been used by over 50 airforces and has seen service in at least 30 wars. The aeroplane shown here is a Yugoslavian Air Force MiG-21UTI, taking off from an airstrip in Kosovo, 1999.

Faring for tailplane actuator

Tailpipe of engine afterburner

Engine bay venting air intake

Communications antenna

Ventral fin

Speed brake

Main undercarriage

PREPARING FOR ACTION
A group of Soviet pilots rush towards a line of waiting MiG-21s during a training scramble in Moscow, July 1965.

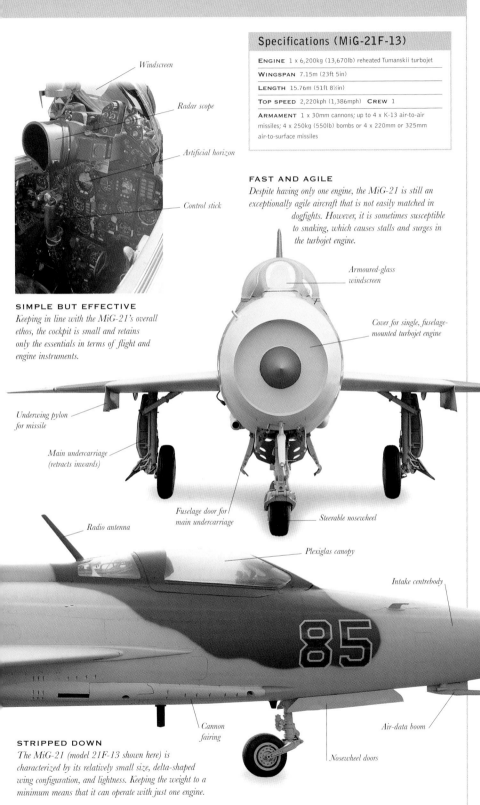

Secretary of Defense Robert McNamara, one of the chief architects of the war, ended up believing that "the picture of the world's greatest superpower killing or seriously injuring 1,000 non-combatants a week, while trying to pound a tiny backward nation into submission on an issue whose merits are hotly disputed, is not a pretty one."

The reference to North Vietnam as "a backward nation" would have had an ironic ring to American airmen at the sharp end of Rolling Thunder. Flying from bases in South Vietnam and Thailand and from carriers of the Seventh Fleet in the Gulf of Tonkin, the fighter bombers had to contend with barrages of anti-aircraft guns, Soviet-supplied SA-2 surface-to-air missiles, and MiG-17 and MiG-21 fighters safely based in the "sanctuary" areas around Hanoi. This was state-of-the-art communist air-defence weaponry and inevitably took its toll. The Americans lost a total of 938 aircraft in the three-year campaign, the majority to anti-aircraft fire or surface-to-air missiles (SAMs). Fifty-six US aircraft are reckoned to have been lost in air-to-air combat, compared with 118 MiGs downed by US fighters. Although this gave the Americans a comfortable 2-to-1 kill-ratio advantage, it was very different from the experience over the Yalu River, when ten MiGs had been shot down for every one American jet.

Strike force

The key aircraft in Rolling Thunder were the F-4 Phantom II as an air superiority fighter and the F-105 Thunderchief for ground attack. Neither was ideal for the tasks it was being asked to perform in Vietnam. The F-105 was a deep penetration fighter-bomber, intended to carry a nuclear bomb through Soviet air defences. It was not designed for precision bombing of targets such as bridges and railroads. The F-4 was a powerful, versatile aircraft, but its designers had not envisaged it in traditional dogfights. It was meant to engage enemy aircraft with radar-guided Sparrow missiles beyond visual range, following up with heat-seeking Sidewinders when it got closer. It did not have a gun for an eyeball-to-eyeball confrontation.

The rules of engagement in Vietnam took away the Phantom II's prime advantage from the outset, by insisting that an aircraft had to be visually identified before it was attacked. This meant that the MiG-17s and MiG-21s could get in close, where their manoeuvrability and their cannon gave them a fair chance. The North Vietnamese pilots consistently showed skill and aggression, although the Phantom pilots soon learned to exploit the

Specifications (MiG-21F-13)

ENGINE	1 x 6,200kg (13,670lb) reheated Tumanskii turbojet
WINGSPAN	7.15m (23ft 5in)
LENGTH	15.76m (51ft 8½in)
TOP SPEED	2,220kph (1,386mph) CREW 1
ARMAMENT	1 x 30mm cannons; up to 4 x K-13 air-to-air missiles; 4 x 250kg (550lb) bombs or 4 x 220mm or 325mm air-to-surface missiles

Windscreen

Radar scope

Artificial horizon

Control stick

SIMPLE BUT EFFECTIVE
Keeping in line with the MiG-21's overall ethos, the cockpit is small and retains only the essentials in terms of flight and engine instruments.

FAST AND AGILE
Despite having only one engine, the MiG-21 is still an exceptionally agile aircraft that is not easily matched in dogfights. However, it is sometimes susceptible to snaking, which causes stalls and surges in the turbojet engine.

Armoured-glass windscreen

Cover for single, fuselage-mounted turbojet engine

Underwing pylon for missile

Main undercarriage (retracts inwards)

Fuselage door for main undercarriage

Steerable nosewheel

Radio antenna

Plexiglas canopy

Intake centrebody

85

Cannon fairing

Air-data boom

Nosewheel doors

STRIPPED DOWN
The MiG-21 (model 21F-13 shown here) is characterized by its relatively small size, delta-shaped wing configuration, and lightness. Keeping the weight to a minimum means that it can operate with just one engine.

SMART BOMBS

GUIDED BOMBS MADE their first appearance in
World War II, when both the Germans and
Americans experimented with attaching radio-
control systems to conventional iron bombs.
They were not ineffective, but the radio link was
too easy to jam and the delivery aircraft was too
vulnerable to anti-aircraft fire while it tracked
the bomb all the way down.

In the late 1950s experiments began with
bombs guided by "electro-optics". The US
Navy's Walleye gliding bomb, first used in 1967
in Vietnam, had a TV camera in its nose, which
transmitted a picture back to the carrier aircraft.
The aircraft's Electronic Warfare Officer could
lock it on to the target or guide it all the way in.

But smart weapons really came of age with
the deployment of laser-guided Paveway bombs
from 1968. The target is selected by a laser beam
and the bomb's guidance system follows the
reflected beam to its source. Because the delivery
aircraft is not itself the target designator, it can
turn away once the bomb is released.

SMART ATTACK
*The Paveway laser-guided bomb demonstrates its accuracy
on a test range. The bomb requires an aircraft or a soldier
on the ground to illuminate the target by
directing a beam of laser light on to it.
The bomb's guidance system does the rest.*

Fixed fins

Laser seeker
head

Moveable
guidance fins

Bomb casing

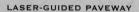

LASER-GUIDED PAVEWAY
*The Paveway bomb is fitted with detectors that acquire
and lock on to the reflected light of a laser-beam
illuminated target.*

superior power of their aircraft in combat
manoeuvres. And by 1967 some F-4s were
sporting a mounted gun; F-4Es, with an integral
gun, were not introduced until the end of 1968.

Initially the Americans employed the tactics
they had developed for nuclear war, penetrating
North Vietnamese air space at low altitude to duck
under radar cover. But they soon found that this
left aircraft too exposed to fire from anti-aircraft
guns and switched to a higher altitude approach,
depending on electronic countermeasures to jam
the radars that guided the SA-2 missiles, and using
Shrike missiles to take out the SAM sites. The
aircraft that carried Shrikes, generally F-105s, were
codenamed Wild Weasel. They had to fly straight
toward the SAMs until they were picked up by the
missile site radar, then fire a Shrike missile, which
homed on the radar emission. Hopefully this would
happen before a SAM was launched back. Despite
all the countermeasures deployed to protect US
aircraft against SAMs, many pilots still found that
survival was best guaranteed by learning to evade
the missiles through sheer skill and speed.

Rolling Thunder was formally ended at the
start of November 1968, in return for North
Vietnamese agreement to join peace talks. By
then American pilots had flown some 300,000
Rolling Thunder sorties, delivering an estimated
860,000 tonnes of bombs. They had caused a lot
of damage, especially
when rules had been
relaxed to allow them
to strike against power
stations and fuel-
storage facilities.
But no one seriously
believed that the
communists had been
"bombed to the
negotiating table". To little evident purpose,
hundreds of American airmen had been killed
and hundreds more delivered into harsh captivity.

When the withdrawal of US ground forces
from Vietnam began in 1969, the Americans also
started handing over responsibility for the air war
to the South Vietnamese. There was something of

a lull until the spring of 1972, when the North
Vietnamese Army invaded South Vietnam in force.
By this time it was politically impossible to commit
American ground troops to battle, but American
air power was redeployed with a vengeance, both
tactically in support of the South Vietnamese
army and against targets in North Vietnam.

Guided weaponry

In the interval between 1969 and 1972, there
had been some notable developments in the US
air forces, partly as a result of reflection on their
unsatisfactory experience in Vietnam. There was a
widespread feeling that fighter pilots had not been
adequately trained for dogfights and that reliance
on missiles had led to neglect of some of the basic
principles of air-to-air combat. In March 1969 the
US Navy established its Post-Graduate Course in
Fighter Weapons, Tactics, and Doctrine at
Miramar Naval Air Station in California – better
known as the Top Gun programme. Meanwhile,
the USAF was re-equipping with the F-4E, which
had an integral gun.

By 1972 American pilots were not only
significantly better prepared for air-to-air combat
but were better armed for ground attack against
precision targets. The evolution of guided
weapons was a gradual process that had already
been well under way by the time the Vietnam
War started, but the introduction of TV and
laser-guided weapons constituted a quantum leap
forwards. For the first time in aviation history, a
target such as a bridge or a single building could
be bombed with a high expectation of success.

Better skills and equipment might have made
little difference but for changes both in the nature
of the war and the rules under which American
airmen operated. The North Vietnamese invasion
replaced guerrilla war with conventional warfare.
This involved the deployment of tanks and other
substantial equipment that provided clear and
valuable targets for air attack. It also required
uninterrupted supplies
of fuel and munitions
on a large scale to
sustain operations.
Intensive American
air strikes – totalling,
for example, more
than 18,000 sorties by
fixed-wing aircraft in
May 1972 – imposed
heavy punishment on North Vietnamese forces
inside South Vietnam. Meanwhile, US air forces
were unleashed against almost the whole of
North Vietnam, with few restrictions in the
selection of targets. US Navy A-7 Corsairs
dropped mines to block North Vietnamese ports,
the entry point for supplies from China and the

> "There's something terribly
> personal about the SAM; it
> means to kill you..."
>
> GENERAL ROBIN OLDS
> USAF PILOT IN VIETNAM

TOP GUNS

IN AIR COMBAT OVER NORTH VIETNAM up to the autumn of 1968 roughly two North Vietnamese MiGs were downed for every US fighter lost. This ratio did not satisfy the US Navy. A Navy inquiry concluded that pilots were not receiving adequate training in close combat and recommended a new

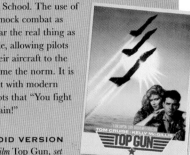

TOP GUN BADGES
Modern fighter combat training encourages competition between pilots, hence the name Top Gun (seen on the badge, bottom right) for the US Navy's Fighter Weapons School.

training programme in which fighter pilots would be pitted against aircraft similar to enemy fighters in realistic air-combat manoeuvres. The result was the establishment of the Top Gun school at Miramar Naval Air Station, California, in 1969.

The progress in combat skills that followed was exemplified by US Navy pilot Lieutenant Randall "Duke" Cunningham and his radar officer Lieutenant William Driscoll. They already had two kills to their credit when, on 10 May 1972, their flight of F-4Js was attacked by MiGs over North Vietnam. As Phantom IIs and MiGs fought a turning dogfight, each pilot trying to get on another's tail, Cunningham downed two enemy fighters with Sidewinder missiles. Heading for the coast, he ran into another MiG-17 whose pilot proved aggressive and tenacious, staying with the Phantom II as it manoeuvred to shake him

off. By suddenly throttling back and applying his airbrakes, Cunningham finally made the MiG overshoot, got on his tail, and brought him down with a Sidewinder. With five kills, Cunningham and Driscoll were hailed as the first American aces of the Vietnam War.

In the 1970s the Top Gun principle was also adopted by the USAF at its Fighter Weapons School. The use of mock combat as near the real thing as possible, allowing pilots to push their aircraft to the limit, became the norm. It is self-evident with modern fighter pilots that "You fight like you train!"

CELLULOID VERSION
The 1986 film Top Gun, set at the US Navy's elite training school, restaged for a new generation the long-established tradition of the fighter pilot as romantic hero. The film was a box-office hit, but the F-14 Tomcats stole the show.

RETURNED FROM COMBAT
Back on the carrier USS Constellation after shooting down three MiG-17s on a single mission, Lieutenant Randall "Duke" Cunningham (centre) treats his colleagues to a graphic account of the day's action.

PHANTOM FIGHTER
Originally designed for the US Navy but also adopted by the USAF, the F-4 Phantom II was an excellent aircraft in its performance and equipment. The ability of the apparently inferior MiGs to hold their own with the F-4s in air combat came as a shock and, in part, led to the founding of the Top Gun school.

THE BIG DROP
A B-52F drops a string of bombs over South Vietnam in 1965. The B-52F could carry 23,500kg (51,750lb) of bombs, but this capacity was dwarfed by the modified "Big Belly" B-52Ds introduced later in the war, whose maximum load was almost 27,000kg (60,000lb).

SCENE OF DEVASTATION
The devastation wrought by military action in South Vietnam, as here in the Cholon district of Saigon after the 1968 Tet offensive, was often the result of artillery fire. But it was the use of air power that became the focus of anti-war sentiment in the United States, with the B-52 an especially demonized aircraft.

"The bombers hide above
the clouds. The whistle
and explosion of bombs
thunder in every corner
of the forest."

TRAN MAI NAM
NORTH VIETNAMESE JOURNALIST

HIGH-ALTITUDE PRECISION BOMBING

DURING THE FIGHTING in 1972, "cells" of three B-52s were used to strike at North Vietnamese ground forces operating inside South Vietnam, dropping their bombs from 11km (7 miles) high, often through dense cloud cover. Although guiding a bomber to release its bombs on a point defined by map co-ordinates had been practised since early in World War II, it still required remarkable skills from aircrews and ground controllers when a five-second delay in bomb release would translate into about a 0.8-km (½-mile) error in targeting. The ground controllers tracking the B-52s on radar worked in threes, checking and double-checking one another's work. The lead B-52 was counted down to bomb release by the voice of a ground controller on the

radio: "Five, four, three, two, one, hack." The radar navigators in the other two planes hit their bomb switches a precise number of seconds later. The remoteness of the aircrew from the effects of their actions was total. Bomb release was felt as a slight shuddering of the aircraft and seen as a series of lights flicking off as each of perhaps 66 bombs fell away. The bombs hit the ground a minute later, by which time the B-52s had turned for home. The pilot might just glimpse flashes lighting up the clouds below. That was all.

BLACK HOLE
The radar navigator in a B-52 was the airman responsible for dropping the bombs. His position was on the lower deck, in the windowless "black hole" beneath the pilot and co-pilot.

Soviet Union. The whole supply system was devastated, from warehouses and fuel depots to roads, railways, and bridges. In October the air-strikes on the North, codenamed Linebacker, were halted after a breakthrough in peace talks. By then, the North Vietnamese had had to admit that they could not conduct a conventional war in the face of American air power.

Christmas bombing

The finale of America's air war in Vietnam brought the B-52s centre stage. They had been extensively used during the fighting in 1972, but in the second half of December that year, with peace talks stalled on the verge of agreement, they were unleashed in a major offensive against Hanoi and Haiphong. President Nixon's typically crude comment on the operation, dubbed Linebacker II, was that "the bastards have never been bombed like they're going to be bombed this time."

Between 18 December and 29 December more than 15,000 tonnes of bombs were dropped on targets in and around the North Vietnamese cities. At the peak of the offensive 120 B-52s attacked in a single night, backed up by fighters, ground-attack aircraft, electronic-warfare aircraft, and helicopter combat-rescue teams. The aim was to destroy North Vietnam's military and industrial infrastructure, including airfields, missile sites, army barracks, power stations, and railroad yards. To achieve accurate bombing and minimize politically unacceptable civilian casualties, the B-52s had to stay on a straight and level course on the approach to their targets, despite facing volleys of SA-2 missiles fired up to greet them. Inevitably there were losses – 15 B-52s were shot

down – and equally inevitably there were bombs that missed their target, including one that struck a hospital. But on the whole Linebacker II achieved its objectives, wreaking destruction on a huge scale.

The Christmas bombing offensive was rapidly followed by an agreement that allowed the final withdrawal of American military forces from Vietnam. But there is little evidence that the bombing forced the North Vietnamese into any concessions. It was a display of strength that alleviated America's pain at having failed to win the war and was meant to warn the communists what might happen if they

broke the peace agreement. In reality, America's will to engage militarily in Vietnam had gone. When the North Vietnamese took over the South in 1975 the United States did nothing. The last use of American aircraft in Vietnam was to fly people out from the roof of the US embassy in Saigon as communist tanks entered the city.

Use of air power on an astonishing scale had failed to decisively influence the course of the Vietnam War, and had in fact become a political liability as the main target for critics of the war. Twenty years earlier, air power had been deployed to immense destructive effect in Korea

LAST HELICOPTER OUT
President Nixon (above) engineered the final withdrawal of American forces from South Vietnam in January 1973. The United States was at the time committed to resume military action if the North Vietnamese broke the peace agreement, but the will to support South Vietnam soon faded. In May 1975, as the communist forces occupied Saigon, people struggled for a chance to escape in helicopters that would fly them to US carriers offshore (right).

WESTERN PRE-1970s COMBAT AIRCRAFT

THE AIRCRAFT THAT SAW ACTION in Vietnam and in the Middle East during the 1960s and early 1970s were mostly designed during the 1950s. Some were extremely long-lived, like the subsonic Douglas Skyhawk, which first flew in 1954 and was still being used by the Argentinians in the 1982 Falklands War. Aircraft such as the F-100 Super Sabre and the Dassault Super Mystère belonged to the first generation of supersonic fighters, while the slightly later F-4 Phantom and Mirage III, show the progression to Mach 2 as a requirement for state-of-the-art fighters. The Dassault Mystère and Mirage series owed their fame to a political quirk which made France Israel's principal arms supplier in the years leading up to the 1967 Six-Day War. The Phantom, originally a US Navy jet, became arguably the foremost fighter of the decade. Both the Vought A-7 Corsair II, a light attack aircraft intended to succeed the Skyhawk, and the Grumman Intruder, were designed in the 1960s and first used in Vietnam. Both were still in front-line service at the time of the Gulf War in 1991.

WARSAW PACT FIGHTER
A lightweight, tactical fighter, the MiG-21 "Fishbed" made an excellent dogfighter. Its widespread export popularity lay in its relative cheapness – about one third of the cost of a Phantom II. See pages 304–5.

Dassault Mirage III C

Development of the French delta-winged fighter was promoted enthusiastically by the French government. The highly successful Mirage III C was exported to many foreign air forces, including that of Argentina, who used them against British forces in the 1982 Falklands War.

ENGINE 6,000kg (13,225lb) thrust SNECMA Atar 9B turbojet	
WINGSPAN 8.2m (27ft)	LENGTH 15.5m (50ft 10in)
TOP SPEED 2,350kph (1,460mph)	CREW 1
ARMAMENT 2 x 30mm DEFA cannon; up to 1,362kg (3,000lb) bombs, rockets, or missiles	

Dassault Super Mystère

Originally designated Mystère IV B, this was the first Western European aircraft to fly at supersonic speed and was retitled Super Mystère. An entirely French design, though taking several ideas from the American F-100, it used a French-developed engine and was in French Air Force service from 1957.

ENGINE 4,500kg (9,920lb) thrust SNECMA Atar 101G-3 turbojet	
WINGSPAN 10.5m (34ft 5in)	LENGTH 14m (46ft 1in)
TOP SPEED 1,200kph (743mph)	CREW 1
ARMAMENT 2 x 30mm DEFA cannon; internal launcher for 35 rockets, 907kg (2,000lb) bombload or weapons on wing pylons.	

Douglas A-4D Skyhawk

The US Navy's first jet attack bomber was so small and light that it was nicknamed "Heinemann's Hot Rod" after its designer. The aircraft was fast enough for the prototype to take the world 500km (311 miles) speed record. Entering service with the Navy and Marines from 1956, Skyhawks were front-line aircraft for 20 years.

ENGINE 3,856kg (8,500lb) thrust Pratt & Whitney J52-6 turbojet	
WINGSPAN 8.4m (27ft 6in)	LENGTH 12.2m (40ft 1in)
TOP SPEED 1,102kph (685mph)	CREW 1
ARMAMENT 2 x 20mm cannon; 3,720kg (8,200lb) bombload	

Grumman A-6E Intruder

This US Navy and Marine Corps all-weather attack bomber went straight into combat in Vietnam from 1965. The bulbous fuselage houses one of the most sophisticated navigation and attack radar systems in the world, including automatic landing on a parent carrier. Intruders were still extremely effective in the Gulf War.

ENGINE 2 x 4,218kg (9,300lb) thrust P&W J52-8A turbojet	
WINGSPAN 16.2m (53ft)	LENGTH 16.7m (54ft 9in)
TOP SPEED 1,037kph (644mph)	CREW 2
ARMAMENT up to 8,165kg (18,000lb) bombs, rockets, or missiles	

McDonnell Douglas F-4J Phantom II

The F-4 was designed as a US Navy air defence fighter, with high supersonic speed, large internal fuel capacity, and powerful radar and missiles. Navy and Air Force Phantoms were in action in Vietnam until 1973. Fitted with diferent equiment, the same aircraft could be used in a variety of roles.

ENGINE 2 x 8,120kg (17,900lb) thrust GE J79-GE-10 turbojet	
WINGSPAN 11.68 m (38 ft 4 in)	LENGTH 17.7m (58ft 2in)
TOP SPEED 2,414kph (1,500mph)	CREW 2
ARMAMENT 4 x AIM-9 Sidewinder and 4 x AIM-7 Sparrow anti-aircraft missiles; or 7,257kg (16,000lb) bombload	

North American F-100D Super Sabre

The prototype Super Sabre flew in 1953, breaking the sound barrier on its first flight. With extensive use of high-strength titanium alloys, it was a radical step forward in aircraft design and soon became one of the USAF's most versatile tactical aircraft, flying over 360,000 combat sorties in Vietnam.

ENGINE 7,688kg (16,950lb) thrust P&W J57-P-21A turbojet	
WINGSPAN 11.9m (39ft)	LENGTH 15m (49ft)
TOP SPEED 1,461kph (908mph)	CREW 1
ARMAMENT 4 x 20mm M-39E cannon in fuselage; up to 3,193kg (7,040lb) of bombs, rockets, or missiles	

Republic F-105D Thunderchief

Famous for their WWII Thunderbolt, Republic built
the largest single-seat, single-engine aircraft ever in this
long-range tactical fighter-bomber. Entering service in
1959, the aircraft carried out more air strikes over North
Vietnam between 1966 and 1971 than any other USAF
aircraft. Losses were heavy and since it was expensive to
maintain, it was gradually phased out.

ENGINE	11,113kg (24,500lb) thrust P&W J75-P-19W turbojet		
WINGSPAN	10.7m (34ft 11in)	LENGTH	19.6m (64ft 3in)
TOP SPEED	2,237kph (1,390mph)	CREW	1
ARMAMENT	1 x 20mm M-61 Vulcan rotary cannon; 3,629kg (8,000lb) bombload, 2,722kg (6,002lb) missiles on wing pylons		

Vought A-7E Corsair II

Derived from but completely different to the Crusader,
this attack aircraft sacrificed supersonic speed for
carrying capacity and range. Its stubby shape earned it
the nickname SLUF – "Short Little Ugly Fella".
Ordered by the US Navy and Air Force in 1966, it was
still in Navy front-line service during the Gulf War.

ENGINE	6,804kg (15,000lb) thrust Allison TF41-A-2 turbofan		
WINGSPAN	11.8m (38ft 9in)	LENGTH	14.1m (46ft 1in)
TOP SPEED	1,123kph (698mph)	CREW	1
ARMAMENT	1 x 20mm M-61 Vulcan rotary cannon, up to 6,804kg (15,000lb) bombs, rockets, or missiles		

Vought (F-8A) F8U-1 Crusader

The US Navy's main fleet defence fighter from 1957,
the Crusader (French F-8E type shown) outperformed
the F-100 using the same engine. The high-mounted
wing tilted up to allow slower landing and take-off on
carrier decks. Effective in combat, the F-8 was gradually
replaced by the better (but more expensive) Phantom.

ENGINE	7,327kg (16,200lb) thrust P&W J57-P-4 turbojet		
WINGSPAN	10.9m (35ft 8in)	LENGTH	16.5m (54ft 3in)
TOP SPEED	1,630kph (1,013mph)	CREW	1
ARMAMENT	4 x 20mm cannon; 2 x Sidewinder air-to-air missiles, 32 rockets in belly pack		

without exciting notable unease in the United
States. But in Vietnam, the more successful
aircraft were in their task of destruction, the more
criticism they provoked. Standards of tolerance
for civilian casualties were falling, and especially,
it seemed, for civilian casualties caused by air
attack. This was a potentially major inhibition on
the use of aircraft in what might be termed "wars
in peacetime".

The Six-Day War

While the Americans were unhappily embroiled
in Vietnam, in 1967 the Israelis gave an
exceptional demonstration of the incisive use of
air power in their Six-Day War with their Arab
neighbours in the Middle East. The Israeli Air
Force (IAF) was a self-conscious elite with only
a few hundred combat aircraft. These were
exclusively French – mostly Dassault Mirage IIIs,
Mystères, Super Mystères, and Ouragans – and
most were fighter-bombers, since Israel did not
have the budget for a separate fighter and bomber
force. For years the IAF had planned to win air
superiority at the very start of a war by destroying
enemy aircraft on the ground. It would then use
its command of the air to provide ground forces
with decisive support. This was a strategy with
clear and achievable objectives, requiring
complete ruthlessness in disregard for such
niceties as a prior declaration of war.

Early on the morning of 5 June 1967, 196
Israeli aircraft, in flights of four, headed at low
altitude across the Mediterranean and the Sinai
desert to deliver co-ordinated strikes against ten
airfields in Egypt. It was the most effective
surprise air attack since Pearl Harbor. The Israeli
pilots maintained complete radio silence, and the
aircraft were not picked up by Egyptian radar.
The Israeli fighter-bombers found Egyptian

aircraft lined up on their hard-standings. They
made a first pass with bombs and a second pass
strafing Egyptian aeroplanes on the ground. Some
aeroplanes were armed with specialized runway-
cratering bombs, which were rocket-propelled
into the runway's concrete surface and exploded,
creating holes about 2m (6ft) deep.

Most of the Egyptians' 600 Soviet-supplied
aircraft were destroyed on the ground. Those that
managed to engage the Israelis in air combat
were outfought and shot down in large numbers.
The low level of the attack went under surface-to-
air missile (SAM) radars and the missiles were
wholly ineffectual, although anti-aircraft fire
claimed a score of Israeli victims. Once the
Egyptians were out of action, the Israelis went on
to attack airfields in Iraq, Syria, and Jordan.
Complete air superiority was achieved in ten hours.
With command of the air, over the following days
the IAF attacked Arab armoured columns and
artillery positions to devastating effect.

Israeli success

The Six-Day War made the Mirage III
momentarily the most famous aircraft in the
world, but the success of the Israelis was not
due to technological superiority. It was partly a
demonstration of the gulf that could exist in air
combat between pilots selected and trained to be
outstanding fighting men and merely competent
adversaries. It was also a sign of how much more
effective air power could be in a theatre of war in
which the weather was usually clear, the terrain
left armies exposed and without the possibility of
concealment, and much of the fighting took place
in areas with little or no civilian population. And
it showed how the military effectiveness of
aircraft could benefit from the absence of
politically imposed rules or limits.

MARCEL DASSAULT

MARCEL DASSAULT (1892–1986) was the French
aeroplane maker responsible for the Mirage and
Mystère fighter-bombers. Born Marcel Bloch, he
was one of the first Frenchmen to take a degree
in aeronautical engineering. During World War
I, in partnership with Henri Potez, he produced
a propeller for the famous SPAD fighters and set
up the aircraft-manufacturing company, SEA.
This went bust after the war and Bloch turned to
making furniture. In the 1930s he set up Avions
Marcel Bloch to build all-metal aircraft.

As a prominent Jewish businessman, Bloch's
position after the Nazi victory was precarious.
He joined the French Resistance, but was
arrested by the Gestapo and sent to Buchenwald
concentration camp. He survived this ordeal and
after the war changed his name to Dassault.

Dassault built up a postwar business empire
centred on aviation. The outstanding jet aircraft
his company produced kept France in the fore
of military aviation
development. He died
shortly after the first flight
of the Rafale fighter.

LONG CAREER
*The long career of
French aviation
pioneer Marcel
Dassault (originally
Bloch) stretched from
the days of wire-and-
strut biplanes to the
era of supersonic jets.*

After 1967, the Soviet Union resupplied Egypt with MiGs and SAMs, while the United States became Israel's arms supplier, providing first Douglas A-4s and then McDonnell F-4 Phantoms. It was the start of a long period in which the Middle East would periodically provide combat trials for the latest Cold War aircraft and air-defence systems.

Electronic warfare

During the Six-Day War, guns and iron bombs had predominated over missiles, and electronic warfare had played only a limited role. But by 1973, when Israeli and Arab forces met again in full-scale conflict in the Yom Kippur war, missiles and the electronic countermeasures (ECM) to them were at the heart of the fighting. Soviet-supplied SAMs and radar-guided anti-aircraft guns initially inflicted heavy losses on Israeli airmen hastily thrown into strike against Egyptian troops who had crossed the Suez Canal and Syrian forces advancing on the Golan Heights. The Israelis lost 40 aircraft in the first few hours of the war – about a tenth of their total air force.

Although the quality of the Israeli pilots ensured that they maintained overwhelming supremacy in air-to-air combat, it was only when the Americans provided them with the latest ECM pods for their F-4s that they were able to overcome the gun-and-missile air-defence systems and once more assert total air superiority over Egypt and Syria. The lesson was clear: even the best-trained and motivated pilots would need the most up-to-date kit if they were to survive on the evolving electronic battlefield.

Vertical take-off and landing

One effect of the 1967 Israeli pre-emptive strike was to make air planners extremely nervous about airfields. The fact that the Egyptian aeroplanes had been lined up on the airfield waiting to be shot up was not any special folly, but largely standard practice in the world's air forces. There was a rush to reorganize airfields so that aircraft were dispersed and to create bomb-proof shelters to house them. Yet the dependence of aircraft on their concrete runways remained a weak point that an enemy could exploit. If runways were cratered, aircraft could not operate even if they survived.

Fears about the vulnerability of airbases created fresh interest in VTOL (vertical take-off and landing) jets, which could operate from any flat space, although some were intended as shipborne fighters. There had been various attempts at creating viable VTOL fixed-wing aircraft. In the mid-1950s, a number of experimental "tail-sitters" were built – aircraft such as the Convair XFY-1 Pogo and Ryan X-13 Vertijet. Their upright

BAe Harrier

THE HARRIER is the only sucessful V/STOL (Vertical or Short Take-Off and Landing) jet aircraft ever produced. Its nearest rival, the Soviet Yak-36 can take off and land vertically but cannot take off conventionally from a short runway. This is an important distinction because using vertical take-off severely limits the amount of fuel and munitions that an aircraft can carry. This means that using a short take-off enables the Harrier "Jump Jet" to increase its range and weaponload. Therefore, the normal profile for a Harrier mission is STOVL – Short Take-Off and Vertical Landing.

The secret to the Harrier's success is its Pegasus vectored-thrust engine, which has four swivelling nozzles instead of the single fixed exhaust pipe found in other jets. For vertical take-off or landing the nozzles are rotated to point downwards. For short take-off they are set at an angle. In normal flight they are swivelled to point backwards. One of the most difficult challenges the Harrier's designers faced was to work out how to control the aircraft during vertical flight, when the usual control surfaces – the rudder, ailerons, and tailplane – would be inoperative. The solution was to place pilot-operated valves in the tail, nose, and wingtips that release jets of high-pressure air to control the aircraft's pitch and roll during hovering flight. However, using this system is by no means easy – one Harrier pilot compared it to balancing on top of four wobbly bamboo poles.

The Royal Navy version of the Harrier, the Sea Harrier, became operational in 1980. Because the Harrier was able to take off from a 180m (600ft) flight deck (with the assistance of an inclined ramp), sea carriers originally considered only big enough for helicopters were able to operate a fixed-wing aircraft. Also, it was Marine pilots who discovered that the swivelling nozzles could be used to produce rapid deceleration and other unexpected jinks unavailable to conventional fighters.

Although the Harrier is too slow to function as a top-class, air-superiority fighter, in the 1990s BAe (British Aerospace) and Boeing developed the Harrier II Plus – a dedicated fighter. This shows that there is still faith in the V/STOL concept in contemporary military aviation.

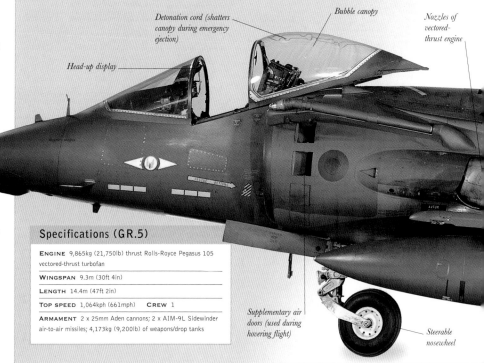

Head-up display

Detonation cord (shatters canopy during emergency ejection)

Bubble canopy

Nozzles of vectored-thrust engine

Supplementary air doors (used during hovering flight)

Steerable nosewheel

Specifications (GR.5)

ENGINE	9,865kg (21,750lb) thrust Rolls-Royce Pegasus 105 vectored-thrust turbofan
WINGSPAN	9.3m (30ft 4in)
LENGTH	14.4m (47ft 2in)
TOP SPEED	1,064kph (661mph) **CREW** 1
ARMAMENT	2 x 25mm Aden cannons; 2 x AIM-9L Sidewinder air-to-air missiles; 4,173kg (9,200lb) of weapons/drop tanks

MAIDEN FLIGHT
The Harrier "Jump Jet" made its first flight in 1966, three years after the first hovering flight made by its prototype, the Hawker P.1127 (shown here). The production Harrier went into full RAF service in 1969.

EYES FRONT
The "head-up" display unit featured in this Royal Navy Sea Harrier (right) shows vital flight information such as altitude and speed on its glass panel. This means that the pilot is able to take in this data without looking down.

Wraparound windscreen

Control stick

Detonation cord (shatters canopy during emergency ejection)

Head-up display unit

Pilot in full flight gear

LANDING AT SEA
This Sea Harrier is about to land on the flight deck of a carrier ship. The ability to land vertically is especially valuable in rough weather, when a conventional aircraft would have difficulty setting on a pitching deck.

Bubble canopy

Probe for refuelling during flight (retracted)

Carbonfibre wing

WEAPONRY
Included in the Harrier GR.5's weaponry are two cannons, which are housed under the fuselage. These are capable of firing over 3,500 shells per minute.

Pylon for air-to-air missile

Stabilizing wheel

Cannon

Engine air intake

External fuel tank

Fin tip

Slotted flaps

ZD408

Underfin

JOINT PRODUCTION
The Harrier GR.5 was produced jointly by British Aerospace (BAe) and McDonnell Douglas. It included several important developments, such as longer wings and an airframe structure incorporating carbonfibre.

Intermediate stores pylon

position made it irritatingly difficult for pilots to climb into the cockpit, and they were alarmingly difficult to land, since the pilot had to make the vertical descent backwards with his feet above his head. The unsatisfactory nature of these aircraft led to experiments with fitting extra lift engines to otherwise conventional aircraft. Dassault fitted the Mirage III V with eight small engines to raise it vertically into the air, after which it flew conventionally. But the extra engines had a severe adverse effect on the aircraft's overall performance.

Jump Jet

The solution was found in the 1960s with the British Aerospace Harrier "Jump Jet", initially developed by Hawker Siddeley. It had the same jet engine for vertical take-off and landing and for conventional flight, using swivelling nozzles to direct the thrust either down towards the ground or towards the rear of the aircraft. It could also

> "I'm not allowed to say how many planes joined the raid, but I counted them all out, and I counted them all back."
>
> **BRIAN HANRAHAN**
> REPORTING ON A HARRIER RAID IN THE FALKLANDS WAR

take off conventionally from a short runway or flight deck. The main function originally envisaged for the Harrier was as a battlefield-support aircraft flying from concealed sites just behind the front line. In part, its thunder was stolen by the development of helicopter gunships, which compete for the same role and also do not need airstrips. The Harrier found its uses in some specific circumstances, and was astonishingly ingenious, but remained essentially marginal. Vertical take-off was simply not compatible with the highest levels of performance in speed, range, or payload. Military planners were stuck with their airfields.

The Falklands War

The Harrier had its moment of glory in the spectacular but idiosyncratic air-sea encounters of the Falklands War, provoked by Argentina's occupation of the British Falkland Islands in 1982. The Royal Navy was sent to the stormy South Atlantic without either a full-size aircraft carrier or an airborne early-warning aircraft – by then an essential element of fleet air defence. Against the British Task Force, defended by a score of Sea Harriers and RAF Harrier GR.3s, the Argentinians deployed aircraft of 1960s vintage, mostly Mirage IIIs and A4 Skyhawks, plus a handful of Super Etendards.

Early in the conflict, an RAF Vulcan became the only member of Britain's one-time independent strategic nuclear bomber force to be used in anger, flying a 6,250-km (3,900-mile) round trip to bomb an airstrip on the Falklands. Delivered by the Super Etendards, the Exocet, an unexceptional French anti-ship missile, leapt to fame through sinking two British vessels. And the Sea Harriers and GR.3s, with their highly trained Navy and RAF pilots, proved more than a match for the aging Argentinian jets. Even more impressive than their combat performance was their ability to maintain constant combat air patrols in often dreadful weather conditions.

The main lesson of the Falklands War was, once again, the vital importance of electronic warfare and missile technology. It was the absence of an early-warning aircraft that made the Royal Navy's ships vulnerable to the Exocet missile. And

it was the improved Sidewinder that above all
gave the Sea Harriers the edge in air combat. For
since the mid-1970s, the American heat-seeking
missile had become "all-aspect" – that is, it no
longer needed to be fired from behind an enemy
aircraft to seek out its target.

In many ways the Falklands War was
anachronistic. By the 1980s air war had already
moved into a new era. In retrospect, the Vietnam
War and the Middle East wars of 1967–73 could
be seen as a transitional phase, in which missiles,
smart bombs, and the ever more complex head-
to-head between guidance systems and electronic
countermeasures had begun to transform aerial
warfare. But it was only with the arrival of a new
generation of astonishingly sophisticated and
expensive fighters and strike aircraft in the 1970s,
backed up by AWACS (airborne warning and
control systems), that an unbridgeable gulf would
open up between the technologically most
adept and the rest.

SEA HARRIER TAKE-OFF
*A Royal Navy Sea Harrier air-superiority fighter launches from
the ramp of HMS* Invincible. *The Sea Harrier's short take-
off capability enabled the Royal Navy to provide air-cover in the
Falklands conflict without a full-size aircraft carrier.*

HI-TECH WARFARE

A NEW GENERATION OF MILITARY AIRCRAFT USING ADVANCED ELECTRONICS AND PRECISION WEAPONS HAS MADE AIR POWER MORE EFFECTIVE THAN EVER

"The jet fighter is one of the great icons of the second half of the twentieth century, a symbol of achievement, of technical excellence, of ultimate modernity, of latent military power."

IVAN RENDALL
FROM HIS BOOK *ROLLING THUNDER*

IN JUNE 1982, THE SYRIAN AIR FORCE sent up its MiG-21 and MiG-23 fighters to engage the Israeli Air Force over the Bekaa Valley in Lebanon. In three days of large-scale air combat, the Syrians lost 50 aircraft shot down by Israel's American-supplied F-15 and F-16 fighters. Not a single Israeli fighter was lost in air-to-air combat. Syria's ground air-defence system was almost equally ineffectual, downing only one piloted Israeli aircraft. As usual the skill and aggression of Israeli pilots had much to do with the outcome, but this could not account for the unprecedented superiority achieved. The Israeli triumph was the sign of revolutionary progress in aviation technology and electronics. This was a revolution that was set to continue through to the 1990s, when the second-order state of Iraq would stand almost in the same relation to US air power as Sudanese tribesmen had to the Maxim gun a century earlier.

The era of the modern fighter can be dated to the introduction of the US Navy's F-14 Tomcat and the USAF's F-15 Eagle in the mid-1970s. These no-expense-spared aircraft were marvels of technological innovation. They had enormously powerful engines, generating around twice the thrust available to any previous fighters, and they were packed with radar and computer equipment that helped the aircraft to new levels of performance in both flying and fighting.

Arguably the first modern fighter, the F-14 had a variable-geometry wing, the degree of sweep automatically adjusted by computer for optimum performance in all situations. The two crew sat in a large glass bubble giving

INSIDE AN F-14 TOMCAT
The view from the two-man cockpit of a US Navy F-14 Tomcat is all-round and unimpeded. The variable-geometry wing is here in the swept-back position that the on-board computers automatically select for supersonic flight.

TORNADO GR.1
The Panavia Tornado, produced as a joint venture by Britain, Germany, and Italy, is a Mach 2 variable-geometry-wing fighter that first flew in 1974. This GR.1 version is an interdictor/strike aircraft.

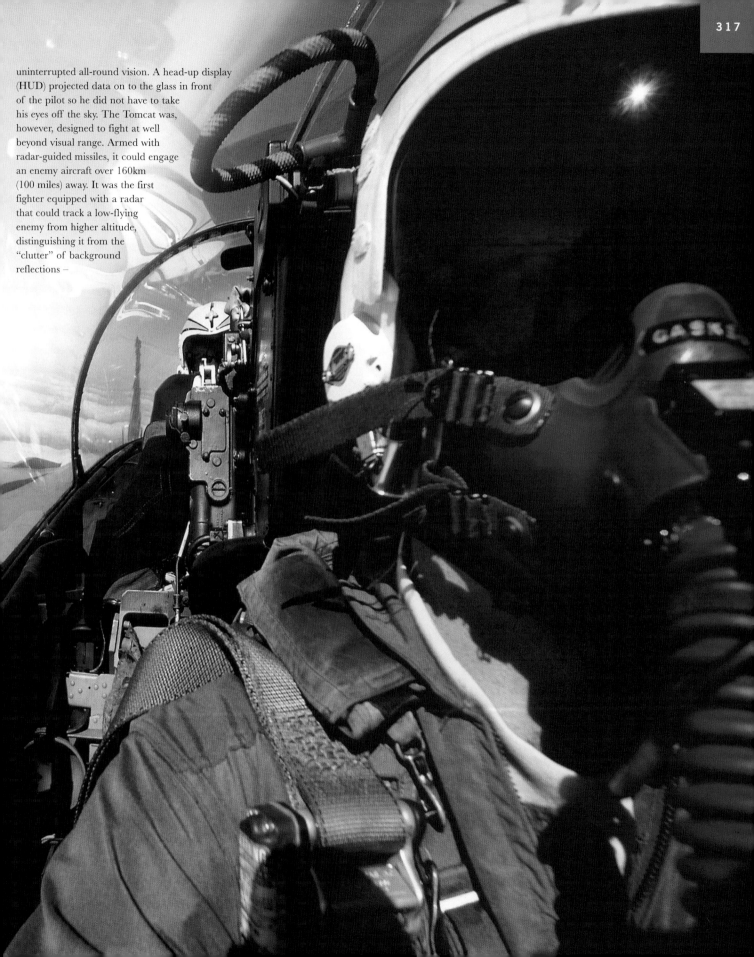

uninterrupted all-round vision. A head-up display (HUD) projected data on to the glass in front of the pilot so he did not have to take his eyes off the sky. The Tomcat was, however, designed to fight at well beyond visual range. Armed with radar-guided missiles, it could engage an enemy aircraft over 160km (100 miles) away. It was the first fighter equipped with a radar that could track a low-flying enemy from higher altitude, distinguishing it from the "clutter" of background reflections –

this gave the Tomcat a famous and coveted "look-down, shoot-down" capability. Its radar could track up to 24 enemy aircraft simultaneously and lock on to six of them as prime targets for its Phoenix missiles. In addition, on-board computers, fed with data from reconnaissance aircraft and satellites, gave the F-14 access to a wide range of ancillary information on activity across the whole battlefield.

New generation of fighters

The F-14 and F-15 (see pages 320–21) were modern weapons systems of astonishing complexity, and, as such, extremely expensive. The next of the US's modern fighters, the Lockheed Martin F-16, was a response to the demand for a lighter, cheaper aircraft that could be afforded in greater numbers and would have an emphasis on manoeuvrability for air-to-air combat. Although it ended up as a complex machine with multirole capability, the F-16 emerged as the most agile of close-combat fighters, setting a standard for aerial manoeuvre that has not yet been surpassed.

Throughout aviation history, there has been a trade-off between stability and manoeuvrability – the more unstable the aircraft, the more agile it will be. The F-16's designers created an aircraft that was inherently unstable, made flyable by sophisticated electronics. With its fly-by-wire controls, a computer linked to the aircraft's control surfaces ensured that the right adjustments would be made at every instant to keep the aircraft under control. The limit on the F-16's high-speed twist-and-turn is human – the amount of g-force that a pilot can withstand without blacking out.

By the time the US Navy's F/A-18 Hornet went into production in 1987, the rest of the world was struggling to keep up. One reason was cost. Even for the US, aircraft such as the F-14 and F-15 were phenomenally expensive, as was their armament – a Phoenix missile cost around $500,000. International co-operation just managed to keep Western Europe in the game. A consortium of British, Italian, and German companies created the Panavia Tornado in the 1970s, and another joint European effort generated the Eurofighter for the third millennium. In France, Dassault produced the Mirage 2000 and the Rafale. However, these aircraft were essentially ancillary to the US air forces, representing primarily an effort to stay in touch with advanced technology and keep aerospace industries active.

The Soviet Union was the country that produced the aircraft the American fighters were likely to meet in combat. The Soviets were prepared to invest the resources to keep up with the Americans, but they had a tradition of producing simpler, more robust aircraft that gave them a potential battlefield superiority in

INSIDE A MODERN FIGHTER COCKPIT

AIRCRAFT DESIGNERS HAVE BEEN well aware that the complexity of modern fighters could overload their pilots, however well trained. The modern "glass" cockpit, which first appeared in the F-18 Hornet and has since been retro-fitted on other fighters, is designed to provide the pilot with all the information he needs in a usable form and to allow him to act upon it with maximum ease and minimum reaction time.

In place of the confusing array of dials and instruments that once confronted the pilot, the modern cockpit has a head-up display (HUD) and, typically, three multifunction display (MFD) colour screens. On these the pilot can call up the information he needs – a moving map overlaid with target and flight information, a range of different radar displays, information on threats identified by the aircraft's ESM (Electronic Surveillance Measures), and so on. The HOTAS (Hands on Throttle and Stick) system pioneered by the F-15 Eagle enables the pilot to respond to a threat or engage a target without taking his hands off either the throttles or the control column. An array of switches and buttons – in the F-15, nine on the throttles and six on the stick – allows the pilot to perform vital tasks such as designating a target, selecting and launching a weapon, and dispensing chaff or flares to distract an incoming missile. Although it requires considerable digital dexterity to operate, the HOTAS system means that a pilot in the heat of combat no longer loses vital seconds groping around for the switch he needs, just as the HUD allows him to absorb essential information without taking his eyes off the sky.

One effect of these innovations has been to make single-seat fighters fashionable again. The accepted wisdom was once that the complexity of a modern fighter aircraft's avionics and weapons systems made it impossible for a pilot to fly alone, but improvements in cockpit design and the use of computers have greatly reduced the workload.

F-16 COCKPIT

The cockpit of an F-16 single-seat fighter shows the colour multifunction visual displays, on which the pilot can call up the information he needs at the touch of a button, and the switches on the throttles and control stick, which enable the pilot to launch weapons or deploy countermeasures with his hands still on the controls.

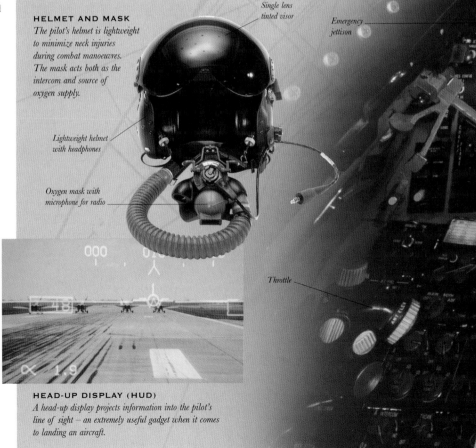

HELMET AND MASK
The pilot's helmet is lightweight to minimize neck injuries during combat manoeuvres. The mask acts both as the intercom and source of oxygen supply.

Single lens tinted visor

Emergency jettison

Lightweight helmet with headphones

Oxygen mask with microphone for radio

Throttle

HEAD-UP DISPLAY (HUD)
A head-up display projects information into the pilot's line of sight – an extremely useful gadget when it comes to landing an aircraft.

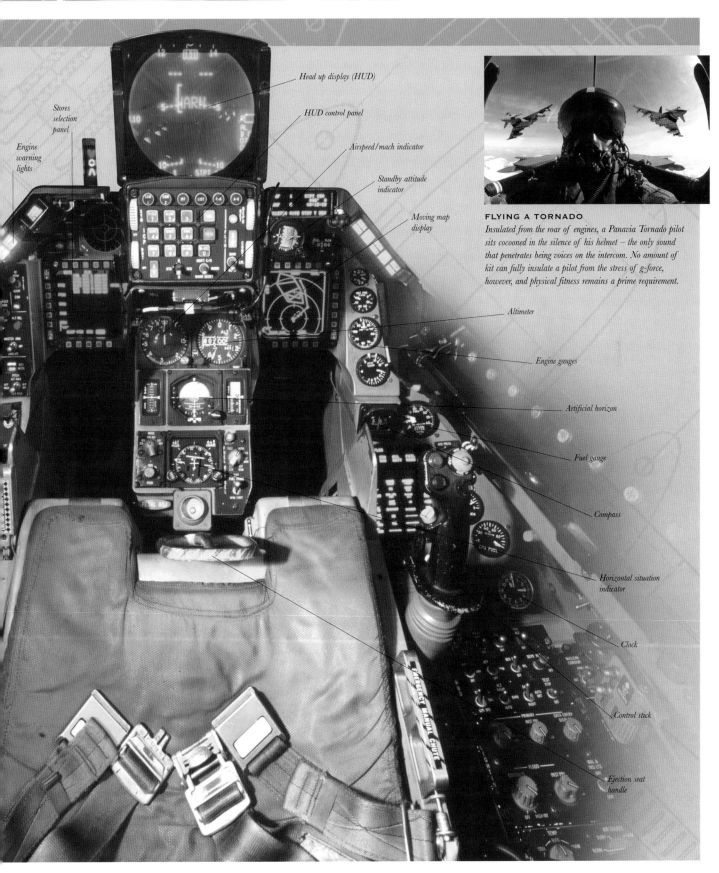

Head up display (HUD)

HUD control panel

Airspeed/mach indicator

Stores
selection
panel

Standby attitude
indicator

Engine
warning
lights

Moving map
display

FLYING A TORNADO

Insulated from the roar of engines, a Panavia Tornado pilot
sits cocooned in the silence of his helmet – the only sound
that penetrates being voices on the intercom. No amount of
kit can fully insulate a pilot from the stress of g-force,
however, and physical fitness remains a prime requirement.

Altimeter

Engine gauges

Artificial horizon

Fuel gauge

Compass

Horizontal situation
indicator

Clock

Control stick

Ejection seat
handle

numbers. It was not until the 1980s, a decade after the US, that they introduced their first truly modern fighters, the MiG-29 and Sukhoi Su-27. These and other subsequent designs showed how brilliant Soviet scientists and engineers could be, but deep-seated problems remained. The Soviet Union was seriously behind the West in computing and microelectronics. And the need to operate such complex aircraft put a strain on every element of the Soviet system, from pilot training to ground maintenance.

The modern fighter pilot

The demands that the new generation of fighter aircraft made on pilots were not entirely novel. In a sense, the new aeroplanes were firmly in a tradition that went back to the World War II Spitfire and beyond, with the aircraft experienced as an extension of the pilot's body – pilots speak of "putting the fighter on" when they climb into the cockpit. The qualities required of a fighter pilot remained much the same as in an earlier era: natural aggression, physical fitness, good eyesight, spatial sense, all-round situational awareness, keen intelligence, and a cool head.

In some ways, modern fighter pilots enjoyed advantages their predecessors would have envied. Virtual-reality simulators largely ended the waste of young lives in training accidents. Overall, being a fighter pilot ceased to be the high-risk occupation that it was in the 1950s and 1960s. Accidental fighter loss rates halved in each decade from 1970 onwards, with more reliable engines, better understanding of jet-aircraft design, and the introduction of computerized controls that to a large degree obviated pilot error. Devices such as terrain-following radar linked to an autopilot allowed aircraft to fly supersonic at altitudes as low as 60m (200ft), even over rough terrain, with reasonable safety.

There was no question, though, that the new aircraft imposed some unprecedented physical demands on the pilot. In earlier times, the main risk of high-speed manoeuvres had been that they would put excessive strain on the airframe. In modern fighters, the risk was excessive strain on the pilot's physiology. The largest g-force that a person can withstand before blacking out is about 5g. But the pilot might be flying an aircraft capable of, say, 12-g manoeuvres. Part of the elaborate kit in which a pilot now steps out to his aircraft is a g-suit; fitting tightly around the lower torso and upper legs, it constricts the bottom half of the body to keep blood in the upper body and brain. This can improve resistance up to 6 or 7g. Beyond this point, pilots have to resort to tensing and

McDonnell Douglas F-15 Eagle

WHEN THE SOVIETS UNVEILED their MiG-25 interceptor in 1967, American defence experts put out an urgent call for a new air-superiority fighter of outstanding performance and agility to match it. The result was the F-15 Eagle fighter-interceptor, whose lightweight construction materials and powerful, specially developed Pratt & Whitney turbofan engines gave it an excellent thrust-to-weight ratio. In a vertical climb the Eagle could reach 8,850m (29,000ft) – the height of Everest – in under a minute.

Most F-15s, which entered service in 1974, are single-seaters, meaning that the pilot has to cope with target identification, spotting and tracking targets, deploying countermeasures against missile attack, and all the other complex business of modern warfare, as well as flying. The designers did an excellent job of making the pilot's task more manageable, and many of their innovations, such as HOTAS ("hands on throttle and stick") became standard on all modern fighters. Later models (the F-15C and D) carried more advanced systems, including the improved APG-70 radar.

Although not as agile in a dogfight as some smaller, lighter aircraft, the F-15's unique ability to accelerate in vertical flight provided the perfect escape route if the pilot was ever in trouble. None has ever been lost in combat.

ACTION STATIONS
On 27 July 1988, two F-15s fulfilled their originally intended fighter-interceptor role by intercepting two Soviet Tu-95 "Bear" long-range bombers off the east coast of Canada.

Rear-facing radar warning receiver

Electronic countermeasures (ECM) jammers

Rudder

Twin tailfin

Missile rail

Starboard main landing gear

Specifications (F-15A)

ENGINE	2 x 10,782kgp (23,770lbst) P&W F100 turbofan
WINGSPAN	13m (42ft 8in)
LENGTH	19.4m (63ft 8in)
WEIGHT	14,515kg (32,000lb) **CREW** 1
TOP SPEED	1,875kph (1,164mph) (Mach 2.5 plus)
ARMAMENT	1 x M-61A1 20mm canon; 4 x AIM-9L/M Sidewinder and 4 x AIM-7F/M Sparrow air-to-air missiles, or 8 x AIM-120 AMRAAMs

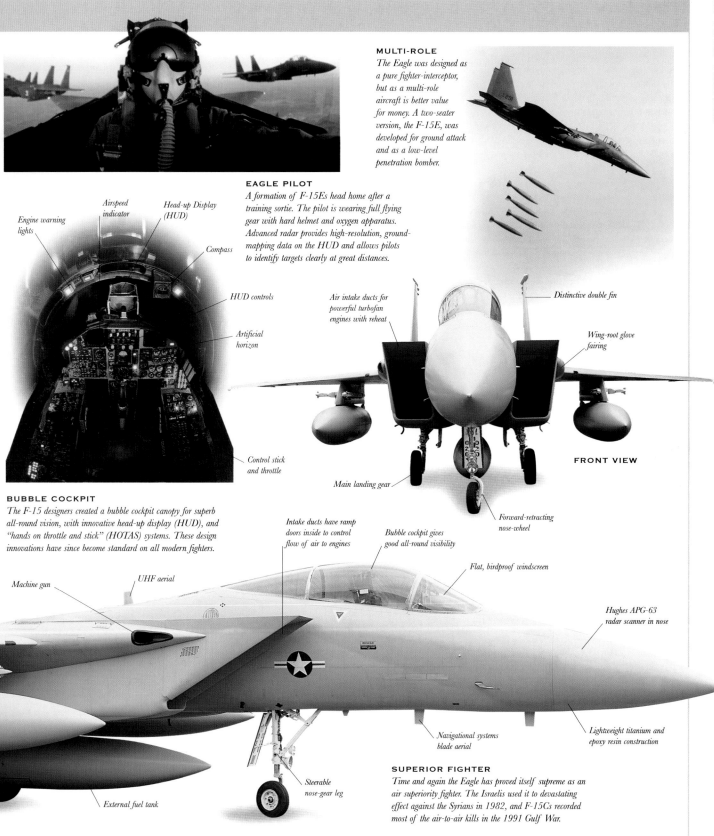

MULTI-ROLE
The Eagle was designed as a pure fighter-interceptor, but as a multi-role aircraft is better value for money. A two-seater version, the F-15E, was developed for ground attack and as a low-level penetration bomber.

EAGLE PILOT
A formation of F-15Es head home after a training sortie. The pilot is wearing full flying gear with hard helmet and oxygen apparatus. Advanced radar provides high-resolution, ground-mapping data on the HUD and allows pilots to identify targets clearly at great distances.

Engine warning lights

Airspeed indicator

Head-up Display (HUD)

Compass

HUD controls

Artificial horizon

Air intake ducts for powerful turbofan engines with reheat

Distinctive double fin

Wing-root glove fairing

Control stick and throttle

Main landing gear

FRONT VIEW

Forward-retracting nose-wheel

BUBBLE COCKPIT
The F-15 designers created a bubble cockpit canopy for superb all-round vision, with innovative head-up display (HUD), and "hands on throttle and stick" (HOTAS) systems. These design innovations have since become standard on all modern fighters.

Intake ducts have ramp doors inside to control flow of air to engines

Bubble cockpit gives good all-round visibility

Flat, birdproof windscreen

Machine gun

UHF aerial

Hughes APG-63 radar scanner in nose

Steerable nose-gear leg

Navigational systems blade aerial

Lightweight titanium and epoxy resin construction

External fuel tank

SUPERIOR FIGHTER
Time and again the Eagle has proved itself supreme as an air superiority fighter. The Israelis used it to devastating effect against the Syrians in 1982, and F-15Cs recorded most of the air-to-air kills in the 1991 Gulf War.

RECONNAISSANCE, AEW, AND AWACS AIRCRAFT

SINCE THE 1950s, when the CIA-operated U-2 spyplane was an important source of Cold War intelligence, military satellites have partly replaced aircraft in the high-altitude reconnaissance role. Airborne Early Warning (AEW) and Airborne Warning and Control System (AWACS) aircraft have, on the other hand, continued to grow in importance. At sea, AEW aircraft are especially vital to fleet defence against submarines or attack by aircraft carrying long-range air-to-surface missiles. Increasingly, the airborne "eyes in the sky" have also become airborne command centres, directing the course of the aerial battle. Since 1991, Joint Surveillance Target Attack Radar System (J-STARS) has extended this concept to the land battlefield. Whereas high-altitude reconnaissance created some technologically extraordinary aircraft, AEW, AWACS, and J-STARS have mostly required only the adaptation of existing airliner or transport aircraft designs, packing them with radar and computer equipment.

HIGH FLIER
The SR-71B Blackbird reconnaissance aircraft depends on Mach 3 speed and very high altitude to stay out of trouble. See pages 290–91.

Boeing E-3A Sentry (AWACS)

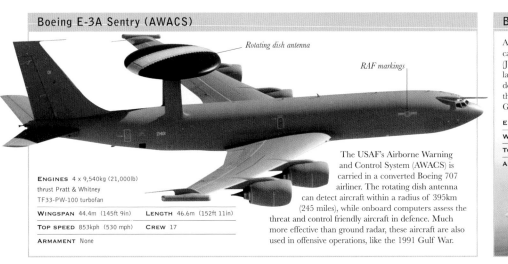

Rotating dish antenna

RAF markings

ENGINES	4 x 9,540kg (21,000lb) thrust Pratt & Whitney TF33-PW-100 turbofan	
WINGSPAN	44.4m (145ft 9in)	LENGTH 46.6m (152ft 11in)
TOP SPEED	853kph (530 mph)	CREW 17
ARMAMENT	None	

The USAF's Airborne Warning and Control System (AWACS) is carried in a converted Boeing 707 airliner. The rotating dish antenna can detect aircraft within a radius of 395km (245 miles), while onboard computers assess the threat and control friendly aircraft in defence. Much more effective than ground radar, these aircraft are also used in offensive operations, like the 1991 Gulf War.

Boeing E-8A J-STARS

A combined US Army and US Air Force programme called Joint Surveillance Target Attack Radar System (J-STARS) is designed to give the same control over land targets as AWACS does over air targets – a downward-looking radar link to army ground stations in the forward battle areas. Prototypes were used in the Gulf War in 1991 and in Bosnia in 1997.

ENGINES	4 x 8,177kg (18,000lb) thrust P&W JT3D-3B turbojet	
WINGSPAN	44.4m (145ft 9in)	LENGTH 46.6m (152ft 11in)
TOP SPEED	853kph (530mph)	CREW 18
ARMAMENT	None	

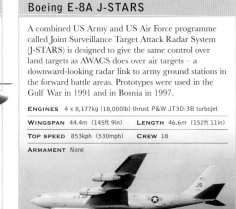

Breguet 1150 Atlantique

A long-range anti-submarine patrol aircraft developed for NATO's European partners, the Atlantique was designed by Breguet in France and is in use by the air forces of France, Germany, Holland, and Italy. On 18-hour patrols, the Atlantique crew searches for submarines using radar and other detection equipment, and can then destroy them with their weaponry.

ENGINES	2 x 6,106hp Rolls-Royce Tyne R Ty 20 Mk 21 turboprop	
WINGSPAN	36.3m (119ft 1in)	LENGTH 31.8m (104ft 2in)
TOP SPEED	657kph (409mph)	CREW 12
ARMAMENT	6,000kg (13,227lb) Exocet or Martel air-to-surface missiles, torpedoes, depth charges, or bombs	

Grumman E-2C Hawkeye

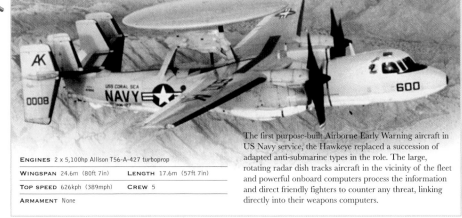

The first purpose-built Airborne Early Warning aircraft in US Navy service, the Hawkeye replaced a succession of adapted anti-submarine types in the role. The large, rotating radar dish tracks aircraft in the vicinity of the fleet and powerful onboard computers process the information and direct friendly fighters to counter any threat, linking directly into their weapons computers.

ENGINES	2 x 5,100hp Allison T56-A-427 turboprop	
WINGSPAN	24.6m (80ft 7in)	LENGTH 17.6m (57ft 7in)
TOP SPEED	626kph (389mph)	CREW 5
ARMAMENT	None	

Hawker Siddeley Nimrod MR.1/2

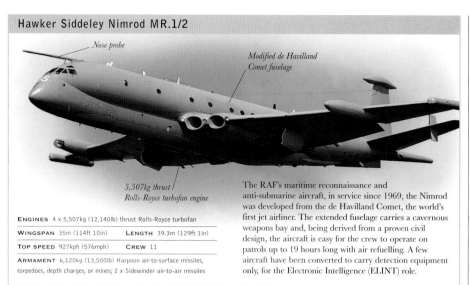

Nose probe

Modified de Havilland Comet fuselage

5,507kg thrust Rolls-Royce turbofan engine

ENGINES	4 x 5,507kg (12,140lb) thrust Rolls-Royce turbofan	
WINGSPAN	35m (114ft 10in)	LENGTH 39.3m (129ft 1in)
TOP SPEED	927kph (576mph)	CREW 11
ARMAMENT	6,120kg (13,500lb) Harpoon air-to-surface missiles, torpedoes, depth charges, or mines; 2 x Sidewinder air-to-air missiles	

The RAF's maritime reconnaissance and anti-submarine aircraft, in service since 1969, the Nimrod was developed from the de Havilland Comet, the world's first jet airliner. The extended fuselage carries a cavernous weapons bay and, being derived from a proven civil design, the aircraft is easy for the crew to operate on patrols up to 19 hours long with air refuelling. A few aircraft have been converted to carry detection equipment only, for the Electronic Intelligence (ELINT) role.

Lockheed P-3C Orion

The Orion has been the US Navy's land-based maritime patrol and submarine hunter/killer aircraft since 1961. It was adapted from the Lockheed Model 188 Electra by shortening the airframe by 2.2m (7ft 4in) and incorporating a weapons bay in the lower fuselage. It can also be used to attack surface warships and to relay information to friendly surface forces. Over 500 have been delivered, and since a proposed replacement has been cancelled, the Orion looks set to continue for many years. It has also been supplied to Australia, Canada, New Zealand, Japan, and Spain, among other countries.

4,910hp Allison turboprop engine

Fuselage modified from Model 188 Electra airliner

ENGINES	4 x 4,910hp Allison T56-A-14 turboprop	
WINGSPAN	30.4m (99ft 8in)	LENGTH 35.6m (116ft 10in)
TOP SPEED	761kph (473mph)	CREW 10
ARMAMENT	Up to 9,072kg (20,000lb) Harpoon air-to-surface missiles, torpedoes, depth-charges, or mines in bay or on wing pylons	

Lockheed U-2A

This high-altitude spy plane was developed in the 1950s and was still in use in the 21st century. From 1956 to 1960, U-2s flew over 30 photographic reconnaissance missions over the former Soviet Union, flying over 22km (14 miles) high. In 1960, a U-2 was shot down and the pilot put on trial in Moscow. In 1962, U-2 photographs revealed Soviet long-range missiles in Cuba.

ENGINE	5,085kg (11,200lb) thrust P&W J75-P-13B turbojet	
WINGSPAN	24.4m (80ft)	LENGTH 15.6m (49ft 7in)
TOP SPEED	795kph (494mph)	CREW 1
ARMAMENT	None	

Myasishchev M-55 Geofizika

The Soviet equivalent of the American U-2 high-altitude spy plane was developed in the 1970s, and its designers claimed their aircraft was far more stable. The original M-17 was developed into the M-55 Geofizika, which the Russians claim is used for atmospheric monitoring. The Americans, of course, claim the same for most U-2 flights.

ENGINES	2 x 5,000kg (11,025lb) thrust Aviadvigadel turbojet	
WINGSPAN	37.4m (122ft 11in)	LENGTH 22.8m (75ft)
TOP SPEED	750kph (466mph)	CREW 1
ARMAMENT	None	

straining of muscles, which can help them withstand much higher gs for short periods.

Pilots also need equipment, up to and including full-pressure spacesuits, for flights at extreme altitude. The design of flight helmets has improved dramatically since the first one was created. They were once heavy enough to cause frequent neck injuries, but are now lightweight while offering the same protection. All modern fighter pilots fly in masks, which, as well as acting as an intercom, provide the oxygen supply. Pilots have to learn to regulate the oxygen appropriately and recognize any sign of hypoxia – oxygen starvation of the brain – almost instantly.

Flying a modern single-seat fighter not only imposes physical strains but also makes intense mental demands on pilots. A combination of complex multimode radars, infrared navigation and targeting equipment, sensors to identify hostile radars or missiles, and data links to AWACS (airborne warning and control system) aircraft can provide more information input than a human brain can readily cope with. At the same time as processing this flow of information, the pilot has to operate his sophisticated targeting and weapons systems for attack, and deploy his defensive array. He also, of course, has to fly the aeroplane.

Improvements in computing have made it possible to organize and analyze the flow of data from different sources so that the pilot is presented with the most relevant information in increasingly easy-to-use graphic form. There is a move towards allowing computers or airborne controllers to take decisions about targeting and weapon launch independently of the pilot. Indeed, with unmanned, remotely piloted vehicles (RPVs) ever more common in combat use, there have been hints that fighter pilots might almost have had their day.

The Gulf War

Most of the general public became aware of the modern revolution in air warfare in 1991 when Operation Desert Storm unleashed the aerial might of the United States and its Coalition allies against Iraq. The conflict displayed, among many other Coalition strengths, the effectiveness of the direction of the battle by airborne controllers, the usefulness of "stealth" technology, and the progress that had been made with missile guidance and smart bombs.

During the Gulf War, four Boeing E-3 Sentry AWACS aircraft were in the air at all times, one for each of the sectors into which the battle area had been divided, and one spare for back-up. The E-3 is a Boeing 707 airliner adapted to carry radars of great power and discrimination. Inside the aircraft, rows of display consoles were manned by highly trained personnel, fighting to maintain absolute concentration through long, exhausting shifts, fixed in front of their screens. The controllers had to interpret data from different sources – including television cameras carried by RPVs over enemy missile sites – and turn this rich and often confusing flood of information into a coherent picture of the air battlefield.

Either through a data link to their on-board computers or by voice over UHF radio, the AWACS controllers kept pilots informed of developments on a minute-by-minute basis. In effect, the controllers could orchestrate the whole battle, giving coherence to the approximately 3,000 sorties a day flown by Coalition forces.

Information from AWACS often allowed Coalition fighters to engage Iraqi aircraft beyond visual range – some Iraqi airmen were downed by missiles fired by aircraft they never saw and of whose presence they were totally unaware. The Iraqis' aircraft shelters proved vulnerable to American precision weapons when their airfields came under attack. The Iraqi air force soon admitted its impotence and gave up the fight, seeking sanctuary in Iran. Iraqi air space was mostly penetrated by the same basic mix of aircraft that had been employed for such missions in North Vietnam, but operating with much improved technology. There might be EF-111s specialized in electronic countermeasures jamming Iraqi radars; F-4s armed with anti-radiation missiles flying Wild Weasel missions to destroy Iraqi SAM sites; a core of strike aircraft for the ground-attack mission; and mostly F-15s escorting the "package" on combat air patrol.

Stealth technology

One aircraft, however, needed no help from others in penetrating Iraqi defences. The F-117 radar-invisible stealth aircraft was used in combat in the Gulf War and resoundingly justified the amount that had been spent on it. Unafraid of radar, it did not have to fly at the ground-hugging altitude adopted by most modern strike aircraft to penetrate air-defence systems. Nor did it need to be accompanied by escort fighters, electronic-warfare aircraft, or anti-radar strike aircraft. Its pilot flew alone through the night in radio silence, eyes fixed on the screens that allowed him to navigate a path into enemy territory and to locate and illuminate his target. Usually the target was destroyed and the F-117 would return as it had come, invisible as a ghost, leaving the enemy guns to blast away blindly into the night sky in the vain hope of a lucky hit.

An aircraft's radar cross-section – the degree to which it shows up on radar – is a function of its shape. Since radar depends on signals striking an

Lockheed F-117 Nighthawk

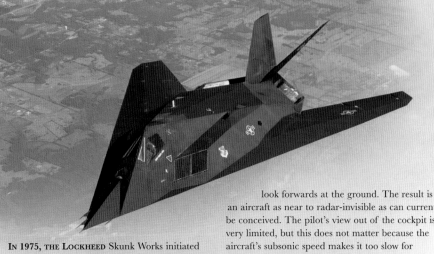

IN 1975, THE LOCKHEED Skunk Works initiated a secret programme to develop a radar-proof "stealth" aircraft. Development was by no means straightforward, and the first F-117As were not delivered to the USAF until 1982. Nicknamed the "Bat Plane" or "Stink Bug", its unique appearance results from the use of flat, faceted surfaces to deflect radar emissions at an angle, especially from the front.

The airframe is covered in radar-absorbent material and the cockpit windows are coated in a gold layer that conducts radar energy into the airframe. There is even a fine mesh over the engine intakes, which might otherwise provide a radar signature for the enemy to track. Any weaponry or fuel tanks are carried internally to maintain the stealth effect. It cannot use radar, because this might give away its position, relying instead on a passive infrared "eye" in the nose to look forwards at the ground. The result is an aircraft as near to radar-invisible as can currently be conceived. The pilot's view out of the cockpit is very limited, but this does not matter because the aircraft's subsonic speed makes it too slow for daylight combat. It only attacks by night, and the pilot never looks outside, keeping his eyes on the multi-function display screens in front of him.

The designation "F" for fighter is a strange one, as the F-117A has no guns and does not usually carry missiles. A night-penetration strike aircraft, it is equipped instead with the most up-to-date smart weapons and targeting systems for precision attack. While the F-117A is a complex weapons system to be operated by a single pilot, who needs a lot of electronic help approaching the target, its strikes proved extremely accurate in the Gulf War in 1991.

The F-117As have established themselves as a central element in America's air armoury. They justify their impressive price-tag – over $40 million each – by their accuracy of attack and near invulnerability, although one was shot down by Serbian artillery during the Kosovo campaign in 1999.

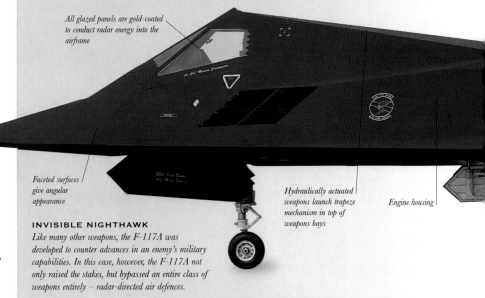

All glazed panels are gold-coated to conduct radar energy into the airframe

Faceted surfaces give angular appearance

Hydraulically actuated weapons launch trapeze mechanism in top of weapons bays

Engine housing

INVISIBLE NIGHTHAWK
Like many other weapons, the F-117A was developed to counter advances in an enemy's military capabilities. In this case, however, the F-117A not only raised the stakes, but bypassed an entire class of weapons entirely – radar-directed air defences.

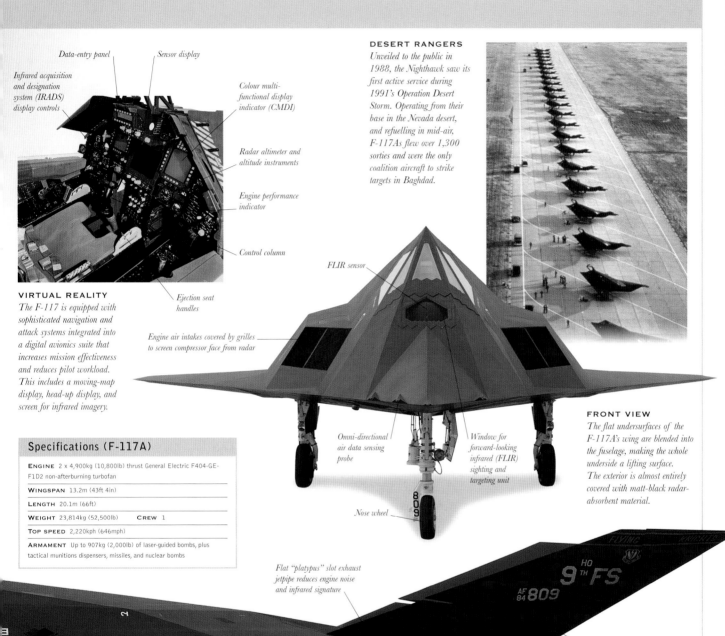

Data-entry panel

Sensor display

Infrared acquisition
and designation
system (IRADS)
display controls

Colour multi-
functional display
indicator (CMDI)

Radar altimeter and
altitude instruments

Engine performance
indicator

Control column

Ejection seat
handles

VIRTUAL REALITY

The F-117 is equipped with
sophisticated navigation and
attack systems integrated into
a digital avionics suite that
increases mission effectiveness
and reduces pilot workload.
This includes a moving-map
display, head-up display, and
screen for infrared imagery.

DESERT RANGERS

Unveiled to the public in
1988, the Nighthawk saw its
first active service during
1991's Operation Desert
Storm. Operating from their
base in the Nevada desert,
and refuelling in mid-air,
F-117As flew over 1,300
sorties and were the only
coalition aircraft to strike
targets in Baghdad.

FLIR sensor

Engine air intakes covered by grilles
to screen compressor face from radar

Omni-directional
air data sensing
probe

Window for
forward-looking
infrared (FLIR)
sighting and
targeting unit

Nose wheel

FRONT VIEW

The flat undersurfaces of the
F-117A's wing are blended into
the fuselage, making the whole
underside a lifting surface.
The exterior is almost entirely
covered with matt-black radar-
absorbent material.

Specifications (F-117A)

ENGINE 2 x 4,900kg (10,800lb) thrust General Electric F404-GE-F1D2 non-afterburning turbofan	
WINGSPAN 13.2m (43ft 4in)	
LENGTH 20.1m (66ft)	
WEIGHT 23,814kg (52,500lb)	**CREW** 1
TOP SPEED 2,220kph (646mph)	
ARMAMENT Up to 907kg (2,000lb) of laser-guided bombs, plus tactical munitions dispensers, missiles, and nuclear bombs	

Flat "platypus" slot exhaust
jetpipe reduces engine noise
and infrared signature

9TH HO FS

AF 809
84

Ruddervators combine functions
of rudders and elevators

Composite construction
leading edge

SLOWING DOWN

The F-117A has a landing speed
of 227kph (172mph) so a drag
chute is needed to reduce the length
of the landing run.

aircraft and returning to a receiver, on the F-117 this was negated by using angled flat surfaces that deflected signals instead of bouncing them back. Beyond this, each element of the aircraft, from the cockpit to the tailfins to the engine air intakes, was designed to minimize radar reflection. The other route to stealth was to cover the airframe with radar-absorbent materials – paint and tiles that would transform the energy of the radar pulse into heat instead of reflecting it.

The F-117 depended absolutely on radar-invisibility for combat survival. It was subsonic, not especially manoeuvrable, and even lacked electronic countermeasures (ECM) equipment. If progress in radar technology or some other tracking method had been able to tear off the Nighthawk's cloak of invisibility, it would have been defenceless. But over Iraq it performed outstandingly – responsible for 30 per cent of targets hit by precisions munitions.

Precision bombing

Essential to the F-117's accuracy of attack was, of course, the improved quality of guidance systems for smart bombs and missiles. There was no essential difference between the munitions being employed in the Gulf War from those used against North Vietnam in the early 1970s, but the effectively presented images of precision attack fed to the media during Operation Desert Storm made a telling impression on the public imagination. And, although subsequent enquiry revealed that many weapons had fallen short of the effectiveness prematurely attributed to them during the war, the laser-guided, TV-guided, and anti-radiation missiles and bombs were unquestionably more accurate than anything that had been seen in action before.

Perhaps the most impressive aspect of air

GROUND STRIKE
First delivered to the US Navy in 1981, the McDonnell-Douglas F/A-18 Hornet was, as its designation indicates, a dual-role aircraft – both an agile air-superiority fighter and a powerful ground-attack aircraft. During the Gulf War in 1991 the Hornets were mostly used in a ground-attack role.

DIRECT HIT
A laser-guided missile is launched at an ammunition depot in Iraq during the Gulf War in January 1991. This was the view of the event as seen through the sights of a French fighter.

power in the Gulf War was its effect on the land battlefield, where it contributed to a victory being achieved with the minimum number of Coalition casualties. The veteran B-52s were once more brought into action, dropping their impressive tonnage of iron bombs on the Iraqis dug in to lines of trenches in the desert. But the crucial damage was done by newer technology. Two Boeing-Northrop Grumman E-8 J-STARS (Joint Surveillance and Target Attack Radar System) aircraft did for the air-land battlefield what AWACS had done for the air-to-air battle. Like the AWACS E-3s, the E-8s were derived from the Boeing 707. J-STARS controllers were able to observe every move of the Iraqi army, directing strike aircraft to destroy armoured columns, truck convoys, fuel dumps, or artillery emplacements. The airborne radars offered such high-resolution imagery that they could even pick out individual vehicles for attack.

The panoply of ground-attack aircraft at the Coalition's disposal ranged from AH-64 helicopter gunships to A-10 Thunderbolt II "Warthog" tank-busters and F-16s or F-18s in ground-attack mode. Linked to J-STARS, this aerial array was deadly. For example, on 22 January 1991, 60 Iraqi tanks were spotted on the move; air-strikes were called in and 58 of the vehicles were destroyed. The carnage of the attempted Iraqi retreat from Kuwait in February was also orchestrated by J-STARS, monitoring convoys crowding the road back to Iraq and calling in wave after wave of F-18s.

Victory by air power

The events of the Gulf War were a challenge to the accepted wisdom – based on the experience of World War II, Korea, and Vietnam – that air power alone could not win a war. USAF Chief of Staff General Merrill A. McPeak asserted bluntly that Desert Storm marked "the first time in air history that a field army has been defeated by air power". There were particular reasons why aircraft were at their maximum effectiveness in this conflict. Iraq presented a range of targets suitable for strategic air attack, from command and control centres to weapons factories, missile sites, and airfields. Similarly, in the ground war, troops in fixed emplacements and tanks moving in open desert terrain both offered

SKY SENTRY
The E-3 Sentry early-warning and control aircraft – essentially a Boeing 707 with a saucer-shaped radar antenna on its back – was the hub of Coalition air operations in the Gulf War.

tempting targets for air attack. The weather was also favourable, with almost guaranteed clear skies. Yet it was difficult to remain cautious about the future potential of air power in the face of such outright triumph.

The first resort

For the key to the political attractiveness of air power as a military option, there is no need to look further than the Gulf War casualty list: only 200 men lost on the Coalition side in a war with the substantial armed forces of a considerable regional power. Thus, when NATO leaders became convinced in 1999 that Serbia had to be driven out of the province of Kosovo, but suspected that this was a cause for which the Western public would not be prepared to see their servicemen die, they fought the first and so far only war to be conducted exclusively by air. Despite a number of notable errors, of which the most diplomatically embarrassing was the destruction of the Chinese embassy in Belgrade, the air-strikes did indeed drive the Serbians out. NATO achieved the extraordinary feat of winning a small war without sustaining a single combat casualty. For the world's most technologically advanced nations, aircraft had become the weapon of first resort.

TANK-BUSTER
The A-10 "Warthog" is an aircraft specifically designed for the close support of ground forces, and in particular for taking on enemy armoured vehicles. It is relatively slow-moving but can stay over the battlefield for long periods of time and withstand a good deal of punishment from ground fire.

SPECTACLE OF DESTRUCTION
Coalition air attacks upon Iraqi ground forces (below) in the Gulf War left an awesome spectacle of destruction in their wake. Attempting to flee Kuwait City by night, the Iraqis found the road blocked by mines dropped by F/A-18s. They were then destroyed (left) by air-strikes orchestrated by controllers of E-8 J-STARS air-to-ground radar aircraft.

STEALTH BOMBER

The Northrop B-2 is a flying-wing design that uses angular faceting to reduce its radar signature. The radar-absorbent coating is so sensitive that the bombers must be kept in spotlessly clean hangers at controlled humidity levels.

The terrorist attacks in America on 11 September 2001 set off a new chain of events in which air power took centre stage. The array of carrier-borne and land-based aircraft that the US brought to bear on Afghanistan in the "war against terrorism" proved effective, once more confounding critics who had suggested that only a considerable ground force could achieve America's objectives. In contrast with the experience of the Soviet Union in Afghanistan in the 1980s, the US was able to depose the Taliban government and destroy al-Qaeda bases with minimal involvement of ground forces. No resistance seemed possible to a military force apparently capable of pinpointing targets however mobile or well hidden they might be, and able to call upon anything from cruise missiles to unpiloted aircraft, fixed-wing gunships, and B-52 bombers to carry out the work of destruction.

Success and limitations

Critics of air power of course pointed out that these were exceptionally one-sided conflicts. NATO's Kosovo campaign faced a Serbian regime that was politically weak and had limited will to resist. In Afghanistan, local opposition forced could be mustered to take over such political power as existed. When the United States and Britain invaded Iraq in 2003, there were pessimistic predictions that this time it would be different. Ground forces, committed in strength, would have to slog it out in costly battle with Saddam Hussein's Republican Guard. Yet once more air power proved its absolute worth. Media attention focussed on the "shock and awe" spectacle of the destruction of Saddam's palaces and government offices in Baghdad, but the great majority of aerial firepower was directed at Iraqi troops and armour. Since Iraq's air defence system had been systematically dismantled by air attacks in the run-up to the invasion, American and British aircraft operated at will. Every movement of Iraqi forces was tracked by Allied reconnaissance, including Global Hawk UAVs, and air strikes were directed accurately on to targets even in adverse weather conditions. In effect, the function of Allied ground forces was to draw Iraqi forces into action so that aircraft could destroy them. Air power brought swift victory in the initial invasion of Iraq, but the aftermath showed its limitations. In the face of a rising tide of insurgency and terrorism, the effectiveness of air power was restricted by the difficulty of identifying targets and by political sensitivity about civilian casualties caused by air attack. However accurate the weaponry, there were always going to be mistakes due to faulty intelligence, equipment malfunction, or simple human error. Unleashed in total war, American air power could have hit any target in Iraq that was deemed worth destroying. But it could not stop suicide bombers.

Next generation

The time-lag involved in development of new aircraft meant that in the early 21st century the world's major air forces were still essentially designed to fight the Cold War. The B-2 stealth bomber, for example, was meant to penetrate Soviet air defences carrying a nuclear payload. The Lockheed Martin F-22 Raptor, which entered service with the US Air Force in 2005, was originally conceived as a fighter that would trounce its Soviet equivalents in air-to-air combat. The future relevance of such aircraft was bound to be called into question, especially as they were immensely expensive – the B-2 programme reportedly came in at around $45 billion, although only 21 of the bombers were built.

Faced with the diffuse threat of global terrorism, the United States looked to a future in which the long reach of air power could be used to strike a vehicle or building anywhere around the world, identified as containing enemies of America. Satellites and unpiloted aircraft were central to this vision of global surveillance and precisely targeted vengeance. Nevertheless, more conventional piloted fighters and bombers looked certain to continue to be produced in large numbers in the immediate future. In the 21st century, whatever its actual defence needs, no country with any pretensions to international status could afford to miss out on a truly modern air force, possession of which had become a prime symbol of technological superiority.

UNMANNED AERIAL VEHICLES

THE USEFULNESS OF DRONES – variously known as remotely piloted vehicles (RPVs) or unmanned aerial vehicles (UAVs) – has been apparent since the early 1980s. Although they can hardly be regarded as disposable, UAVs can carry out reconnaissance missions at a risk level that would be considered unacceptable for manned aircraft. "Flown" by a technical team on the ground, they transmit back television images or other data – for example, from radar and infrared sensors – often staying over enemy positions for many hours. During the US intervention in Afghanistan in 2001–02, RQ-1 Predator UAVs were used for the first time in an attack role, firing Hellfire air-to-ground missiles. Also introduced in Afghanistan were high-altitude RQ-4A Global Hawk UAVs, equipped for all-weather reconnaissance.

RQ-1 PREDATOR

The General Atomics Predator UAV is 7.9m (26ft) long and flies at 134–224kph (84–140mph). It is controlled by an air-vehicle operator and three sensor operators at a ground station.

Sensor package including cameras and radar

Stabilizing fin

RAPTOR AND FALCON

An F-22 Raptor, a fighter designed for the 21st century, flies on the left of an F-16 Falcon. The differences between the two are subtle, although they add up to a major advance. The F-22 has stealth features including the use of composite materials; it can cruise at around Mach 1.4, rather than going supersonic only in short bursts; it has vectored thrust to increase its agility; and has astonishingly sophisticated electronics and computer systems.

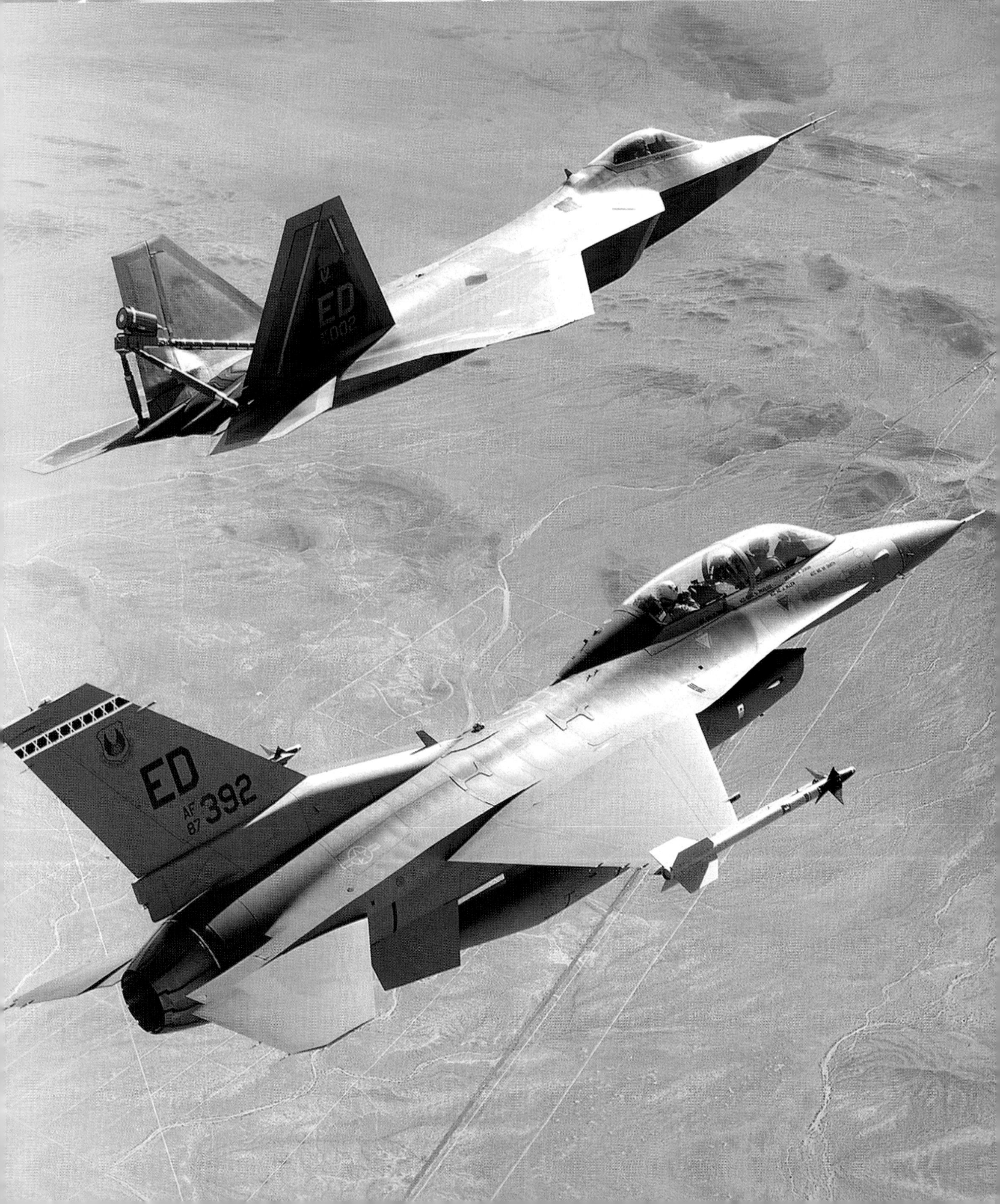
ED
ED
AF
87
392

MODERN FIGHTING AIRCRAFT

MODERN FIGHTERS ARE AMONG the wonders of modern technology. These astonishing machines can typically reach speeds well in excess of Mach 2; climb around 10,000m (30,000ft) a minute to an operational altitude of about 16km (10 miles); execute high-speed jinks and twirls in close combat; and, of course, operate by day and night in all weathers. Reliable engines, improved design, and fly-by-wire controls have made these aircraft much safer to fly than earlier high-performance jets. Immense engine power – up to 22,700kg (50,000lb) total thrust – means that they have been able to grow heavier without losing out on performance. Fitted with an array of radar and infrared devices able to identify enemy aircraft at distance and warn against incoming missiles, their own missile systems are able to engage targets well beyond visual range. They are also very expensive. Economics dictate that even the most dedicated air-superiority fighters end up being used in the ground-attack role too.

STEALTH FIGHTER
The F-117 is totally distinct from other "F"-designated modern aircraft. This subsonic night flier depends on its stealth features for survival in hostile airspace. See pages 324–25.

Dassault Mirage 2000C

Dassault's third generation of tail-less delta interceptor uses fly-by-wire controls to give far better turning ability than previously available with this wing form. The aircraft entered French Air Force service in 1988, and 14 were deployed to Saudi Arabia in 1991 as part of Operation Desert Storm, although they did not see combat. There is also a two-seat nuclear or conventional attack version. Mirage 2000s have been exported to India, Egypt, Peru, and Greece.

ENGINE	9,715kg (21,385lb) thrust SNECMA M53-P2 turbofan		
WINGSPAN	9.1m (29ft 11in)	LENGTH	14.3m (47ft 1in)
TOP SPEED	2,336kph (1,452mph) (Mach 2.2)	CREW	1
ARMAMENT	2 x 30mm DEFA cannon, 4 x Matra air-to-air missiles; up to 6,300kg (13,890lb) attack weapons		

Eurofighter (EFA) Typhoon

Studies for a replacement for the Panavia Tornado began in the late 1970s. Development, by a four-nation consortium of Britain, Germany, Italy, and Spain, was long and slow, but the first Eurofighters entered service in 2003. It is a "swing role" aircraft, able to switch from air-to-air to air-to-ground mission configuration in flight.

ENGINES	2 x 9,086kg (20,000lb) thrust Eurojet EJ 200 turbofan		
WINGSPAN	11m (35ft 11in)	LENGTH	16m (52ft 4in)
TOP SPEED	c.2,124kph (c.1,320mph) (Mach 2)	CREW	1
ARMAMENT	1 x 27mm IWKA-Mauser cannon; 13 weapon carriage points under fuselage and wings		

General Dynamics F-111E

The F-111 was the world's first swing-wing aircraft, entering service with the USAF in 1967. Variable geometry enables the aircraft to take off and land with straight wings, but to fly at supersonic speeds with swept wings. The F-111 was also first with automatic terrain-following radar, making it a powerful low-level strike aircraft. Used in Vietnam in 1972–73, against Libya in 1986, and then in Operation Desert Storm in 1991. F-111s were finally retired in 1993.

ENGINES	2 x 8,392kg (18,500lb) thrust P&W TF30-P-3 turbofan		
WINGSPAN	19.2m (63ft)	LENGTH	22.4m (73ft 6in)
TOP SPEED	2,655kph (1,650mph) (Mach 2.5)	CREW	2
ARMAMENT	2 x 340kg (750lb) nuclear/conventional bombs; or 1 x 20mm cannon and 11,340kg (25,000 lb) of bombs or missiles		

Grumman F-14A Tomcat

The US Navy's fleet defence fighter – in service since 1974, and set to continue well into the 21st century – is the first carrier-based aircraft to feature variable-geometry wings. The Tomcat is claimed to be unsurpassed as a dogfighting aircraft, partly due to automatic wing-sweep variation, which adjusts according to the manoeuvre being undertaken.

ENGINES	2 x 9,480kg (20,900lb) thrust Pratt & Whitney turbofan		
WINGSPAN	19.5m (64ft 1in)	LENGTH	19.1m (62ft 8in)
TOP SPEED	2,517kph (1,564mph) (Mach 2.34)	CREW	2
ARMAMENT	1 x 20mm Vulcan rotary cannon; 6 x air-to-air missiles, or up to 6,577kg (14,500lb) attack weapons		

General Dynamics F-16C Fighting Falcon

The US Air Force's perennial demand for a lightweight fighter produced, in the 1980s, the F-16. Fast, extremely manoeuvrable and relatively cheap, over 2,500 were produced. The F-16A was limited to the daylight interceptor role and most of these aircraft have now been transferred to the Air National Guard. The F-16C has greater all-weather and attack capability and is regarded as a fighter-bomber. The F-16's many export customers, who include Belgium, Holland, and Israel, use it as an attack aircraft.

ENGINE	12,538kg (27,600lb) thrust GE F110-GE-100 turbofan		
WINGSPAN	10m (32ft 9in)	LENGTH	15m (49ft 4in)
TOP SPEED	2,124kph (1,320mph) (Mach 2)	CREW	1
ARMAMENT	1 x 20mm Vulcan rotary cannon; 2 x Sidewinder air-to-air missiles, up to 9,276kg (20,450lb) attack weapons		

Designation shows an AFTI (Advanced Fighter Technology Integration)-F16, the product of a joint programme with NASA, the US Army, and the US Navy

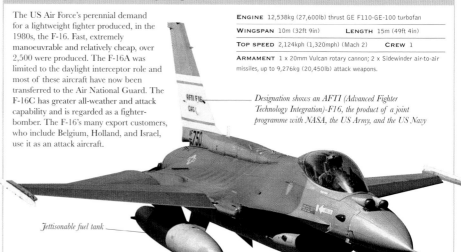

Jettisonable fuel tank

McDonnell Douglas F/A-18C Hornet

The US Navy's and Marine Corps' strike aircraft since 1981, the Hornet (NASA safety support aircraft shown) has been given the unusual dual designation "F/A", as it can be used as both a fighter and an attack aircraft. Though slower than the F-14, its small size makes it extremely manoeuvrable. Formations of attack F/A-18s can defend themselves en-route to their target and chase enemy fighters after they have dropped their bombload.

ENGINES	2 x 7,257kg (16,000lb) thrust General Electric turbofan	
WINGSPAN	11.4m (36ft 6in)	LENGTH 17.1m (56ft)
TOP SPEED	1,915kph (1,190 mph) (Mach 1.8)	CREW 1
ARMAMENT	1 x 20mm Vulcan cannon; 2 x Sidewinder air-to-air missiles, up to 7,711kg (17,000lb) attack weapons (bombs or missiles)	

Mikoyan-Gurevich MiG-23M "Flogger"

The MiG-23 was the first Soviet swing-wing aircraft, and its main purpose was to take on Phantoms and other Western attack aircraft. To do this, it carried more interception radar and so was bigger than its predecessor, the MiG-21. Fitting variable-geometry wings reduced the take-off and landing run, so the aeroplane could still use small front-line airfields in the traditional Soviet manner. MiG-23s served with the Soviet Union and its allies from 1973 until the 1990s.

ENGINE	12,500kg (27,512lb) thrust Khachaturov turbojet	
WINGSPAN	14m (45ft 9in)	LENGTH 15.7m (51ft 7in)
TOP SPEED	2,490kph (1,546mph) (Mach 2.35)	CREW 1
ARMAMENT	1 x 23mm twin-barrel cannon; 10 x air-to-air missiles	

Mikoyan-Gurevich MiG-25P "Foxbat"

The MiG-25 was developed in response to the American Lockheed A-11 (which became the SR-71). The prototype flew in 1964, and although by then the American aeroplane was only a reconnaissance project, development of the Russian interceptor continued. It remains the world's fastest combat fighter.

ENGINES	2 x 11,200kg (24,651lb) thrust Tumanskii turbojet	
WINGSPAN	14m (45ft 11in)	LENGTH 19.8m (64ft 9in)
TOP SPEED	3,000kph (1,868mph) (Mach 2.83)	CREW 1
ARMAMENT	4 x AA-6 air-to-air missiles	

Mikoyan-Gurevich MiG-29 "Fulcrum"

The MiG-29, with the now familiar layout of twin tails with twin underslung engines, uses the fuselage between the engines as part of the lift area, giving it amazing manoeuvrability. Like the Su-27, it was designed to counter the newest American aircraft – F-15, F-16, and F-18. In service since 1984, the MiG demonstrated its abilities by performing "tail-slides" at air shows during the 1990s – something no Western aircraft could do.

ENGINES	2 x 8,300kg (18,268lb) thrust Klimov RD-33 turbofan	
WINGSPAN	11.4m (37ft 3in)	LENGTH 17.3m (56ft 10in)
TOP SPEED	2,450kph (1,521mph) (Mach 2.3)	CREW 1
ARMAMENT	1 x 30mm cannon; 6 x AA-10 air-to-air missiles	

Panavia Tornado GR 1

The first European multi-role combat aircraft, the Tornado was designed and built by a consortium of British, German, and Italian aircraft manufacturing companies. In service with the air forces of these three countries since 1980, the Tornado uses a variable geometry or swing-wing to give good manoeuvrability and supersonic attack capability. RAF Tornados undertook the most dangerous missions of the 1991 Gulf War – low-level attacks on Iraqi runways against fierce anti-aircraft gunfire.

ENGINES	2 x 7,178kg (15,800lb) thrust Turbo Union turbofan	
WINGSPAN	13.9m (45ft 8in)	LENGTH 16.7m (54ft 9in)
TOP SPEED	2,336kph (1,452mph) (Mach 2.2)	CREW 2
ARMAMENT	2 x 27mm IWKA-Mauser cannon; up to 8,172kg (18,000lb) of ordnance, including free fall or guided bombs, unguided rockets, JP 233 runway-cratering weapons, Sidewinder air-to-air missiles, air-to-ground missiles, and electronic warfare pods	

*Swing wings can sweep through 40°
and carry an array of high-lift devices
on their leading and trailing edges*

*Advanced
nose sensor*

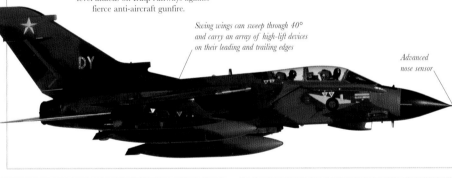

Saab 39 Gripen

Using the same delta wing with canard (a second wing fitted near the nose) layout as its Viggen predecessor, the Gripen is significantly smaller yet performs even better. It can also operate as an interceptor, ground-attack, or sea-attack aircraft, simply by fitting different armaments and selecting the appropriate mission characteristics from the advanced computer system. Design began in 1982 but the aircraft was not declared operational until 1997. As well as being used by the Swedish Air Force, it has been bought by Hungary, the Czech Republic, and South Africa, and serves as a jet trainer with Britain's Empire Test Pilots' School. There are both single- and dual-seat versions.

ENGINE	8,223kg (18,500lb) thrust Volvo RM12 turbofan	
WINGSPAN	8.4m (27ft 6in)	LENGTH 14.1m (46ft 3in)
TOP SPEED	2,124kph (1,320mph) (Mach 2)	CREW 1
ARMAMENT	1 x 27mm Mauser cannon; 2 x Sidewinder air-to-air missiles, plus up to 8 x air-to-air missiles, air-to-ground, or anti-ship, or cruise missiles, or bomb dispenser	

Sukhoi Su-27 "Flanker"

While the Sukhoi is considerably larger than the MiG, this, surprisingly, does not compromise its agility, but allows instead a smoother blending of wing and fuselage lift areas. The aircraft entered service with the Soviet (now Russian) Air Force in 1986. The Su-27 also forms the core of China's People's Liberation Army Air Force.

ENGINES	2 x 12,533kg (27,588lb) thrust Saturn/Lyul'ka turbofan	
WINGSPAN	14.7m (48ft 2in)	LENGTH 21.94m (71ft 11in)
TOP SPEED	2,495kph (1,550mph) (Mach 2.35)	CREW 1
ARMAMENT	1 x 30mm cannon; 10 x AA-10 air-to-air missiles	

SPACE TRAVEL

6

WHEREAS WINGED FLIGHT WAS DEVELOPED mostly through the efforts of individual enthusiasts, inventors and entrepreneurs, the exploration of space demanded the resources that only large nations could command. The basics of flight into space were simple enough. Given a sufficiently powerful thrust, any object could theoretically be flung either into orbit or beyond the pull of the Earth's gravity. But in practice such a venture posed daunting problems – requiring rockets with awesome power, space vehicles capable of withstanding extreme forces and temperatures, communication and control systems of great precision, and support systems to sustain life outside the atmosphere. That space travel has become a reality is a supreme example of organization and technological innovation, as well as of individual human courage.

HEADING FOR THE MOON
The Apollo 11 mission blasts off for the Moon on 16 July 1969. The manned Moon landing was the fulfilment of a dream that had long inspired rocket and science fiction enthusiasts, but that hard-headed practical people had mostly dismissed as mere fantasy.

> "First comes, inevitably, the idea, the fantasy, the fairy tale. Then comes scientific calculation. Ultimately, fulfillment crowns the dream."
>
> KONSTANTIN TSIOLKOVSKY
> EARLY THEORIST OF ROCKET FLIGHT
> INTO SPACE, 1926

JOURNEY TO THE MOON

IN HALF A CENTURY MANNED SPACE TRAVEL GREW FROM THE OBSESSION OF A FEW DREAMERS INTO A DESPERATE RACE BETWEEN COLD WAR RIVALS

IN 1903, AS THE Wright brothers were approaching success at Kitty Hawk, an unknown Russian schoolteacher, Konstantin Tsiolkovsky, published a paper entitled "The Exploration of Space with Reaction Propelled Devices" in a small-circulation scientific journal. Although the world at large paid no attention, Tsiolkovky had theoretically solved the basic problem of manned space flight – how to propel human beings out of the Earth's atmosphere without killing them.

Fantasy about space flight had a long history. In the seventeenth century, when the moon and planets had only recently been identified as bodies moving in space, astronomer Johannes Kepler fantasized about a journey to the Moon in one of the first science fiction stories, *Somnium*. By the nineteenth century, moon travel had developed into a familiar element of fantasy fiction. Most influential were the works of French author Jules Verne, who imagined his Moon-bound astronauts being fired from a canon sufficiently powerful to overcome the Earth's gravitational pull.

Tsiolkovsky had been inspired by Verne's stories to begin research into space travel in the 1880s. He worked out that, although it was theoretically possible to fire a projectile from a canon into space, it could not be done without subjecting its passengers to forces that would kill them. Working in provincial obscurity, Tsiolkovsky lighted on the correct solution: a rocket. Following Newton's third law of motion, that every action

RUSSIAN PIONEER
Konstantin Tsiolkovsky cracked many of the problems of space flight in theory, but not in practice.

produces an equal and opposite reaction, it would operate as a propulsion system both in and outside the atmosphere. Tsiolkovsky also worked out what fuel would power the rocket engine, suggesting liquid hydrogen and liquid oxygen to provide the required thrust. In the early 1900s, he went on to design spacecraft with prescient refinements such as steering vanes and multi-stage launchers, as well as a winged proto-Shuttle capable of gliding back to Earth.

Tsiolkovsky's eccentric interest in space rockets was shared by a number of isolated individuals in different countries who, up to the 1920s, pursued the subject with little or no reference to one another's work. In France, aviation pioneer Robert Esnault-Pelterie gave lectures on the possibility of space travel in 1912. In the US, by 1909 physicist Robert Goddard, then a postgraduate student at Clark University, had calculated the velocities required to reach Earth orbit. He carried out research into solid-fuel rocket engines and in 1919 attracted press notice by stating that a rocket could reach the Moon. Goddard said nothing of manned space flight, but in 1923 German teacher Hermann Oberth published *Die Rakete zu den Planetenräumen* (The Rocket into Interplanetary Space), assessing the practicality of human space travel, and giving plans for a liquid-propellant two-stage rocket.

LUNAR MODULE
The Apollo 13 astronauts practiced their Moon landing in a training module similar to this replica. However, their module would never land on the Moon, but was instead used as a "life raft" in space (see page 356).

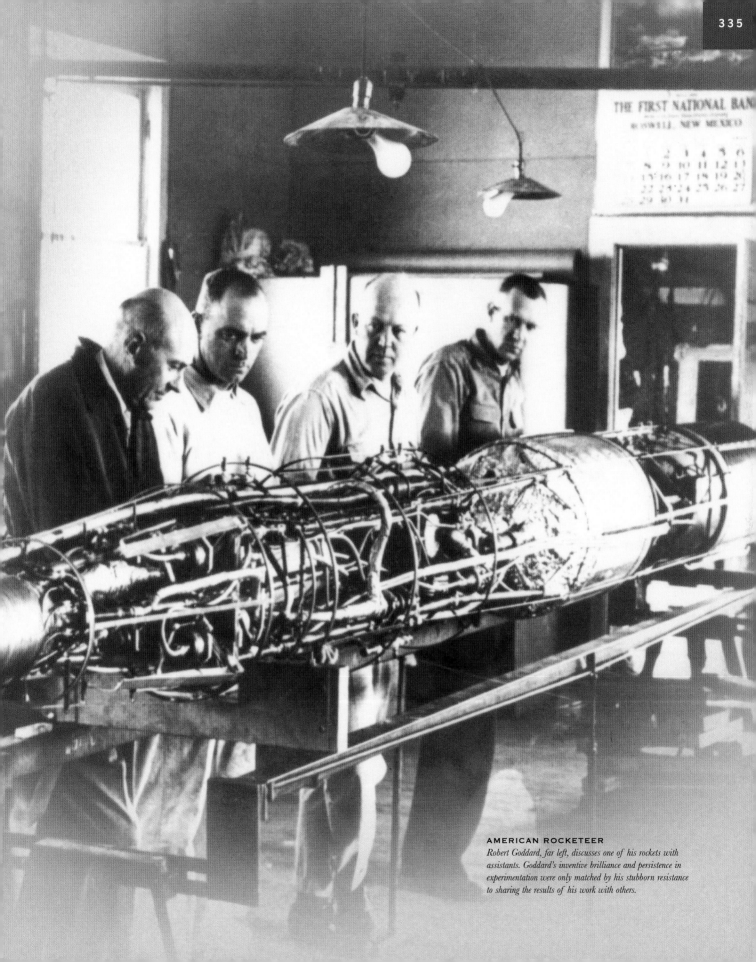

AMERICAN ROCKETEER
Robert Goddard, far left, discusses one of his rockets with
assistants. Goddard's inventive brilliance and persistence in
experimentation were only matched by his stubborn resistance
to sharing the results of his work with others.

ROBERT GODDARD

BORN IN WORCESTER, MASSACHUSETTS, Robert Goddard (1882–1945) was inspired by reading science fiction as a youth to a lifelong obsession with space travel. He pursued research into rocket motors as a student and then professor at Clark University, and in 1926 launched the first rocket fuelled by a liquid propellant. Goddard's work drew the attention of Lindbergh, and through him funding from the Guggenheim Foundation. This allowed Goddard to experiment with bigger and faster rockets in the 1930s, working at Roswell, New Mexico. He also developed sophisticated control systems.

SOLITARY GENIUS
By nature a loner, Robert Goddard mostly worked in isolation, keeping details of his experiments secret. As a result, the practical development of rocketry in Germany occurred without reference to his pioneering work.

CAPTURED ROCKET
Wernher von Braun's V-2 rocket was the starting point for both the American and Soviet space and missile programmes after World War II. This one had been brought to the United States.

By the 1920s the theoretical possibility of sending rockets into space had been established, but it was still a topic more likely to capture the imagination of oddball enthusiasts than mainstream scientists. Goddard, Oberth, and the ageing Tsiolkovsky – elevated to membership of the Soviet Academy of Sciences by the new Bolshevik regime in 1919 – were revered by groups of youths who set up rocket clubs and societies dedicated to interplanetary travel. Space fiction flourished, from the frivolous adventures of Buck Rodgers and Flash Gordon to the fantastic prophesies of Fritz Lang's 1929 movie *Frau im Mond* (Woman in the Moon). But the association of rockets with fantasy fiction and utopian projects for space colonies deterred more level-headed people from taking the subject seriously.

Still, practical progress was made, initially with Goddard in the lead. His first liquid-propelled rocket, launched in 1926, reached only a modest height of 12.5m (41ft), but by 1930 he fired a rocket up to 600m (2,000ft), and in 1935 one of his rockets rose to 2,300m (7,500 ft).

Also, at the California Institute of Technology in the second half of the 1930s, Theodore von Karman, a professor of aeronautics, encouraged the work of rocket experimenters such as Frank Malina and John Parsons. In the following decade the work at Caltech led to, among other things, development of rocket-assisted takeoff for aircraft. Another American group traced a remarkable path from fantasy to reality: the American Interplanetary Society, founded in 1930 mostly by science fiction writers, turned into the American Rocket Society in 1934, dedicated to practical experimentation, and in the 1940s gave birth to Reaction Motors Inc., the company that built the engine that powered the Bell X-1.

Military interest

The US was far from leading the world in rocket development. Through the 1920s and 1930s, it was in the Soviet Union and above all Germany that the idea of space exploration excited most serious interest, and that military authorities came to view rocket weapons as potential war-winners.

The initial development of liquid-propellant rockets in Germany was largely the work of the Verein für Raumschiffahrt (Society for Space Travel). In 1931–32, at their Raketenflugplatz (rocket flying field) in a Berlin suburb, the German enthusiasts carried out around 100 rocket launches, reaching altitudes up to 1,552m (4,922ft). Their work was being monitored by the German army, dreaming of escape from the severe restrictions placed upon it by the Versailles peace treaty. Suspicious of what they saw as the "humbugs, charlatans, and scientific cranks" of the rocket group, the army decided to recruit them into its own weapons programme, where they could be kept under supervision. The only

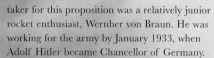

taker for this proposition was a relatively junior rocket enthusiast, Wernher von Braun. He was working for the army by January 1933, when Adolf Hitler became Chancellor of Germany.

The development of a large rocket capable of striking distant targets attracted ever-increasing resources from the Nazi state. By 1936 von Braun had 80 workers at his command and a top secret research and development facility was under construction at Peenemünde in a remote area on the Baltic. It eventually included a rocket factory, launch pads, and a liquid oxygen plant. By 1943 some 6,000 scientists, technicians, and engineers were employed there, along with countless forced labourers and prisoners of war.

The product of their efforts was the A-4, a liquid-propellant missile that, when fired vertically, reached an altitude of 176km (110 miles). Renamed by Nazi propaganda chief Joseph Goebbels as the V-2 (V for *Vergeltungswaffe*, or "vengeance weapon"), it was used against the Allies from the summer of 1944. It fell far short of the decisive impact the Nazis had hoped

> "A failed weapons system, the product of horrific slave labour, the V-2... opened the door to the universe."
>
> TOM D. CROUCH
> AIMING FOR THE STARS

for – each V-2 launched against Britain killed on average 1.76 people – but it revealed the existence of a revolutionary new technology to the Western Allies and the Soviet Union. Getting their hands on that technology was a major priority for both in 1945.

Von Braun's position, as Germany's defeat loomed, was on the face of it a desperate one. He was a member of the SS and his secret weapon had been produced by slave labour under murderous conditions. But he knew that the knowledge he and his team possessed was a powerful bargaining counter. By the end of 1945, along with captured V-2s, the Americans had carried von Braun and some of his colleagues back to the United States to work for their former enemies as "prisoners of peace". Over the following years about 120 of the 6,000 from Peenemünde reassembled, eventually creating an expatriate colony at Huntsville, Alabama. Their first collective task was to develop America's first nuclear ballistic missile, the Redstone, which was test launched from Cape Canaveral, Florida, in 1953.

FILMING THE LAUNCH
Recorded by movie cameras, a test launch of the Bumper rocket, using a captured V-2 as first stage, takes place at Cape Canaveral on 24 July 1950. Although rocket enthusiasts were inspired by the idea of space exploration, funding for rocket research depended on its military potential.

WERNHER VON BRAUN

BORN INTO THE PRUSSIAN landed aristocracy, Wernher von Braun (1912–77) was caught up in the "rocket craze" of the 1920s and in 1932 agreed to work for the German army. He always claimed that he had no interest in weapons research, but was "milking the military purse" to develop rockets that would eventually travel into space. After heading the Nazi programme that produced the V-2 rocket, at the end of World War II he engineered the transfer of his key staff to the United States. There he emerged in the 1950s as a high-profile advocate of travel to the Moon and Mars, fronting Disney TV shows and writing influential articles for Collier's magazine. His team played a key role in the development of US ballistic missiles and of launch rockets for the space programme.

CLEAN IMAGE
His past in Nazi Germany was rarely allowed to cast a shadow over von Braun's reputation as an inspired rocket pioneer.

REDSTONE ROCKET
The US Army Redstone rocket, created by Wernher von Braun's team at Huntsville, Alabama, was a logical step forward from the V-2. Although designed as a nuclear missile, it became an important early US space booster.

> "We gawked at what he showed us as if we were a bunch of sheep seeing a new gate for the first time."
>
> **NIKITA KHRUSHCHĒV**
> ON THE POLITBURO SEEING KOROLĒV'S ROCKET

By that time, both the development of long-range missile weaponry and the placing of satellites in Earth orbit had been adopted as priority goals by the United States and the Soviet Union, locked in Cold War confrontation. At stake were military security and the prestige of the competing ideological systems. The Soviets had a long tradition of rocket research to build on. As early as the 1920s the Soviet regime had embraced space exploration as a suitable goal for a society building a new future. Soviet inventors formed rocket clubs such as MosGIRD – the Moscow Group for Study of Reaction Motion – which sent its first rockets aloft in 1932, one reaching 400m (1,300ft).

Along with other rocket clubs in the Soviet Union, MosGIRD was soon absorbed into the state bureaucracy. The Red Army commander responsible for armaments, Marshal Mikhail Tukhachevsky, was especially interested in rockets as a potential form of long-range artillery. By 1937 most Soviet rocket experts, including Sergei Korolēv, were working for Tukhachevsky, which was unfortunate since

SERGEI PAVLOVICH KOROLĒV

BORN IN UKRAINE, Sergei Korolēv (1906–66) was a recent graduate in aeronautical engineering from Moscow Higher Technical College when he became a founder member of the MosGIRD rocket club in 1931. His talent was recognized by the Soviet state and he led a team at the Rocket Research Institute (RNII) under Marshal Tukhachevsky. In 1938, in the Stalinist purges, Korolēv was arrested, charged with subversion, and sent to the infamous Kolyma mines. Like many others, he was rescued from almost certain death by being recalled to work, still as a prisoner, on aircraft design and then rocket programmes. Released in 1944, he directed postwar Soviet rocket development. He was responsible for the first satellite launch and the first manned space flight, among other firsts. His death, of cancer, in 1966 was a severe blow to the Soviet space programme.

SOVIET HERO
Korolēv died as a hero of the Soviet state which had once sent him to a prison camp. His design genius was allied to remarkable willpower and astuteness that enabled him to operate within the brutal Soviet bureaucracy.

SPUTNIK 1

EXTERIOR

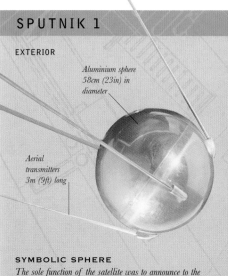

Aluminium sphere 58cm (23in) in diameter

Aerial transmitters 3m (9ft) long

SYMBOLIC SPHERE

The sole function of the satellite was to announce to the world a triumph of Soviet technology. Its signal, transmitted while in orbit, announced the start of the space age.

THE FIRST ARTIFICIAL EARTH SATELLITE, Sputnik 1, was launched from the Baikonur Cosmodrome near Tyuratam, Kazakhstan, late on the evening of 4 October 1957. To minimize the chances of failure, the Soviet team had chosen to keep the satellite simple, far lighter than the maximum load that the Soviet R-7 rocket was capable of lifting into space. Watching from a bunker as the huge rocket launched, Sergei Korolëv was almost incredulous at his own success. "Is this really all?" he said. "Have we really done it?" As Sputnik 1 circled some 800km (500 miles) above the Earth, emitting beeps from its transmitters, the Soviet people were euphoric. As for the American reaction, future president Lyndon B. Johnson recalled "the profound shock of realizing that it might be possible for another nation to achieve technological superiority over this great country of ours."

EXPLODED VIEW

INSIDE SPUTNIK

Sputnik 1 was an aluminium sphere containing two radio transmitters, a series of silver-zinc batteries, and a thermometer. It weighed 84kg (184lb). Travelling at 30,000kph (18,000mph), it orbited the Earth every 96 minutes.

---

the marshal then fell victim to Stalin in the great purge that decimated the Red Army officer corps. Korolëv was among many scientists sent to the Gulag on absurd charges of subversion. Soviet rocket research went into temporary decline.

The end of World War II left the Soviets in control of the V-2 production site at Nordhausen. They also kidnapped a number of generally lower-level scientists and technicians who had worked on the German rocket programme. Information from these sources was used by Korolëv, now released from the Gulag, and other Soviet rocket experts such as Valentin Glushko to advance the development of the Soviet rocket programme from artillery expertise

SPACE DOG

The ill-fated Laika, the first living creature to travel into Earth orbit, demonstrates its space pod. The dog paid for worldwide celebrity status with its life, dying when Sputnik 2 overheated.

initially begun in World War II. By the early 1950s the Soviets had rocket motors capable of generating over 90,000kg (200,000lbs) of thrust – the Redstone's motor gave 35,000kg (75,000lbs).

First satellite

Whereas the Americans concentrated much of their energy on developing a nuclear bomber force, from 1952 the Soviets put the lion's share of their resources into building a nuclear-armed intercontinental ballistic missile (ICBM). The goal, as defined by Soviet Air Force chief Marshal Zhigarev, was to make "long-range, reliable rockets capable of hitting the American continent."

In 1955–56 a new launch complex, known as the Baikonur Cosmo-drome, was built in the Soviet republic of Kazakhstan. The Soviets set out to test a rocket that would astonish the world. By grouping five rocket motors together, Korolëv gave their R-7 Semyorka almost 455,000kg (1 million lb) of thrust. It had a range of 6,400km (4,000 miles), making it the world's first ICBM, and could potentially lift a one-ton payload into space.

After many delays, test launches of the R-7 began in May 1957. The first eight attempts either resulted in an aborted launch or explosion some time after lift-off. Korolëv was close to nervous collapse by the time a fully successful flight was achieved,

on 21 August 1957. After one repeat performance, Soviet leader Nikita Khrushchëv gave the order for a satellite mission. The launch of Sputnik 1 on 4 October had a propaganda impact that could only be described as sensational. The Soviet leadership immediately gave instructions for a follow-up. The launch of a dog into orbit in Sputnik 2 came exactly one month later.

SPUTNIK 3

Launched on 15 May 1958, Sputnik 3 was a conically-shaped vehicle measuring 3.57m x 1.73m (11ft 8in x 5ft 8in), and weighing 1,327kg (2,919lb). It was powered by solar panels embedded around the base of the main body.

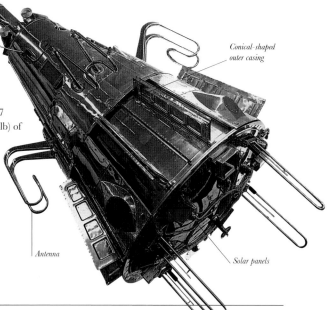

Conical-shaped outer casing

Antenna

Solar panels

SPACE TRAVEL

As Soviet propagandists crowed over the Soviet Union's technological achievement, claiming it as evidence of the superiority of the communist system, the United States raced to catch up in what had instantly become a "space race". Both the USSR and the US had announced a satellite launch for International Geophysical Year 1957–58, developing a launcher named Vanguard. It was in no way comparable to the Soviet R-7 in thrust or payload, but its first launch in December 1957 was eagerly awaited as an urgently needed riposte to the Sputniks. But Vanguard blew up on liftoff – the explosion seeming to one observer "as if the gates of hell had opened up". The delighted Soviets sneeringly enquired whether America might like to benefit from technical aid for backward countries.

Fortunately Vanguard was not the only satellite project the United States had in hand. Wernher von Braun's team at Huntsville, working for the US Army, had developed a more powerful version of their Redstone rocket, the Jupiter-C, which could be adapted to launch a satellite.

Although interservice rivalries had caused the Army's long-range rocket programme to be curtailed, when the shock of the Sputnik launches struck, von Braun and his boss, General John Medaris, were ready to step forward with their own proposal for a satellite launch. The Jupiter-C rocket was adapted into a Juno 1 launcher, while the Jet Propulsion Laboratory, California, headed by William Pickering, put together a satellite to go on top of it. Explorer 1 was successfully launched into orbit on 31 January 1958. Unlike the first Soviet satellites, it carried scientific instruments that achieved an immediate step forward in human knowledge, identifying the regions of

SPACE AIRCRAFT
The North Anmerican X-15 rocket plane could fly to an altitude of 108km (67 miles) – high enough to qualify as sub-orbital space flight. But it was not capable of going into orbit, and even if it had been, it would not have been able to re-enter the atmosphere safely.

66671

"From a nation of 175 million they stepped forward... seven men cut of the same stone as Columbus..."

TIME MAGAZINE
AFTER THE FIRST MERCURY 7 PRESS CONFERENCE

AMERICA'S FIRST ASTRONAUTS
The Mercury 7 pose for a formal portrait in their space gear in 1962: (left to right) Wally Schirra, Alan Shepard, Deke Slayton, Gus Grissom, John Glenn, Gordon Cooper, and Scott Carpenter. Selection for the space programme had made these pilots among the most famous men in the world.

radiation trapped by the Earth's magnetic field that are now known as the Van Allen Belts – after University of Iowa physicist James Van Allen, who set up the experiment conducted by Explorer 1.

But the pursuit of scientific knowledge or of practical technological goals such as improved communications were to remain for a long time marginal to the main thrust of both American and Soviet space programmes. The space race was a contest for prestige between two superpowers, each seeking to prove the superiority of its own political and economic system on a cosmic stage. Both sides knew that in the eyes of the world, the winner (at least of round one) would be the first to put a man in space. The imperative of the moment was not long-term planning or practical utility, but simply to put a man in space as soon as possible.

During 1958, the United States sorted out the basic organization of its space effort into three programmes. Determined that one programme should be under civilian control, President Eisenhower chose to transform NACA, the long-established federal agency for flight research, into NASA – the National Aeronautics and Space Administration. The second program under the US Air Force was to continue its own reconnaissance satellite programme with the CIA, and the third program was controlled by the US Air Force in a military role, while the Army was soon told to end its involvement in space. In 1960 von Braun's team transferred from the Army to NASA, becoming the George C. Marshall Space Flight Center.

NACA had been involved with the USAF in the X-15 experimental rocket plane programme, which was regarded by many engineers as a stepping stone on the right road to space travel. The X-15 could travel almost as fast as a rocket launcher and reach the edge of the atmosphere. NACA engineers had argued that the next generation of X-planes might be able to fly into orbit and fly back down to land. Committed to getting a man into space as quickly as possible, however, NASA had no time to wait and find out whether a winged aircraft could really achieve this. While ready to back von Braun's plan to produce a giant launch rocket for the future, what was needed at this point was a space capsule and a flight plan that would take an American into space and back using existing rocket technology.

By the time NASA was formed, engineer Maxime Faget had already sketched out the requisite vehicle. Small and light enough to be lifted into suborbital flight by a Redstone missile, it would be a cone-shaped capsule fitted with rockets to slow it before re-entry to the atmosphere, and with a heat shield to protect its blunt base. It would come down with the blunt end first and drop into the ocean on parachutes. But for all its limitations, if it meant an American was the first astronaut into space, it would do.

Circus performers were one group initially considered as possible candidates for America's first corps of astronauts, but Eisenhower sensibly laid down that they should be military test pilots.

SPACE APE
NASA sent Ham the chimpanzee into space in January 1961. More fortunate than Soviet space dog Laika, Ham survived the experience.

ASTRONAUT TRAINING

WEIGHTLESS MOVES
Trainee Mercury astronauts undergo the experience of weightlessness in the cargo hold of a transport aircraft. Flown in a vertical parabola, the aeroplane could create gravity-free conditions for up to 45 seconds as it passed over the top of the arc.

OCEAN EXERCISES
Gemini astronauts come to terms with a "splashdown", the clumsy but fully practical solution to landing without wings.

THE TRAINING THE FIRST American astronauts underwent consisted primarily of exposure to repeated simulations of the experience of space flight, including weightlessness and the g forces produced by rapid acceleration. The experience of weightlessness was created by taking the astronauts up in aircraft that were flown in a parabolic curve, or hump, a trajectory that created shortlived zero-gravity conditions. They were also whirled around at the end of the giant arm of the US Navy's centrifuge at Johnsville, Pennsylvania, which was computer programmed to recreate the different g forces they would experience at various stages of a space flight.

In other exercises the astronauts were made to lie for hours in a simulator at Cape Canaveral, sitting in a seat that had its back on the floor, staring up at a replica of the Mercury spacecraft console. There they were put through the routines of the flight and confronted with a variety of emergencies.

All highly skilled test pilots, the astronauts were initially discontented with a program that seemed to place them in the role of "redundant components" in an automatic system. They wanted to fly their space craft, not just serve as passive guinea-pigs to test the effects of space flight on the human body. From the start, the astronauts fought successfully to win more control than the early Soviet cosmonauts enjoyed. They were soon to find themselves playing a vital hands-on role, as automatic systems proved considerably less than one hundred percent reliable.

The Mercury 7

A rigorous selection procedure was set in motion to find the men who would represent the strength and pride of the United States. Candidates were required to have educational qualifications, be in peak physical and psychological condition, and have the sort of lifestyle and personal qualities that would fit them to be American heroes. The seven eventually chosen were all married men with children, between 32 and 37 years old. The project that they joined was dubbed Mercury, and they were the Mercury 7.

The Mercury astronauts did not, of course, possess the uniform qualities that their selectors had perhaps hoped for. Marine Colonel John Glenn, the oldest of the group and a veteran of World War II and Korea, was a man of modest manners and strong religious principle. Alan Shepard and Walter Schirra were wilder characters, with a taste for practical jokes and fast cars. Some of the men's family lives were not quite as near the ideal as their *Life* magazine profiles suggested. But all had the necessary courage and patriotism for the risky task at hand.

FIRST MANNED SPACE FLIGHT

PRESENTED TO THE WORLD as a smooth and trouble-free triumph of Soviet technology, the first manned space flight was in fact made using equipment that came very close to disaster.

On the morning of 12 April 1961, cosmonaut Yuri Gagarin climbed into the cramped, bare capsule of the Vostok spacecraft on top of its towering launcher. He was left there for 50 minutes, listening to Russian love songs, while preparations for the launch were completed. Then the engines roared and Gagarin felt the g-load build as the vehicle accelerated upwards. Minutes later the rocket shut down and he was in orbit, undisturbed by weightlessness and enjoying the astonishing view.

DAWN OF AN ERA

The era of manned space flight begins as Vostok 1 blasts off from the Baikonur Cosmodrome on 12 April 1961. Cosmonaut Yuri Gagarin flew for just 1 hour 48 minutes, long enough to circle the globe.

Heading east over the Pacific he plunged precipitately into darkness, crossed South America, and emerged to a new dawn over the Atlantic.

An hour and a quarter into the flight, over Africa, Gagarin heard and felt the retro-rocket ignite to begin re-entry. But the capsule failed to separate from the rest of the spacecraft. Gagarin found himself tumbling into the atmosphere, "head, then feet, rotating rapidly". This continued for 10 minutes until a cable burnt through and the capsule broke free.

Plunging through the atmosphere, Gagarin heard a cracking noise as the heat shield reached high temperatures and felt the g-load intensify. But all was now well. According to plan, at 7,000m (23,000ft) the capsule hatch blew off and Gagarin's ejector seat fired. He parachuted down into a ploughed field near the city of Saratov.

BRAVE PASSENGER
Gagarin was essentially a passenger in the Vostok spacecraft. Although the controls of the spacecraft were locked, a key had been placed in a sealed envelope in case an emergency situation made it necessary for Gagarin to take control.

> "The noise was... like the noise in an aircraft. Then the rocket smoothly, lightly rose..."
>
> **YURI GAGARIN**
> ON HIS FIRST SPACEFLIGHT

Crew module

Command control antenna

Oxygen-nitrogen pressure bottles

Equipment module

Launch vehicle third stage

VOSTOK 1
The cosmonaut travelled in the small spherical capsule on top of the Vostok spacecraft. The rest of the craft contained the environmental control system and the retro-rocket which ignited to send the capsule back from orbit.

ROCKET POWER
Directly derived from the R-7 which had launched the first Sputnik in 1957, the launcher used to propel Vostok into orbit consisted of a core vehicle and four boosters that dropped off.

The risks that the Mercury astronauts faced were soon clear enough. The programme was being rushed through – the original goal was for a first manned flight in mid-1960 – but rocket technology remained dangerously unreliable. The plan was for a sub-orbital flight using a Redstone booster to be followed by an orbital flight using a more powerful Atlas launcher. But in July 1960 an Atlas test-fired with a Mercury capsule on top blew up spectacularly just after launch, and the following November the Redstone-Mercury combination proved a dud, lifting barely 15cm (6in) off its launch pad before settling back to rest.

Pipped at the post

There was thus no end of relief at NASA when on 31 January 1961, a chimpanzee called Ham was successfully sent into space strapped into a Mercury capsule. The event went far from smoothly – Ham eventually splashed down 210km (130 miles) off target and nearly drowned before being picked up. Nor were the proud and skilful Mercury pilot-astronauts pleased to be reminded that the flight they were training for had been successfully carried out by a monkey. But NASA was ready for a human launch in the spring, only pushing it back from March to May in order to iron out the glitches revealed by Ham's space adventure. It was a small delay, but it let the Soviets win.

In complete contrast to the United States, where the glare of publicity surrounding the space programme was dazzling, the Soviet Union proceeded in secrecy. The appearance was of an unbroken series of triumphs, each emerging without prior announcement – with a suddenness that in practice enhanced their impact. And the triumphs were real. In 1959 they won the first Moon race with their Luna spacecraft, crashing Luna 2 on to the Moon's surface and sending Luna 3 around the Moon to take the first photos of its invisible dark side. But there were also

TARGETTING THE MOON
On 25 May 1961 President Kennedy told Congress that the USA should commit itself to "landing a man on the Moon and returning him safely to Earth" before the end of the decade.

setbacks that were hidden from the world. In October 1960 an attempt to send an unmanned spacecraft to Mars failed, and this was followed by the worst single disaster of the space age, when a new rocket booster on its launch pad at Baikonur unexpectedly ignited while being examined by senior officers, engineers, and technicians. Over 100 people died, engulfed in flames, yet the event went totally unreported.

Like the Americans, the Soviets were hell-bent on being first to send a man into space, and selected a team of trainee cosmonauts from their pool of air force fighter pilots. The 20 Soviet cosmonauts were significantly younger than their American counterparts, ranging in age from 24 to 34. In contrast to the instant fame of the Mercury 7, the cosmonauts were unknown to the public, their very existence a secret. As a result, they were dispensible. In March 1961, cosmonaut Valentin Bondarenko, aged 24, carried out a series of tests that involved living in a pressurized oxygen chamber. He had an electric hot plate in the chamber for cooking. At the end of the tests, he carelessly tossed a piece of alcohol-soaked cotton on to the hot plate, causing the oxygen-soaked atmosphere to burst into flames. Badly burned, Bondarenko died shortly afterwards.

Because of its secrecy, such an event was no hitch in the progress of the Soviet manned space programme. From May 1960, the Soviets had begun test launches of their Vostok space capsule, sending a veritable menagery of animals into orbit, including dogs, mice, rats, frogs, and literal, rather than figurative, guinea pigs. By April 1961 they were ready for the real thing. Only three weeks after the death of Bondarenko, Gagarin went into orbit and became, for a time, the most famous man in the world.

There was no doubt of the political motive behind the Soviet space effort, or of its effectiveness. Across the Third World, where new nations were becoming independent of colonial rule every year, leaders and their peoples saw the Soviet Union as the world's leading technological

power and Soviet-style communism as the way of the future. Losing the space race was causing real damage to the global interests of the United States at a time of intense Cold War confrontation – this was the year of the failed Bay of Pigs invasion of Cuba and the building of the Berlin Wall.

Once Gagarin had won the race to be first in space, the only way the Americans could hope for victory was by raising the stakes – by setting a more ambitious target, distant enough to give the Americans time to catch up with and overtake the Soviets. The new US president, John F. Kennedy, decided to challenge the Soviet Union to a race to put a man on the Moon. His

> ## "This is the new age of exploration; space is our great New Frontier."
>
> **JOHN F. KENNEDY**
> WHILE US PRESIDENTIAL CANDIDATE IN 1960

YURI GAGARIN

YURI ALEXEYEVICH GAGARIN (1934–68) was born on a collective farm near Smolensk. As a child he survived the Nazi occupation during World War II, as well as the harshness of life under Stalin's regime. In the 1950s he went to technical college, and was expected to become a factory supervisor. But instead he found a place on an air force training course and in 1957 graduated as a pilot officer. In March 1960, at the age of 26, he was selected to train as a cosmonaut. This was no passport to glamour. Recently married and starting a family – his second child was born a month before the first space flight – Gagarin lived in a modest Moscow apartment and commuted to cosmonaut training in the city suburbs.

Becoming a Soviet hero turned his life into a round of public ceremonies and official foreign visits. Gagarin always longed to return to flying and to space. He had his wish in 1968, resuming cosmonaut and jet fighter training. But in March of that year he was killed when his MiG15UTI trainer crashed.

CHARMING HERO
Raised to the rank of colonel in the Soviet air force, Yuri Gagarin proved an excellent ambassador for his country, displaying confident charm in his public role as a Soviet hero.

predecessor, Eisenhower, had baulked at the cost of a long-term manned space programme, seeing the Mercury project as an end rather than a beginning. Kennedy was no more keen on the high price tag, but he was more ready to identify with a bold technological project that would restore America's belief in itself and its international prestige.

On 25 May 1961, Kennedy told Congress: "I believe that this nation should commit itself to achieving the goal, before this decade is out, of landing a man on the Moon and returning him safely to Earth." It was the opinion among some US space experts that, in the time scale laid down, the goal was unlikely to be achieved. But the die was cast.

> ## "Why don't you fix your little problem... and light this candle."
>
> ### ALAN B. SHEPARD
> TO GROUND CONTROL BEFORE 1961 LAUNCH

By the time Kennedy made his famous speech to Congress, Americans had already found something to cheer about as the Mercury programme produced its first success. On 5 May 1961, astronaut Alan B. Shepard was, with considerable difficulty, shoehorned into the Freedom 7 Mercury space vehicle on top of a Redstone launcher. There, sitting with his back to the ground facing the sky, barely able to move, he had to wait for four hours as delay followed delay. This led to an acutely embarrassing problem with his bladder, as no thought had been given to providing a means of urinating in what was planned as a 15-minute mission. At last, approaching 9.30am, the final countdown began. Across America, millions followed the tense moments on television or radio. The Redstone lifted smoothly from its launch pad and rose through the atmosphere. The rocket separated from the capsule and Shepard sailed through space up to an altitude of 186km (116 miles), his voice on the radio link relayed to the nation.

Safe return

After taking an opportunity to experiment with manual control of the spacecraft in pitch, yaw, and roll, so important to pilots who did not want to be reduced to passengers, Shepard made a successful re-entry and splashed down in the Atlantic off Bermuda, some 480km (300 miles) from his starting point in Florida.

In some ways, Shepard's quarter-hour loop through space only emphasized how far the Americans lagged behind the Soviets – it could not compare with Gagarin's 108-minute orbital flight. But it also showed how far the United States was ahead of the Soviet Union in the use of mass media. The American side of the space race began as it was to continue – as a television spectacular. It was to be a gripping drama enacted live that would engage the emotions of viewers worldwide, and stand as an ongoing demonstration of the openness of American society.

Both the Americans and the Soviets continued a programme of manned flights with their Mercury and Vostok craft over the next two years. Following in Shepard's footsteps with a suborbital flight in July 1961, Virgil Grissom narrowly avoided disaster when his emergency hatch blew off after splashdown. Grissom made a hasty exit

as his cockpit filled with water and survived with a drenching, although the spacecraft sank and was lost. The following month cosmonaut Gherman Titov put the American efforts in perspective by staying in orbit for over 24 hours in Vostok 2 (the Soviets numbered their Vostoks sequentially, while all the Mercury craft were given the number 7).

By 1962 the Americans had at last overcome the many difficulties experienced with the Atlas launcher and were ready to put an astronaut into orbit. On 20 February John Glenn lifted off from the Cape aboard Friendship 7 on top of a perfectly functioning Atlas. This was not to be another human cannonball loop through space,

AMERICAN FIRST

On 5 May 1961 Alan B. Shepard became the first American to enter space. The Redstone launcher (left) was not powerful enough to propel the Freedom 7 Mercury space vehicle into orbit. Shepard rose to a maximum altitude of 186km (116 miles). During his 15-minute flight, the astronaut's facial expressions were monitored by an on-board camera (below left). Shepard splashed down in the ocean and was rescued by a US Marine helicopter (below) which flew him to USS Lake Champlain.

INSIDE FRIENDSHIP 7
Compared with a Soviet Vostok capsule, John Glenn's Friendship 7 was crammed with dials and switches – in part a response to the insistence of the Mercury astronauts on exercising as much manual control as possible over their space vehicles.

HEADING INTO ORBIT
The Atlas launcher blasts off from Cape Canaveral on 20 February 1962 carrying the Friendship 7 space vehicle and John Glenn in its nose. Developed by the US Air Force, Atlas was significantly more powerful than Redstone. Glenn circled the globe three times in a flight lasting almost five hours.

but a genuine orbital space flight. Glenn went round the Earth three times in a flight that lasted 4 hours and 55 minutes. It was far from being a sightseeing trip. There were unnerving problems with some automatic systems and at one point ground control began to think that the capsule's heat shield might have become loose. As Friendship 7 re-entered the atmosphere, the whole of America held its breath. Even Glenn went through a moment when he believed his vehicle was burning up. But splashdown was successfully accomplished and the astronaut returned to a tickertape parade down Broadway believed by some to be the biggest since Lindbergh.

The Soviets continued to execute the longest space flights, extending to an hour short of five days for Valery Bykovsky in Vostok 5 in June 1963. They also went on scoring political points. While Bykovsky was still circling the Earth, a former cottonmill worker, Valentina Tereshkova, was sent up in Vostok 6. The Soviets not only had two people in orbit at once – a feat they were performing for the second time – but the first woman in space. In this respect, as it turned out, they were two decades ahead of the United States.

The longest flight under the Mercury programme was 34 hours 20 minutes, by Gordon Cooper in May 1963. He was the sixth of the Mercury 7 to go into space, and the last under the banner of that programme. Only Deke Slayton had missed out, grounded with a minor heart problem. Slayton finally made it into space in 1975.

GEMINI 4 SPACECRAFT

The two-man Gemini spacecraft was a major step forward from the Mercury capsules. Weighing 3,200kg (7,000lb), it could be controlled by its pilots to perform elaborate manoeuvres in space. Hatches swung open to provide an easy exit for spacewalks.

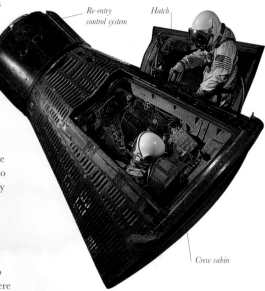

Re-entry control system

Hatch

Crew cabin

The announcement of the goal for a manned Moon mission increased the importance of the Mercury project; it was really the first step in learning about human spaceflight. But there was no straight line from the Mercury capsules and their launchers to a Moon voyage. The lunar mission would require not just a far more powerful booster rocket, but also spacecraft with onboard power systems and computers capable of elaborate precision manoeuvres such as docking with another craft in space. It would also require astronauts to be trained to spend much longer periods in space and to operate outside their space vehicles – so they could walk on the airless Moon. It was to develop these skills and techniques that NASA embarked on the Gemini project, while longer-term work on the Moon programme proper, dubbed Apollo, gathered momentum.

By 1964 the US space programme had turned into one of the largest-scale endeavours in the country's history. Yet the Soviets still gave the impression of holding their lead. As NASA worked towards

GLENN IN ORBIT

John Glenn prepares to view the sun through a photometer during one of the three sunsets he witnessed in his five-hour journey through space in February 1962. The astronaut saw what looked like fireflies during his flight, but the "spacebugs" turned out to be particles of ice shaken from the surface of the capsule.

WALKING IN SPACE

"I felt red, white, and
blue all over."

EDWARD H. WHITE
ON HIS SPACE WALK, *LIFE* JUNE 1965

LIVING HIGH
*Astronaut Ed White, pilot of the Gemini 4 mission, floats
above the Earth during his historic space walk on 3 June
1965. In his right hand White holds the self-manoeuvring
gun which allowed him to propel himself through space,
while remaining attached to the Gemini craft by an 8m
(25ft) umbilical line.*

THE SOVIET UNION'S DETERMINATION to remain
ahead in the propaganda battle in space forced
Sergei Korolev to rush into extra vehicular activity
(EVA) – a space walk – before the right technology
was in place. Aware that America's new Gemini
project would probably lead to a space walk in
1965, Korolev set about adapting his Voskhod
spacecraft to the purpose. Because the Soviets
were still using vacuum-tube technology instead
of circuit boards, the tube-filled interior of the
Voskhod cabin could not be exposed to space.
Instead, an inflatable airlock was fitted over the
spacecraft's main hatch. A crew member would
crawl into this, close the hatch behind him, and
open the airlock hatch to float out into space.

On 18 March 1965 cosmonaut Alexei Leonov
duly exited into space from Voskhod 2, tethered to
the craft by an umbilical line that provided his air
supply and a communications link to his colleague
Pavel Belyayev who stayed behind to command the
craft. Unfortunately, Leonov soon found himself
spinning uncontrollably as his tether, which had
become tangled, untwisted itself. Worse, when he
attempted to re-enter the airlock, he found his
spacesuit had ballooned and he could no longer
fit into the opening. He finally squeezed in after
lowering the pressure in the suit. Even though
Voskhod 2 had an eventful journey, completed by

COSMONAUTS
On 18 March 1965 cosmonaut Alexei Leonov exited Voskhod 2 to perform the first space walk (right). He was protected by his pressurized space suit and his eyes were shielded from the Sun's glare by mirrored goggles worn inside his helmet. Soviet manned missions continued after the Voskhod flights with the Soyuz spacecraft (top).

landing in a forest 3,200km (2,000 miles) from their target, Leonov had become the first man to walk in space.

America's first space walk was quite different to that which the cosmonauts experienced. Astronaut Edward H. White floated out of the main hatch of his Gemini 4 spacecraft on 3 June 1965. Unlike Leonov, White had a hand-held device that enabled him to move around at will. It consisted of two small tanks of compressed gas which, released in small bursts, produced enough thrust to propel the astronaut around. He was, of course, attached to the spacecraft by an umbilical line. White was so euphoric at his weight-free acrobatics that he jokingly threatened to refuse to come back in at the end of his planned 10 minutes outside.

White and fellow crewman James McDivitt stayed in space for four days. This was also a major landmark. It was longer than the total of all previous US space flights added together. Although the Soviets had made longer flights, there were fears that the astronauts might suffer serious ill-effects on returning from such a time in zero-gravity conditions. When White and McDivitt were retrieved from splashdown evidently fit and happy, another possible barrier to long-distance space flight had evaporated.

sending two astronauts into space on board a Gemini spacecraft, Korolëv went one better. In October 1964 he put three men into orbit in a craft called Voskhod. It was a desperately risky venture. Voskhod was no more than an improved version of the old Vostok spacecraft. The only way to fit three people inside was to have them fly without space suits and to strip out the ejection seats. But Boris Yegarov, Vladimir Komorov, and design engineer Konstantin Feoktistov made a successful day trip into space and landed inside the capsule.

Because of Soviet secrecy, the Americans had no idea of the desperate straits to which Korolëv had been driven in order to carry off the Voskhod 1 mission. Nikita Khrushchëv was deposed as Soviet leader on 12 October 1964, not long after Voskhod 1 was launched, although the reasons had nothing to do with space. His replacements, the stolid Leonid Brezhnev and Alexei Kosygin, were much less reliable supporters of an extravagant space programme than the ebullient Khrushchëv. In the United States, by contrast, the death of President Kennedy in 1963 had only brought the equally space-minded Lyndon B. Johnson to the White House. While the American space effort was concentrated on a single politically determined goal – putting a man on the Moon – the Soviet programme increasingly lacked direction. Yet with less resources at their disposal, the Soviets could ill-afford to disperse them.

The last moment when the Soviets appeared seriously to be ahead of NASA was with the first space walk during the Voskhod 2 mission in March 1965. Yet this again was a risky improvisation rather than a carefully considered step in a planned programme. The death of Korolëv in January 1966 – which brought his instant transformation from an anonymous engineer to a publicly hailed Soviet hero – was another important setback. With their experience of powerful rockets and of space vehicles, the Soviets were in no way out of contention in the race to the Moon, but by 1965–6 the United States was making all the running.

SPACE RENDEZVOUS
Astronauts Wally Schirra and Tom Stafford aboard Gemini 6 had this close-up view of Gemini 7 as they drew near during the first Gemini orbital rendezvous on 15 December 1965. This was a vital preparation for an eventual manned Moon mission.

In preparation for the Gemini and Apollo programmes, NASA took in new recruits to the astronaut corps. This time there were few hesitations about exchanging the status of test pilot for that of astronaut. Being fired into space with limited control of the craft had not seemed an attractive option to many men used to piloting the latest jet fighters or X-planes. Now there was not only the chance of setting foot on the Moon, but also spacecraft that could genuinely be controlled in space by their pilots.

Gemini missions

The first manned Gemini mission, Gemini 3, blasted off on 23 March 1965. The crew consisted of Gus Grissom from the original Mercury 7 and rookie John Young. The craft they journeyed in was made up of two parts: the spacecraft proper and a support module containing, among other things, an oxygen supply and propellant for thrusters. This module would be jetisoned before re-entry. Much larger than a Mercury capsule, Gemini 3 was lifted into space by a Titan II, a two-stage rocket delivering 196,000kg (430,000lb) of thrust at lift-off.

The spacecraft was still small for two men to occupy on missions that would eventually last for weeks. Sitting side-by-side, they each had less room than you would find in a phone booth. But for pilots it was engagingly like the cockpit of a high-performance fighter, with ejection seats and a complicated set of controls designed to allow the pilots to change their orbit and eventually dock with another vehicle in space.

Not all the Gemini missions went smoothly. The object was to learn, and learning was often from mistakes. The most important goal of the programme was to achieve docking between two spacecraft. However, the Gemini 6 mission to dock with an unmanned Agena vehicle was called off when the Agena broke up during launch. Another use was found for Gemini 6 when it was sent up to rendezvous with Gemini 7 in December 1965. The craft manoeuvred to within a few feet of one another, and Gemini 7's crew of Frank Borman and James Lovell went on to stay in space for a fortnight. Four months later Gemini 8, crewed by Neil Armstrong and Dave Scott, docked with an Agena vehicle achieving the first successful docking in space. And even with their success, all was not well. The joined spacecraft began to tumble end over end. The pilot managed to control the craft by activating thrusters intended for use during re-entry. It worked, and the astronauts achieved the first successful docking in space, but it had been a close call.

A smoother docking was achieved by Gemini 10 in July 1966, crewed by John Young and Michael Collins. They not only rendezvoused and docked with an Agena vehicle, but carried out manoeuvres with the two spacecraft joined together.

Space walking also ran into problems after the successful start with Gemini 4. During the Gemini 9 mission, astronaut Gene Cernan experienced severe difficulty trying to work outside the spacecraft and had to curtail his EVA. But all difficulties were triumphantly overcome during the last Gemini mission, Gemini 12, in November 1966, when Edwin "Buzz" Aldrin carried out five hours of space walks without a problem.

Apollo gets under way

With the successful completion of the Gemini programme, the path to the Moon was open. NASA had long decided on the best route to follow. Back in 1959, Wernher von Braun had chosen the name Saturn for a new series of powerful launchers. Like all his generation of rocket pioneers, von Braun had always assumed that a launcher of such power would be required for a manned Moon landing, because the space vehicle would have to be large and packed with heavy propellant to blast it off the Moon's surface for the return to Earth. When President Kennedy announced his time-scale for the Moon

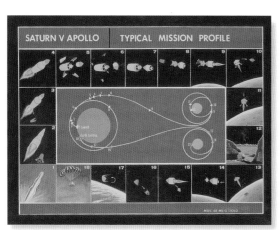

MOON MISSION PROFILE
This depiction of the various stages of the planned Apollo Saturn mission gives a graphic representation of the complexity of the operation. The key to the mission plan was the use of a lunar module to land on the surface of the Moon and then blast off to meet up with the command module in lunar orbit.

programme, however, von Braun was clear that developing Nova would take too long.

He suggested instead a scheme involving the use of two smaller Saturn launchers. One would lift the crew into Earth orbit, while the other would lift the propellant. They would link up in orbit and proceed to the Moon. But this was not the scheme adopted. Instead, in 1962, NASA opted for a "lunar orbit rendezvous" plan. This involved using a spacecraft consisting of a command and service module (CSM) and a lunar module (LM). Only the lunar module would land on the Moon's surface, later blasting off to rejoin the CSM in lunar orbit before the return journey to earth. The advantage of this scheme was that relatively little propellant would be needed, allowing the spacecraft to be smaller and lighter, and thus lifted by a less powerful launcher.

The scale of the effort required to put this plan into practice was awesome. While von Braun's Marshall Space Center co-ordinated the work of literally hundreds of contractors and subcontractors developing the Saturn series of launchers, Grumman took on building a lunar module and North American wrestled with the problems posed by the command module. NASA's additional facility, the 1961 Manned Spacecraft Center in Houston, Texas, today, the Johnson Space Center, took over as the nerve centre for America's space flights from 1965. By the mid-1960s NASA's budget had risen to over $5 billion a year and around half a million people in the United States were carrying

ILL-FATED CREW
The crew of the Apollo 1 mission, Gus Grissom, Ed White, and Roger Chaffee, were the first casualties of the American space programme. They died when fire broke out in their command module (left) during a routine ground test in January 1967.

out work related to the space programme, about 36,000 of them being employed by NASA.

Yet a poor contractor, North American, and a lack of attention by NASA to the problem of filling the spacecraft with pure oxygen making the interior atmosphere at normal pressure highly combustible led to corner-cutting and a lack of adequate attention to safety. Astronauts Gus Grissom, Ed White, and John Chaffee paid the price. On 27 January 1967, they were victims of a flash fire in the pressurized pure-oxygen atmosphere of their command module during a ground test in preparation for the first manned Apollo flight. A shocked NASA put a temporary stop to manned missions and embarked on a scrutiny of the Apollo spacecraft that led to over a thousand design changes being made.

Yet 1967 ended on a high for the space agency after the successful first test of the Saturn V launcher. This astonishing machine, the crowning glory of von Braun's career, consisted of three

SATURN TRANSPORTATION
Transporting the three stages of the giant Saturn V rocket, from the different factories, ultimately reaching their destination – Cape Canaveral (then Cape Kennedy), was an achievement in its own right. The second stage was 10m (33ft) in diameter and had to be carried on a special vehicle to a barge for transportation by sea.

SATURN V ASSEMBLY

The three stages of the Saturn V launcher were joined together in the Vehicle Assembly Building at the Cape. At 111m (363ft), the launcher was taller than a 40-storey building. The thrust of its stage one engines (right) was 100 times that of the Redstone used for the first US manned space flight.

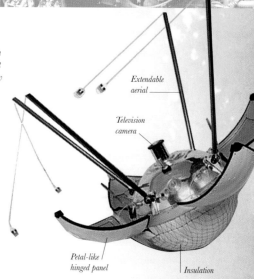

LUNA 9
The first successful landing on the surface of the Moon was made by the unmanned Luna 9 spacecraft in February 1966. It was an achievement that masked the growing problems besetting the Soviet space programme.

stages, the first powered by five F-1 engines using a propellant mix of kerosene and liquid oxygen, and the other two with J-2 engines using a mix of liquid oxygen and liquid hydrogen – the propellant originally proposed by rocket pioneer Tsiolkovsky.

The three stages were joined together in the vast Vehicle Assembly Building at the Cape, on top of a platform with a 115m (380ft) high support tower. The rocket, platform, and tower were then all carried on a massive "crawler" to the launch pad three miles away. The first launch, on 9 November 1967, astounded everyone who witnessed it. The 3.4 million kg (7.5 million lb) of thrust created a shock wave like a minor earthquake – astronaut Mike Collins said it felt "as if a giant had grabbed my shirtfront and started shaking." It was the heaviest object ever to fly, and the most powerful flying machine in history – and it worked.

As well as carrying on the Apollo Saturn missions, NASA was pursuing unmanned exploration of the Moon, partly as a preparation for a manned landing. Five Surveyor spacecraft set down on the Moon between May 1966 and January 1968, while five Lunar Orbiter spacecraft photographed almost the entire Moon surface between August 1966 and August 1967. These missions enabled NASA to identify sites where the lunar module would be able to land safely.

There was a prolonged interruption to manned flights while the problems in the command module revealed by the Apollo 1 fire were ironed

LUNAR SIMULATOR
Lunar Landing Research Vehicles (LLRV) were designed to simulate the descent of a lunar module onto the Moon. Like giant flying bedsteads, they could be piloted down from around 500m (1,500ft) to the ground using the thrust from a jet engine. In 1967 they were converted into trainers for the astronauts.

out. There were also delays in building the lunar module – not surprisingly as it contained around a million parts, each of which had to be as near perfection as humanly possible. And the second test launch of Saturn V, in April 1968, revealed serious glitches that fortunately were rapidly fixed. It was not until October 1968 that American astronauts returned to space, Apollo 7 going into Earth orbit with Walter Schirra, Donn Eisele, and Walt Cunningham in the command module.

Soviet developments

A constant fear for NASA at this time was, of course, that the Soviets might leap in and put a man on the Moon with a one-off shot while the Apollo programme was proceeding systematically towards its goal. In February 1966 the Soviet Union had landed the spacecraft Luna 9 on the Moon three months head of America's first Surveyor mission. It could be assumed that they would attempt similar coups to pre-empt the United States in the next two "firsts" – the first manned flight around the Moon and first manned Moon landing.

The Soviets were indeed planning for both types of mission, even though their space programme had long ceased to compare with that of the United States in scale and resources. The new unmanned Soyuz spacecraft, first put into orbit in 1966, was to be the basis for a circumlunar flight vehicle, the Zond, which would be launched by a three-stage Proton rocket. Soyuz was also to be the basis for a spacecraft to ferry two cosmonauts into lunar orbit, from where one of them would descend in a lunar lander to the surface of the Moon. This mission would require a more powerful launcher, the four-stage N-1. The chief difference between the Soviet plan for a manned Moon landing and its American equivalent was that the cosmonaut

would have to perform a spacewalk to transfer back and forth between the main spacecraft and the lunar lander.

The Soviet government did not authorize a manned Moon programme until 1964, and having started later than the Americans the Soviets soon fell even further behind. Initial unmanned test flights of Soyuz revealed problems that should have been sorted out before a manned flight was attempted. They were not. In April 1967, 40-year-old cosmonaut Vladimir Komarov, who had survived the incredibly risky three-man spacesuit-free Voskhod mission in October 1964, was sent up into orbit on board the bug-ridden Soyuz. He was supposed to rendezvous and dock with another Soyuz craft, but the second launch was quickly abandoned as Komarov's problems mounted. After only a day in space he was ordered to return to Earth. Partially out of control, Komarov had great difficulty in achieving re-entry. When he did, the parachute on his spacecraft failed to deploy and he plunged into the ground at high speed. Komarov was the first person to die carrying out a spaceflight.

Race to orbit the Moon

After Komarov's death the Soviets carried out no further manned spaceflights for 18 months – a break that roughly paralleled the interruption of American manned missions after the Apollo 1 fire. A cosmonaut returned to space a week after the Apollo 7 mission, in October 1968, a manned Soyuz craft manoeuvring close to an unmanned craft in Earth orbit. At the same time the powerful N-1 booster – the Soviet equivalent of Saturn V – was being readied for a test launch. More worrying for NASA, in September 1968, an unmanned Zond spacecraft, Zond 5, was sent around the Moon and brought back to Earth. Cosmonauts Alexei Leonov and Oleg Makarov began training for a circumlunar flight in the hopes that the Soviets could claim another first. With the loss of Komarov, the fact that the N-1 rocket had never tested successfully, and the return of Zond 5 at such a high trajectory, which would have killed any crew, the Soviets were more cautious. Another unmanned flight, Zond 6, was staged in November to improve the chances of bringing the cosmonauts back alive. As the Soviets hesitated, the Americans jumped in.

NASA had planned to follow Apollo 7 with another mission in Earth orbit, to test the lunar module and docking procedures. Apollo 9 would then go into lunar orbit, carrying out all the procedures for the ultimate Moon mission short of an actual Moon landing. But the desire not to be upstaged by the Soviet Union led to a change of plan. Because the lunar module for Apollo 8 was unavailable, NASA would proceed directly to a manned flight into lunar orbit.

Frank Borman, Jim Lovell, and William Anders blasted off on board Apollo 8 on 21 December

HEAVENLY PHOTOS
This breathtaking photograph of the Earth seen from lunar orbit, with the surface of the Moon in the foreground, was taken by astronaut Bill Anders from the Apollo 8 module on Christmas Eve 1968. The photograph below, showing Apollo 8 reentering the Earth's atmosphere like a meteor, was taken from a tracking aircraft.

1968. They were the first crew to experience launch by the mighty Saturn V – Apollo 7 had been lifted into orbit by a Saturn 1B. Shaken and stirred by the power of the booster, within 11 minutes they had reached Earth orbit. Three hours later the Saturn's third-stage J-2 engine reignited to send them off into space. The first humans to go beyond Earth orbit watched their home planet dwindle behind them through the three-day 400,000km (250,000-mile) voyage to the Moon.

The course they took through space was accurate, and communications with the ground remained good until the spacecraft disappeared behind the Moon. There a blast from the module's main rocket brought the craft into lunar orbit, where it stayed for

the following 20 hours. The event was capped by a media spectacular. Millions of viewers watched on television as black-and-white images of the astronauts and of the surface of the Moon passing below their spacecraft were broadcast live to the world. They heard the crew give their impressions of what Lovell described as "the vast loneliness" of space. As it was Christmas Eve, the astronauts all read from the Book of Genesis, and Borman wished a Happy Christmas to "all of you on the good Earth". Apollo 8 then returned safely to splash down on 27 December, bringing with it what became one of the most famous photographs

> "The earth from here is a grand oasis in the big vastness of space."
>
> JIM LOVELL
> ASTRONAUT ON APOLLO 8 SPEAKING LIVE ON TELEVISION, 24 DECEMBER 1968

APOLLO 9 ORBIT
The Apollo 9 lunar module floats in Earth orbit, photographed from the nearby command module. The main purpose of the Apollo 9 mission was to try out the operation of the lunar module in space. The module has its landing gear deployed as it would when approaching the Moon.

in history – the blue ball of the Earth rising above the bleak horizon of the Moon, set against the darkness of space. This image is widely credited with helping to create a new vision of the Earth as a beautiful but fragile planet, whose capacity to support life amid the inanimate vastness of space was a miracle that humankind would do well to treat with much greater respect.

SATURN V LIFTOFF
A frame-by frame record of the launch of Apollo 11 gives a vivid impression of Saturn's power. The first stage engines used up 150,000 litres (40,000 US gallons) of propellant a minute.

One-horse race

After the triumph of Apollo 8, there was only one contestant left in the Moon race. A Soviet circumlunar flight would now be pointless, while the chances of a cosmonaut being first to set foot on the Moon shrank to zero as test launches of the N-1 booster repeatedly failed. The Soviets continued with some spectacular Soyuz space rendezvous and docking exercises in Earth orbit, while beginning to hint that they had never really wanted to go to the Moon at all.

Bursting with optimism, NASA now had a manned Moon landing in its sights. In March 1969, the Apollo 9 mission, crewed by James McDivitt, David Scott, and Russell Schweickart, successfully put the lunar module through its paces in Earth orbit. The following May, Apollo 10 repeated these exercises in lunar orbit. Mission commander Thomas Stafford and pilot Eugene Cernan flew the lunar module to an altitude of 15,000m (50,000ft) above the surface of the Moon while their colleague John Young awaited their return on board the orbiting command module. Once again, the trip to the Moon and back was accomplished without drama.

ONE SMALL STEP
Ever aware of the importance of television coverage, NASA had arranged for a camera to film Neil Armstrong emerging from the Eagle landing module and climbing down the ladder onto the Moon's surface. Under the stress of the occasion Armstrong slightly fluffed the lines that he had been mentally rehearsing, coming out with: "That's one small step for man, one giant leap for mankind".

Big moment approaches

All was now ready for a manned landing on the Moon. Apollo 11 would be the mission to achieve the goal that President Kennedy had laid down, with more than five months to spare before the end-of-decade deadline. The crew would consist of Neil Armstrong, the mission commander; Edwin "Buzz" Aldrin; and Michael Collins. Apollo crews were assigned on a strict rotation basis to minimize possible conflict between a notoriously competitive set of individuals. That Armstrong's team landed the mission was therefore pure chance, not a reflection of any

LUNAR MODULE
This replica shows the layout of the interior of the lunar module in which Neil Armstrong and Buzz Aldrin descended to the Moon. They stood side by side, Armstrong on the left and Aldrin on the right, each with a restricted view through a triangular window.

special distinction that was supposed to mark them out from the other astronauts. Collins was there because a spinal problem had forced him to withdraw from the Apollo 8 mission, swapping with Jim Lovell who was originally slated for Apollo 11.

Armstrong, Aldrin, and Collins lifted off from launch pad 39A at Cape Kennedy (Canaveral) on 16 July 1969. During their three-day voyage into lunar orbit inside the command module Columbia, they had plenty of time to take in the astonishing visual spectacle. For Armstrong, the most dramatic moment was passing through the lunar shadow, with the Sun eclipsed and the Moon lit blue-grey by reflected light from the Earth. As they drew near

to the Moon, Collins was struck most vividly by its barrenness – he described it as a "monotonous rock pile" starkly contrasted to the "verdant valleys" of the planet that the astronauts had left behind.

The Moon landing took place on 20 July, though from the point of view of Europe, Africa, and Asia, Armstrong did not walk on the Moon until 21 July. The landing was not without incident. Armstrong and Aldrin climbed into the lunar module *Eagle* and, leaving Collins to his lonely vigil, separated from the command module, which was orbiting 100km (60 miles) above the Moon's surface. They then dropped to an orbit of 13km (8 miles) altitude. The final powered descent from orbit to the surface of the Moon was the tensest phase of the entire mission. Each move was scrupulously monitored by ground controllers in Houston, ready to abort the landing at any moment if it

"Here men from the planet Earth first set foot upon the Moon, July 1969 A.D. We came in peace for all mankind."

INSCRIPTION FROM THE PLAQUE LEFT BY THE APOLLO 11 CREW ON THE MOON

MEN ON THE MOON
Edwin "Buzz" Aldrin walks on the Moon in July 1969. Reflected in his visor – which was goldplated to protect the astronaut's eyes against the Sun – you can see the photographer, mission commander Neil Armstrong, and one leg of the Eagle lunar module. Beyond stretches the dusty plain of the Sea of Tranquillity, marked for the first time by human footprints.

APOLLO 11 CREW

APOLLO 11 MISSION COMMANDER
Neil Armstrong (b. 1930) grew up in Wapakoneta, Ohio, and learned to fly at the age of 15. He fought in the Korean War, flying off the carrier USS *Essex*, and went on to become one of America's top test pilots at Edwards Air Force Base, pushing the X-15 to almost 6,500kph (4,000mph). Armstrong transferred to the NASA space programme in 1962 from the USAF's Dyna-Soar project for winged flight beyond the atmosphere. He had been into space once before, surviving the uncomfortable Gemini 8 mission in 1966.

Buzz Aldrin (b. 1930), the mission's lunar landing module pilot, came from a military family, and his father had flown with the likes of Charles Lindbergh and Jimmy Doolittle. Like Armstrong he was a Korean War veteran, but instead of going on to become a test pilot he took a doctorate at the Massachusets Institute of Technology (MIT), for which he wrote a thesis on orbital rendezvous. Aldrin was regarded as an especially competitive man even by his colleagues – people not noted for self-effacement. He lobbied hard to be the first man to exit the landing module. Fellow Apollo 11 astronaut Michael Collins later wrote of Aldrin that he

"resents not being the first man on the Moon more than he appreciates being second."

Mike Collins (b. 1930) was another scion of a military family – his father was an army general and his brother was a colonel. An Edwards AFB test pilot, he made a successful application to become a NASA astronaut in 1962. He seems to have accepted with equanimity his role as the sole crew member not to set foot on the Moon.

MOON TRAVELLERS
The Apollo 11 crew pose for an official photo, from left to right, mission commander Neil Armstrong, command module pilot Mike Collins, and lunar module pilot Buzz Aldrin.

seemed to be going awry. The astronauts had enough propellant to fuel the descent engine for 12 minutes. Despite worries about overload to the onboard computer which was flying the module, the descent continued until Armstrong could see where they were going to land – in a boulder-strewn area at the edge of a crater. Taking manual control, he set off in search of a gentler landing place. The module was about 20 seconds away from running out of propellant when Armstrong brought it down, through a duststorm whipped up by the engine, coming to rest on the Sea of Tranquillity. Armstrong then spoke the famous words: "Houston, the *Eagle* has landed",

before he and Aldrin turned to one another and shook hands.

It was another six hours before Armstrong opened the hatch and climbed down the ladder on to the Moon. He had planned to say that it was "one small step for a man, but one giant leap for mankind". Instead, he forgot the indefinite article, calling it "a small step for man", which made the sentence meaningless. Not that it mattered, for here was a human being standing on the Moon, witnessed by an estimated 600 million viewers on television.

Aldrin joined Armstrong in a two-and-a-half-hour moonwalk. They spent some time setting up

RENDEZVOUS IN LUNAR ORBIT
With a half-Earth in the background, the lunar module ascent stage with Moon-walking astronauts Neil Armstrong and Buzz Aldrin aboard approaches for a rendezvous with the Apollo 11 command module Columbia *(below) manned by Michael Collins.*

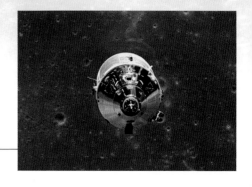

scientific instruments and collecting rock samples, but there was no serious pretence that the mission was about anything other than getting there and being there. The astronauts proudly planted the Stars and Stripes, offsetting this patriotic gesture with a plaque, attached to the leg of the lunar lander, expressing the noble sentiment: "We came in peace for all mankind". Like any tourists that have arrived somewhere interesting, they took photographs of one another, and they also phoned home – talking to President Richard Nixon through a special telephone link.

Having returned to the lunar module, the astronauts hooked up to the spacecraft's oxygen supply before shedding their backpacks and overshoes. The following day, after a sleep period, the launch back to the command module would take place. Igniting the engine on the module's ascent stage, they blasted off; the bottom half of the module – the descent stage – was left behind. The ascent stage then docked with the command module still in lunar orbit, reuniting Armstrong and Aldrin with Collins. The return journey to Earth, re-entry to the atmosphere, and splashdown followed in textbook fashion. Apollo 11 landed in the Pacific off Hawaii 8 days, 3 hours and 9 minutes after the launch. The splashdown was some ten seconds behind schedule.

Fears had been expressed before the mission that the astronauts might bring back with them Moon viruses or bacteria that would pose a threat to life on Earth, which would have no natural resistance to such alien infections. Consequently the Apollo 11 crew were transported to Houston in a sealed container, known as the Mobile Quarantine Facility. The rock samples they had collected were also quarantined. The astronauts were finally declared clear of contamination on 10 August, and catapulted from confinement into a blaze of public exposure, including a traditional Broadway tickertape parade and an exhausting celebratory tour of 25 countries in 35 days.

Keeping the momentum

NASA had no intention of resting on its laurels. There were enough Saturn V boosters and Apollo spacecraft for nine more missions. Beyond that, there were visions of a permanent Moon base, a space station, and manned landings on Mars.

Yet masked by the temporary euphoria, difficulties were already mounting. Many American politicians were inclined to see the Moon landing as a culminating triumph, rather than a first step on the way to further projects. There was pressure to bring NASA's budget down as the American economy faltered and national finances suffered from the rising cost of the Vietnam War. The optimistic vision that had underpinned the space programme through the

1960s was becoming blurred in a growing national mood of doubt and self-criticism.

It did not help NASA's cause that the material brought back from the Moon proved short on the sort of excitement that makes headlines outside the scientific press. Certainly, the public queued round the block to see Moon rocks displayed at the Smithsonian Institution in Washington D.C. Equally certainly, geologists and astronomers found much to ponder and enlighten in the study of these rocks. But only traces of life could have supplied the drama that NASA desperately needed to hold public attention.

The first follow-up mission to the Moon, by Apollo 12 in November 1969, was another

TRIUMPHANT HOME-COMING
After Apollo 11's punctual splashdown in the Pacific (left), Armstrong, Aldrin, and Collins were quarantined inside a sealed container, cramped quarters in which they received a visit from President Nixon (below). Their tickertape reception on a motorcade down Broadway (bottom) was one of the most enthusiastic on record.

APOLLO 13

APOLLO 13 was almost 56 hours into its journey to the Moon and some 330,000km (205,000 miles) from Earth when, on the evening of 13 April 1970, the crew heard a sudden bang. An explosion in an oxygen tank had knocked out their command and service module's electricity supply and their remaining oxygen was leaking into space. For mission commander James Lovell, Fred Haise, and Jack Swigert, the chances of survival looked slim. Swigert must have been tempted to think himself one of the world's unluckiest men, having at the last moment taken the place of Ken Mattingly, believed to be at risk of German measles.

The crew's one hope of returning to Earth was the undamaged condition of the lunar module, which had its own oxygen, water, power systems, and engine. By transferring to the lunar module they could survive. Mission control at Houston worked out a flight plan to bring the astronauts to safety by swinging around the Moon onto a course back to Earth. But this would take four days and the power supply on board the lunar module was only good for two.

Mission controllers and the crew working together found ways of closing down equipment

ORIGINAL CREW
The original Apollo 13 crew were, left to right, Jim Lovell, Ken Mattingly and Fred Haise. Jack Swigert replaced Mattingly two days before the launch.

to save electricity and, in a feat of improvisation, rigged up a system for extracting the carbon dioxide the crew were exhaling from the module's unventilated atmosphere. The small cabin soon became cold and damp, temperatures falling close to freezing. Unable to sleep, with no hot food and short of water, the three astronauts clung on. Meanwhile, on Earth their progress was followed by hundreds of millions – prayers were said, and hushed groups clustered in front of televisions in shop windows.

Early on the afternoon of 17 April, Apollo 13 splashed down in the Pacific just 6 km (4 miles) from its recovery ship. Shown live on television, this event may have been watched by more people than the first Moon landing. *Newsweek* described it as "the most amazing rescue operation of all time", and few would disagree.

MISSION CONTROL
The Houston Manned Spacecraft Center acted as the mission control center throughout the Apollo programme. During the Apollo 13 mission it was a scene of great tension as the drama unfolded. Instructions from controllers helped find the solutions the astronauts needed for survival.

STRICKEN SERVICE MODULE
The Apollo 13 service module floats dead in space, with a panel blown off its side by an explosion, exposing battery fuel cells, tanks, and other components. This was the view of the service module that the astronauts had as they prepared to re-enter the Earth's atmosphere.

staggering display of technological prowess, with Pete Conrad and Alan Bean landing within a few hundred metres of their target, the remains of the Surveyor 3 spacecraft. But their mission was marred by the misfunction of the camera that was to broadcast live pictures of their moonwalk. For most of the world, without the television pictures it was hard to believe or even imagine.

Ironically, it took a failed mission to raise public interest back to fever pitch. The only manned Moon shot that did not reach its goal, Apollo 13 became a drama of survival against the odds that touched the human heart in a way no perfectly executed feat of technology could (see panel, left).

The final missions

Despite possessing the hardware for more Moon landings, NASA was forced by ever-tightening budgetary constraints to limit the Apollo programme to four more shots after the "successful failure" of Apollo 13. The emphasis was on scientific investigation, although only one scientist went to the Moon, geologist Harrison Schmitt, who accompanied Gene Cernan in the lunar module on the Apollo 17 mission.

Improvements in the techniques of lunar exploration continued to be made. When Apollo 14 went to the Moon in February 1971, Mercury 7 veteran Alan Shepard and his lunar module pilot Ed Mitchell were provided with the first lunar vehicle, a two-wheeled cart – rather pompously designated as a "modularized equipment transporter" – to carry equipment and samples. Apollo 15, in August 1971, brought the debut of the battery-powered Lunar Roving Vehicle, which extended the area of exploration on the Moon to a radius of 10km (6 miles) around the landing site.

Also equipped with improved pressurized suits, the crews of the last three Apollo missions were able to get through a great deal of work on the Moon. The total moonwalk time rose on each mission, reaching over 22 hours on Apollo 17. This was far from easy for the astronauts, operating in Moon gravity, which is one-sixth that of the Earth. Harrison Schmitt described how the simple operation of climbing into the Lunar Rover became a complex piece of acrobatics, involving a two-footed jump with a sideways twist, after which you "wait until you slowly settle into the seat, ideally in the correct one". It was no small feat to collect, as Schmitt and Cernan did, about 115kg (250lb) of rock samples when simply tipping a scoop of material into a bag required total concentration.

Apollo 17 was the last manned Moon mission. After three days on the Moon's surface, Schmitt and Cernan departed on 14 December 1972. They left behind a plaque that drew a line under

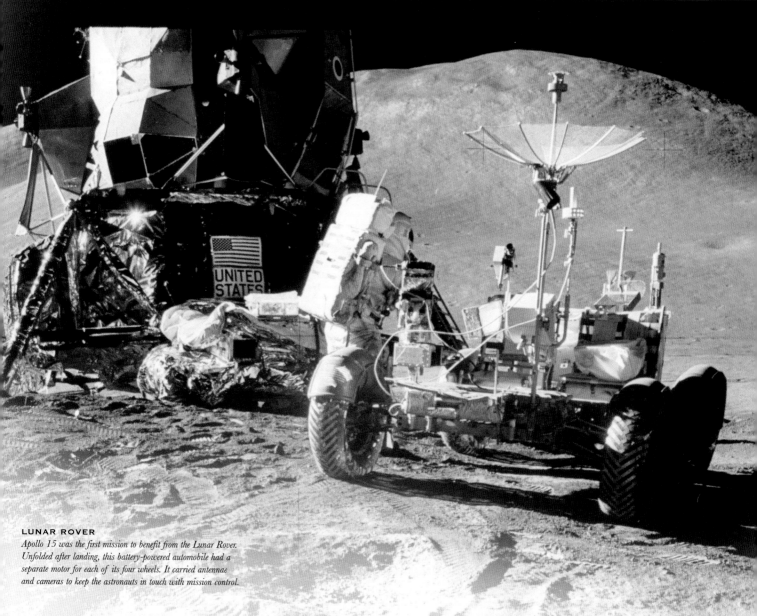

LUNAR ROVER

Apollo 15 was the first mission to benefit from the Lunar Rover. Unfolded after landing, this battery-powered automobile had a separate motor for each of its four wheels. It carried antennae and cameras to keep the astronauts in touch with mission control.

the Apollo programme. It read: "Here man completed his first exploration of the moon... May the spirit of peace in which we came be reflected in the lives of all mankind."

As the Apollo programme drew to a close, with waning public interest, there was an inevitable feeling of anticlimax. Many Americans at the time were inclined to be critical of the

amount of money and effort that had been expended on a project with so little relevance to the pressing problems confronting life on Earth.

Yet the achievements of the Apollo project were incontravertible. Twelve men had walked on the Moon. They had brought back to Earth a total of 379kg (842lb) of lunar rock for geologists to study. Each mission had left scientific

instruments on the Moon that contributed to our understanding of the Moon's structure and its history. Moreover, the Apollo programme had led to valuable technological progress in many areas.

And it had been more than simply a great feat of technological innovation and organization. It had been an heroic venture, an affirmation of the human will, always to go further, always beyond.

FAREWELL TO THE MOON

Filmed by a television camera left behind on the Lunar Rover, the ascent stage of the lunar module Challenger lifts off from its descent platform, carrying astronauts Gene Cernan and Harrison Schmitt to rendezvous with the Apollo 17 command module on 14 December 1972. After that day, no humans set foot on the Moon's surface for the rest of the twentieth century.

SHUTTLE TO SPACE

AFTER THE ENDING OF THE APOLLO PROGRAMME, THE FOCUS OF MANNED SPACEFLIGHT SHIFTED TO REUSABLE SPACECRAFT AND SPACE STATIONS

> "Aiming at the stars ... is a problem to occupy generations, so that no matter how much progress one makes, there is always the thrill of just beginning."
>
> ROBERT GODDARD
> LETTER TO H.G. WELLS, 1932

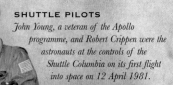

SHUTTLE PILOTS
John Young, a veteran of the Apollo programme, and Robert Crippen were the astronauts at the controls of the Shuttle Columbia on its first flight into space on 12 April 1981.

IN THE 1970s manned space flight programmes had to adapt to a rapidly changing situation that threatened their survival. The political heat had gone out of the competition between the superpowers which had energized the space race. Caught up in post-Vietnam and Watergate traumas, the United States was in one of its least bullish and assertive national moods. The world economy was also in crisis after decades of rapid expansion. The new demand was for a focus on cost-effectiveness and practical applications of space technology. In this context, the growing everyday usefulness of satellites and the spectacular triumphs of unmanned space probes inevitably led to a questioning of the need for manned flights at all.

Practical progress

In this less supportive context, space agencies continued with manned programmes that were far less intensive in the short term, but still based on an ambitious long-term vision of a developing human presence in space. The logic of the development of a reusable space vehicle seemed overwhelming in the light of the expected cost advantage it should bring, and its practicality – it was hard to imagine space travel ever growing into a phenomenon on a substantial scale if every return to Earth involved being picked up from the sea. Not surprisingly, then, the focus of the American manned space programme became the Shuttle. Meanwhile the Soviets took the lead in the development of space stations and exploration of the effects of long-duration space flights.

The idea of permanently inhabited space stations in orbit around the Earth had occurred to some of the earliest space-flight visionaries. In 1923 German rocket enthusiast Hermann Oberth wrote of an orbiting "observing station" from which astronauts with "powerful instruments" would be able to see "fine detail on Earth" and give warning of icebergs on the oceans – thus averting any repetition of the Titanic disaster. Oberth imagined this information being flashed to Earth in code by reflecting sunlight off a hand mirror.

As a disciple of Oberth, Wernher von Braun picked up an early enthusiasm for space stations that endured throughout his life. He drew up a blueprint for one in the 1950s and promoted the idea as a future follow-on to the Moon programme during the second half of the 1960s. But it was the Soviets who first put a functioning space station in orbit, and who were to make space stations their most distinctive contribution to late twentieth-century space exploration.

By 1969, with the goal of being first to the Moon beyond their grasp, and no desire

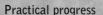

HUBBLE TELESCOPE
The Hubble Space Telescope (HST) was launched in April 1990. It is capable of observing objects far more clearly than most of Earth's largest optical telescopes. The HST was launched and has been repaired by Shuttle astronauts.

BOOSTED TAKEOFF
The Shuttle's two solid-fuel rocket boosters, strapped on to the massive external fuel tank, deliver 11.8 million newtons (2.6 million lb) thrust each to help lift the craft off its launch pad. The Shuttle was designed for an operational life of 100 missions.

DOOMED CREW
Vladislav Volkov, Viktor Patsayev, and Georgi Dobrovolsky, the crew of Soyuz 11, became the first men to die outside the Earth's atmosphere when the air leaked from their descent module.

to be second, the Soviets were turning hardware originally intended for manned lunar flight to another purpose. Their Soyuz spacecraft carried out a series of manoeuvres and dockings in space, with at one time three of the craft in orbit at once. In June 1970 the crew of Soyuz 9 set a new record of 18 days in orbit. This was the prelude to the launch of the first space station, Salyut 1, on 19 April 1971.

Troubled debut

Salyut 1 was designed to accommodate three cosmonauts, who would have a range of scientific and observation instruments to keep them usefully employed on board. As it turned out, no one was ever to recount their experience of this space novelty. The first crew sent up, on Soyuz 10, docked with Salyut 1 but could not get inside it. The next crew, on Soyuz 11, spent three weeks on board Salyut 1 but died during their return to Earth after depressurization of the descent module.

This tragedy was followed by other setbacks for the Salyut programme which meant that it was not until June 1974 that a Soviet crew successfully travelled to an orbiting station (Salyut 3), spent some time on board, and returned to Earth. By then, the United States had carried out its own first space station experiment.

Looking beyond the Moon landings, leaders of the US

manned space programme had always seen space stations as a part of their future plans, along with a reusable space vehicle and the more distant goal of bases on the Moon and Mars. The barrier was cost. Although NASA presented space stations and a reusable space vehicle as interdependent parts of the same package, budgetary constraints dictated that they had to choose between them. As a result, Skylab, launched on 14 May 1973, was a one-off, a relatively cheap orbiting station using hardware adapted from the Moon flights. Despite the undoubted success of Skylab, there was no follow-on to a sustained US space station programme.

The last US astronauts left Skylab in February 1974. Only one American crew went into space in the next seven years. They were on board an

Apollo launched for the Apollo-Soyuz Test Project, a highly publicized symbolic gesture of international co-operation between Cold War adversaries engaged in a policy of detente. The Apollo and Soyuz spacecraft rendezvoused in space on 17 July 1975. Soviet and American spacemen exchanged gifts and friendly phrases, spent two days working together, and went back to Earth.

From then until the spring of 1981, Soviet cosmonauts had space to themselves. A succession of military and civilian Salyuts were sent up into orbit in a programme that stubbornly overcame difficulties and setbacks. The early Salyuts were crude, uncomfortable, and shortlived. Even the most successful, Salyut 4, which was inhabited by two crews for sustained periods, was reportedly cold and damp. However, Salyut 6, launched in September 1977, marked a significant step forward. Refuelled by unmanned space tankers, it was able repeatedly to use its rocket engine to counter orbital decay, staying in orbit until 1982. Its successor, Salyut 7, lasted from 1982 to 1991. Ferried to these space stations by rapidly improving Soyuz craft, cosmonauts broke space endurance records time and again – Valery Ryumin and Leonid Popov stayed aloft for 185

APOLLO-SOYUZ LINK-UP
These badges celebrate the Apollo-Soyuz space rendezvous in 1975, an unprecedented gesture of co-operation between the USA and the USSR.

APOLLO-SOYUZ TEST PROJECT (ASTP)
America's Apollo command and service module (replica shown) carried the docking adaptor to connect with the Soviet Soyuz 19 spacecraft. The space rendezvous was the last Apollo mission.

> "All the inexperienced cosmonauts agitate for longer space flights, but the veterans of long-term flights don't."
>
> OLEG GAZENKO
> DIRECTOR OF THE INSTITUTE FOR BIOMEDICAL PROBLEMS

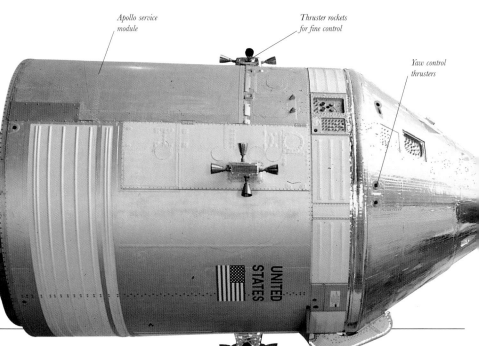

Apollo service module

Thruster rockets for fine control

Yaw control thrusters

Engine nozzle

UNITED STATES

THE SKYLAB

SKYLAB WAS THE FIRST AMERICAN space station. Coming at a time when NASA was under pressure over its budget, the emphasis was on economy and practicality. Creating a space station by adapting the third stage of a Saturn 5 rocket was much cheaper than starting from scratch, and to emphasize the practicality of the project it was dubbed an Orbital Workshop. There were to be medical tests of the effects of weightlessness, observation of the Sun and stars, photography of the Earth's surface, and experiments concerning possible industrial activities in space.

Skylab was launched into orbit on 14 May 1973, the crew following 11 days later in an Apollo module. The project began with a crisis. An accident during the ascent left Skylab with no thermal shield and only one of its two solar panels operational. The crew – Charles Conrad, Paul Weitz, and Joseph Kerwin – were fortunately able to repair the damage to the rapidly overheating craft by passing a sunshade through an airlock.

The initial crew stayed on Skylab for 28 days, but two subsequent three-man crews stayed for 59 days and 84 days respectively. Skylab was about as spacious as a small house, with three sleeping areas, a kitchen and eating space, a bathroom with a gravity-free air-suction toilet, and a shower – which had to be an enclosed container, to stop the water floating away. The astronauts were committed to an intensive programme of experiments and medical tests. Some of the effects of weightlessness were strange – for example, the Skylab crew members grew in height by about an inch during their stay on board. Conrad quipped: "At last I'm taller than my wife." Constant exercise maintained muscle tone, which would otherwise have tended to deteriorate. Although they complained about relentless work schedules and tasteless food, the astronauts had the sensational ever-changing view of the Earth to cheer their spirits. In this they were somewhat fortunate, as Saturn engineers had originally intended not to put in a window.

SKYLAB IN ORBIT

Skylab, photographed in orbit from the Apollo command and service module, was created by modifying the third stage of a Saturn 5 rocket. This provided quite roomy accommodation for three astronauts.

SKYLAB BADGE

The Skylab project was rated as a major success. The spacecraft remained in orbit for six years, although astronauts were aboard for only a total of 171 days.

INSIDE SKYLAB

Skylab had a kitchen and dining area, and facilities for medical tests. Astronauts found that prolonged weightlessness had little adverse effect on health, apart from an embarrassing excess of gas after eating.

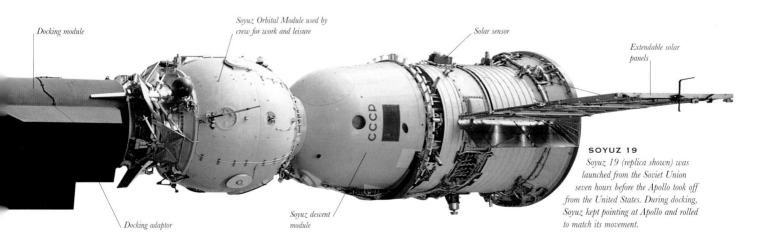

Docking module

Soyuz Orbital Module used by crew for work and leisure

Solar sensor

Extendable solar panels

Docking adaptor

Soyuz descent module

SOYUZ 19

Soyuz 19 (replica shown) was launched from the Soviet Union seven hours before the Apollo took off from the United States. During docking, Soyuz kept pointing at Apollo and rolled to match its movement.

days in 1980. Conditions on Salyut 6 and 7 were far better than they had been on Salyut 4, but life in space was still a test of endurance rather than an enjoyable adventure – the cosmonauts' accounts of their long-duration flights are full of nostalgia for Earth. Still, the Soviet Union had unquestionably succeeded in establishing a potentially permanent human presence in space.

Wings into space

The American Space Shuttle programme could be said to have started with a step backwards, to rejoin a line of development that had been elbowed aside by the drive to reach the Moon. Von Braun had always had a reusable winged space vehicle in mind as central to the opening up of space. Also, the USAF had confidently pursued the notion of an aircraft that would fly out of the atmosphere and back in again, before shelving such plans in the early 1960s. The air force was involved with NASA in developing what was soon dubbed the Space Shuttle Orbiter from 1969. In the new atmosphere of belt-tightening and political questioning of technological extravaganzas, NASA sold the Shuttle idea to the politicians by stressing its role as an economical and trusty workhorse. The project was given the go-ahead in 1972, but NASA was left in no doubt that the financial aspect of the programme would be under close scrutiny.

In its original concept the Shuttle was to have been fully reusable. One proposal, from North American Rockwell, was for a piloted winged launch vehicle that would carry the orbiter into the upper atmosphere and release it at 1,200kph (7,400mph) before turning back to land. This combination of two winged vehicles would have been economical to operate, but development costs promised to be astronomical. NASA was forced to fall back on the use of a jettisonable external fuel tank – to keep the orbiter small – and reusable solid-propellant rocket boosters.

The first Shuttle, named *Enterprise* in homage to the spaceship in Star Trek, made its maiden experimental unpowered flight in 1977, lifted into the air by a Boeing 747 and gliding down to land

POWERFUL COMBINATION
The Shuttle orbiter was built by Rockwell International, the huge external liquid-propellant tank was made by Martin Marietta, and the solid-fuel rocket boosters were produced by Thiokol. Total thrust from the orbiter's three main engines and the boosters is 30.4 million newtons (6.9 million lb).

CHALLENGER

CONCERNED WITH drumming up public support for the Shuttle programme, NASA announced in 1985 that a teacher would be selected for a space flight. More than 11,000 applications were received from teachers across America. Christa McAuliffe of Concord, New Hampshire, won the competition amid a blaze of publicity that was a triumph for the space agency's public relations.

McAuliffe joined the crew for a flight on board *Challenger* scheduled for January 1986, along with mission commander Dick Scobee,

CHALLENGER CREW

The crew of the Shuttle Challenger pose cheerfully for the camera before their ill-fated flight. High school teacher Christa McAuliffe is second from the left in the back row.

TRAGIC IMAGE

Fragments of Challenger *drag smoke trails across the blue sky over Cape Canaveral after the explosion of the Shuttle on 28 January 1986.*

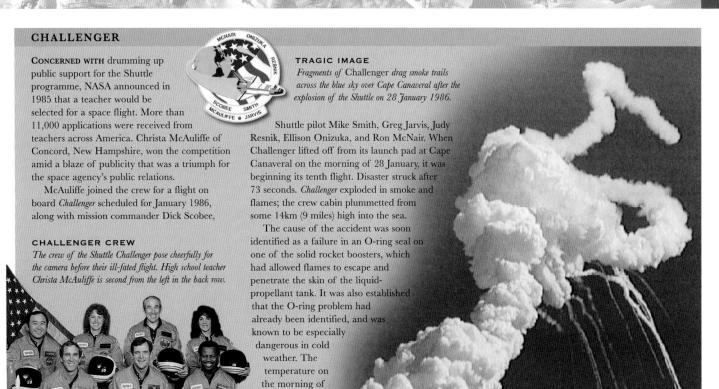

Shuttle pilot Mike Smith, Greg Jarvis, Judy Resnik, Ellison Onizuka, and Ron McNair. When Challenger lifted off from its launch pad at Cape Canaveral on the morning of 28 January, it was beginning its tenth flight. Disaster struck after 73 seconds. *Challenger* exploded in smoke and flames; the crew cabin plummetted from some 14km (9 miles) high into the sea.

The cause of the accident was soon identified as a failure in an O-ring seal on one of the solid rocket boosters, which had allowed flames to escape and penetrate the skin of the liquid-propellant tank. It was also established that the O-ring problem had already been identified, and was known to be especially dangerous in cold weather. The temperature on the morning of 28 January was below freezing.

at Edwards Air Force Base. The first Shuttle flight into space had originally been projected for 1979, but problems, especially with the main rocket engines, produced delays and helped push the programme $1 billion over budget. Nonetheless, on 12 April 1981, exactly 20 years after Gagarin's epoch-making space flight, the Shuttle *Columbia* lifted off from Cape Canaveral with John Young and Robert Crippen at the controls. Despite a nerve-racking moment when the astronauts discovered that some of the heat-protective tiles that would prevent the vehicle burning up on re-entry had shaken loose, *Columbia* returned to the Earth's atmosphere without mishap and

smoothly glided to land at Edwards AFB. Further test flights confirmed the success of the Shuttle, which went operational in November 1982.

Technologically, the Shuttle was one of the wonders of the world. But as a programme that had promised to make space flight routine and, in the words of President Richard Nixon, to "take the astronomical costs out of astronautics", it was soon in serious trouble. The Shuttle, it turned out, was neither cheap nor easy to operate. NASA had worked out costings on the basis of 24 Shuttle flights a year. Yet even with engineers and technicians working overtime, it proved difficult to achieve

a launch every two months. The Shuttle was supposed to cover its costs by luring paying customers who needed satellites launched or other space services. But even charging only a fraction of the real cost of a launch, the Shuttle was undercut by the European Space Agency's Ariane rockets. Even the USAF deserted the Shuttle as a launcher for reconnaissance satellites, after *Challenger*, reverting to disposable launch rockets.

NASA continued to promote the Shuttle as above all a safe means of travel – a spaceliner usable by anyone. For the first time, people who were not pilots voyaged into space, first scientists or engineers who conducted experiments on board, and then selected passengers whose presence might boost the programme. One of these, teacher Christa McAuliffe, was on board when *Challenger* exploded on 28 January 1986. Further Shuttle missions were postponed as an investigation began into the causes of the accident. It was NASA's darkest hour.

SHUTTLE PIGGYBACK

A Boeing 747 was adapted to transport the spacecraft from its landing strip at Edwards Air Force Base, California, to its launch point at Cape Canaveral, Florida. Initial flight tests of the Shuttle's gliding ability were also carried out by air launch from the back of the 747.

Space Shuttle Orbiter

THE SHUTTLE IS A SPACE VEHICLE shaped by compromises enforced by budgetary constraints. NASA chose the basic configuration in 1972 as above all affordable to develop. The orbiter would be a delta-winged craft capable of gliding to land after re-entering the atmosphere. For takeoff there would be an external jettisonable liquid-propellant tank to fuel the main engines and two solid-fuel boosters strapped on to the tank to increase thrust. This somewhat ungainly arrangement slashed development costs, but produced a vehicle far removed from dreams of an aircraft that would be flown on wings into space and back.

As well as the three main engines, which proved a formidable technical challenge, the Shuttle was designed with two additional manoeuvring engines and 44 thrusters. To protect it from temperatures that would reach 1,659°C (3,000°F) on reentering the atmosphere at around 30,000kph (18,000mph), the orbiter's skin was covered with around 30,000 silica tiles. It took one worker three weeks to glue on four of these tiles.

START OF A MISSION
The Space Shuttle Columbia *is slowly rolled out of its hangar to the launch pad where it will await the countdown. A typical Shuttle mission usually lasts about a week – this one was a service mission to the Hubble Space Telescope.*

> "It's an exciting and dynamic moment. The whole vehicle starts to shake and rattle so much you can barely read the instruments."

CREW-MEMBER JAMES VOSS
REFLECTIONS ON THE LAUNCH AT LIFT-OFF

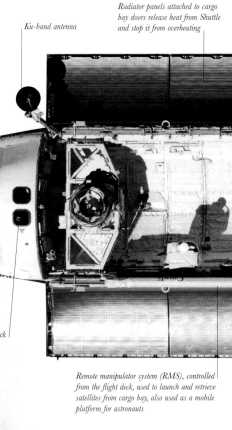

Ku-band antenna

Radiator panels attached to cargo bay doors release heat from Shuttle and stop it from overheating

Control thrusters situated in nose (and at rear of Shuttle) are used to change position in space

Two-storey cabin, with upper flight deck and lower mid-deck crew quarters

Flight deck windows

Rudder and speed brake

Remote manipulator system (RMS), controlled from the flight deck, used to launch and retrieve satellites from cargo bay, also used as a mobile platform for astronauts

COLUMBIA MAKES FIRST SHUTTLE FLIGHT
On 12 April 1981, the Space Shuttle Columbia *made its maiden flight, lasting just over 54 hours. At the end of its twenty-eighth mission, on 1 February 2003,* Columbia *was lost during re-entry, killing its seven-person crew.*

First launched into space on 12 April 1981, the Shuttle was another triumph of American technology and can-do spirit. Once described as a "space truck", it has proved a versatile spacecraft, carrying satellites into orbit, acting as a mini space station and laboratory, carrying out space repairs, and latterly helping in the construction and operation of space stations. Two of the six Shuttles built have been lost in accidents, *Challenger* in January 1986 and *Columbia* in February 2003.

Wings used only when Shuttle glides in to land – they have no function in space

Elevon

Cargo bay measures 18.3 x 4.6m (60 x 15ft)

Three main engines are supplied with liquid hydrogen and oxygen from external tank during first eight minutes of flight

Orbital manoeuvring system engines

Aft reaction control system, used to manoeuvre Shuttle into correct position once it reaches space

Thermal Protection (TPS) tiles protect the underside and other surfaces of the Shuttle from burning up on re-entry into the Earth's atmosphere

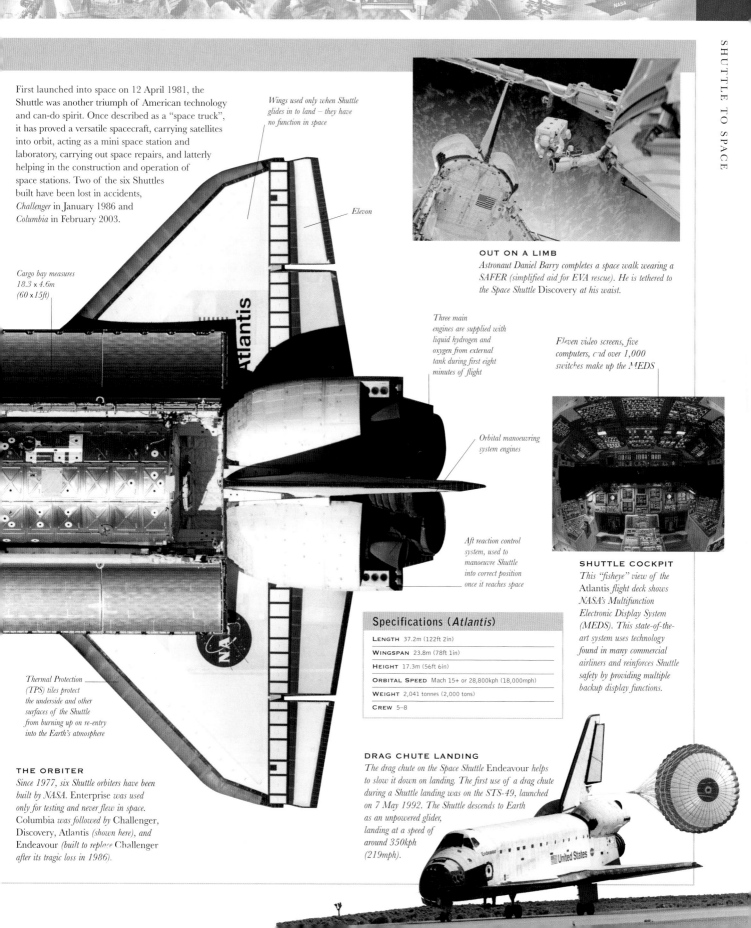

OUT ON A LIMB
Astronaut Daniel Barry completes a space walk wearing a SAFER (simplified aid for EVA rescue). He is tethered to the Space Shuttle Discovery *at his waist.*

Eleven video screens, five computers, and over 1,000 switches make up the MEDS

SHUTTLE COCKPIT
This "fisheye" view of the Atlantis *flight deck shows NASA's Multifunction Electronic Display System (MEDS). This state-of-the-art system uses technology found in many commercial airliners and reinforces Shuttle safety by providing multiple backup display functions.*

Specifications (*Atlantis*)

LENGTH	37.2m (122ft 2in)
WINGSPAN	23.8m (78ft 1in)
HEIGHT	17.3m (56ft 6in)
ORBITAL SPEED	Mach 15+ or 28,800kph (18,000mph)
WEIGHT	2,041 tonnes (2,000 tons)
CREW	5–8

THE ORBITER
Since 1977, six Shuttle orbiters have been built by NASA. Enterprise *was used only for testing and never flew in space.* Columbia *was followed by* Challenger, Discovery, Atlantis *(shown here), and* Endeavour *(built to replace* Challenger *after its tragic loss in 1986).*

DRAG CHUTE LANDING
The drag chute on the Space Shuttle Endeavour *helps to slow it down on landing. The first use of a drag chute during a Shuttle landing was on the STS-49, launched on 7 May 1992. The Shuttle descends to Earth as an unpowered glider, landing at a speed of around 350kph (219mph).*

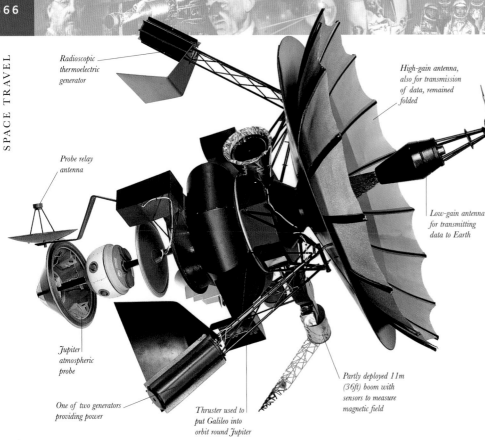

Radioscopic thermoelectric generator

Probe relay antenna

Jupiter atmospheric probe

One of two generators providing power

Thruster used to put Galileo into orbit round Jupiter

Partly deployed 11m (36ft) boom with sensors to measure magnetic field

High-gain antenna, also for transmission of data, remained folded

Low-gain antenna for transmitting data to Earth

The HST went on to prove itself one of the greatest science success stories of the modern era, supplying astonishing images of distant galaxies. There could no longer be any question of the Shuttle's importance in launching and servicing satellites. But the very success of such unmanned scientific missions as the *Galileo* probe and the HST called into question the need for a manned space programme, except as a subsidiary service to maintain unmanned satellites.

Continuing goal

Yet the goal of manned space flight continued to be pursued, if fitfully as arguments against its high cost intermittently blocked the way forward. In favour of manned flight it was urged that scientists in space could perform experiments that could not be done remotely. It was said that putting human beings in space was in itself a valuable experiment – when 77-year-old John Glenn travelled in the Shuttle in 1998 this was presented as a test of the effects of space travel on an older body, that might ultimately benefit senior citizens on Earth. And beyond all practical arguments was the sense of a final frontier that must always be pushed further back, the exploratory imperative – if humans were capable of visiting Mars, how could they resist doing it?

The early 1980s brought a revival of tension between the superpowers that favoured prestigious space projects as it had in the 1960s. President Ronald Reagan not only committed himself to the

The report of the Rogers Commission, set up to inquire into the *Challenger* accident, concluded that NASA had been under such pressure to maintain the schedule of Shuttle launches that the agency had ceased to give safety and quality control the priority they required. While reaffirming the importance of NASA and the space programme as "symbols of national pride and technological leadership", the commission called for a thorough safety review before Shuttle launches resumed at a more manageable rhythm.

Return of the Shuttle

The Shuttle returned to space in September 1988 and the program worked up to a launch every six to eight weeks. But the vision of the Shuttle as a self-financing part of the commercial space business was at an end. It would now carry only loads specifically designed for it or essential military space hardware.

Nor was the Shuttle any longer expected to replace disposable launch systems for satellites. The traditional rocket launchers had themselves gone through a difficult period – in 1986 a USAF Titan rocket blew up eight seconds after launch and a NASA Delta launcher also failed early in flight. But the rockets were there to stay, working alongside the reusable space vehicle.

Projects delayed by the interruption of the Shuttle programme included the *Galileo* space

probe designed to fly to Jupiter. It had been scheduled for launch in 1986 and had actually been transported to Cape Canaveral when the *Challenger* accident occurred. *Galileo* returned there in 1989 and was taken into space by the Shuttle *Atlantis*.

Also waiting for Shuttle launch in 1986 had been the Hubble Space Telescope (HST). Taking its place in the queue, the HST went into orbit in April 1990. At first, the mission appeared to be another expensive embarrassment for NASA. The great mirror designed to gather light from near to the origin of the universe did not work well because of a minor grinding error. Out of adversity, however, came a resounding triumph. In December 1993, the Shuttle *Endeavour* met up with the HST in orbit and grabbed it with a manipulator arm. Members of the Shuttle crew then exited the vehicle to carry out a gruelling repair job, their space-suited figures shown to the world in another space spectacular.

HUBBLE REPAIR MISSION
*The repair of the Hubble Space Telescope by Shuttle
astronauts in December 1993 was a spectacular
demonstration of the usefulness of the vehicle and
crew in servicing satellites in orbit. Here one astronaut
is raised on the manipulator arm of the Shuttle while
another works inside the payload bay.*

Strategic Defense Initiative that
would put weapons in space,
but also announced in 1984 that
the United States would have a
permanently manned space station in
orbit within a decade.

This was, of course, moving into territory
already solidly occupied by the Soviet Union. The
Soviet space station programme had continued to
progress through the 1980s – during the post-
Challenger interruption of the
Shuttle programme cosmonauts
were the only people travelling into space. In
February 1986 Salyut 7 was joined in orbit by a
third generation orbital station, Mir. Once Mir
was occupied, the ageing Salyut was left
to fly on unmanned until 1991.
Mir was a considerable step forward from

SHUTTLE DOCKING

The Shuttle Atlantis, *docked with the Mir space
station, is photographed from a Russian Soyuz-TM
space craft in 1995. The Shuttle missions to Mir
were a preparation for collaboration on the
International Space Station.*

SPACE CO-OPERATION

*A badge celebrates co-operation in
space between America and Russia
in the post-Soviet era.*

MIR SPACE STATION

*"Mir" in Russian means both "peace" and "world".
The space station of that name launched in 1986
was designed with six docking ports, which could be
used to attach special modules for scientific research.*

the Salyuts. The key improvement was Mir's
design as a modular station, with more
comfortable living conditions and more
sophisticated on-board life-support systems, and
the practicality of being able to attach extra
modules to the original core. Additional modules
sent up to dock with Mir included an astrophysics
observatory and laboratories for research in zero-
gravity, metallurgy, and crystallography.

The Soviets also had a much improved
manned spacecraft to service Mir, the Soyuz-TM.
In response to the Americans' Space Shuttle
programme, the Soviets began building the
Shuttle Buran in 1980. However, the Buran only
made one unmanned flight – lifted into space by
the powerful Energyia liquid-hydrogen booster
introduced in 1988 – before it was cancelled due

LIFE ON MIR

IN ORBIT FROM 1986 TO 2001, the Mir space station was the site for the most extensive experiments in living outside the Earth's atmosphere that have yet been attempted. Russian cosmonaut and physicist Valeriy Polyakov established the world record for space endurance by spending 437.75 days on board Mir in 1994–95, while his colleague Sergei Avdeyev spent a record cumulative total of 747.6 days orbiting in space.

Of the American astronauts who joined the cosmonauts on Mir from 1995 to 1998, the longest stay was made by Shannon Lucid, who spent 188 days in orbit. It was found that the human body adapted quite quickly to gravity-free conditions. For example, the heart, veins, and arteries adjusted to the relative ease with which blood would circulate. Adapting back to conditions on Earth posed more serious problems. Some cosmonauts found that they were too weak to walk or even sit up unaided when returned to gravity after months of weightlessness. To counter the disorientating effect of an environment in which the Sun rose and set 16 times in

MICHAEL FOALE
By the time that British-born American Mike Foale spent time on Mir in 1997, the ageing station was becoming seriously unsafe.

SHANNON LUCID
American biochemist Shannon Lucid broke a string of records with her 188 days on board Mir in 1996, a trip prolonged by a problem with the Shuttle that delayed her return.

24 hours, a strict Earth-based time routine was imposed, with cosmonauts woken by a wrist alarm at 8.00am Moscow time every day. The crew slept in floating sleeping bags attached to a wall. There was no real sense of "up" or "down" – astronaut Jerry Linenger described sleeping with his feet to the ceiling and his head near the floor.

Life on board was inevitably cramped and nostalgia for friends and family affected all the crew. Mir was increasingly cluttered with objects brought up from Earth and never returned, including personal effects such as books and videos. Still, Lucid recalled the small pleasures of communal life – for example, the crew eating meals together "floating around a table in the Base Block", with Jello as a Sunday treat. Later Americans on Mir had a darker experience as the station deteriorated, with systems breakdowns, power failures, and other life-threatening crises.

FLOATING WORLD
Every task in space presented cosmonauts and astronauts with special problems related to weightless conditions. To sit on a chair, for example, they would have to hook their feet into rings on the floor.

---

to economic and technological difficulties, and a lack of scientific and military support.

The failure of Soviet President Mikhail Gorbachev's attempts to modernize the communist state led to the political disintegration of the Soviet Union. At the end of 1991, the state that had once taken on America in the space race ceased to exist.

Running out of cash
The personnel and facilities of the Soviet space programme were primarily inherited by the Russian Federation, although the Baikonur Cosmodrome was in independent Kazakhstan. Even if the successor states of the Soviet Union had the will to continue the space programme – and to a degree they undoubtedly did – money

was in desperately short supply. With budgets slashed to a small fraction of those in the Soviet era, plans for a Mir 2 space station were binned and unfinished Burans were dismantled.

Even before the fall of the Soviet Union, the new spirit of enterprise sweeping the communist bloc had led the Glavkosmos space agency, set up in 1985, to sell journeys to Mir for millions of dollars at a time. For example, one place was bought by the Japanese for $7 million in order to put one of its nationals into space. Instead, the American space agency incorporated Mir into its own programme for building a new space station.

President Reagan's project for Space Station Freedom had, like so many bold manned space projects, run aground on the issue of cost. A way forward was found in international co-operation,

with Canada, Japan, and the countries of the European Space Agency soon signing up to take part. Adding Russia to the consortium promised a financial lifeline for the beleaguered ex-Soviet space researchers, engineers, and cosmonauts, while the use of Russian facilities had much to offer in terms of economy to their American counterparts. Co-operation across the previous Cold War divide also had a political payoff, symbolizing aspiration to a new world order.

As part of the International Space Station project, in 1994 the Russians and Americans began a series of joint missions in which cosmonauts travelled on Shuttles, Shuttles docked with Mir, and American astronauts spent time on board the Russian space station. Over five years, the old rivals learned to work together.

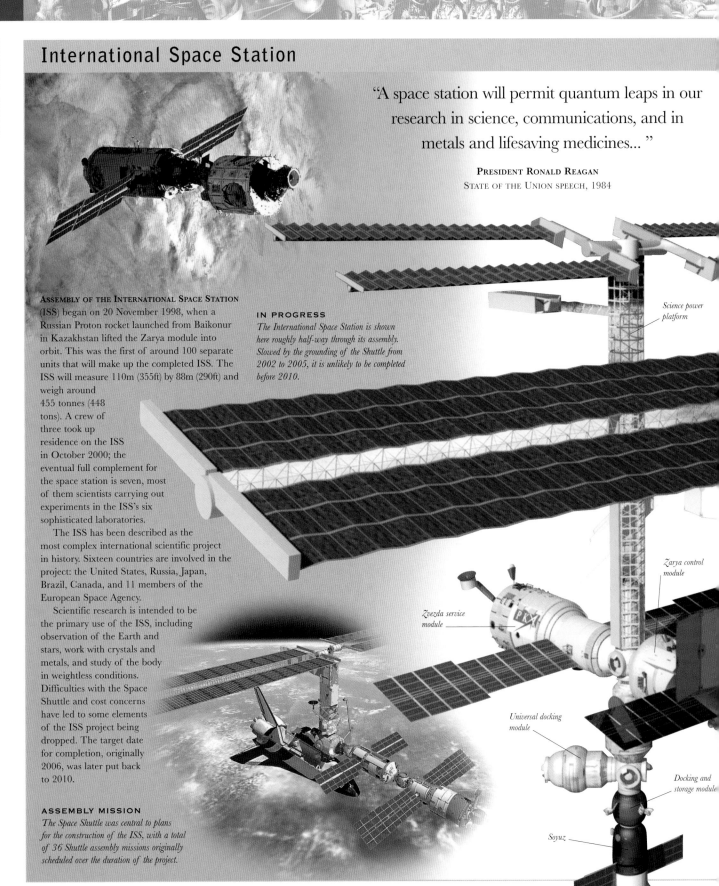

International Space Station

"A space station will permit quantum leaps in our research in science, communications, and in metals and lifesaving medicines... "

PRESIDENT RONALD REAGAN
STATE OF THE UNION SPEECH, 1984

ASSEMBLY OF THE INTERNATIONAL SPACE STATION (ISS) began on 20 November 1998, when a Russian Proton rocket launched from Baikonur in Kazakhstan lifted the Zarya module into orbit. This was the first of around 100 separate units that will make up the completed ISS. The ISS will measure 110m (355ft) by 88m (290ft) and weigh around 455 tonnes (448 tons). A crew of three took up residence on the ISS in October 2000; the eventual full complement for the space station is seven, most of them scientists carrying out experiments in the ISS's six sophisticated laboratories.

The ISS has been described as the most complex international scientific project in history. Sixteen countries are involved in the project: the United States, Russia, Japan, Brazil, Canada, and 11 members of the European Space Agency.

Scientific research is intended to be the primary use of the ISS, including observation of the Earth and stars, work with crystals and metals, and study of the body in weightless conditions. Difficulties with the Space Shuttle and cost concerns have led to some elements of the ISS project being dropped. The target date for completion, originally 2006, was later put back to 2010.

ASSEMBLY MISSION
The Space Shuttle was central to plans for the construction of the ISS, with a total of 36 Shuttle assembly missions originally scheduled over the duration of the project.

IN PROGRESS
The International Space Station is shown here roughly half-way through its assembly. Slowed by the grounding of the Shuttle from 2002 to 2005, it is unlikely to be completed before 2010.

Science power platform

Zarya control module

Zvezda service module

Universal docking module

Docking and storage module

Soyuz

Specifications

SIZE	About 110 x 88m (360 x 290ft)
WEIGHT	About 455 tonnes (448 tons)
SPEED	28,000kph (17,500mph)
EARTH ORBIT TIME	90 minutes
ALTITUDE	350km (220 miles)
CREW	7

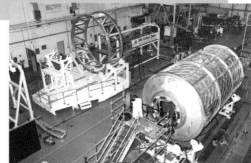

ORBITAL LABORATORY
A US Laboratory Module, Destiny, one of the major elements of the ISS, is seen under construction at the Marshall Space Flight Center. Experiments to be conducted in space include the growing of protein crystals for medical research.

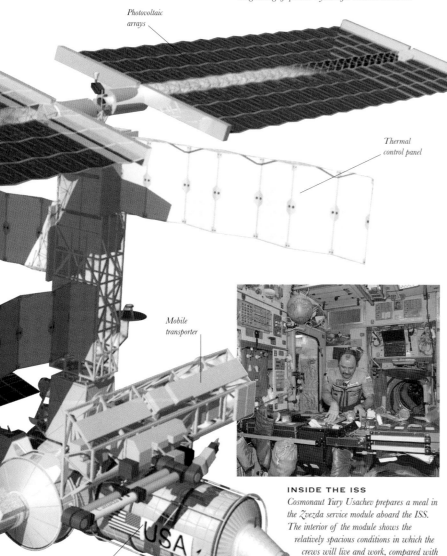

Photovoltaic arrays

Thermal control panel

Mobile transporter

Spacewalking airlock

US laboratory

INSIDE THE ISS
Cosmonaut Yury Usachev prepares a meal in the Zvezda service module aboard the ISS. The interior of the module shows the relatively spacious conditions in which the crews will live and work, compared with the cramped quarters experienced by the crews of the Mir space station.

The first American astronaut on board Mir was Norm Thagard, who stayed for 115 days from March to July 1995, and the last was Andy Thomas, from January to June 1998. It was, in the words of one NASA official, a critical exercise in "getting to know how to work with an international partner, getting to understand the Russian way of doing business". But it also brought warnings that Russian technology and operational expertise might not meet the standards that the Americans were used to.

By the time the joint American-Russian programme began, Mir had been in orbit for almost a decade and much of its equipment was deteriorating, and in need of replacement. The overall quality of the Russian space programme on the ground was also under threat from budgetary constraints. Mounting problems with Mir posed a serious risk to the lives of astronauts and cosmonauts. In June 1997 a cargo craft crashed into the space station, making a hole in its skin and taking out some solar panels. The following month the station lost power after one of the crew accidently disconnected a cable. Other incidents in the same year included oxygen system failures, computer breakdowns, coolant leaks, and a fire on board which filled the space station with smoke.

Fiery end for Mir

Despite the dangers, NASA persisted with the joint Mir programme until it was considered complete in the summer of 1998. The Americans then urged the Russians to abandon Mir and devote themselves singlemindedly to the incipient International Space Station project. But Mir had won a place in Russian hearts and was not to be so readily ditched. Cosmonauts continued to man it up to August 1999. Even after they left, there was a stay of execution when a consortium of Western businessmen put in a bid to buy Mir and use it as a space hotel. Whether this deal was a serious proposal for consideration, or not, Mir was given the death sentence. In March 2001 it was nudged lower by a Progress supply ship until it was close enough to the Earth to be pulled down out of orbit. Mir burnt up in the atmosphere, fiery fragments crashing into the Pacific, ending a journey of over 3 billion km (2 billion miles).

By the time Mir hurtled from the sky, parts of the International Space Station had already been assembled in orbit and were providing a new platform for a permanent human presence in space. The project literally got off the ground when a Russian Proton rocket lifted the Zarya ("sunrise") module into orbit. The second module, Unity, linked up with Zarya the following month, carried into orbit by the Shuttle *Endeavour*. The

FIRST SPACE TOURIST

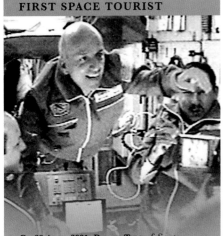

ON 28 APRIL 2001, DENNIS TITO of Santa Monica, California, became the first person to enter space on a trip paid for out of his own pocket. The space tourist, whose fare for a journey to the International Space Station and back was reputed to be $20 million, lifted off from the launch pad at Baikonur, Kazakhstan, on board a Soyuz-TM rocketcraft.

Tito had originally hoped to pay to visit the Mir space station, but when that prospect evaporated in 2000, his ambitions shifted to the ISS. The Russian space authorities, keen for cash from any direction, saw no problem in fulfilling Tito's dream. NASA, by contrast, was openly hostile, but unable to block the trip. Tito was evidently delighted with his experience of eight days in space. On his return he described himself as "a serious man who had a dream and pursued it in the face of great difficulty."

Faced with the inevitability of further space tourism, NASA withdrew its objection in February 2002. South African Mark Shuttleworth became the Russians' second paying passenger the following May.

next waymark was the arrival of the first three-person crew to begin the permanent occupation of the ISS in October 2000.

As the demanding schedule of flights and spacewalks went forward, the ISS was justly described as the most extraordinary building site in existence. But the cost of the ISS programme naturally continued to escalate from an originally agreed $17 billion. Estimates were soon putting the eventual price at between $50 billion and $100 billion. And cost was not the worst of the ISS's problems. On 1 February 2003, the shuttle programme suffered its second fatal accident when *Columbia* broke up on re-entry, killing its crew of seven. An investigation revealed that foam broken off the main propellant tank on lift-off had struck the leading edge of *Columbia*'s left wing, damaging the ceramic tiles that protected against the

> "I spent 60 years on Earth and I spent 8 days in space. From my viewpoint, it was two separate lives."
>
> **DENNIS TITO**
> FIRST SPACE TOURIST

extreme temperatures experienced during re-entry. All remaining Shuttles were grounded while safety issues were addressed. It was not until July 2005 that flights resumed with the launch of *Discovery* on mission STS-114, but even then the problems continued. Foam was again shed at launch. Although this time no damage was caused to the orbiter, Shuttle flights were again suspended while the problem was fixed. The ISS project stumbled forward in the absence of the shuttle. Russian Soyuz and Progress spacecraft were able to keep a crew of two in occupation of the space station and limited construction work continued. But only the Shuttle had the cargo capacity to carry major components of the ISS.

In the immediate aftermath of the loss of *Columbia*, President Bush had hastened to assure the world that the future of the manned space

VENTURESTAR
Intended as the next step on from the Space Shuttle, the Lockheed Martin X-33 VentureStar was to be a fully reusable space vehicle – essentially an aircraft that could fly into space and back. Begun in the late 1990s, the project soon ran into apparently insuperable difficulties and was cancelled in 2001.

NASA X33 LOCKHEED MARTIN

BFGoodrich Aerospace BOEING Rocketdyne AlliedSignal Sverdrup

USA

flight programme remained secure. The official line was that the momentum behind its continuation and expansion was unstoppable. Many observers felt, though, that NASA's manned flight programme was lacking in direction and focus. The main justification for continuing with Shuttle flights was to build the ISS, but was the key purpose of the ISS simply to give the Shuttle something to do? The sense of drift in the manned space programme ended in January 2004, when President Bush announced his "vision for space exploration". There was to be a return to the Moon to establish a permanent base, and a future journey onward to Mars.

New direction

This fresh clarity of goal was welcomed by NASA, but required a substantial change of direction. There was soon evidence of cutbacks in the number of planned Shuttle missions to complete the ISS, with a reduction in the number of elements making up the space station. This was necessary to cut costs and divert resources to the renewed Moon programme.

It had long been obvious that the ageing Shuttle fleet would need to be phased out, but the assumption had always been that a Shuttle replacement would be produced. This time it would be a fully reusable launch system, flying under its own power into space and back. The pursuit of this concept, which had originally inspired the Shuttle programme, was renewed in 1994 with a directive from President

Bill Clinton. It brought a new crop of experimental X-planes with the ambition of creating a "space liner" that would ferry people back and forth between bases on Earth and the ISS – and transform the economics of journeys into space. From 2004, however, NASA switched to development of a Crew Exploration Vehicle to go to the Moon and beyond. Replacing the Shuttle was no longer a priority. The "space liner" seemed likely to be left to private entrepreneurs to develop instead.

Manned versus unmanned

Many scientists remained openly sceptical of the value of manned, as opposed to unmanned, ventures into space. The achievements of unmanned space probes and satellites in the first 50 years of the space era were indeed impressive. They had increased scientific knowledge of the Universe to a degree that manned missions could not begin to match. Since the 1960s, solar observatories, astronomical observatories, and geophysical observatories have orbited the Earth, providing a vast quantity of new information on phenomena such as black holes and solar flares, with no need of scientist-astronauts to crew them. Nor does exploration of the solar system require humans to travel through space. The achievements of robot planetary explorers have been extraordinary. Thanks to NASA's Magellan spacecraft of 1989, for example, we have an accurate map of the surface of Venus, permanently hidden by cloud but revealed by radar. Since the Viking mission of 1976, we know what the surface of Mars looks like close up and have analysed the planet's soil and atmosphere. Voyagers 1 and 2 gave new insight into Saturn and remote Uranus, and Galileo in 1996 produced many surprising revelations about Jupiter and its moons. This process of unmanned exploration continued into the 21st century.

The practical impact of unmanned satellites on everyday human life has been all-pervasive. They have, among other things, revolutionized

NASA'S HYPER-X
The X-43A scramjet was created by Micro Craft for NASA to explore the possibility of hypersonic flight at speeds up to Mach 10 using an air-breathing engine. One of the possible uses for such an aircraft would be as a reusable space launch vehicle.

communications, meteorology, cartography, and navigation. Thanks to satellites an individual can, at little cost, know his or her location on the Earth's surface at all times with great precision, and contact another person instantaneously almost anywhere on the globe.

The impetus to manned space flight comes, however, from a different impulse to either scientific research or practicality. Its essence was perhaps most clearly revealed in the beginning of space tourism in 2001. Individuals who were rich enough would, it turned out, be ready to pay a fortune for a short stay on the ISS. President Bush, in his project for a return to the Moon as a prelude to a manned journey to Mars, was hoping that, by the same token, Congress would be prepared to pay the price for a human adventure. Space travel had been an inspirational dream since the days of Tsiolkovsky and Goddard, and so it remains.

PLANETARY EXPLORATION
Impressions of the future: below, a remote-controlled NASA aeroplane to fly survey missions on Mars; right, an orbiter powered by a tether – a long wire that generates an electric current as it passes through a planet's magnetic field.

Star™

SHRINKING WORLD

7

UNTIL THE 1940S, AIR TRAVEL was a growing but marginal alternative to travel by ship and train. By 1957, however, more people were crossing the Atlantic by plane than by boat. The triumph of jet airliners in the following decade carried the transformation of travel patterns to a new level. People made journeys they would never previously have made to places they would never have visited. Boeing executive Tex Boullioun said in 1983, "Great cities and great resorts now stand in places where neither ship nor train could have caused them to be." While passenger flight became commonplace – without losing, for some people, associations of anxiety – thousands of individuals recovered the early excitement of flight by piloting their own aircraft or experimenting with small but technologically sophisticated flying machines.

QUEUING FOR TAKE-OFF
By the year 2000, more than four million people worldwide were taking to the air every day. The familiar scene of airliners queuing for take-off at a busy airport was the natural result of technological progress that had made flight fast, safe, reliable, and affordable.

"The modern airplane creates a new geographical dimension… There are no distant places any longer: the world is small and the world is one."

WENDELL WILLKIE
ONE WORLD (1943)

JET PASSENGER TRAVEL

DURING THE SECOND HALF OF THE 20TH CENTURY AIR TRAVEL BROUGHT EVERY PART OF THE GLOBE WITHIN EASY REACH OF ANY TRAVELLER

AT THE START OF THE 1940s, American airline bosses and aircraft manufacturers sensed that passenger air travel was on the brink of a quantum leap forwards. The Boeing 307 Stratoliner, with its pressurized cabin and supercharged engines, was already flying above the weather, offering a new level of comfort to passengers. Millionaire aviation enthusiast Howard Hughes had bought into TWA and, with fellow executive Jack Frye, was pushing Lockheed into developing the Constellation, a four-engined airliner capable of transporting people non-stop across the United States. At Douglas, the four-engined DC-4 was almost ready as a successor to the twin-engined DC-3, and a pressurized version, later to become the DC-6, was already under discussion.

Then came the war. After the Japanese attack on Pearl Harbor in December 1941, all production of commercial aircraft was stopped and resources devoted exclusively to building military aircraft. Before it could enter service as an airliner, the DC-4 became the C-54 transport plane. Work on the Constellation was halted, and then the aircraft was rather tardily ordered by the armed forces as the C-69 transport. Like the manufacturers, the airlines were drafted into war service.

Although in a sense this was an interruption of commercial aviation, the war actually generated a rapid acceleration in the growth already under way. There were important technological advances, chiefly in navigation, radar, and communications, but above all there was a transformation in scale.

DOUGLAS DC-4
Marketed as the "First with Four", the DC-4 was a four-engine version of the DC-3. It could carry between 50 and 80 passengers and had the range to offer a transatlantic service with a single fuel stop. However, it lacked a pressurized cabin for high-altitude flight. Shown below is the military adaptation, the C-54 Skymaster.

MIGHTY JUMBO
The sheer size of the Boeing 747, introduced into service in 1970, transformed the scale of long-distance passenger flight. Carrying three or four times as many passengers as previous jet airliners, it ushered in the age of mass air travel. Although the 747 was a remarkable technological achievement, its success inevitably made flight more banal.

In 1941 the US airline fleets totalled 322 aircraft, mostly DC-3s; over the next four years, more than 11,000 transport versions of the DC-3 were built, and 1,600 transport versions of the DC-4. Although there was an inevitable collapse in production once the war ended, the expanded industrial base was available to supply an expanding commercial market. Equally important was the development of a network of international air routes, with massive investment in infrastructure, including concrete runways from which four-engined aircraft could safely operate. By the time peace came, the airlines and their personnel had amassed an impressive experience of operating worldwide, including transoceanic services.

During the war a clear vision emerged of a future in which aviation would shrink the globe. For example, American politician Wendell Willkie pronounced: "There are no distant places any longer: the world is small and the world is one."

CRUISING THE STRATOSPHERE
The 377 Stratocruiser was Boeing's last propliner. Its name drew attention to the fact that its pressurized cabin allowed it to cruise in the stratosphere, above the discomfort of the weather. Unfortunately, however, it suffered frequent engine failures.

HOWARD ROBARD HUGHES

WHEN MILLIONAIRE Howard Hughes (1905–76) became a major force in passenger aviation by buying into TWA in 1939, he was already an aviation legend. After first investing his inherited fortune and eccentric talent in Hollywood movies, including the air-combat epic *Hell's Angels*, in 1934 Hughes had sought out the excitements of air racing. He set up Hughes Aircraft to design his own racer, the H-1, in which he established a new world land-plane speed record of 563.7kph (352.3mph) in 1935. Other aviation feats followed, culminating in a record 91-hour round-the-world flight in a Lockheed Super Electra in 1938.

ECCENTRIC MILLIONAIRE
Howard Hughes was a daring pilot but often a poor judge of aircraft design. In later years his eccentricities got the upper hand over his undoubted talents, and he died a recluse.

As an aeroplane manufacturer, though, Hughes was not a success. Convinced that plywood, not aluminium, was the material for the aircraft of the future, he proposed a wooden-framed bomber to the US War Department. They turned it down, as they had the H-1, offered as a fighter.

Hughes' lobbying strength in Washington prevented immediate refusal of his next project, for a huge military flying boat. The H-4 Hercules, made mainly of birch but euphoniously dubbed the "Spruce Goose", had eight engines and, at 97.5m (320ft), a larger wingspan than any other aircraft before or since. But only a prototype was ever built, and that two years too late for the war in which it was meant to fight. It rose into the air just once, on 2 November 1947, with Hughes at the controls, reaching an altitude of about 20m (70ft) and covering less than a mile.

By that time Hughes' flying career was largely at an end. In 1946 he was almost killed piloting the XF-11, an all-metal reconnaissance version of his wooden D-2 bomber. After that crash he rarely flew. He lost hands-on control of Hughes Aircraft in 1953 and was elbowed out of the financially ailing TWA in 1960.

SPRUCE GOOSE
Howard Hughes' gigantic H-4 Hercules flying boat, known as the "Spruce Goose", was intended as a military transport aircraft that would carry a large enough cargo in its huge hold to act as a replacement for some of the U-boat-threatened merchant ships plying the Atlantic in World War II.

In 1944, a conference met in Chicago to lay the political and administrative foundations for a new era of international air travel. The US representative, Adolf Berle, saw the Chicago meeting as a chance to write "the charter of the open sky". This did not quite happen. Countries were not ready to give up control of their airspace or to stop promoting the interests of their national airlines. But enough agreement was reached to let the expansion of international air travel happen.

Increased competition

The international airline industry did not even approximate to a model of free competition. Most airlines either had government subsidy or were state-owned, and the tight regulation of routes and operations was the accepted rule. The International Air Transport Association (IATA) fixed fares and co-ordinated the schedules of different companies. The US was the leader in world aviation, and by 1951 more Americans were travelling by air than on long-distance train services. But airlines based in other parts of the developed world held their own. Although Germany was totally excluded from aviation, and

BOEING STRATOCRUISER

NX 1039V

Britain and France continued to concentrate much of their energy on their imperial connections, there were some surprising success stories, including the Australian airline Qantas, Swissair, and the Scandinavian SAS.

The expansion of the US's international aviation followed the model of "controlled competition" that developed its domestic industry in the 1930s. Pan American lost the monopoly of overseas flights it had enjoyed in the 1930s, and lobbying in Washington determined which airlines got what slice of the action. It was American Overseas Airways that flew the first scheduled transatlantic land-plane service, from Boston to London, in October 1945. But Pan American's chief global rival was TWA, which had developed a flourishing Intercontinental Division during the war. In 1950 the meaning of the abbreviation was appropriately changed from Transcontinental and Western Airlines to Trans World Airlines.

Rivalry between airlines was hooked into a competition between manufacturers Lockheed and Douglas. Although Pan American also began its postwar services with Lockheed Constellations, it was TWA that was identified with the "Connie" from the outset. In 1943, in a calculated publicity coup, Jack Frye and Howard Hughes flew the first Constellation from California to Washington D.C.

in under seven hours – painted in TWA colours. The airline stuck with the successive Constellation upgrades throughout the 1950s – Super Constellations, then the ultimate Starliner from 1957. Douglas countered with the DC-6 in 1947 and DC-7 from 1953, repeatedly matching and surpassing the Lockheeds' range and speed. The DC-6B and DC-7 were the aircraft that Pan American flew in the 1950s, along with the Boeing 377 Stratocruiser, a propliner intended to match the pre-war flying boats for luxury.

ELEGANT CONNIE

A TWA Lockheed 1049 Super Constellation looks its best flying over New York City in the 1950s. TWA was the source of the original request for Lockheed to produce a large pressurized airliner with transcontinental range. The aircraft that Kelly Johnson and the Lockheed team came up with was a design classic, marked out by its extraordinary triple tail. Flying as a passenger in a "Connie" was a comfortable and stylish experience.

TWA TRANS WORLD AIRLINES

Air travel in the "golden age" of the propliner came far closer to matching the airlines' publicity image of comfort and glamour than had ever been possible before. Turbulence was much reduced by flying at high altitude and soundproofing cut down engine noise. Journey times were progressively shorter – the San Francisco-to-New York route, for example, came down from around 11 hours in an early-model Constellation to about eight hours in a mid-1950s DC-7. Although the pressure towards economy flight was always there, propliners often provided very spacious accommodation. Many had bunk beds, and the Stratocruiser had a downstairs lounge and bar reached by a spiral staircase.

However, for those who could afford it, flying remained something of an adventure. The passengers on postwar transatlantic services, with their refuelling stops at the bleak outposts of Gander, Newfoundland, and Shannon in Ireland, could not regard their experience as mundane. Flights were still subject to delays or enforced stops for bad weather, and many of the propliners were plagued by engine trouble. Cockpit technology was advancing fast – for example, on-board radars came into general use during the 1950s – but flying was still very hands-on.

Advent of jet airliners

The propliners of the 1940s and 1950s were machines of great power and their achievements were impressive. Through the 1950s they developed the range to offer non-stop transoceanic and transcontinental flights. The DC-6B allowed SAS to start the first scheduled transpolar flights in 1954, linking northern Europe to America's west coast. By 1957 the Lockheed 1649 Starliner was flying from Los Angeles to London non-stop in 19 hours. But the 1950s propliners represented a technology that had been pushed as far as it would go.

The transfer of jet technology to commercial aviation was not straightforward. Economy and reliability were the essence of passenger operations, but early jet engines could provide neither. They used up enormous quantities of fuel and made heavy demands on maintenance staff. No wonder American airlines and manufacturers, with their heavy commitment to piston-engined airliners, ignored the jet for so long after the war. The British, by contrast, found themselves world leaders in jet-engine design after the forced withdrawal of Germany from the field. They had every interest in pursuing jet aviation and soon set the pace, first with turboprops and then with turbojets.

The turboprop was, in retrospect, clearly a transitional technology. It used the hot gas inside a jet engine to drive propellers, giving greater power and speed than a piston engine, but much better fuel economy than contemporary turbojets. The first turboprop airliner, the Vickers Viscount,

POSTWAR PROPLINERS

AFTER THE WAR, piston-engined, propeller-driven airliners were not a serious challenge until the late 1950s. The propliners were able to hold off the challenge of turboprops and jets for more than a decade, not only because of the difficulty of developing a reliable jet airliner, but also because they were such excellent aeroplanes. Far superior in comfort, range, and speed to most pre-war airliners, aeroplanes such as the Constellation and the DC-7 could operate transcontinental services that outclassed the service offered by ships and trains. By 1957, propliners were carrying more people across the Atlantic than ocean liners. Keys to their success included pressurized cabins and more powerful, efficient engines.

BRITISH AMBASSADOR
One of the most elegant aircraft ever built, the 30-seat Airspeed Ambassador, which entered service in 1952, was intended to revive the UK's civil airline network.

Boeing Model 377 Stratocruiser

The bulbous Stratocruiser was deceptively fast and its size evoked a pre-war atmosphere of spacious elegance. The new double-decked fuselage was grafted onto the wings, engines, and tail of the B-29 Super Fortress bomber. As usual Pan American was the first customer, putting the 377 into service in February 1949 but it was not destined to be a commercial success for Boeing. Only 55 Model 377s were built, due to their high operating costs and also because each cost $1.75 million against $1 million for the competing Lockheed Constellation and Douglas DC-6.

ENGINE 4 x 3,500hp P&W R-4360B Wasp Major 28 radial	
WINGSPAN 43.1m (141ft 3in)	LENGTH 33.6m (110ft 4in)
CRUISNG SPEED 547kph (340mph)	
PASSENGERS 55–112	CREW 5

Breguet 763 Deux-Ponts/Provence

Sadly the story of the Deux-Ponts is typical of early postwar attempts to build a successful airliner. The first bulbous double-decker Deux-Ponts flew in February 1949, following a French government order in 1947 for 15 aircraft. To improve its commercial potential, the following aircraft had American engines in place of the prototype's French power plants. Despite lengthy military trials, no more B-761s were built, although 12 of the improved B-763 Provences were flown by Air France.

ENGINE 4 x 2,100hp P&W R-2800 Double Wasp 18 radial	
WINGSPAN 41.7m (136ft 8in)	LENGTH 28.7m (94ft 2in)
CRUISING SPEED 320kph (286mph)	
PASSENGERS 101	CREW 4

Bristol Brabazon (Type 167)

The Bristol Brabazon was named after Lord Brabazon, who led the committee which recommended the types of postwar civil aircraft Britain should build. The enormous Brabazon was, in fact, the least successful and the only model finally flew on 4 September 1949. It was powered by eight engines coupled in pairs to four counter-rotating propellers. Sadly it never went into production, and the aircraft was scrapped in 1953, having flown less than 400 hours.

ENGINE 8 x 2,500hp Bristol Centaurus 18-cylinder radial	
WINGSPAN 70.1m (230ft)	LENGTH 54m (177ft)
CRUISING SPEED 402kph (250mph)	
PASSENGERS 100	CREW 12

Convair 240 Convair-Liner

A direct competitor to the Martin 404, the Convair family of medium-sized airliners was far more successful. The first Convair Model 110 prototype flew in 1946, but could only carry 30 passengers. The enlarged 240, with more powerful engines, flew in March 1947. When production ended in 1958, a total of 571 Convair 240s had been manufactured, 176 as airliners and 395 as military C-131 and T-29 variants.

ENGINE	2 x 2,400hp P&W R-2800 Double Wasp 18 radial	
WINGSPAN	28m (91ft 9in)	LENGTH 22.3m (74ft 8in)
CRUISING SPEED	432kph (270mph)	
PASSENGERS	40	CREW 3–4

Douglas DC-6B

The Douglas DC-6 was a natural development of the DC-4, much of the work being undertaken on behalf of the US military who wanted an improved C-54. Although the war ended before a military version was complete, the initial commercial version was immediately successful with 175 built. Pan Am's first transatlantic all-cargo service was operated by DC-6s. With constant improvements in range and capacity leading through to the DC-6C capable of carrying 107 passengers, over 700 DC-6s had been built when production closed in 1958. A few continued to fly as freighters in the 21st century.

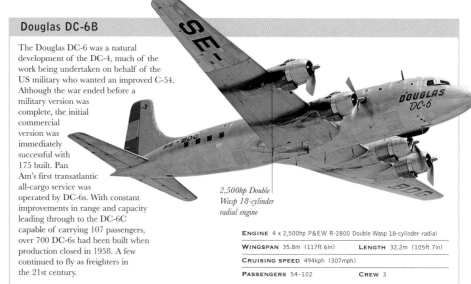

2,500hp Double Wasp 18-cylinder radial engine

ENGINE	4 x 2,500hp P&EW R-2800 Double Wasp 18-cylinder radial	
WINGSPAN	35.8m (117ft 6in)	LENGTH 32.2m (105ft 7in)
CRUISING SPEED	494kph (307mph)	
PASSENGERS	54–102	CREW 3

Douglas DC-7C Seven Seas

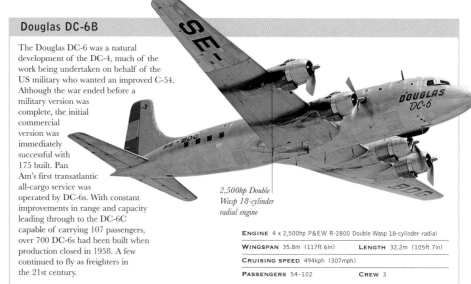

ENGINE	4 x 3,400hp Wright R-3350 Turbo 18-cylinder radial	
WINGSPAN	38.8m (127ft 6in)	LENGTH 34.2m (112ft 3in)
CRUISING SPEED	580kph (360mph)	
PASSENGERS	65–105	CREW 7–8

The intense postwar competition between Douglas and Lockheed for the long-range airliner market produced some remarkable aircraft and engines. The DC-7 was the ultimate member of the four piston-engined family which began with the DC-4. Using energy efficient Wright Turbo Compound engines, the first DC-7 flew in May 1953. The production line closed in late 1958 after 338 had been built including 128 of the final variant, the superb DC-7C which BOAC flew as the Seven Seas.

Ilyushin Il-12 "Coach"

Designed to replace the Douglas DC-3 which had been built in the Soviet Union as the Lisunov Li-2, the Il-12 was produced in large numbers to rival the quantity of its famous predecessor. Conceived in 1943, the first flight of the unsuccessful prototype was on 15 August 1945. The re-designed aircraft, with the NATO reporting name of Coach, finally went into service with Aeroflot in August 1947 and was supplied in large numbers throughout the Warsaw Pact nations in both military and civil variants. Production ceased in 1949 in favour of the improved Il-14 (shown below).

ENGINE	2 x 1,650hp Shvetsov Ash-82FN 14-cylinder radial	
WINGSPAN	31.7m (104ft)	LENGTH 21.3m (69ft 11in)
CRUISING SPEED	350kph (217mph)	
PASSENGERS	27–32	CREW 4–5

Lockheed 1049G Super Constellation

One of the most graceful airliners of all time, the elegant L-49 was first conceived in 1939, only to be requisitioned for military use as the C-69. After the war, the 81-seat L-649 became the L-1049, after the fuselage was lengthened to accomodate up to 109 passengers

ENGINE	4 x 3,250hp Wright R-3350 Turbo 18-cylinder radial	
WINGSPAN	37.5m (123ft)	LENGTH 34.7m (113ft 7in)
CRUISING SPEED	526kph (327mph)	
PASSENGERS	81–109	CREW 6

Martin Model 404 (4-0-4)

The Martin Model 202 was the first postwar, twin-engined airliner to fly. However, an accident in 1948, attributed to structural weakness of the wing, caused the temporary grounding of the type. An improved, pressurized Model 404 "Skyliner" overcame the accident's stigma to become a 1950s classic. Refinements included a longer fuselage and more powerful engines, making its cruising speed nearly 160kph (100mph) faster than the DC-3. A total of 103 were built between 1951 and 1953, mostly split between Eastern Airlines and Trans World Airlines (TWA).

ENGINE	2 x 2,400hp P&W R-2800 Double Wasp 18 radial	
WINGSPAN	28.4m (93ft 3in)	LENGTH 22.7m (74ft 7in)
CRUISING SPEED	460kph (286mph)	
PASSENGERS	34–42	CREW 3

COMET PROTOTYPE

The prototype of the de Havilland Comet made its maiden flight in 1949. The first jet airliner to enter service, the Comet suffered a spate of crashes caused by metal fatigue.

arrived on the scene in 1950. It proved an immensely popular aircraft, offering a smoother, quieter ride than any piston-engined airliner. Although it only carried 50 passengers, it was run at a profit by British European Airways, which in the 1950s was responsible for a quarter of all passenger traffic in Western Europe. Britain went on to make the larger Bristol Britannia and Vickers Vanguard turboprops, while America belatedly came up with the Lockheed turboprop Model 188 Electra, an aircraft whose reputation was severely damaged by a series of shocking crashes.

The country that produced the most remarkable of all turboprop airliners was the Soviet Union. The Tupolev Tu-114 was a spin-off from Soviet military aviation, developing out of the Tu-95 long-range turboprop bomber. Entering service with Aeroflot in 1957, the Tu-114 was a giant – the largest airliner in the world before the Boeing 747. It could carry 170 passengers over 8,800km (5,500 miles) between stops, cruising at around 770kph (480mph). In 1959 a Tu-114 flew from Moscow to New York in 11 hours, and in the 1960s the Tupolevs operated scheduled services linking the Soviet capital to Cuba, Canada, West Africa, India, and Japan.

de Havilland's Comet

In the course of the 1950s, however, it became obvious that the future lay not in hybrid turboprops, but in pure turbojets. Here, too, it was Britain that took the first step. The de Havilland Comet, which began flight testing in

1949, was a bold gamble on advanced technology. Company chairman Sir Geoffrey de Havilland, a veteran of the pioneering age of aviation who had designed aircraft that flew in World War I, staked his fortune on the success of the world's first jet airliner. Initially, the gamble seemed set to come off. The Comet entered service on BOAC's London to Johannesburg route in 1952 and was an instant hit with the elite market at which it was aimed. The new airliner was fast, smooth, and stylish. Passengers fought for the limited number of tickets – the aircraft initially only seated 36 – and Comets were soon being flown by other airlines, including Air France. The American business magazine *Fortune* declared 1953 "the year of the Coronation and the Comet".

Disaster struck in January 1954, when a Comet flying out of Rome exploded at 9,000m (30,000ft). An earlier Comet crash in India had been blamed on bad weather, but this time there seemed no obvious explanation. Flights were resumed, but three months later another Comet disintegrated over the Mediterranean. While experts sought a cause for these disasters, all Comets were grounded. Lengthy investigation finally revealed that the aircraft had a structural weakness around the windows. A redesigned Comet resumed commercial flights four years later, stretched to carry 72 passengers. It was just in time to make BOAC the first airline to operate a scheduled jet service across the Atlantic (with one stop at Gander), but by then the technological and commercial lead had been lost.

During the Comet's period out of service, the only jet airliner operating anywhere in the world was the Soviet Tu-104, another product of Andrei Tupolev's design bureau and another derivative of the Cold War nuclear bomber programme. From 1956 onwards it successfully established a number of scheduled services, including Moscow to Beijing via Irkutsk, and had an excellent safety

> "At jet speed you could circle the globe in 40 hours. The world was shrinking to half the size."
>
> HAROLD MANSFIELD
> *VISION: THE STORY OF BOEING*

record. The main beneficiary of the Comet debacle, however, was the American aircraft industry, which grasped the opportunity to catch up with and far surpass the foreign competition.

Boeing breakthrough

Until the 1950s Boeing had no great reputation as a manufacturer of commercial aircraft. Its strength was large military aeroplanes. But this provided the pathway to the aircraft that truly founded the era of the passenger jet: the Boeing 707. While Douglas, the world's leading manufacturer of civilian passenger aircraft, reflected the scepticism of most American airlines about the commercial feasibility of jet air travel, Boeing was already heavily involved in the development of large military jet aircraft such as the B-47 and B-52 bombers. Boeing boss Bill Allen reckoned that if he could make a jet aeroplane that would be saleable both to the Air Force and to civilian airlines, he would be able to make a profit.

In 1954 Boeing rolled out the prototype designated 367-80 and universally known as the "Dash 80". It was pitched to the USAF as primarily an in-flight refuelling aircraft to service the new jet bomber fleet, a role it eventually fulfilled under the military designation KC-135. For airlines, Boeing found it had to offer a significantly

TURBOFAN 707

Introduced into airline service in 1958, the Boeing 707 was the aircraft that made jet travel a commercial success. This 707 prominently displays the fact that it is fitted with turbofan engines, which became standard on the airliner from 1962.

different aircraft, increasing development costs, but orders soon began to flood in. Here was a four-engined jet with a range and payload that would transform long-distance travel.

The Boeing 707 came into service in October 1958, flying the Pan Am transatlantic routes from Idlewild Airport, New York, to Paris and London. Although configurations varied, the 707 carried more than twice as many passengers as the new model Comets and substantially more than the latest propliners. Flying at 960kph (600mph), it was far quicker than a propliner or even a turboprop. At a stroke, long-distance journey times were almost halved.

The 707 was not without its problems. Like all early jets, it may have been quiet for its passengers but it was howlingly noisy for people living around airports. Also, the first 707s entered

service with inadequate engines. These JT-3 turbojets had to be water-injected to produce enough thrust for take-off and only allowed a range of around 4,800km (3,000 miles). The 707s were soon, however, being fitted with more powerful and fuel-efficient fan-jet engines, as well as having a larger wing. The turbofan-engined 707-320B, which became the standard model after 1962, had a range of around 7,200km (4,500 miles), still far short of today's

long-haul airliners but perfectly adequate for intercontinental flights.

With the introduction of the Boeing 707, jet air travel took the world by storm. People simply did not want to fly with propellers any longer. Operators of turboprops began advertising them (accurately) as jet-powered, and TWA even dubbed its latest Constellation the "Jetstream" – presumably hoping to associate the propliner with some of the glamour of the jet age.

JET NETWORK
This route map shows Pan Am's worldwide passenger jet network in the 1960s. Note the massive blank on the map represented by the Soviet Union and China, no-go areas for American airlines.

Sud-Aviation Caravelle

LARGER CAPACITY
The Caravelle 12 was able to carry up to 140 passengers (the original Caravelle accommodated only 52). The passenger cabin, shown above, was comfortable and well protected from external noise (it is said that a whisper at the back could be heard in the front).

THE WORD MOST OFTEN ASSOCIATED with the Caravelle is "elegant": many aircraft enthusiasts consider that the design team led by Pierre Satre at Sud-Est Aviation was responsible for one of the most aesthetically pleasing aeroplanes ever produced. The Caravelle is emblematic of the "glamour era" of jet flight in the 1960s.

In 1951 the French government decided to promote the development of the country's first jet airliner. They issued a specification for an aeroplane with a range of 2,000km (1,242 miles) and capacity for about 60 passengers. Because the French did not possess a forceful enough powerplant for a two-engine design, the Sud-Est team initially submitted a three-engine design. Only after the Rolls-Royce Avon turbojet became available was a twin-engine design settled on.

The Caravelle prototype first flew in 1955, and by the time the first Caravelles entered service in 1959, Sud-Est had merged with Sud-Ouest to form Sud-Aviation. Initially, the omens for the aircraft were favourable. It was effectively the only short-to-medium haul jet airliner available, and its stylish looks provided strong passenger appeal. The first Caravelles were bought by Air France, the Scandinavian airline SAS, and the Brazilian airline Varig. In 1961, the aircraft broke into the valuable but notoriously patriotic American market with an order from United Airlines for 20 of the upgraded Caravelle VI model. Furthermore, Douglas, desperate for a product to run against the Boeing 727, seriously considered building Caravelles under licence. In 1962, Douglas signed an agreement to market Caravelles in the US.

This, however, turned out to be the high point of the Caravelle's fortunes. American airline operators

POPULAR IN THE SIXTIES
Although production of the Caravelle halted in 1972 (after 282 had been made), it was the most successful jet airliner developed by any country in western Europe. It paved the way for the later success of the Airbus series.

Stylish, sweptback wings free of engines

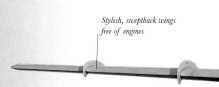

were thirsty for jet airliners to operate on shorter routes, and it turned out that the French plane simply was not economical enough – its operating costs were too high for the number of passengers it could carry. When the British BAC 1-11 entered the market in the early 1960s, it could offer a better ratio of passengers and cargo carried to fuel consumption. Douglas, meanwhile, decided to produce its own aircraft, developing the immensely successful DC-9, and Boeing followed up with the 737. In a now crowded marketplace, the Caravelle was elbowed out.

Passenger windows

Two flight crew in cockpit

Flightdeck well forward in nose

Port forward door

AIR PROVENCE

Nose-gear doors

Longer fuselage accommodates more passengers

PROTOTYPE IN PRODUCTION
Under construction here is the first Caravelle prototype (completed in 1955). In the background is the nose, and various other parts, of the second prototype.

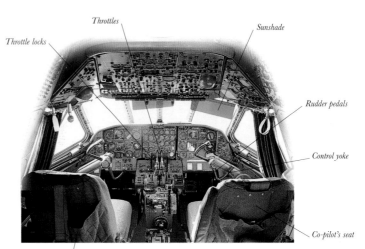

Throttles

Throttle locks

Sunshade

Rudder pedals

Control yoke

Co-pilot's seat

Pilot's seat

SIMILAR STYLINGS
Because the British De Havilland company gave Sud-Est access to information about its Comet, the Caravelle has a very similar flight deck layout (and an identical nose section) to the British airliner. The Caravelle's instrumentation is obsolete in modern airliners.

Toughened flightdeck windscreen

6,577kgp thrust Pratt & Whitney turbofan engine

Nose section the same as the de Havilland Comet

Nose-gear

ENGINES AT THE BACK
The Caravelle was the first airliner with its engines in pods at the rear of the fuselage. This had a number of advantages: it left the wings uncluttered, reduced noise inside the passenger cabin, and limited fire risk by placing the engines far from the fuel tanks.

Air Provence colours

Specifications (Caravelle 12)

ENGINE	2 x 6,577kgp (14,500lbst) Pratt & Whitney JT8D-9 turbofan
WINGSPAN	34.3m (112ft 6in)
LENGTH	36.24m (118ft 9in)
WEIGHT	31,800kg (70,107lb)
CRUISING SPEED	785kph (487mph)
CREW 2 (plus 4 cabin crew)	**PASSENGERS** 140

F-GCVL

AIR PROVENCE INTERNATIONAL

Tail cone

Main undercarriage

Engines situated at rear of fuselage

LONGER MODEL
Sud Aviation became part of Aérospatiale in 1970, and this was the company that built the last version of the airliner, the Caravelle 12 (shown here). The Caravelle 12 is a stretched model able to carry more passengers than previous models. It also has more powerful engines.

Other manufacturers rushed to compete with Boeing in the jet market. Douglas had hoped to trump the 707 with its jet-powered DC-8, which entered service only 11 months after the Boeing. But the DC-8 ran into production difficulties, and by the time these were ironed out the 707 had an unshakeable lead in the market. The success of the large jets led to a demand for smaller, shorter-range jet airliners that was quantitively far greater. Through the 1960s Boeing continued to show infallible judgment, scoring with the three-engined 727 and the two-engined 737.

European manufacturers chipped in with aircraft such as the Caravelle and the highly successful British Aircraft Corporation BAC One-Eleven. In 1966 Douglas produced the impressive short-haul DC-9, a resounding commercial success, but too late to save the cash-strapped company from a takeover by McDonnell the following year.

Shrinking world

One of the major worries that had made airline executives hesitant about jet airliners was the threat of overcapacity. The new airliners not only had more seats but, because of their faster journey times, could make twice as many flights a day. This offered a tempting economic prospect, as long as the seats could be filled. Yet there was no evidence that people would travel in such quantities. In 1957, before the jets were introduced, for the first time more passengers

SEXUAL EXPLOITATION
In the late 1960s and early 1970s, the blatant use of air hostess sex appeal to sell seats on airliners became universal practice in the industry. Here Southwest Airlines cabin staff adopt a thigh-revealing pose.

NOT A PLAIN PLANE
From 1965 Braniff decided to shake up its reputation as a reliable but dull airline by painting its aircraft in bold colours. The campaign pulled in extra customers by the thousand. This aircraft, Alexander Calder's DC-8 Flying Colors, *dates from 1973.*

crossed the Atlantic by air than by sea. But even if all the sea passengers converted to air travel, there would not be anything like enough of them to fill the fleets of jets rolling off the production lines. In the event, jet travel generated a large-scale movement of people that, without jet airliners, would not have existed at all.

Previously airliners had offered travellers an alternative to ships and trains for journeys they would have made in any case. Now people were tempted into making journeys when previously they would have stayed where they were, or at least not have travelled as far, especially since ticket prices were becoming more affordable. The growth of passenger miles flown on long-distance routes in the first decade of jet travel was astonishing – for TWA, for example, the figure rose from 4.6 billion in 1958 to 19.1 billion in 1969, and for Pan Am from 3.8 billion to 17.1 billion. In effect, jets did not just transform air travel; they changed the world. As Pan Am boss Juan Trippe, one of the first to sign up for the 707, proclaimed: "In one fell swoop we have shrunken the earth."

Jet-set era

The growth in passenger numbers inevitably required airports to expand and improve their facilities. Air-traffic control had to be upgraded to cope with the increasingly crowded skies. Fortunately, on-board systems were making notable progress in the 1960s – the first autopilot landing on a scheduled flight was achieved by a Caravelle in 1964. On the whole, jets could fairly claim to increase safety and reliability as well as speed and comfort. Passengers were amazed at how quiet and smooth air travel had become. Advertising the first coast-to-coast jet service in the United States in 1959, American Airlines boasted that "engine noise and vibration, the two factors that contribute most to travel fatigue, have all but vanished".

The new experience of jet flight became part of the wider phenomenon of the 1960s, with its

ebullient popular culture and its worship of youth, sex, and fun. There was still an air of glamour surrounding long-distance flight, but it was far from the propliner era of well-groomed sophisticates sipping cocktails in an airborne bar. Jet travel was slick, brash, and totally modern. The press invented the term "jet set" for the bright young things who partied one night in New York and the next in Swinging London. The steps down from a 707 provided a natural photo opportunity for starlets, and airports turned into arenas for welcoming pop stars – a phenomenon that reached its peak in the astonishing scenes

FUTURISTIC STYLE
Paris' Charles de Gaulle airport was one of a new generation of facilities built to cope with the rapid expansion of air travel brought about by the introduction of jet airliners. Its futuristic appearance reflected the new importance of airports as architectural style-setters – the cathedrals of the late 20th century, as train stations had been for the 1800s.

that attended the Beatles' arrival at Kennedy Airport, New York, in February 1964.

Airlines wanted flight to be sexy and fun. The alluring "air hostess" became a key weapon in the battle to fill seats. American company Braniff captured the mood in 1965 when it announced "the End of the Plain Plane". All its aircraft were painted in one of seven garishly bright colours – the press dubbed them the "jelly bean fleet" – and female flight attendants were fitted out in outrageous outfits designed by Italian fashion guru Emilio Pucci. One Braniff advertising punchline was: "Does your wife know you are flying with us?" The exploitation of sex appeal reached its apogee in a National Airlines advertising campaign in 1971, which grabbed the attention of potential passengers with a picture of a young flight attendant and the slogan: "I'm Linda. Fly me." This was extreme, but all airlines agreed that flight attendants should be female, young, and have a good figure. It was, no doubt, the spirit of the age.

THE BEATLES DISCOVER AMERICA
Arriving at New York's Kennedy Airport on Pan Am Flight 101 in February 1964, the Beatles were greeted by thousands of hysterical fans. Given the glamour and novelty of jet flight, airports seemed an apt location for the celebration of star status.

EARLY JET AND TURBOPROP AIRLINERS

THE BASIS OF THE AIRLINE BUSINESS is not technical innovation or elegance of design, but the economic equation of passenger numbers against operating costs. Jet engine technology did not at first seem an attractive proposition to airlines because of high fuel and maintenance costs and low passenger payload. Turbojets were developed in the postwar period to provide some of the comfort and speed advantages of jets, but with more economical operating costs. The Boeing 707 was the breakthrough jet that could carry enough passengers fast enough to make it profitable on long-distance routes. Its success in the 1960s left plenty of room at the bottom for less powerful airliners that could use smaller airports or operate on shorter routes where it was important to perform economically at slower speeds. The BAC 111, one of the few commercially successful British aircraft, catered for this market, as did Douglas's DC-9, but once again it was Boeing that tailored airliners most precisely to the needs of its customers with the 727 and 737.

CARAVELLE
The Caravelle was a trend-setting design – its rear-mounted engines were much imitated – but it did not carry enough passengers to operate profitably. See pages 384–85.

BAC 111

The BAC 111, which entered service in 1965, was based on a similar design to the Sud-Aviation Caravelle, the world's first short-haul jet airliner. An unusual feature of the design, copied by Hunting aircraft (part of BAC) was the tail mounting of the twin engines.

ENGINE 2 x 5,690kgp (12,000lbst) thrust Rolls-Royce Spey turbofan		
WINGSPAN 28.5m (93ft 6in)	**LENGTH** 32.6m (107ft)	
CRUISING SPEED 871kph (541mph)		
PASSENGERS 97	**CREW** 2	

Boeing 707-120

The development of the 707 was a major gamble by Boeing. Determined to break the Lockheed/Douglas monopoly, Boeing bet half the total value of the company on building the first dual-role military tanker/civil airliner. In 1955, in the face of intense competition from the DC-8, the first 707-120s were ordered by Pan Am.

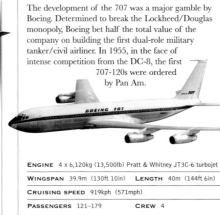

ENGINE 4 x 6,120kg (13,500lb) Pratt & Whitney JT3C-6 turbojet		
WINGSPAN 39.9m (130ft 10in)	**LENGTH** 40m (144ft 6in)	
CRUISING SPEED 919kph (571mph)		
PASSENGERS 121–179	**CREW** 4	

Boeing 727

Although Boeing built a shorter ranged 707, it could not use the smaller airports where passengers also demanded the speed and comfort of jet travel. The answer appeared in 1963 as the rear-mounted tri-jet 727, with its complex multi-flap wing married to a spacious fuselage with the same diameter as the 707. The design proved incredibly successful, becoming the world's best-selling jet transport with 1,831 built when production ended in 1984.

ENGINE 2 x 7,260kgp (16,000lbst) P&W JT8D-17 turbofan		
WINGSPAN 32.9m (108ft)	**LENGTH** 46.7m (153ft 2in)	
CRUISING SPEED 964kph (599mph)		
PASSENGERS 134–189	**CREW** 3	

Boeing 737-200

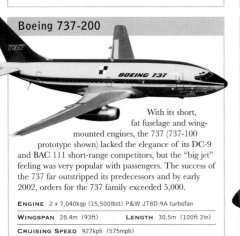

With its short, fat fuselage and wing-mounted engines, the 737 (737-100 prototype shown) lacked the elegance of its DC-9 and BAC 111 short-range competitors, but the "big jet" feeling was very popular with passengers. The success of the 737 far outstripped its predecessors and by early 2002, orders for the 737 family exceeded 5,000.

ENGINE 2 x 7,040kgp (15,500lbst) P&W JT8D-9A turbofan		
WINGSPAN 28.4m (93ft)	**LENGTH** 30.5m (100ft 2in)	
CRUISING SPEED 927kph (575mph)		
PASSENGERS 115–130	**CREW** 2	

de Havilland D.H.106 Comet 4B

The Comet 1 was the first jet airliner. In 1952, 10 were delivered to British Overseas Airways Corporation. They cut flying times by half, but a series of disastrous accidents led to the grounding of all Comet 1s by 1954. Lengthy investigations showed the accidents were caused by metal fatigue, a little known phenomenon at that time. The improved Comet 4 had longer range. On 4 October 1958, two Comets, one eastbound and another westbound, made the first jet-powered passenger flights across the Atlantic.

ENGINE 4 x 4,760kgp (10,500lbst) thrust R-R Avon 524 turbojet		
WINGSPAN 35m (114ft 10 in)	**LENGTH** 34m (111ft 10in)	
CRUISING SPEED 809kph (503mph)		
PASSENGERS 60–81	**CREW** 4	

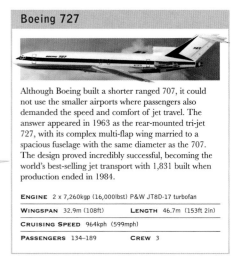

Douglas DC-9

Competing in the short-to-medium-range market with the BAC 111, which had a two-year head start, and the later Boeing 737, the DC-9, with over 2,200 sold, was much more commercially successful than the British aircraft. This was largely because the DC-9, like the 737, evolved into a large family of aircraft providing the right combination of range and payload for each airline customer. The original DC-9 Series 10 of 1965 was developed up to the 1977 DC-9 Series 80, which was evolved further as the McDonnell Douglas MD-80 series.

ENGINE	2 x 5,450kgp (12,000lbst) P&W JT8D-5 turbofan		
WINGSPAN	27.3m (89ft 5in)	LENGTH	32.8m (104ft 5in)
CRUISING SPEED	797kph (495mph)		
PASSENGERS	90	CREW	2

Fokker F.27 Mk200 Friendship

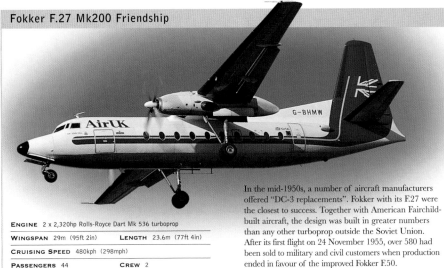

In the mid-1950s, a number of aircraft manufacturers offered "DC-3 replacements". Fokker with its F.27 were the closest to success. Together with American Fairchild-built aircraft, the design was built in greater numbers than any other turboprop outside the Soviet Union. After its first flight on 24 November 1955, over 580 had been sold to military and civil customers when production ended in favour of the improved Fokker F.50.

ENGINE	2 x 2,320hp Rolls-Royce Dart Mk 536 turboprop		
WINGSPAN	29m (95ft 2in)	LENGTH	23.6m (77ft 4in)
CRUISING SPEED	480kph (298mph)		
PASSENGERS	44	CREW	2

Ilyushin Il-18

ENGINE	4 x 4,190hp Ivchenko AI-20M turboprop		
WINGSPAN	37.4m (122ft 9in)	LENGTH	35.9m (117ft 9in)
CRUISING SPEED	674kph (419mph)		
PASSENGERS	100–120	CREW	4–5

Throughout the Cold War, the Soviet Union made a number of commercial aircraft which were based on Western designs. The Il-18 resembled the Lockheed Electra and the Vickers Vanguard. However, with over 550 built between 1957 and 1979, production of the Il-18 was more numerous than its western counterparts.. The "Moskva", as it was called by its manufacturer, was exported to most of the Warsaw Pact countries and many are still in use with central and eastern European carriers today. A military version (Il-20) also exists.

4,190hp Ivchenko turboprop engine

Tupolev Tu-104A

The arrival of the prototype Tu-104 at London Airport on 22 March 1956 was the first the West knew of the Soviet Union's first jet airliner. With the early Comets grounded and Boeing's 707 still two years from service, the impact was immense. Developed from theTu-16 "Badger" medium bomber, the Tu-104 began the first Soviet internal jet service in September 1956 carrying just 50 passengers. The improved 104A set a number of commercial records and was the most numerous out of all the Tu-104s built.

ENGINE	2 x 9,710kgp (21,385lbst) Mikulin AM-3M turbojet		
WINGSPAN	34.5m (113ft 4in)	LENGTH	38.9m (127ft 6in)
CRUISING SPEED	770kph (478mph)		
PASSENGERS	70	CREW	4

Vickers Super VC10

The VC10 was designed in 1958 as a new long-range airliner for BOAC. Though extremely popular with passengers, the VC10 series failed to win large orders because they were less economical than their 707 and DC-8 rivals. By 1969, VC10s were in use on all BOAC's long-distance routes, but were phased out in the early 1970s with the introduction of Boeing 747s.

ENGINE	4 x 10,205kg (22,500lb) Rolls-Royce Conway 550 turbofan		
WINGSPAN	44.6m (146ft 2in)	LENGTH	52.3m (171ft 5in)
CRUISING SPEED	914kph (568mph)		
PASSENGERS	174	CREW	9–11

Vickers Viscount 701

The Viscount was the first turboprop aircraft to operate a commercial passenger service and the first British airliner to achieve substantial sales in North America. The small 32-passenger prototype made its maiden flight in 1948 and entered BEA service in April 1953. The last aircraft was delivered in 1964, bringing the total number built to 444. The Viscount's pressurized and air-conditioned cabin, comparatively short landing run, and quiet, smooth turbine engines made it popular with passengers and airlines alike.

ENGINE	4 x 1,550hp Rolls-Royce Dart 506 turboprop		
WINGSPAN	28.6m (93ft 9in)	LENGTH	24.7m (81ft 2in)
CRUISING SPEED	517kph (323mph)		
PASSENGERS	59	CREW	3

1,550hp Rolls-Royce Dart propeller turbine

British Aerospace/Aérospatiale Concorde

THERE WAS ARGUABLY NOTHING quite as remarkable in the history of flight as a Concorde scheduled service, which saw over 100 passengers – without flying suits or helmets – travelling 17.7km (11 miles) above the Earth. They moved faster than a rifle bullet – yet, as a journalist observed, their coffee, "moving a mile every 2.7 seconds, doesn't even ripple."

When the French and British governments signed an agreement to design and produce a supersonic transport in 1962, they were convinced that the future of passenger aviation lay in ever-increasing speed and shorter journey times. The challenge was to produce an airliner that could cruise at Mach 2 – a speed only achieved by fighter aircraft for brief periods. Much ingenuity went into the task of designing Concorde. For example, because the craft flew steeply (tail-down) during take-off and landing, the pilots would not have been able to see forwards in a conventional plane. This difficulty was overcome by hingeing the nose to drop during these phases of the flight. The nose was raised hydraulically for supersonic flight and a visor lifted to streamline it to the windscreen.

The first prototype did not fly until March 1969; the first production aircraft appeared in 1973. Air France and British Airways put their first Concordes into service on the same day in 1976. But by then conditions for the launch of a supersonic airliner were abysmal. Attacks from the ecological lobby (responding to the threat of high levels of pollution and noise) meant that Concorde was banned from many of the world's airports. Spiralling fuel prices also deterred prospective buyers. As a result, only 16 production Concordes were built. Yet Concorde did establish its niche as a luxury or once-in-a-lifetime experience. And many agree with pilot Christopher Orlebar, who described Concorde as "an engineering dream, a beautiful aircraft, way ahead of anything... conceived since."

"It is not unreasonable to look upon Concorde as a miracle."

BRIAN TRUBSHAW
CONCORDE TEST PILOT

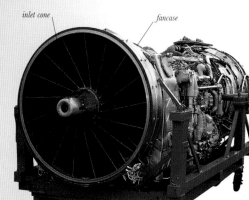

inlet cone

fancase

PUTTING IT ALL TOGETHER
An engineer is seen working on the forward fuselage of the 002 Concorde prototype, in 1968. The 002 was the first Concorde assembled in Britain.

AIRFLOW CONTROL
Concorde was powered by four Rolls-Royce/SNECMA Olympus turbojet engines. An ingenious feature of these engines was that no matter how fast the craft was moving, the speed of airflow to the engines remained less than 483kph (300mph).

Antenna for radio
navigation aid

Dorsal
fin

Tail cone

Titanium and
steel skin

G-BOAG

Engine cowling

Landing-gear door

Main landing gear

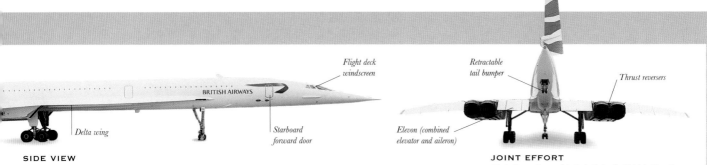

Flight deck windscreen

BRITISH AIRWAYS

Delta wing

Starboard forward door

SIDE VIEW

Retractable tail bumper

Thrust reversers

Elevon (combined elevator and aileron)

JOINT EFFORT
Each Concorde was jointly built by the British Aircraft Corporation (now British Aerospace) and Aérospatiale. For example, the British made most of the fuselage and the vertical tail; the French made the wings and control surfaces.

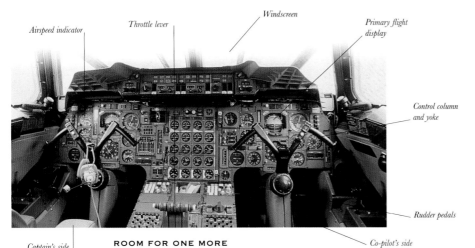

Airspeed indicator

Throttle lever

Windscreen

Primary flight display

Control column and yoke

Rudder pedals

Captain's side

Co-pilot's side

ROOM FOR ONE MORE
While the usual Concorde flight crew numbered three (the pilot, co-pilot, and another person behind the pilots looking after the systems-management panel), there was space for a fourth member, also behind the pilots. Despite the impressive technology that went into Concorde, the cockpit looks oddly antiquated.

CAREFUL INSPECTION
After the crash near Paris in 2000 (where all 109 on board, plus four on the ground, were killed), Concorde maintenance came under close scrutiny. Modifications were also made, including improved tyres and the addition of a Kevlar-based lining to strengthen the fuel tanks.

Passenger windows

Aerodynamic strake

Cockpit screen

Long, pointed nose (in raised position)

SH AIRWAYS

Standby pitot head

Narrow fuselage

Twin-wheel nose-gear

BUILT FOR SPEED
Everything about Concorde was optimized for high-altitude supersonic flight, including the elegant delta wing, the narrow fuselage, and the high-powered Rolls-Royce/SNECMA Olympus turbojet engines. The unavoidable consequences of this approach were relatively little room to carry passengers or cargo and high levels of engine noise and pollution.

Specifications

ENGINE 4 x 17,259kg (38,050lb) thrust reheat Rolls-Royce/SNECMA Olympus 593 Mk.610 turbojet

WINGSPAN 25.56m (83ft 10in)

LENGTH 61.7m (202ft 4in)

WEIGHT 79,265kg (174,750lb)

TOP SPEED 2,179kph (1,354mph)

CREW 3   **PASSENGERS** 128

In the 1960s everyone in the airline business assumed that supersonic commercial jets would be the next big thing. The history of airliners seemed to show that increase in speed was a constant factor. In the 1930s the DC-3 had flown at 290kph (180mph); in the 1940s the four-engined propliners flew at upwards of 480kph (300mph); the turboprops flew at around 640kph (400mph) in the 1950s; and the Boeing 707 was carrying passengers at 960kph (600mph) in the 1960s. Why should this acceleration of commercial flight cease? It seemed a fair assumption that, given the choice, passengers would always opt for the shortest journey time, and thus the fastest aeroplane. As with turbojets, there were daunting problems to overcome, especially in the commercially crucial relationship between fuel consumption, payload, and range. But with the supersonic transport (SST) apparently the new holy grail for the aircraft industry, designers and engineers bent to the task.

Supersonic showdown

Three SST projects developed through the 1960s: the Anglo-French Concorde; the Soviet Tupolev Tu-144; and, in the US, the Boeing 2707. They were so costly that only government money could cover the expense – even Boeing depended almost entirely on federal funding. In many ways the SST projects were close in spirit to their contemporary, the American Apollo moon-landing programme, driven by the same technological imperatives and heightened national pride, rather than by considerations of profit or practical advantage.

The Anglo-French and Soviet projects got under way more quickly, but the Americans were more ambitious, setting their sights on Mach 3 flight – around 3,200kph (2,000mph) – while their rivals settled for Mach 2.

The Concorde and Tu-144 developed into such superficially similar, slender, delta-wing designs that in the West there were inevitable rumours of espionage. In fact, the number of possible solutions to the problem of designing an SST was very limited, and it is hardly surprising that two teams independently came up with a similar shape – no more surprising than that all three SSTs had a movable nose (to give better visibility on landing). Boeing pursued a radically different design, with a variable-geometry wing. It was not only intended to fly faster than its rivals, but also to carry more than twice as many passengers.

The Tu-144 made its first flight in December 1968, followed by Concorde in March 1969. Meanwhile, the Boeing SST was in trouble: the variable-wing concept had had to be abandoned and the projected passenger payload scaled down.

DELTA DESIGNS
Various shapes of delta wing suggested for Concorde are lined up on display. The one eventually chosen, furthest from the camera, had elegant curves that were aerodynamically complex, contrasting with the more simple angular delta adopted by the Tupolev bureau for the Soviet SST.

Shattered dreams

By 1971 the US economy was in trouble and prestigious high-tech projects out of favour. Badgered by residents groups from airport districts and by the ecology movement, US Congress withdrew funding from the Boeing SST. With the American authorities no longer having any interest in promoting supersonic flight, Concorde faced an uphill struggle against protesters. A further blow came with the tripling of oil prices in 1973–74, which fatally undermined any prospect of operating fleets of fuel-guzzling SSTs as an economic proposition.

The Tu-144 project suffered a shattering setback at the Paris Air Show in June 1973, when one of the Tupolev SSTs disintegrated pulling out of a steep dive. The Soviets nonetheless pressed ahead with production and the aircraft entered service with Aeroflot in 1975, initially flying cargo. It proved uneconomical and impractical to operate. After another Tu-144 crashed in May 1978, production was halted and the attempt at supersonic services discontinued.

Concorde was far more successful. It went into service with British Airways and Air France in 1976 and, after a stiff fight, won the prized access to New York. Concorde cut the transatlantic journey time to three and a half hours, but it could only operate at a profit by charging an astonishing price for a ticket – £6,000 ($9,000) return by 2000. Concorde had become a prestigious curio, an elegant relic of a lost dream of supersonic passenger flight, long before commercial services ended on 24 October 2003.

ILL-FATED SOVIET SST

The Soviet Tupolev Tu-144 was the first supersonic airliner to fly, beating Concorde by two months. The Tu-144 programme suffered a severe setback when one disintegrated in the air at the 1973 Paris Air Show. In operation it suffered similar problems to Concorde, including uneconomically high fuel consumption.

BOEING SUPERSONIC

The Boeing SST was conceived in the 1960s, when funding was readily available for prestige projects that promised to push technology to new limits. Immensely ambitious, the SST was originally planned to fly at three times the speed of sound with 200 passengers on board.

But these technical setbacks were a relatively minor matter compared with the political storm that now confronted all the SSTs. No one in the aircraft industry had foreseen the scale of popular opposition that now awaited any innovation that would increase noise and pollution. When the first turbojets entered the scene in the 1950s, New York noise regulations were changed to accommodate them. But by the late 1960s times were changing. There was no way that the engines required for supersonic flight could avoid being loud on take-off and landing. They would also inevitably drag a sonic boom along their supersonic flight path. And the newly ascendant ecology movement claimed they would damage the ozone layer.

CONCORDE CRASHES

Concorde had an impeccable safety record until 25 July 2000, when one of Air France's fleet crashed after take-off from Charles de Gaulle airport, Paris, killing 109 passengers and crew. All Concordes were grounded. It was later established that one of the airliner's tyres had hit a strip of metal on the runway, shredding off rubber that struck fuel tanks in the bottom of the wing; kerosene poured out of a hole and caught fire. The Concordes returned to the air in 2001, with modifications that included lining some fuel tanks with Kevlar, the material used in bulletproof vests. But in spite of these improvements, three years after the crash all Concordes were retired from service.

CONCORDE IN FLAMES

The pilot of the Air France Concorde tried to nurse his aircraft as far as another Paris airport but the intensity of the fire gave him no chance.

Boeing 747 ("Jumbo Jet")

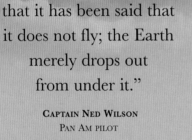

A NEW DAWN
Over 64m (210ft) long and with a tail reaching six storeys high, the first 747 (right) towers over several thousand guests at the roll-out ceremony in Everett on 30 September 1968. It made its maiden flight in February 1969.

THE BOEING 747 was the last product of the visionary era of commercial aviation. In 1965, legendary Pan Am boss Juan Trippe and Bill Allen of Boeing gambled the future of their respective companies on an airliner that would have over double the passenger capacity of any existing jet airliner. Trippe ordered 25 at a cost of \$550 million, allowing Boeing to finance its \$2 billion development costs. Boeing's starting point was an aborted project for a giant military transport aircraft, which it had lost out on after the contract went to Lockheed. The 747 was to be a passenger airliner on the same unprecedented scale. It would also be convertible to a freighter, since both Trippe and Allen suspected the future of passenger flight might lie with supersonic aviation, which would make the 747 redundant as a passenger carrier.

At Everett, north of Seattle, Boeing built an entirely new factory (larger in volume than any other building in the world) to make the giant airliner. Lack of time and capacity meant that much of the design and manufacture had to be contracted out to other firms, with Boeing responsible for the nose and wings. It was soon obvious that the aircraft was going to be too heavy to be powered by existing Pratt & Whitney turbofan engines. Great ingenuity was put into shedding weight, including a cutback on intended passenger load, pending an engine upgrade.

The 747 prototype was due to roll out in the autumn of 1968, but did not make its maiden flight until February 1969, seven weeks behind schedule. By the time it entered service in January 1970, over 150 747s had already been ordered by airlines across the world. Its teething troubles would not prevent it being a huge commercial success for Boeing, eventually making flight available to more people than ever before.

> "The Boeing 747 is so big that it has been said that it does not fly; the Earth merely drops out from under it."

CAPTAIN NED WILSON
PAN AM PILOT

Fin built up on two-spar box structure

Rudder divided into upper and lower segments

TESTING A TURBOFAN
Test engineers place a prototype turbofan engine into a test stand at the Pratt & Whitney Aircraft Plant in Connecticut. Many of the problems during the development of the 747 concerned its engines.

Rearmost passenger exit/entry door

City of

VH-OJE

JUMBO CARRIER
The 747 has undergone substantial upgrading since first entering service. Currently the world's largest, heaviest, and most powerful airliner, the 747-400 is the only model still in production. Its wingspan is wider than earlier models and the fairings, engine pylons, and nacelles have all been redesigned.

When wing is full of fuel, outer portions bend down and winglets are angled further outwards, increasing span by 48cm (19in)

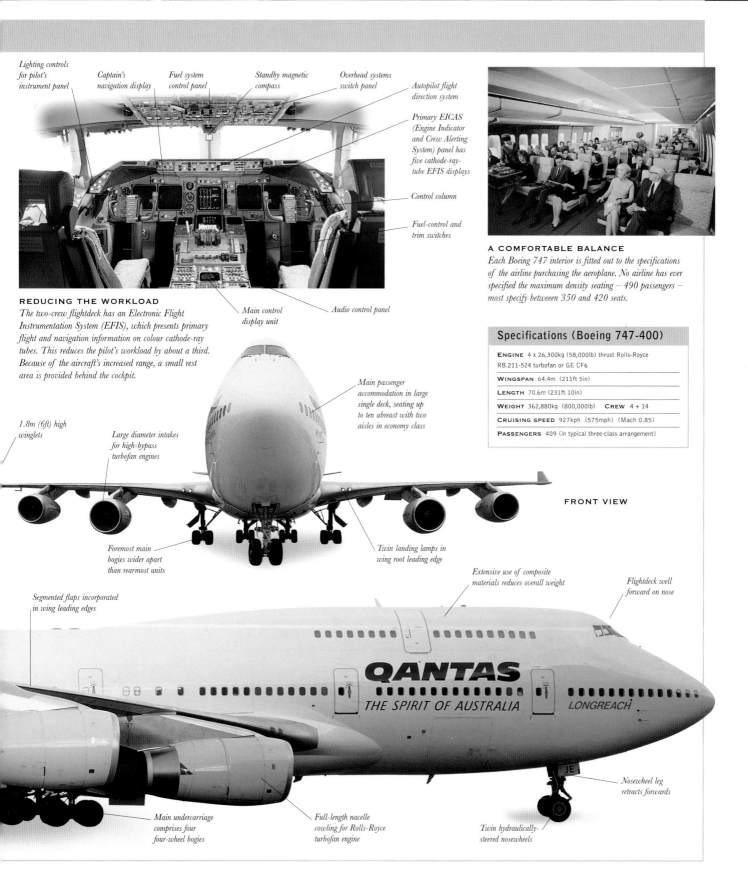

Lighting controls for pilot's instrument panel

Captain's navigation display

Fuel system control panel

Standby magnetic compass

Overhead systems switch panel

Autopilot flight direction system

Primary EICAS (Engine Indicator and Crew Alerting System) panel has five cathode-ray-tube EFIS displays

Control column

Fuel-control and trim switches

A COMFORTABLE BALANCE
Each Boeing 747 interior is fitted out to the specifications of the airline purchasing the aeroplane. No airline has ever specified the maximum density seating – 490 passengers – most specify betweeen 350 and 420 seats.

REDUCING THE WORKLOAD
The two-crew flightdeck has an Electronic Flight Instrumentation System (EFIS), which presents primary flight and navigation information on colour cathode-ray tubes. This reduces the pilot's workload by about a third. Because of the aircraft's increased range, a small rest area is provided behind the cockpit.

Main control display unit

Audio control panel

Specifications (Boeing 747-400)

ENGINE 4 x 26,300kg (58,000lb) thrust Rolls-Royce RB.211-524 turbofan or GE CF6	
WINGSPAN 64.4m (211ft 5in)	
LENGTH 70.6m (231ft 10in)	
WEIGHT 362,880kg (800,000lb)	**CREW** 4 + 14
CRUISING SPEED 927kph (575mph) (Mach 0.85)	
PASSENGERS 409 (in typical three-class arrangement)	

Main passenger accommodation in large single deck, seating up to ten abreast with two aisles in economy class

1.8m (6ft) high winglets

Large diameter intakes for high-bypass turbofan engines

FRONT VIEW

Foremost main bogies wider apart than rearmost units

Twin landing lamps in wing root leading edge

Extensive use of composite materials reduces overall weight

Flightdeck well forward on nose

Segmented flaps incorporated in wing leading edges

QANTAS THE SPIRIT OF AUSTRALIA LONGREACH

Nosewheel leg retracts forwards

Main undercarriage comprises four four-wheel bogies

Full-length nacelle cowling for Rolls-Royce turbofan engine

Twin hydraulically-steered nosewheels

Instead of supersonic speed, the way forward for passenger airliners turned out to lie in size and economy. Even in the 1960s, the widespread adoption of jet airliners brought downward pressure on seat prices. More people were travelling further than ever before, partly because it had become quicker, easier, and more pleasant to fly, but also because it was becoming cheaper.

Wide-bodied jets

The entry of the wide-bodied Boeing 747s into service in 1970 carried this process to a new level. The message of the 747's success was that air travel was going to be mass travel. Airports had to reinforce runways and expand passenger and baggage handling facilities, initially

swamped by 300 or 400 people disgorging at once from a single aircraft. In-flight caterers had to adjust to supplying their fare in previously undreamed of quantities. Hotels had to be built to cope with the rising tide of travellers.

The 747 represented no great technological breakthrough, just better engines and business daring. Other widebodies inevitably came in its wake as rival manufacturers sought to break Boeing's increasing domination of the market. They could not challenge the giant 747 head-on, but sought out a market share among airlines for which the 747 was just too big. McDonnell Douglas came up with the DC-10, Lockheed produced the TriStar, and a new consortium of European manufacturers, Airbus Industrie, built the Airbus A300. The DC-10 and TriStar were

SPACIOUS INTERIOR
The spacious interior of a 747 is a familiar sight today, but when the airliner was first brought into service it astonished passengers used to narrow cabins with a single aisle.

three-engined airliners competing for the same market niche; they inevitably ran into commercial difficulties because there was not enough room for both of them. Lockheed, who came off worse in this contest, never made an airliner again. The Airbus A300, on the other hand, was a major success. With only two engines and two crew members in the cockpit, it marked a significant step forwards in economy of operation. By the 1980s Airbus Industrie had established itself as Boeing's most vigorous competitor on the world market.

Increased competition

The introduction of the wide-bodied jets was not an immediate commercial success. From 1973, economic recession slowed the rise in passenger numbers just as oil price hikes made the airliners vastly more expensive to fly – a 747 uses more than 13,500 litres (3,000 gallons) of fuel an hour. The new 747s and DC-10s often flew three-quarters empty, giving passengers the temporary advantage of impressively spacious accommodation. Some desperate airlines used the cavernous space to provide airborne piano bars or restaurants. Once the recession was over, of course, airlines went back to packing in passengers, sacrificing space and comfort to hold down prices.

The difficult years of the 1970s left some airlines with precarious finances – in 1975 TWA

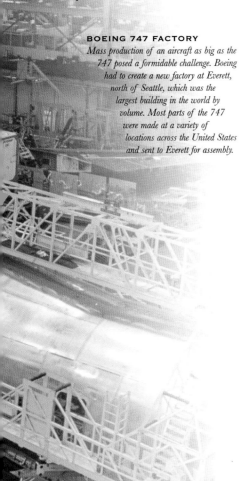

BOEING 747 FACTORY
Mass production of an aircraft as big as the 747 posed a formidable challenge. Boeing had to create a new factory at Everett, north of Seattle, which was the largest building in the world by volume. Most parts of the 747 were made at a variety of locations across the United States and sent to Everett for assembly.

had to sell some of its 747s to Iran to raise the cash to pay its staff's wages. Another blow to the long-established major carriers was the progress of deregulation from the late 1970s. This was primarily an American phenomenon, although mirrored to a degree worldwide. Free (or at least freer) competition on air routes forced prices down further and soon brought some famous casualties, notably Braniff in the 1980s, and then PanAm, which stopped services in 1991. Even in Europe, where national support for flagship airlines continued, the irrepressible rise of charter flights undercut the official carriers. Small operators were able to buy perfectly viable second-hand airliners, as the major airlines felt forced to renew their fleets long before their existing aircraft had exhausted their usefulness. The airline business had also long ceased to be largely the preserve of Europe and North America, with companies such as Singapore Airlines and Cathay Pacific becoming among the world's most prestigious airlines.

ECONOMICAL AIRBUS
Airbus Industrie shook the airliner market with the introduction of the A300 into service in 1974. It was the first widebody airliner to fly with only two engines and two cockpit crew. This made all its variants impressively economical to operate.

From the 1970s jet flight became an experience open to everyone. As such it inevitably lost its connotations of glamour, romance, or excitement. The European consortium's choice of the name "Airbus" for its product spoke volumes about the prosaic nature of air travel in the age of the wide-bodied jets. Most passengers on a 747 did not even have a window. Much effort was devoted to distracting passengers from the fact that they were flying at all, insulating them with in-flight entertainment and reducing them to passive consumers of duty-free goods, snacks, and drinks. One jaundiced journalist referred to modern air travel as "the most constrained form of mass transport since the slave ships".

DC-10 ENGINE
The fan-jet engines used to power widebody jetliners are astonishing pieces of equipment. A turbofan such as this one from a DC-10 can generate over 25,000kg (50,000lb) of thrust – almost three times the thrust of each engine on a 1960s Boeing 707. The extra power translates into increased aircraft size and payload.

FLYING A MODERN AIRLINER

A MODERN JET IS TYPICALLY FLOWN by a flight deck crew of two – the pilot, and a co-pilot who acts as navigator and flight engineer. The number of aircrew has been reduced through the use of computers and improvements in instrumentation and navigational equipment. In the latest airliners, colour screens have mostly superseded dials for the display of data on altitude, airspeed, position, and other essentials.

Navigation has achieved a high degree of accuracy. It still partly relies on radio beacons that provide waypoint markers along airways, although these now operate on VHF. The copilot tunes in a radio compass to each successive beacon along the marked route. On-board inertial navigation systems (INS) will plot the aircraft's movements to within about a mile tolerance on a long-haul flight, if its position is set correctly at the start of the journey. Sophisticated airports have VOR (VHF omnidirectional range transmitters) with DME (distance measuring equipment), allowing the aircrew to track their position relative to their destination. In recent years, GPS satellite

A320 COCKPIT
The Airbus A320 was the first airliner with fly-by-wire controls and a "glass cockpit", with a small number of coloured display screens replacing the dozens of potentially confusing dials and gauges found in earlier airliner cockpits.

DISPLAYS
The modern instruments system centres around multi-functional screens, including a navigation display (right), and primary flight display (below), which includes an artificial horizon.

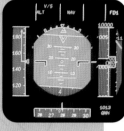

positioning has been added to the array of navigational improvements. Satellites also provide superb weather forecasting, a major contribution to safety and comfort.

For most of a flight, the pilot's main task is to supervise the work of the autopilot and other on-board computers. ILS (instrument landing systems) are now of high quality at many airports. Responding to signals from a transmitter on the runway, on-board ILS tells the pilot if he is too high or too low, to the left or right of the correct approach path. ILS linked to an autopilot can deliver automatic landing in poor visibility. But safety still depends on pilots' alertness, judgement, and flying skill.

The power of the world-shrinking jets to promote social and economic change was unquestionable.

As early as the 1960s, for example, the tranquil Mediterranean island of Majorca was abruptly transformed into a package-holiday destination through the building of an airport – its shores quickly becoming lined with high-rise hotels. The same fate befell the Portuguese Algarve, the Canary Islands, and some parts of Greece. By the 1990s, the falling cost of long-haul flights had extended mass tourism to Thailand and Goa, Sri Lanka and Gambia, transforming the lives of local inhabitants.

Tourism also grew into a dominant economic activity in the world's major cities: the centres of London, Paris, and New York became as much destinations to visit as places to work or live in. Nor was all non-business travel for pleasure: millions of Muslims were able to make the pilgrimage to Mecca thanks to aircraft. Meanwhile freight carried in jet transports altered patterns of consumption. Soon no one was surprised to find fresh fish from an African lake in a British supermarket or fresh flowers from Mauritius decorating a restaurant table in Chicago.

While airliners carried on changing the world, after 1970 the world of airliners changed comparatively little. The last 30 years of the 20th century brought no further revolution in speed or size. But progress in engine design made aircraft quieter and more fuel efficient, and increased operating range – the Boeing 747-400 could fly almost 11,000km (7,000 miles) with more than 400 passengers on board. There was also great progress in avionics and navigation systems. In the late 1980s the Airbus A320 introduced digital fly-by-wire controls to civil aviation, as well as the so-called "glass cockpit". The Boeing 777, introduced into service in 1995, was the first American fly-by-wire airliner.

Prospects for passenger travel
It was hard to feel that the chief executives of aircraft manufacturers or airlines in the 1990s had much in common with the flight-besotted empire builders of an earlier era. This tended to be a business like any other, dominated by marketing gurus and accountants. The kind of vision that produces entirely new aircraft ideas was in short

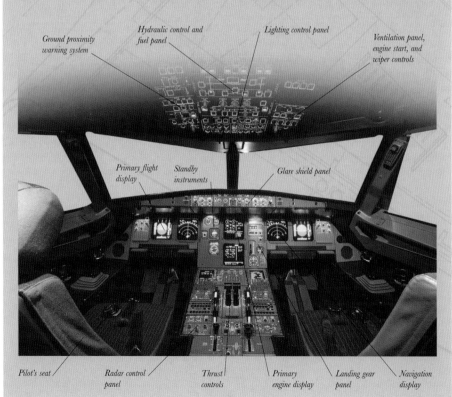

Ground proximity warning system

Hydraulic control and fuel panel

Lighting control panel

Ventilation panel, engine start, and wiper controls

Primary flight display

Standby instruments

Glare shield panel

Pilot's seat

Radar control panel

Thrust controls

Primary engine display

Landing gear panel

Navigation display

supply. The accepted wisdom was that revisiting the notion of supersonic flight was unlikely to prove the way forwards, but that a further leap in aircraft size might make economic sense. From the mid-1990s, Airbus began touting the idea of a super-jumbo, the A3XX, which would be capable of carrying 600 to 800 people. Redesignated the A380 and shrunk somewhat to a typical payload of 550, it made its maiden flight in April 2005 and was set to enter airline service in 2006.

Wrong-footed by the successful development and marketing of the A380, Boeing struggled to catch up with the launch of the 787 Dreamliner, due to enter service in 2008.

The key selling point of the 787 was neither speed nor size, but fuel efficiency. At the start of the new millennium, projections were for future air passenger traffic tripling in volume to four billion journeys a year by 2020. There were constraints that might affect this expansion, particularly concerns about the effect of aircraft on global warming. But this did not mean that any serious question mark hung over the future of mass air travel. A triumph of technology and organization, it had given human beings an undreamed-of mobility, and it was here to stay.

FUTURE INTERIOR
A mock-up of the interior of the Airbus A3XX giant widebody gave an impression of the comfort to which its designers aspired. Experience suggested that ample legroom and champagne bars would, in practice, give way to demand for maximum passenger numbers.

AIRBUS A3XX
In the 1990s, Airbus Industrie took the bold decision to challenge the aging Boeing 747 by developing an even bigger widebody. Originally designated the A3XX, this was to be the largest airliner ever built, and was intended to carry more than 600–800 passengers. The A3XX eventually evolved into the A380. As pictured here, the A3XX differed from the A380 in significant details such as the shape of the fin and nose and the size of the wings.

MODERN AIRLINERS

THE INTRODUCTION OF THE BOEING 747 in 1970 marked the start of the contemporary era in air travel. The fiercely competitive nature of the airliner market since then has led to the fall of several great names. Lockheed pulled out of airliner production after unsuccessfully running the L-1011 TriStar against the DC-10, and Douglas itself became McDonnell-Douglas, before being absorbed into Boeing. In Europe, Airbus Industrie emerged as a serious rival to Boeing, successfully challenging the American company's dominance of the world market. Once it became clear that the future did not lie with supersonic travel, the aircraft themselves were inevitably committed to a process of evolution rather than revolution. Jet engines became progressively more powerful, but also more fuel-efficient and quieter. Avionics improved rapidly, with fly-by-wire controls and improvements in all forms of electronic equipment.

AIR FORCE ONE
"Air Force One" is a 747-200B, extensively modified to meet the US President's needs. See pages 394–95.

Airbus A320-200

Aimed at the 150-seat market, the advanced A320 first flew in February 1987. With its fully computerized cockpit and fly-by-wire control system, the design set the standard by which all future airliners were judged. Based on the A320, Airbus developed a further two jets: the A319 (up to 120 passengers) and the A321 (up to 220 passengers). All three remain in production.

ENGINE	2 x 11,350kgp (25,000lbst) CFM56-5A1 turbofan	
WINGSPAN	33.9m (111ft 3in)	LENGTH 37.6m (123ft 3in)
CRUISING SPEED	903kph (550mph)	
PASSENGERS	150–179	CREW 2

Airbus A330-200

The combined A330/A340 programme was launched in June 1987, with both designs sharing the same wings, airframe, cockpit, and systems. Only the number and type of engines differed. The first to fly was the long-range, four-engined A340 in October 1991, followed by the medium-range, twin-engined A330 in November 1992. Airbus now offer many permutations of passenger capacity and range.

ENGINE	2 x 30,870kgp (68,000lbst) Rolls-Royce Trent 768	
WINGSPAN	60.3m (197ft 10in)	LENGTH 59m (193ft 7in)
CRUISING SPEED	880kph (547mph)	
PASSENGERS	256–293	CREW 2

Boeing 737-700

Boeing offered its "Next Generation" family of medium-range 737s (737-600 onwards) with a new wing permitting a higher cruise speed and more efficient CFM International engines giving longer range. The 737-700 and –800 first flew in 1997, and the –900 in 2000. By summer 2006, Boeing had delivered more than 2,000 737 NGs and over 3,000 other 737s.

ENGINE	2 x 10,990kgp (24,200lbst) CFM56-7B turbofan	
WINGSPAN	34.3m (112ft 7in)	LENGTH 33.6m (110ft 4in)
CRUISING SPEED	853kph (530mph)	
PASSENGERS	126–149	CREW 2

Boeing 757-300

The narrow-body (single-aisle) 757 was developed in parallel with the wide-body 767. Conceived as a replacement for the 727, the first 757 flew in 1979, a few months after the initial 767. Offered with a choice of engines from either Pratt & Whitney or Rolls-Royce, sales were rather slow at first but improved into the late 1980s, as the Rolls-Royce version became the preferred choice. Production ended in 2005, with 1,050 built.

ENGINE	2 x 19,750kgp (43,500lbst) R-R RB211-535E4B turbofan	
WINGSPAN	38.1m (124ft 10in)	LENGTH 54.5m (178ft 7in)
CRUISING SPEED	853kph (530mph)	
PASSENGERS	240–289	CREW 2

Boeing 767

The 767 was developed as a medium- to long-range, wide-body (twin-aisle) airliner in parallel with its narrow-body sister, the 757, and first took to the skies in September 1981.
Sharing an identical cockpit to the 757, a pilot trained on one could fly the other without additional training, offering considerable savings to airlines buying both. The largest of the family is the stretched -400ER (Extended Range) which entered service in September 2000. By 2006, Boeing had manufactured 926 of the 767 family and 767s were estimated to have made almost 8 million flights.

28,830kgp General Electric turbofan engine

ENGINE	2 x 28,830kgp (63,500lbst) GE CF6-80C turbofan	
WINGSPAN	51.9m (170ft 4in)	LENGTH 61.4m (201ft 4in)
CRUISING SPEED	853kph (530mph)	
PASSENGERS	245–375	CREW 2

Boeing 777-300

Designed to fit in passenger capacity between the Boeing 767 and the renowned 747, the 777 programme was announced in October 1990. The design featured many firsts. It was the first airliner to be digitally designed using 3-D computer graphics, eliminating the need for a costly, full-scale mock-up; and the first Boeing product to have fly-by-wire flight controls and an advanced computerized cockpit. The first of the 777 family, the 777-200, went into service in July 1995. The latest, the 777-200LR, first flew in 2005.

ENGINE	2 x 38,140kgp (84,000lbst) Rolls-Royce Trent 884 turbofan	
WINGSPAN	60.9m (199ft 11in)	LENGTH	73.9m (242ft 4in)
CRUISING SPEED	905kph (564mph)	
PASSENGERS	368-550	CREW	2

Douglas DC-10-30

The last in the line of Douglas Commercials, the tri-jet DC-10 entered service with both American and United Airlines in August 1971. Despite being in direct competition with the Lockheed TriStar, and early catastrophic accidents, the DC-10, particularly the longer-range -30 version, was the more successful of the two. When production ended in 1989, 386 civil DC-10 had been built plus 60 military KC-10 tanker transports for the USAF. The design evolved into the MD-11.

ENGINE	3 x 23,840kgp (52,500lbst) General Electric CF6-C1	
WINGSPAN	50.4m (165ft 5in)	LENGTH	55.5m (182ft 1in)
CRUISING SPEED	982kph (610mph)	
PASSENGERS	250-380	CREW	3

Fokker F.50

Building on the outstanding success of the F.27, Fokker announced the F.50 in November 1983. Although based on the F.27-500 fuselage, the design had new engines, advanced six-bladed propellers, a computerized cockpit, and other improvements. In March 1996, Fokker collapsed due to financial problems, after some 220 F.50s had been delivered or were on order.

ENGINE	2 x 2,250shp Pratt & Whitney PW 125B turboprop	
WINGSPAN	29m (95ft 2in)	LENGTH	25.3m (82ft 19in)
CRUISING SPEED	532kph (331mph)	
PASSENGERS	58	CREW	4

Ilyushin Il-96M

Although resembling the earlier Il-86, the Il-96 was virtually a new design which first flew in September 1988. Equipped with Soloviev PS-90 turbofans, some 15 were built before the appearance of the IL-96M in 1993. With a stretched fuselage, western computerized avionics, and American engines, the new variant also came in a IL-96T freight version. Despite this attempt to widen the commercial appeal, the future of the IL-96 remains uncertain. By 1999, a number of unsold airframes were awaiting engines at the Voronezh Aircraft Plant of the Ilyushin Association.

ENGINE	4 x 16,800kgp (37,000lbst) P&W PW2337 turbofan	
WINGSPAN	60.1m (197ft 3in)	LENGTH	63.94m (209ft 9in)
CRUISING SPEED	880kph (547mph)	
PASSENGERS	312-375	CREW	2

Lockheed L-1011-500 TriStar

The last Lockheed airliner, the TriStar would lose out to its DC-10 rival because of the choice of Rolls-Royce engines. The development costs of the highly advanced RB211 engine bankrupted Rolls-Royce in 1971, just four months after the L-1011's first flight in November 1970. With the RB211's technical and financial problems overcome, the TriStar joined Eastern and TWA in April 1972. Depite introduction of the longer-range -500 version, total production was limited to 250.

50,000lbst thrust Rolls-Royce turbofan engine

Shortened fuselage for longer-range -500 model

Increased-span wings for longer-range operations

ENGINE	3 x 22,700kgp (50,000lbst) R-R RB211-524 turbofan	
WINGSPAN	50.1m (164ft 4in)	LENGTH	50.1m (164ft 3in)
CRUISING SPEED	960kph (597mph)	
PASSENGERS	246-330	CREW	3

McDonnell Douglas MD-11

Following the merger of McDonnell and Douglas, the DC-10 was re-launched in December 1986 as the modernized, stretched, and re-engined MD-11. Offered as a passenger airliner, a freighter, or combination of both, the first MD-11 was delivered in December 1990. Production continued after the 1997 takeover by Boeing and it remained in the Boeing sales list in early 2002.

ENGINE	3 x 27,240kgp (60,000lbst) P&W PW4460 turbofan	
WINGSPAN	51.7m (169ft 6in)	LENGTH	61.2m (200ft 10in)
CRUISING SPEED	945kph (588mph)	
PASSENGERS	323-405	CREW	2

McDonnell Douglas MD-87

The original DC-9 Series 10 of 1965 became the DC-9 Series 80 by 1977, and then the MD-80 series in 1983, following the McDonnell merger. It was very successful with over 1,400 delivered by December 1999. The MD-80-87, abbreviated to just MD-87, featured a shortened fuselage. The MD-90 is a re-engined version of the MD-80, the MD-95 variant of which was relaunched as the Boeing 717.

ENGINE	2 x 9,080kgp (20,00lbst) P&W JT8D-217C turbofan	
WINGSPAN	32.9m (107ft 10in)	LENGTH	39.8m (130ft 5in)
CRUISING SPEED	811kph (504mph)	
PASSENGERS	139	CREW	2

Tupolev Tu-154

By the mid-1990s, the Tu-154, with approximately 900 built, was the standard medium-range airliner throughout the former USSR. Designed in the mid-1960s with the ability to operate from the more rural areas of Russia, the first tri-jet 154 flew in October 1968. The improved Tu-154M appeared in 1982, with quieter and more economical engines and continued in production into the 21st century. The type is slowly being replaced by the twin-jet Tu-204.

ENGINE	3 x 10,610kgp (23,380lbst) Aviadvigatel turbofan	
WINGSPAN	37.6m (123ft 3in)	LENGTH	47.9m (157ft 2in)
CRUISING SPEED	950kph (590mph)	
PASSENGERS	180	CREW	3-4

FEAR OF FLYING

STATISTICS PROVE THAT FLYING IS BY FAR THE SAFEST WAY OF TRAVELLING LONG DISTANCES, BUT AIR ACCIDENTS INSPIRE A MORBID FASCINATION

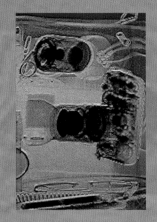

"There had been no deliberate plan, but in elaborating the aeroplane, and in doing all they could to calm those that flew in it, they had created… the most infernal conditions in which to die."

JULIAN BARNES
FROM THE NOVEL *STARING AT THE SUN*

THE AIRLINE INDUSTRY has always known that its success depends on convincing the public that air travel is safe. This has never been an easy task. The drama of major air disasters impresses itself so intensely on the public consciousness – partly, no doubt, precisely because they are rare – that flying is often inextricably associated in people's minds with sudden and violent death. Yet measures to reduce the number of air accidents and aviation-related deaths may undermine the image of air travel as a normal, safe, everyday experience. The more safety procedures air

passengers are subjected to, the less secure they are likely to feel. Surely flying cannot be that safe if we are searched before boarding and flight attendants insist on telling us where the oxygen masks and emergency exits are? For the nervous,

DOUGLAS DC-10
All DC-10s were grounded in 1979 after a crash at Chicago killed 271 people. A badly designed cargo door was blamed for another DC-10 accident near Paris in 1974 when 346 people died.

there is nothing quite so disquieting as constant reassurance "for your comfort and safety".

Yet the figures are unequivocal. Although accident statistics fluctuate from year to year, flying on a commercial airliner always emerges as by far the safest way of travelling long distances. In 1996, for example, a relatively bad year for aviation deaths, a total of 1,187 people were killed on commercial jet flights worldwide. This compares with over 40,000 people killed that year in road accidents in the United States alone, and worldwide probably a quarter of a million road-accident deaths. Flying is far from being equally safe in different parts of the world: in a typical year, the United States might have one flight fatality for every two million passenger-hours flown, while Africa might have 13 fatalities per million flight hours. But even in Africa you are more likely to be killed or injured driving to and from the airport than on board the aeroplane. The risk of a fatal accident each time you board an airliner has been calculated at roughly three in a million. This means that if an otherwise immortal individual made a flight every day, he or she could expect, on average, to survive for over 900 years before dying in an air accident. (Flying in a private aircraft carries a quite different risk – it is almost 50 times more dangerous than flying in a commercial jet.)

Progress on safety has been the necessary condition for the development of mass air travel. In the early 1930s, there was a fatality for every 4.8 million passenger-miles flown in the United States. In a single, admittedly exceptional, period in the winter of 1936–37, there were five fatal air crashes in the US in 28 days. Translated into the contemporary world of widebody jets, a 1930s-style accident rate would have produced a totally unacceptable mass of fatalities. By the 1980s, American airlines flying major routes had reduced the death rate to around one for every 300 million passenger-miles. Even so, recent decades set all the records for air disasters, because of the large numbers of passengers on a single flight. The worst year for air-accident fatalities worldwide was 1985, with 2,129 people killed – although 1,105 of the victims died in just three incidents. In 2005, a worse than average year for aviation in recent times, there were 1,050 deaths worldwide. To put this in perspective, there were by then some 1.2 billion passengers being carried

LOCKERBIE DISASTER
A terrorist bomb was responsible for the destruction of the Pan Am Boeing 747 that exploded above the small Scottish town of Lockerbie in December 1988. Despite constant efforts to upgrade security at airports, it remains extremely difficult to prevent the smuggling of bombs or weapons on to airliners.

each year. Many of the world's major airlines had not had a fatality in the 21st century.

The safety of commercial flying is a triumph of organization and regulation, and a tribute to the professionalism of all involved in the aviation business, from those who make the airframes, engines, and avionics, through the ground maintenance staff and flight crews to administrators and air-traffic controllers. The volume of traffic that air-traffic control has to cope with has, of course, increased dramatically in the jet age. By the late 1990s there were some 7,000 flights a day into and out of New York. To look at it another way, controllers at Chicago's O'Hare airport were responsible for the safety of around 70 million passengers a year. But despite occasional panics about overstretched air-traffic controllers being overwhelmed by numbers, the system has continued to cope well. So has the system of periodic checks and overhauls designed to ensure that aircraft are fit to fly, with faultless engines and free of structural weaknesses. Considering what amazingly complex machines modern aircraft are – a Boeing 747 has about 4.5 million moving parts – it is astonishing how rarely they suffer serious faults. A modern jet may have ten hours ground maintenance for every hour it spends in the air.

Human error

Humans cannot be engineered to operate as faultlessly as machines. Airline pilots of course undergo rigorous training. They have mandatory medical check-ups every six months and their flying time is limited to prevent dangerous levels of stress and fatigue building up. Yet pilot error is still

WORLD'S WORST CRASH
In the aftermath of a collision between two Boeing 747s at Tenerife airport in 1977, dazed survivors try to help the injured. Images such as this, shown in the media worldwide, inevitably have more impact than the recitation of safety statistics.

held to be a contributory cause of about three quarters of air accidents. These may range from simple mistakes in reading instruments – for example flying into the ground after misreading an altimeter – to confused reactions in an emergency, as in the British Midlands Boeing 737 crash in 1989, when, with one engine failed, the crew shut down the other one, which was working.

Improvements in cockpit equipment have done a lot to aid pilots and to counter the human factor. The presentation of flight information on-screen in the new "glass cockpits", for example, should bring an end to the simple misreading of indicators. As well as a whole gamut of navigational aids and instrument landing systems, there are now ground-proximity

CROWDED AIRPORT
Air-traffic controllers are responsible for ensuring that take-off and landing at a busy airport do not result in a collision. Airliners make a quick initial climb to carry the noise from their engines as high as possible over surrounding housing.

warning devices that tell the pilot to "pull up" when he is too close to the ground, and others that warn of an impending stall or a collision course with another aircraft. There is an increasing tendency to go one step further and make the avionics automatically override the pilot if he is committing an error – pull the aircraft up when it is on a collision course with the ground, for example, or block a movement of the controls that might make the aircraft stall.

Creating a catastrophe

When accidents do happen, it is more often through a combination of factors than a single error or fault. The worst aviation disaster of the 20th century occurred in 1977 when two Boeing 747s, one operated by KLM and the other by Pan Am, collided on the ground at Tenerife in the Canary Islands, killing 582 people. The accident happened in poor visibility. Like many airports, Tenerife had no ground radar, so aircraft on the runways could only be tracked by visual observation, but the two 747s were invisible from the control tower, as well as to one another. The American aircrew, who had difficulty understanding the Spanish controller's English, were trying to follow instructions for taxiing, searching for the right turn off the runway. The Dutch KLM pilot, in a hurry and confused, assumed he had clearance to take off when he did not, and accelerated down the runway into the taxiing Pan Am 747. Weather conditions, substandard air-traffic control equipment, poor communications, and pilot error had all conspired to create a catastrophe.

AIR TRAFFIC CONTROL

IN ITS BASICS, THE SYSTEM for organizing air traffic devised in the 1930s and 1940s remains valid in the jet age. Before take-off a pilot files a flight plan and is cleared for a certain route by air-traffic control (ATC), which lays down the time and altitude at which he must pass waymarks along an airway. Each airliner has a "squawk" transponder, which emits a signal that identifies it to ATC radar, allowing a controller to track the aircraft. Controllers ensure that aircraft are separated vertically and horizontally – typically at least ten minutes flying time apart when on the same level, or 300m (1,000ft) apart by altitude.

The ATC centres that control the airways each cover a wide area, and may not even be situated at an airport. When an airliner draws

MODERN ATC ROOM
Computers have helped air-traffic controllers cope with the density of traffic, although air safety still depends on controllers' intelligence and steadiness under pressure.

near to its destination, it is handed over to approach controllers at the airport. They either place it in a holding pattern or in the line of landing aircraft. Approach controllers are in the airport control tower, but normally behind tinted glass because they work solely with radar screens. Aerodrome controllers, at the top of the tower behind clear glass, depend on radar and visual observation to deal with aircraft during taxiing, take-off, and landing. They take over as the aircraft makes its final approach and talk it down.

TRAGIC COLLISION
A Boeing 727 with 135 people on board plunges in flames after colliding with a Cessna light aircraft over San Diego, California, in September 1978. It was a clear day and the Cessna pilot was operating on Visual Flight Rules. All on board the 727 were killed, plus two people in the Cessna and two on the ground.

Common hazards

Some of the hazards of the jet age are as old as flying itself. In October 1960, for example, a Lockheed Electra turbojet taking off from Boston flew into a dense swarm of starlings, which brought it down, killing 62 people. Slush on runways has been the cause of a number of notorious crashes, including the Munich air disaster that killed so many of the young Manchester United football team in 1958. Flying by instruments has taken much of the sting out of low cloud and fog, but it is still safer to fly when you can see where you are going. This is especially true around the more hazardous airports – pilots do not look forward to instrument landing where air-traffic control is not of the highest standard and flight paths bring the aircraft in low over or between mountains or hills.

Aircraft still need to avoid thunderstorms or typhoons, although improved weather forecasts and on-board weather radar have made this far

easier to achieve. Violent winds, notably the downward waves of air sometimes encountered in the lee of mountains, can still be treacherous for modern planes – one such caused a Uruguayan aircraft to come down in the Andes in October 1972, a crash that became notorious when it was revealed that some passengers survived until rescue by eating the dead. Wind shear, a sudden change in wind direction or velocity that occurs where different bands of air meet in the jet stream, can cause airliners to drop vertically for hundreds of feet, a terrifying and potentially fatal experience for passengers.

Mid-air collisions

As has been mentioned, the first mid-air collision between airliners happened in 1922. Fears about such events are stoked by scary statistics about the number of reported close encounters between aircraft around busy major airports. The world's worst ever collision, between a Saudi Boeing 747 and a Kazakh Airways Ilyushin Il-76 outside Delhi, India, in 1996, killed 350 people. But even if something has gone wrong, it requires a lot of bad luck to actually hit another moving object in what is actually quite a large three-dimensional space. Tragic incidents from the propliner era, such as the collision of a United Airlines DC-7 and a TWA Constellation over the Grand Canyon in 1956, and of a TWA Super Constellation and a United DC-8 over Staten Island in 1960, were examples both of flaws in air-traffic control and of extreme ill fortune. The presence of private aircraft in the same airspace as commercial jets

creates the greatest potential for collisions – as was demonstrated in September 1978, when a Pacific Southwest Airlines 727 collided with a much slower light aircraft over San Diego, an accident that cost more than 150 lives.

Mechanical faults

Serious flaws in aircraft can elude the usually rigorous testing procedures of manufacturers and regulatory authorities. In 1959–60, for example, two of the newly introduced Lockheed Model 188 Electra turbojets crashed through a wing falling off as a result of a weakness in the engine nacelles. America's biggest single air accident, a DC-10 crash at Chicago in May 1979, which killed all 271 people on board, was caused by the failure of a bolt, leading to the separation of an engine from the pylon assembly – all

RIVER CRASH-LANDING
Air-accident investigators often have to work under difficult conditions. This aircraft operated by Indonesia's Garuda airline crash-landed in the Bengawan Solo river in Klaten, Central Java, in January 2002. Indonesian investigators can be seen inspecting the wreckage and searching for debris in the river.

OVERSIZE LOAD

DC-10s were judged prone to the same problem and temporarily grounded. The fatal vulnerability of Concorde's fuel tanks to a tyre blow-out was revealed in the Paris crash in 2000.

Air-crash investigations, which can identify such faults, are obviously a vital part of the safety system. Few human activities are pursued quite as rigorously and painstakingly as the collection and examination of every fragment of a crashed aircraft. This sometimes occurs under the most trying circumstances. When a ValuJet DC-9 crashed in the alligator-infested Florida Everglades in May 1996, it was swallowed by the swamp leaving not a trace on the surface. Two months later, Boeing 747 Flight TWA 800 exploded in the air off Long Island, and investigators had to search the Atlantic seabed for pieces and reassemble the airliner like a giant jigsaw puzzle. But, finally, conclusions were reached in both these cases: the ValuJet DC-9 had been carrying an inflammable cargo; an electrical fault on the TWA 747 had probably ignited fuel tanks.

BLACK BOX RECORDERS

IN-FLIGHT RECORDERS have been mandatory on airliners since 1957. Although habitually referred to as black boxes, they are bright orange with reflective stripes. At first the boxes contained only a Flight Data Recorder (FDR), keeping a record of the aircraft's airspeed, altitude, compass headings, and so on. From 1966 a Cockpit Voice Recorder (CVR) was added. This has traditionally been a 30-minute magnetic tape loop – self-erasing, so that it always contains a record of the half hour before any crash. CVRs are being upgraded to solid-state two-hour recorders. The black boxes are virtually indestructible. None has ever been crushed by force of impact, although a very few have been disabled by fire. If an aircraft crashes in the ocean, divers can still find the in-flight recorder by homing in on a locator beacon that emits an ultrasonic signal.

IN-FLIGHT RECORDER
The so-called black box is fitted in the tail of the aircraft and linked to microphones in the cockpit and flight instruments by cables running along the fuselage.

Carrying handle · Connections to aircraft systems · Recorder motor · Eight-track magnetic tape · Kevlar lining insulates recorder against heat of fire

RECONSTRUCTED FUSELAGE
This length of Boeing 747 fuselage was pieced together from fragments of TWA Flight 800, which crashed in the Atlantic shortly after take-off from JFK Airport, New York, in August 1996. Thousands of pieces of the jet were retrieved from the ocean and painstakingly reassembled.

The importance of accident investigators' work was underlined by a notorious case in which their findings were not sufficiently heeded. A report into a near-fatal incident involving a DC-10 at Detroit in 1972 indicated that the aircraft's cargo door represented a hazard, since it could appear to be securely locked when it was not. Two years later a DC-10 flying from Paris to London crashed into the forest of Ermenonville, killing all 346 people on board – the first widebody jet crash and the worst air accident at that date. Investigation revealed that once again the

DAWSON'S FIELD HIJACKING
Hijacked by Palestinian terrorists en route from Bahrain to London in September 1970, a BOAC Vickers VC-10 waits on the desert airstrip at Dawson's Field in Jordan, while negotiations decide the fate of the aircraft, passengers, and crew.

cargo door had not been securely fastened and had disastrously burst open at an altitude of 3,500m (11,500ft). An official report later concluded that the "risks had already become evident 19 months earlier… but no efficacious corrective action had followed".

In our heavily armed and deeply divided world, the hazards of flight include military action. Such disasters can be purely accidental, as when an airliner was shot down over the Black Sea by a stray Ukrainian missile during military exercises in 2001. But they are more likely to be deliberate acts committed in an atmosphere of tension and paranoia. As early as 1955, an El Al Constellation that inadvertently crossed the Iron Curtain into Bulgarian airspace was shot down by interceptors. In 1973 a Libyan Boeing 727 that strayed over the Sinai desert – part of Egypt but then occupied by Israel – was brought down by Israeli interceptors, killing 106 people. Ten years later a Korean Air Lines 747, Flight 007, inexplicably wandered into Soviet airspace

on a night trip from Anchorage in Alaska to Seoul in South Korea, and was shot down by a Sukhoi Su-15 with air-to-air missiles. All 269 people on board were killed. During a tense period in the Gulf in 1988, the US Navy cruiser *Vincennes* shot down an Iran Air Airbus with a SAM missile, killing 286 people, after somehow interpreting the radar profile of a scheduled airline flight as that of a jet fighter on an attack mission.

Hijacking

Such actions on the part of regular armed forces have been rare, however, compared with the depredations of hijackers and terrorists. From the start hijacking was encouraged by an often ambivalent attitude of many states to the perpetrators. The phenomenon began in the 1950s with individuals escaping from communist-ruled Eastern Europe – three airliners were hijacked from Czechoslovakia to West Germany on a single day. The hijackers received a hero's

SYMBOLIC DESTRUCTION
The Palestinian terrorists begin the destruction of the three airliners they had hijacked to Dawson's Field – a VC-10, a Boeing 707, and a Douglas DC-8. With their crew and passengers removed, the airliners were "sacrificed" as symbols of Western technology and capitalism.

welcome in the West. In the early 1960s the focus of hijacking was Cuba, recently taken over by Fidel Castro's guerrillas. In May 1961 a man armed with a knife and a gun ordered the pilot of a National Airlines service from Miami to Key West to fly him to Havana. It was America's first hijacking and made "Take this plane to Cuba" a catchphrase. In response, hijacking became a federal offence in the United States and air authorities worldwide agreed to counter it. But in practice little was done. Until the 1970s boarding an aircraft anywhere in the world was a casual matter, involving no more than, where appropriate, the long-established formalities of passport control and customs checks.

In these lax circumstances, by the late 1960s hijacking had become a growing epidemic, with 33 occurrences in 1968 and 91 in 1969. Many were carried out by disturbed individuals with no clear rational motivation. Some were straight-

SCATTERED WRECKAGE
Accident investigators had an awesome task after the Lockerbie bombing in 1988. Because the explosion took place at high altitude, the wreckage was spread over 2,200 square km (850 square miles) of countryside. Evidence from the wreckage eventually helped to convict two Libyans.

forwardly mercenary – as in the case of the hijacker who registered as D.B. Cooper on a Boeing 727 flight in Washington state in 1971, and parachuted out of the aircraft with $200,000 ransom money in a sack, never to be seen again. But the most frightening trend was set in the summer of 1968, when members of the Popular Front for the Liberation of Palestine diverted an El Al airliner to Algiers and held the Israeli passengers hostage, negotiating the release of Arab prisoners in Israel. An impressively successful operation from the Palestinian point of view, it revealed that airliners constituted a vulnerable point at which otherwise well-defended states – and world capitalism – could be attacked, with potential for maximum propaganda impact on the media. As the world entered the era of international terrorism, airliners and airports found themselves in the front line.

Terrorist operations
By 1970 three types of terrorist operation had become established: hijacking and hostage-taking; attacks on passengers and airliners on the ground, initiated at Athens airport in December 1968; and the planting of bombs on board airliners set to explode in mid-air. Palestinian groups soon extended their attacks from Israeli

targets to all Western airlines, both because America had emerged as Israel's main backer in the Middle East, and because they had learned to identify their local cause with a worldwide struggle against "Western imperialism". In September 1970, an astonishing terrorist spectacular was staged involving the hijacking of four airliners (a fifth hijack attempt was foiled). One of the planes, a Pan Am 747, was flown to Cairo and blown up on the ground after the passengers and crew had been moved off. The three others – a Swissair DC-8, a TWA Boeing 707, and a BOAC VC-10 – were diverted to the Dawson's Field airstrip in Jordan. After lengthy negotiations, the hostages were also removed from these aircraft and they too were blown up. Staged in front of the world's TV cameras, this blazing funeral pyre of prized Western technology was a symbolic gesture skilfully crafted to appeal to the powerless and poverty-stricken peoples of the Middle East and beyond.

Security measures
The response of governments and air authorities to the rising tide of hijackings and terrorism was patchy and often laggardly. Israel and its airline, El Al, understandably introduced the most rigorous security measures and were scornful of the reluctance of other countries and airlines to spend sufficient money or impose the delays and restrictions on their passengers that safety required. Yet much was done. In the United

ENTEBBE RETURN
A crowd at Lod airport, Tel Aviv, welcomes back Israeli soldiers and freed hostages after the raid on Entebbe in July 1976. In the background is one of the C-130 Hercules transports used in the raid.

States, for example, pre-flight checks on baggage were finally introduced in 1972. Armed air marshals were placed on selected flights. A rash of D.B. Cooper copycats parachuting to the ground with ransom money was stopped by simple design changes that made it impossible to jump out of an airliner with a parachute. In 1974 the ICAO, the governing body of international aviation, agreed minimum security standards that all airlines had to meet, including checks on baggage loaded in the hold and on passengers and their hand luggage, plus measures to prevent unauthorized access to aircraft.

Israeli and Western governments took an increasingly hard line over hijackings, leading to spectacular incidents that were not necessarily reassuring for nervous passengers. In July 1976, the Israelis flew paratroops into Entebbe airport, Uganda, where Palestinian and German Baader-Meinhof terrorists were holding Jewish passengers from an Air France Airbus hostage, with the complicity of Idi Amin's Ugandan government. Thirteen terrorists were killed in the assault; three hostages were also killed, but the rest were taken back to Israel and safety. In October 1977, another combined Palestinian–Baader-Meinhof hijacking ended with an assault on the hijacked aircraft by German and British special forces at Mogadishu, Somalia, in which three terrorists were shot dead and the hostages were freed.

Tightening security and a tougher response definitely made it harder for hijackers and terrorists to operate, but too many loopholes existed in the system for them to be stopped completely. In June 1985, for example, TWA Flight 847 was hijacked by Islamic Hezbollah terrorists from Lebanon, who murdered one American on board and held 39 others hostage for 17 days, while Sikh extremists blew up an Air India jumbo jet over the Atlantic, killing all 329 people on board.

Stopping explosive devices being hidden in baggage was a challenge that seemed to defy solution. Just before Christmas 1988 a Pan Am 747 flying from Frankfurt to New York via Heathrow exploded at 9,400m (31,000ft) over the small Scottish town of Lockerbie, killing 259 people in the air and 11 on the ground. The cause was a bomb in the baggage hold.

11 September 2001

To a few individuals this sorry catalogue of misery and disaster was an inspiration. Islamic extremists such as Osama bin Laden, obsessed by the goal of humiliating the United States with spectacular acts of terror, saw airliners as both a vulnerable and a symbolic target, just as the Palestinians had in 1970. What represented Western civilization more completely than the aeroplane, so often a key element in Western dominance? In the 1990s, the American public was informed that the FBI had uncovered an extraordinary plot by Islamic extremists to hijack 12 airliners simultaneously and destroy them in an airborne holocaust. It sounded the stuff of fantasy, but was not.

"The pictures of airplanes flying into buildings, fires burning, huge structures collapsing, have filled us with disbelief, terrible sadness, and a quiet, unyielding anger..."

PRESIDENT GEORGE W. BUSH
IN HIS ADDRESS TO THE NATION, 11 SEPTEMBER 2001

On 11 September 2001, terror in the air reached its apogee. As had happened before, terrorists exploited carefully researched weaknesses in the system, targeting American domestic flights, on which security was relatively lax, and relying on blades rather than guns – much easier to pass through security checks. They also exploited the received wisdom that governs response to hijacking – the well-drilled pattern of non-resistance and co-operation with the hijackers in which aircrews are trained, and the fearful but calm compliance of passengers. No one had thought of hijackers having trained as pilots, turning airliners into guided missiles. Of the four besieged aeroplanes, one was slammed into each of the Twin Towers of New York City's World Trade Center and one struck the Pentagon building in Washington, D.C. The fourth aircraft fatally crashed in a field in rural Pennsylvania when the heroic passengers and crew overpowered their assailants. By the time it was clear that these hijackings were different, it was, in all but one case, too late to resist.

The impact of the events of 11 September on air travel was severe but shortlived. According to pollsters, in the aftermath of the terrorist attacks 40 per cent of Americans said they would not fly. Airlines already under pressure from a world economic downturn and rising oil prices faced bankruptcy. But a combination of tighter security measures and the simple passage of time without further disaster soon restored public confidence in air travel. From 2002 to 2005, the number of fatal air accidents averaged 32 a year. The presence of armed air marshals on US flights and time-consuming security checks before boarding probably did not reassure everyone. But passenger flight had become as safe as it could get.

SECURITY MEASURES
In the wake of the 11 September terrorist hijackings a tightening of security measures was in evidence at the world's airports. Above are items confiscated from travellers at Beijing's main airport, and, right, a new scanner is demonstrated at Orlando International.

TRADE CENTER TERROR
As one tower of the World Trade Center burns, a second airliner heads for the tower's twin. Using airliners as human-guided missiles, with their fuel to wreak destruction in place of a warhead, the terrorists were able to inflict vast damage while armed with nothing more than sharp blades.

SMALL IS BEAUTIFUL

SOME OF THE MOST SPECTACULAR FLYING FEATS OF MODERN TIMES HAVE BEEN ACHIEVED IN LIGHTWEIGHT AIRCRAFT OR EVEN BALLOONS

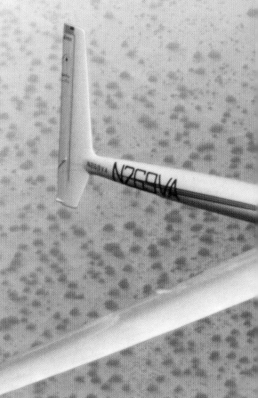

O**N 14 DECEMBER 1986,** at 8.00am, former fighter pilot Dick Rutan and his partner Jeana Yeager set out from Edwards Air Force Base in California to fly non-stop around the world, without in-flight refuelling. Their propeller-driven aircraft, *Voyager*, had been designed by Dick Rutan's younger brother, aeronautical engineer Burt Rutan. The pilots were to make the journey of over 42,000km (26,000 miles) at an average speed of only 175kph (110mph), in a cockpit and cabin that were barely 0.6m (2ft) wide.

Voyager was made of low-density high-strength carbon composites, so light that the airframe weighed only 426kg (939lb). That weight was virtually doubled by the two engines, but dwarfed by the fuel load, over 3,200kg (7,000lb) at take-off. The wing was 34m (111ft) long – longer than that of a Boeing 727 – but only 1m (3ft) across, a shape offering maximum lift and minimum drag. Like the fuselages, it was packed with fuel.

Dick Rutan has said that before the flight he was not looking forward to "being inside this flailing carbon structure, a long way from home, listening to engines and hoping they keep running". *Voyager* had never taken off with a full fuel load before the round-the-world attempt, and only just made it, using almost all of Edwards Air Force Base's 4,500m (15,000ft) runway, with the tips of the fuel-filled wings dragging on the ground. It was very fragile in its fuelled-up state and could not have landed if it had got into trouble.

Rutan and Yeager benefited from excellent state-of-the-art instruments, including an autopilot, weather radar, and the latest navigational aids, but once they headed out over the Pacific they were on their own, at grips with the unknown. Weather proved a terrifying hazard as they dodged thunderheads and a typhoon. Over Africa on the fifth day, they had to fly up to 6,000m (20,000ft) to go above a line of thunderstorms, and both suffered from hypoxia through not breathing enough of their oxygen supply.

"To fly! To live as airmen live! Like them to ride the skyways from horizon to horizon, across rivers and forests... Ah! That is life."

HENRI MIGNET
L'AVIATION DE L'AMATEUR, 1934

PEGASUS XL SE MICROLIGHT
Microlights are often in effect powered hang-gliders, controlled by pivoting the wing, although some have an aeroplane-style control system. Flying one of these machines creates an experience that is in some ways similar to that of the early flight pioneers.

VOYAGER PILOTS

*Jeana Yeager and Dick Rutan
attend to Voyager, the
aircraft that they flew non-stop
around the world without
refuelling in December 1986.
They travelled at only 175kph
(110mph), the speed that
would give the maximum
distance for every gallon of
fuel. Although they alternated
at the controls, in principle
allowing enough rest, fatigue
became a serious problem
during the nine-day journey.*

VOYAGER IN FLIGHT

*Like the Wright brothers, Voyager designer Burt Rutan
favoured canard designs, with stabilizers at the front
instead of on the tail. In the Voyager, these canard
surfaces were extended to link twin side fuselages that were
in effect long fuel tanks. Between them was the central
fuselage with a cockpit and a tiny cabin, and two engines
front and back driving pusher and tractor propellers.*

GOSSAMER CONDOR
*Created by a team headed by Dr Paul MacCready in
California, the* Gossamer Condor *was the first aircraft to
demonstrate sustained, controlled, human-powered flight,
winning the Kremer prize in August 1977. It was
made of the lightest available materials and designed
to maximize lift at very low speeds.*

The worst moment came on the
last night of the flight, when the
engine stopped over the Pacific north
of Mexico and they glided silently in
the darkness for five minutes before
regaining power. The following morning they
reached Edwards AFB, 9 days, 3 minutes, and
44 seconds after taking off, completing a journey
of 42,120km (26,178 miles). There were 48kg
(106lb) of fuel left in the tanks.

Return of the pioneers

In the era of widebody jets and fly-by-wire
fighters, the *Voyager* flight was a project that
harked back to more romantic times, when
pioneering individuals could produce innovative
designs in a small workshop or make record-
breaking flights in private planes. It showed that
the spirit of the Wright brothers and Lindbergh
was still very much alive – that there were still
gifted and courageous experimenters who
regarded flight as an adventure and sought out
areas to explore that had been neglected in the
onward march of the big battalions.

PEDAL POWER
The Gossamer Albatross *takes
off on a test flight pedalled by
Bryan Allen, the cyclist who would
fly the aircraft across the English
Channel in June 1979. Whereas
dreamers before the era of flight
had imagined people flying like
birds, with flapping wings,*
Gossamer Albatross *used the
power generated by leg muscles.*

The latter part of the 20th century saw a return to the fascination with "flying like a bird" that had inspired the early flight pioneers. The growth of hang-gliding as a leisure activity allowed thousands of people to, in effect, repeat the experiences of Otto Lilienthal, while the microlight, essentially a hang-glider with an engine attached, is a remote descendant of the powered glider Percy Pilcher was working on when he died in 1899. In the 1970s, it even proved possible to fulfil the dream of human-powered flight, long dismissed by most as a physical impossibility.

Human-powered flight

As early as 1680, Italian mathematician Giovanni Borelli had calculated that human muscles could not generate enough power for bird-like flapping flight. "It is impossible", he wrote, "that men should be able to fly craftily, by their own strength." Until very recently, Borelli seemed to be right. Several cash prizes were put up in the 1930s for the first controlled, human-powered flight, but none was claimed. A human athlete is capable of generating around 0.4hp for sustained periods; the Wright brothers' first aero-engine had generated 12hp. Manpower, it seemed, was just not power enough.

In 1959 a British industrialist, Henry Kremer, set out to reignite interest in human-powered aircraft with a substantial cash prize for anyone who could fly a figure-of-eight course around two pylons at least 800m (2,625ft) apart. Regularly increased to keep pace with inflation, the prize stimulated numerous experiments, but remained unclaimed for 18 years. However, in 1976 a team led by aeronautical engineer Dr Paul MacCready of Pasadena,

IMITATING DAEDALUS

ACCORDING TO LEGEND, the first human-powered flights were made by Daedalus and his son Icarus, escaping from the island of Crete. In 1984, professors and students at the Massachusetts Institute of Technology (MIT) decided to see if they could build an aircraft to match this legendary feat. They assumed that the goal of the flight had been the smaller island of Santorini, about 118km (74 miles) north of Crete. Their design followed a pattern by then well established – lightweight materials, a long narrow wing, and a propeller driven by pedal-power. It still took four years and millions of dollars to create an aeroplane that might manage the distance. *Daedalus* took off from Crete's Heraklion airbase early on the morning of 23 April 1988, with Greek champion cyclist Kanellos Kanellopoulos in the cockpit.

FLYING WING
Daedalus, *the human-powered aircraft that made the record-breaking flight from Crete to Santorini, was created by staff and students at MIT.*

Fortuitously aided by a following wind, he headed across the sea at about 32kph (20mph), flying some 5m (16ft) above the waves. The journey to Santorini took 3 hours 54 minutes and proceeded without mishap until the final approach to land on a beach, when a sudden gust of headwind broke up the aircraft a few metres off shore. A gleeful Kanellopoulos emerged unscathed to celebrate a record for distance and endurance.

LIGHTWEIGHT DAEDALUS
Daedalus *weighed just 31kg (69lb) empty, despite having the wingspan of an airliner, at 34m (112ft). Like the Gossamer series, it used pedal-power to drive a propeller.*

California, decided to take up the challenge. Starting from scratch, with no reference to current powered-aircraft designs, they built an aircraft called the *Gossamer Condor*. It was made of light, plastic-covered aluminium tubes and its long, narrow wing was braced with stainless-steel wires, the leading edge reinforced with corrugated cardboard and styrene foam. Intriguingly, it shared some essential features with the Wright Flyer – a canard stabilizer at the front, a pusher propeller, and wing-warping for lateral control. The pilot powered the aircraft by pedalling.

By the summer of 1977, MacCready's team was ready for an attempt at winning the Kremer prize, which by then stood at £50,000. The pilot was to be Bryan Allen, a champion cyclist and hang-gliding enthusiast. At Shafter, California, on 23 August 1977, Allen powered the *Gossamer*

Condor along a figure-of-eight course for a distance of 2.15km (1⅓ miles) at a speed of around 17.5kph (11mph). The MacCready team was able to carry off the prize because the aircraft was so light – the pilot weighed more than the machine – and so perfectly designed to maximize lift and minimize drag. When Kremer put up an even larger prize for the first human-powered flight across the Channel – shades of Blériot – the MacCready team won that too. With Allen again as pilot, on 12 June 1979 the *Gossamer Albatross* flew from Folkestone, England, to Cap Gris-Nez in France, a distance of 35km (22 miles).

MacCready subsequently turned his attention to improving the performance of his human-powered aircraft with a little help from the sun. By placing solar cells on the surface of the wing, he was able to generate electricity to assist the pedalling pilot in his gruelling task. In July 1981, driven by a blend of electric power and the leg muscles of pilot Steve Ptacek, MacCready's

ROUND THE WORLD IN 19 DAYS

BREITLING ORBITER 3
Orbiter 3 is the balloon that won the race to fly non-stop around the world. The silver aluminized coating on the envelope acted as insulation, helping keep the gas inside at an even temperature.

GLOBAL CHALLENGERS
Left to right, Steve Fossett, Richard Branson, and Per Lindstrand pose in front of the balloon ICO Global Challenger, *in which they failed to fly around the world in 1998. Both Fossett and Branson were rich men who repeatedly risked their lives in pursuit of the round-the-world record.*

Solar Challenger flew 261km (163 miles) from just outside Paris to Manston, Kent, in the south of England. Battery-assisted manpower became an important area of experimentation, with aircraft achieving speeds in excess of 50kph (30mph). But human-powered flight remained the preserve of the extremely fit. It was also sensitive to weather conditions – any strong wind would destroy the fragile aircraft in no time.

Record-breaking ballooning
Another activity that attracted adventurers and experimenters in the late 20th century was ballooning. Although interest in the first form of manned flight had never totally faded, it had long appeared a technology incapable of much further development. However, in 1931 Swiss physicist Auguste Piccard invented a pressurized gondola that allowed balloonists to ascend to previously fatal heights. Balloon ascents became both a scientific means of exploring the upper reaches of the atmosphere and a field of international competition to establish altitude records.

In 1935, the American *Explorer II* balloon, crewed by Albert Stevens and Orville Anderson, reached an altitude of 22,056m (72,395ft), a record that was not broken for two decades.

More exciting from a media and sporting point of view was long-distance ballooning. All the spectacular record-breaking achievements of the aeroplane pilots of the 1920s and 1930s – first transatlantic flight, first non-stop transcontinental flight, and so on – stood to be repeated in unpowered lighter-than-air craft. By the 1970s, balloon technology and, crucially, the understanding of winds and weather patterns, had

BERTRAND PICCARD and Brian Jones became the first humans to circle the earth in a balloon on 20 March 1999, as the *Breitling Orbiter 3*, travelling at 210kph (130mph), crossed an invisible finishing line 11,000m (36,000ft) above west Africa.

Orbiter 3 was a hybrid of a helium and hot-air balloon, a type known as a Rozier. The towering silver envelope was filled with helium, while at the bottom a cone-shaped bag was filled with hot air. This was used as variable ballast, allowing the balloonists to gain or lose altitude easily. The gondola was far removed from the wicker basket of early balloonists. It was more like a space capsule, fully pressurized and packed with high-tech equipment.

Like all aspirant round-the-world balloonists of the 1990s, Piccard and Jones took advantage of the jet stream, which from December to March flows eastwards in the mid-latitudes at speeds of up to 320kph (200mph). Starting from Chateau d'Oex, Switzerland, they flew south across Spain to pick up the jet stream over north Africa. From then on, manipulating the hot air to make the balloon rise and fall, they were able skilfully to locate winds that carried them on a remarkably direct route east, at around 20 degrees north of the equator. They spent much of the journey at above 9,100m (30,000ft), enduring very low temperatures.

Hoping to finish the voyage in style, Piccard and Jones tried to reach the Pyramids at Giza, but ended up coming down in the Sahara and waiting hours for a helicopter to pick them up. Their epic journey had taken 19 days, 21 hours, and 55 minutes.

INSIDE VIEW
Inside the gondola of the Breitling Orbiter 3 most of the available space was devoted to highly advanced communications and navigation equipment.

SOLAR CHALLENGER
The use of solar energy to power flight has been explored in aircraft such as the Solar Challenger. This used the sun's rays, transformed into electricity, to boost the efforts of a pedalling pilot, giving better speed and range than purely human-powered flight. Aircraft have since been produced that are entirely solar-powered (see page 425).

improved to the point where such flights were feasible, although still extremely dangerous. The first 17 attempts to fly by balloon across the Atlantic all failed, with the loss of seven lives. Then, on 11 August 1978, Ben Abruzzo, Maxie Anderson, and Larry Newman, hailing from Albuquerque, New Mexico, set off from Presque Isle, Maine, bound for France in the helium balloon *Double Eagle II*. They named their gondola the *Spirit of Albuquerque*, in reference to Lindbergh's

Spirit of St Louis, and took Paris' Le Bourget airport, Lindbergh's landing point, as their goal. On the evening of 17 August, short of ballast and with daylight fading, they set down in a wheatfield 96km (60 miles) short of Paris. They had covered 4,960km (3,100 miles) in 137 hours 6 minutes.

Over the next two decades other records fell – the first non-stop flight across the North American continent in 1981, the first solo transatlantic balloon flight (by Joseph Kittinger) in 1984, the first solo transpacific flight (by Steve Fossett) in 1995. There remained the ultimate goal of a non-stop flight around the world.

An extremely expensive contest, in the late 1990s round-the-world ballooning turned into a battle between adventurous millionaires. American

financier Steve Fosset and British entrepreneur Richard Branson set much of the running, Fossett making the first of three solo attempts in 1996. The American was one of the crew on Branson's third round-the-world attempt in 1998, which lifted off from Morocco and ended in the Pacific on Christmas Day. But these high-profile personalities were trumped by Bertrand Piccard (grandson of the inventor of the pressurized gondola) and Brian Jones, who made it round the globe in *Breitling Orbiter 3* in March 1999 (see panel, left).

Personalized flight

For many thousands of people, just flying is exciting enough, without the need for record-breaking ambitions. Gliding has developed as its own fascinating area of flight, the silent communion with the winds and the weather appealing to many aerial romantics with no taste for the roar of engines. The emergence of gliding as a skilful and satisfying sport began in German gliding clubs of the 1920s, which also provided flight training for future Luftwaffe pilots. Gliders soon proved an important testing ground for

aerodynamic improvements, and found military uses during World War II. But "soaring flight" has remained perhaps the purest of aerial activities, only marginally involved with commerce or war.

The idea of flying an aircraft as an activity open to millions has remained one of the frustrated dreams of aviation. As early as 1924 American automobile manufacturer Henry Ford envisaged a future in which aircraft would be produced in similar numbers to cars. In the 1930s American air administrator Eugene Vidal was a prominent campaigner for a "poor man's airplane", a Model T Ford of the air, that would transform flying from "a rich man's hobby to a daily utility or inexpensive pleasure for the average American citizen". From the 1920s onwards, private aviation in light aircraft did become a popular sport open to the moderately affluent. But the notion of an aircraft parked in every driveway never came off.

There were several attempts to produce flying cars – vehicles that could be driven on roads as well as fly through the air. Perhaps the most promising was Molt Taylor's Aerocar, produced in the 1950s. The flying surfaces folded up to turn the aeroplane into an automobile; the engine drove a propeller when in aircraft mode and the wheels when it was being driven on the ground. But the Aerocar's dual function involved too many compromises to perform well enough in either genre.

There were also attempts at making aviation very cheap, of which perhaps the most memorable was Frenchman Henri Mignet's Pou du Ciel, or Flying Flea. Brought out in 1933, this tiny aircraft was sold in kit form, to be assembled at home. After some 30 days' hard work, the purchaser would have a machine capable of reaching 130kph (80mph), but with a landing speed of only 30kph (19mph). The Flying Flea was greeted with a rush of enthusiasm, but despite Mignet's claim that it

FLYING CAR

Molt Taylor's Aerocar, marketed in the 1950s, was one of the most ambitious efforts to create an aircraft for people to keep in their garage. It is shown here in both flight and automobile mode – the wings are folded up in the trailer.

FLIGHT ACROSS THE CURTAIN

ONE OF THE MOST FAMOUS flights in a light aircraft was made by Matthias Rust, a 19-year-old German with about 50 hours' flying experience. In May 1987, with the Cold War still in progress, Rust decided to go for a jaunt across the Iron Curtain. Flying solo in a four-seater Cessna, he took off from Helsinki, Finland, and flew east to Moscow, totally eluding the supposedly imposing Soviet air-defence system – apparently, the aircraft was simply too slow to register. Rust cheerfully buzzed the Kremlin before landing alongside Lenin's Mausoleum in Red Square. He emerged from his aircraft to sign autographs for bemused Russian passers-by before being put under arrest. His exploit led to the resignation of the Soviet minister of defence and made a small but real contribution to undermining confidence in the communist system before the fall of the Berlin Wall.

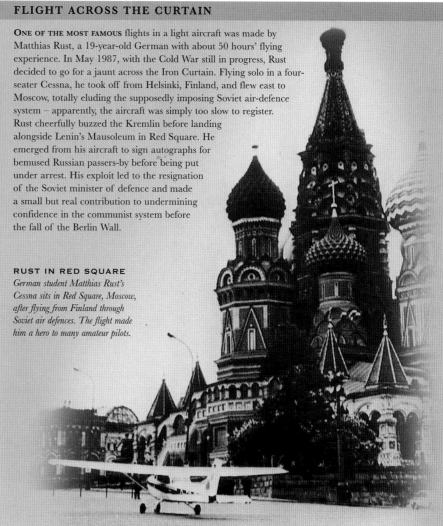

RUST IN RED SQUARE
German student Matthias Rust's Cessna sits in Red Square, Moscow, after flying from Finland through Soviet air defences. The flight made him a hero to many amateur pilots.

POPULAR PIPER
A Piper Cherokee flies over the desert in California. The Cherokee series, the first of which appeared in 1960, is typical of Piper's successful light aircraft in its reliability, simplicity, and adaptability to different roles.

involved "definitely less risk" than driving, deaths in Flying Flea accidents soon mounted and it was banned by the French government.

Light aircraft

Although there will no doubt always be a self-assembly market for hobbyists, in the end the major genre in private flying became the mass-produced light aircraft. This had to be easy to fly, economical to operate, and robustly reliable, but at the same time offer enough speed and manoeuvrability to satisfy the leisure flier's aspiration to excitement and fulfil a wide variety of practical functions. State-sponsored flying clubs proliferated in the 1920s in Germany and Britain, where the de Havilland Moth of 1925 quickly established itself as a much-loved favourite. In the United States, the three companies that were to dominate light aircraft

FLYING FLEA
Henri Mignet's Pou du Ciel – accurately translated as the Sky Louse, but generally known as the Flying Flea – enjoyed short-lived popularity in the 1930s. Mignet was an eloquent propagandist for private flying, publishing a bestselling book, L'Aviation de l'Amateur, in 1934.

manufacture, Cessna, Beech, and Piper, were founded between 1927 and 1937, in the trough of the Great Depression. The early Cessna A, marketed as "everyman's airplane", proved to be something that everyman, under the economic circumstances of the day, could not afford. But Piper had enormous success with

the Cub. Very stripped down and economical – it had no airspeed indicator or compass – the Piper Cub superseded such veterans of the barnstorming era as the Curtiss Jenny to become the bestselling light aircraft in the United States in the 1930s.

Cessna and Beech, with other manufacturers, profited in the 1950s when the American private flying market expanded dramatically. One of the most popular aircraft was the appropriately named Beechcraft Bonanza. Cessna's successes included the ubiquitous Cessna 172 of 1955. But Piper long headed the field in quantity production – when the company rolled out its 100,000th machine in 1977, it was reckoned to have manufactured one tenth of all the aircraft ever built in the world. Piper Cherokees and Comanches became so common around busy airports that commercial pilots dubbed their low-altitude airspace "Indian territory".

Bombardier Learjet

WILLIAM P. LEAR DEVELOPED the original Learjet Model 23 from the Swiss American Aviation Corporation's projected P-16 fighter-bomber. Unsurprisingly, given its military origins, when the Learjet entered service in 1964 it exhibited performance characteristics unprecedented for a private aircraft. In 1965 a Learjet flew from Los Angeles to New York and back in under 11 hours. The eight-seat Learjet 24, introduced in 1966, flew around the world in 50 hours and 20 minutes. As well as being able to fly long distances at impressive speeds, Learjets could also operate at higher altitudes than even jet airliners – in 1977 the Model 25 was authorized to fly at up to 15,550m (51,000ft).

The early Learjets had wingtip fuel tanks, but during the 1970s these were replaced by drag-reducing winglets. Also, in newer models turbofans took the place of the turbojet engines.

Although the executive jet market became increasingly competitive in the 1980s and 1990s, Learjets held their own by constantly evolving. For example, the range on some models increased to 4,000–5,000km (2,500–3,100 miles), and the passenger capacity rose from six to nine or ten.

While the original Learjet Corporation has been through many changes (and is presently part of the Bombardier aviation empire), the Learjet aircraft remains a classic.

> "To the public, the Learjet came close to being the ultimate status symbol."
>
> **T. A. HEPPENHEIMER**
> *A Brief History of Flight*

RECOGNIZABLE PROFILE
The distinctive front-on profile of the Learjet 45 is characterized by the two turbofan engines set at the rear of the fuselage, the tailplane positioned high on the fin, and the upturned winglets at the end of the wings (used to reduce drag).

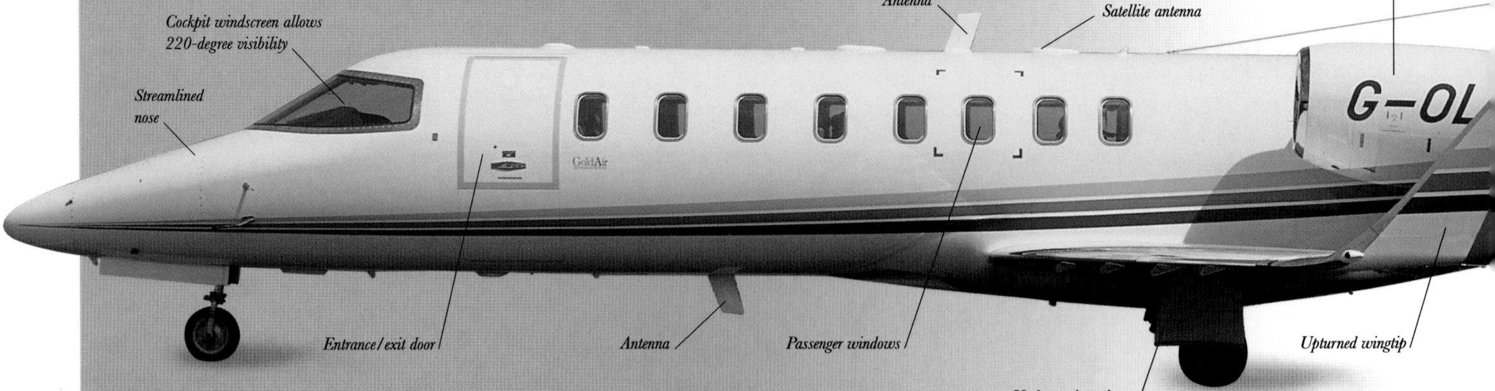

ADVANCED SYSTEMS
The compact cockpit of the Learjet 45 is populated with precision flight-management systems. In addition, the wide, electrically heated windscreen allows the pilots 220-degree visibility.

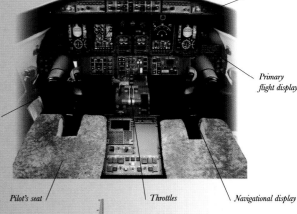

Engine-instrumentation and crew-alerting systems

Radio-management units

Primary flight display

Electric power control panel

Pilot's seat

Throttles

Navigational display

Tailplane set high on fin

AlliedSignal turbofan engine

Seamless aluminium wing skin

Nose-gear

Drag-reducing winglet

Cockpit windscreen allows 220-degree visibility

Streamlined nose

Antenna

Satellite antenna

Engines at rear of fuselage

G-OL

Entrance/exit door

Antenna

Passenger windows

Undercarriage doors

Upturned wingtip

LUXURY FLIGHT
The Learjet 45 provides truly luxurious flight conditions for its nine passengers. Features include a spacious, flat-floor cabin; comfortable, fully reclining chairs; folding tables; and swivelling television monitors.

Specifications (Learjet 45)

ENGINE	2 x AlliedSignal TFE731-20 turbofan
WINGSPAN	14.57m (47ft 10in)
LENGTH	17.6m (52ft 8in)
WEIGHT	6,146kg (13,550lb)
TOP SPEED	867kph (538mph)
CREW	2
PASSENGERS	9

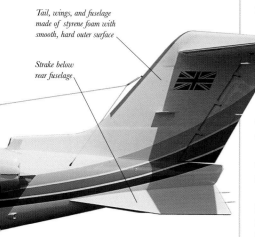

Tail, wings, and fuselage made of styrene foam with smooth, hard outer surface

Strake below rear fuselage

TOP EXECUTIVE
The uncluttered, streamlined design of the Learjet 45 was realized with optimum aerodynamic efficiency in mind. One example of this is the 13-degree sweep of the wings.

The secret of the success of light aircraft has been their adaptability. Their functions have ranged from pilot training to aerial battlefield observation, from crop-spraying to salting icy roads, from light commercial transport to leisure flying.

Flying in the wild
In some of the world's wildest places, such as Alaska, the Amazon basin, or New Guinea, they have provided the main link between population centres and the only access to trackless wilderness. Although rarely recorded, the experiences of the pilots who fly Cessnas into the jungle airstrips of New Guinea's Bismarck Mountains, for example, have almost certainly been as demanding and hazardous as anything else in the entire history of flight.

There was a constant tendency for light aircraft to become more complex and more powerful, a response to the ready supply of private customers willing to pay the price for the latest in aviation technology. It was inevitable that this would eventually stretch to the private jet. Aimed largely at corporations, although also bought for personal use by the seriously rich, business jets have proliferated since the 1960s, freeing their owners from the straitjacket of airline timetables and from reliance on large commercial airports.

Status symbol
The trigger that set off the explosion of the executive-jet market was the Learjet, which became the ultimate 1960s status symbol, surpassing a luxury yacht or Manhattan penthouse for both glamour and desirability. The Learjet's main rival in the early years was the Dassault Falcon series; by the mid-1970s more than 500 Learjets and 300 Falcons had been sold. By the 1990s, the running was being made by the Gulfstream Aerospace Corporation's Gulfstream V, which had a range of 12,000km (7,450 miles) and could fly at 15,500m (51,000ft). The Gulfstream has state-of-the-art avionics, to make it as safe as a commercial airliner.

This marks a clear distinction from private flying as practised by the amateur pilot, which shows up in statistics as far more dangerous than travelling in an airliner. The figures are underlined by the

NINETIES PACESETTER
The Gulfstream Aerospace Corporation's Gulfstream series set new performance levels for executive jets, as well as raising safety standards to match the airlines.

WILLIAM LEAR

WILLIAM POWELL LEAR (1902–78), the creator of the Learjet, was born in Hannibal, Missouri. A high-school dropout from an unsettled family background, he turned out to have a genius for electronics. He invented the first practical car radio in 1932 and subsequently, turning to aviation, created a wide range of electronic cockpit instruments. In 1947 Lear produced the first autopilot for jet aircraft, winning the prestigious Collier Trophy and making his fortune. After a spell in Switzerland in the 1950s, he returned to the United States and founded the Learjet Corporation in 1962. The success of the Learjet won him immense public renown, but he was not good at business. The firm was taken over by Gates Rubber Company in 1967. Lear subsequently frittered away a great deal of time and money on designing supposedly ecologically friendly steam-powered cars, as well as on the unsuccessful Learfan jet.

KING LEAR
The Learjet made William Lear a celebrity in the 1960s, able to pull in investment for any innovative technological project, however impractical it might be.

media impact of crashes involving high-profile individuals – for example, the death of John F. Kennedy Jr piloting a Piper Saratoga in 1999.

Knowing the risks involved is unlikely to deter any aviation enthusiast, whether in an expensive light aircraft or a hang-glider. The adventurous have always found occasional adrenaline-pumping danger part of the appeal of flight. Aviation is destined to retain a twin nature, as a central part of the practical business of the world, but also as one of those frontiers where individuals will sometimes go to test and prove themselves.

PRIVATE AIRCRAFT

INITIALLY ALL AIRCRAFT WERE private aircraft, made for sportsmen and women or for the amusement of wealthy enthusiasts. The growth of military and then commercial aviation led to the evolution by the 1930s of a distinctive category of light aeroplane aimed at leisure fliers or small-scale practical uses such as crop-dusting or passenger transport in remote areas. To succeed, these aircraft had to be cheap and sturdy, and relatively easy to fly. Around the same time, home-assembly aircraft kits began to gain a popularity they would never entirely lose. The use of light aircraft expanded massively after World War II, with the Piper, Beech, and Cessna companies as the main beneficiaries. The introduction of the private jet in the 1960s created a new category of aircraft for corporations and the super-rich, which rivalled commerical airliners in sophistication and performance. Meanwhile the development of ultralights and microlights brought a new form of relatively inexpensive leisure flying.

LEARJET LUXURY
The top-of-the-range Learjet 60 series combined the Longhorn wings with the fuselage and engine of the 35 series to give it the fastest speed and longest range of all the Learjet planes. See pages 420–21.

Beech Bonanza D35

The Bonanza, with its distinctive V-tail, all-metal construction and retractable undercarriage, set a new standard for light aircraft. However, nobody could have forecast that over 10,000 would be built since its debut in 1945 or that production would continue until 1982.

ENGINE	205hp Continental E-185 flat 6 piston engine	
WINGSPAN 10m (32ft 10in)		LENGTH 7.7m (25ft 2in)
CRUISING SPEED 281kph (175mph)		
PASSENGERS 3		CREW 1

Cessna 172R Skyhawk

A tricycle undercarriage version of the popular Cessna 170, the first four-seat Cessna 172 flew in November 1955. The type was immediately successful, and this popularity has continued to the present day with over 42,000 sold, making the Cessna 172 the world's most successful light aircraft.

ENGINE	160hp Textron Lycoming IO-360 flat 4 piston engine	
WINGSPAN 11m (36ft 1in)		LENGTH 8.3m (27ft 2in)
CRUISING SPEED 226kph (140mph)		
PASSENGERS 3		CREW 1

Cessna 340

When the Cessna 340 was first offered for sale in 1971, it fitted in between the long established 4–6 seat Cessna 310 and the 6–8 seat Cessna 414/421. With a pressurized cabin and an integral airstair door, the aircraft was aimed at the business market. A lighter, unpressurized version, the Cessna 335, was also built in small numbers from 1979. When the production line closed in 1984, 1,267 Cessna 340s had been built.

ENGINE	2 x 310hp Continental TSIO-520 Turbocharged flat 6 piston	
WINGSPAN 11.6m (38ft 1in)		LENGTH 10.5m (34ft 4in)
CRUISING SPEED 425kph (264mph)		
PASSENGERS 5		CREW 1

Dassault Mystère Falcon 900C

Derived from the earlier tri-jet Falcon 50 to provide greater range, the Falcon 900 was announced at the June 1983 Paris Air Salon with the first prototype flying in September 1984. The second prototype demonstrated its intercontinental range by flying non-stop from Paris to Little Rock, Arkansas. Deliveries to customers started in December 1986 with over 180 orders. The current standard model is the Falcon 900C, but an even longer-range version, the 900EX, was launched in October 1994.

ENGINE	3 x 4,750lbst Honeywell TFE731-5BR-1C turbofan	
WINGSPAN 19.3m (63ft 5in)		LENGTH 20.2m (66ft 4in)
CRUISING SPEED 887kph (551mph)		
PASSENGERS 8–15		CREW 2

Eipper Quicksilver E (Ultralight)

Eipper Formance manufactured the Quicksilver hang-glider, designed in 1972 by Bob Lovejoy, and some 2,000 had been sold when production ended in 1981. When fitted with an engine, the aircraft became one of the first foot-launched ultralights (microlights).

ENGINE	46hp Rotax 503 piston engine	
WINGSPAN 9.8m (32ft)		LENGTH 5.5m (18ft 1in)
CRUISING SPEED 66kph (41mph)		
PASSENGERS None		CREW 1

Europa XS

In the early 1990s, Ivan Shaw had a dream. He conceived an aeroplane – the Europa – which could be built and kept at home, easily transported to the nearest airfield, filled up at a petrol station along the way, and assembled in under five minutes, before carrying two people for extended touring.

ENGINE	80hp Rotax 912S	
WINGSPAN 8.3m (27ft 2in)		LENGTH 5.8m (19ft 2in)
CRUISING SPEED 259kph (161mph)		
PASSENGERS 1		CREW 1

Gulfstream Aerospace Gulfstream V-SP

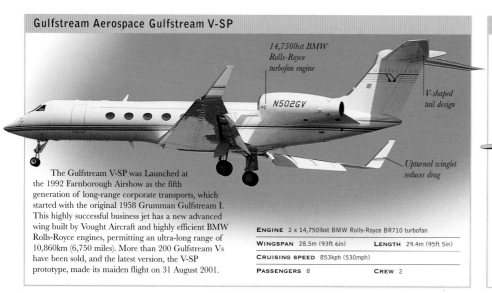

14,750lbst BMW Rolls-Royce turbofan engine

N502GV

V-shaped tail design

Upturned winglet reduces drag

The Gulfstream V-SP was Launched at the 1992 Farnborough Airshow as the fifth generation of long-range corporate transports, which started with the original 1958 Grumman Gulfstream I. This highly successful business jet has a new advanced wing built by Vought Aircraft and highly efficient BMW Rolls-Royce engines, permitting an ultra-long range of 10,860km (6,750 miles). More than 200 Gulfstream Vs have been sold, and the latest version, the V-SP prototype, made its maiden flight on 31 August 2001.

ENGINE	2 x 14,750lbst BMW Rolls-Royce BR710 turbofan	
WINGSPAN 28.5m (93ft 6in)	LENGTH 29.4m (95ft 5in)	
CRUISING SPEED 853kph (530mph)		
PASSENGERS 8	CREW 2	

Piper PA-18-150 Super Cub

More powerful than any of its predecessors, the Super Cub was the final development of the original 1937 Piper J-3 Cub, first flying in November 1949. When continuous production ended in 1981, nearly 7,500 had been built. The most powerful version, the PA-18-150, first appeared late in 1954. Many thousands of Super Cubs remain in private ownership worldwide.

N4155R

ENGINE	150hp Lycoming 0-320 flat 4	
WINGSPAN 10.7m (35ft 3in)	LENGTH 6.9m (22ft 6in)	
CRUISING SPEED 206kph (128mph)		
PASSENGERS 1	CREW 1	

Piper PA-46-350P Malibu Mirage

The six-seat Piper Malibu, which first flew in August 1982, survived the late-1980s product-liability laws, which decimated much of the US general aviation industry. The design, powered by a Continental turbocharged engine, offered many of the features of more expensive business twins, such as cabin pressurization and an airstair door. The improved Malibu Mirage, with its 350hp Lycoming engine, entered production in October 1988.

ENGINE	350hp Textron Lycoming TIO-540 flat 6 piston engine
WINGSPAN 13.1m (43ft)	LENGTH 8.7m (28ft 7in)
CRUISING SPEED 398kph (247mph)	
PASSENGERS 5	CREW 1

Robin DR 400/120

The Jodel ancestry of the Robin DR 400 is immediately apparent in its characteristic "bent wing". The Robin company had its origins in 1957 when Pierre Robin and Jean Delemontez (the "del" of Jodel) got together to develop the original wooden Jodel design. The initial designs evolved up to the DR 400 which first flew in June 1972. Powered by a choice of engines and power outputs, the DR 400 can provide two seats or a full four-seater. Over 1,200 DR 400s have been built.

G-BRNU

ENGINE	112hp Textron Lycoming 0-235 flat 4 piston engine.
WINGSPAN 8.7m (28ft 7in)	LENGTH 7m (22ft 10in)
CRUISING SPEED 215kph (134mph)	
PASSENGERS 1 + 2 children	CREW 1

Rutan Model 33 Vari-eze

The Rutan Aircraft Factory Inc., formed in 1969 by the two Rutan brothers, Burt and Dick, has produced some incredible aircraft, the most famous being the "round-the-world" Voyager. The first composite-structure, canard (tail-first) Varieze appeared in 1975, and was followed by a second prototype a year later. The design, with its retractable nose-wheel and fixed main wheels, achieved an incredible performance on just 100hp. It was immediately successful with home builders such that, by 1984, over 600 had flown and more than 1,400 were under construction.

ENGINE	Various, including 100hp Continental 0-200 flat 4 piston
WINGSPAN 6.8m (22ft 3in)	LENGTH 4.3m (14ft 2in)
CRUISING SPEED 313kph (195mph)	
PASSENGERS 1	CREW 1

G-TIMB

Slingsby T.67C Firefly

In 1981, the renowned Yorkshire sailplane manufacturer bought the rights to the French Fournier RF-6B which they rebuilt as the Firefly. With their extensive experience of glassfibre-reinforced plastics (GRP), the redesigned aircraft made maximum use of the material in its structure. It was built in both civilian and military versions.

G-FFSM

ENGINE	160hp Textron Lycoming 0-320 flat 4 piston engine.
WINGSPAN 10.6m (34ft 9in)	LENGTH 7.3m (24ft)
CRUISING SPEED 215kph (134mph)	
PASSENGERS None	CREW 2

Socata TB10 Tobago

In the mid-1970s, the French company Socata designed a family of four- or five-seat light aircraft to replace its highly successful Rallye. The TB series were all given Caribbean names and ranged from the basic fixed undercarriage TB9 Tampico Club to the turbocharged, retractable gear TB21 Trinidad TC. The first TB10 flew in February 1977 and became the most popular of the range, with a faired, fixed undercarriage and more powerful engine than the basic Tampico.

180hp Textron Lycoming piston engine

G-VMJM

Fixed undercarriage

ENGINE	180hp Textron Lycoming 0-360 flat 4 piston engine
WINGSPAN 9.8m (32ft)	LENGTH 7.7m (25ft 3in)
CRUISING SPEED 240kph (150mph)	
PASSENGERS 3	CREW 1

THE FUTURE OF FLIGHT

AT THE END OF A HUNDRED YEARS IN WHICH FLIGHT HAS CHANGED THE WORLD, THERE ARE STILL NEW FRONTIERS FOR AVIATION TO EXPLORE

"This conquest of the air will prove, ultimately, to be man's greatest and most glorious triumph."

CLAUDE GRAHAME-WHITE
AVIATION PIONEER, 1914

JOINT STRIKE FIGHTER
The Joint Strike Fighter (JSF) is intended to give NATO an affordable, mass-produced, air-superiority and strike aircraft for the 21st century. It is smaller and more versatile than the Lockheed Martin F-22 Raptor, which in many ways it closely resembles.

IN THE GREEK MYTH that so fascinated many of the earliest pioneers of flight, Daedalus' son Icarus died after flying too close to the sun, which melted the wax on his feathered wings. In 2001, ironically inverting the mythical experience, a NASA *Helios* ultralight flying wing powered by the sun's rays flew to the outer edge of the earth's atmosphere.

Surely one of the most extraordinary aircraft yet built, *Helios* was "piloted" by a controller on the ground and travelled at a sedate 32kph (20mph). Its wing, measuring 75.25m (247ft) and thus longer than that of a Boeing 747, was covered in solar panels that generated the electricity to drive its 14 motors. Storing electricity in fuel cells during the day allowed it to continue to operate through the night. Totally ecologically friendly, *Helios* was destined for sustained flight at the edge of space. On 13 August 2001 it set an altitude record for a propeller-driven aircraft, rising to 29,511m (96,863ft). The earth's atmosphere at that altitude is similar to the atmosphere of Mars, so the flight allowed NASA scientists to learn about the feasibility of a flying machine that might cruise the skies of the "red planet". Helios also had the potential to serve many of the functions of a satellite – in communications or weather observation, for example – at a fraction of the cost. With no need to refuel, NASA believed Helios would eventually be able to fly for months at a time – in effect until its parts wore out.

Helios's brief experimental career came to an end in June 2003, when it went out of control flying over the Pacific and was lost. Yet it remained an indicator of some of the key directions that flight was taking in the 21st century. The development of unmanned aerial vehicles (UAVs) did not quite threaten to make pilots obsolete, but it was set to revolutionize many areas of aviation. And the idea of using solar power was the kind of solution designers might be forced to adopt if the pollution produced by conventionally powered aircraft in the end proved politically and socially unacceptable.

Distance travelled

Helios was above all a superb example of the constant power of aviation to amaze with unexpected feats of technological innovation, revealed time and again through the 20th century. Looking back at the distance flight advanced in its first 100 years offers a vertiginous perspective. Any measure of aircraft performance reveals dizzying progress – speed, for example, accelerating from Glenn Curtiss's record-breaking 75kph (47mph) in 1909 to top speeds passing 640kph (400mph) in the 1930s; the breaking of the the sound barrier by the end of the 1940s; aircraft reaching Mach 2 and Mach 3 in the 1960s; the X-15 reaching Mach 6.7 in 1967; and X-43A scramjet hitting Mach 9.6 in 2004.

Similar progress could be traced in range and payload. The propeller-driven airliners of the 1930s and 1940s could fly 1,000–3,000km (600–1,800 miles); jet airliners of the late 1950s had a maximum range of around 6,000km (3,720 miles); in the early 1970s the Boeing 747 raised this figure to over 10,000km (6,210 miles); by the early 21st century, the latest passenger aircraft could fly 14,000–18,000km (8,700–11,200 miles) non-stop. The Lockheed C-5 transport introduced at the end of the 1960s could carry about 100 times the payload of a World War I bomber, and has itself been far surpassed by transports such as the extraordinary Airbus Beluga series, capable of carrying cargoes well in excess of 50 tonnes.

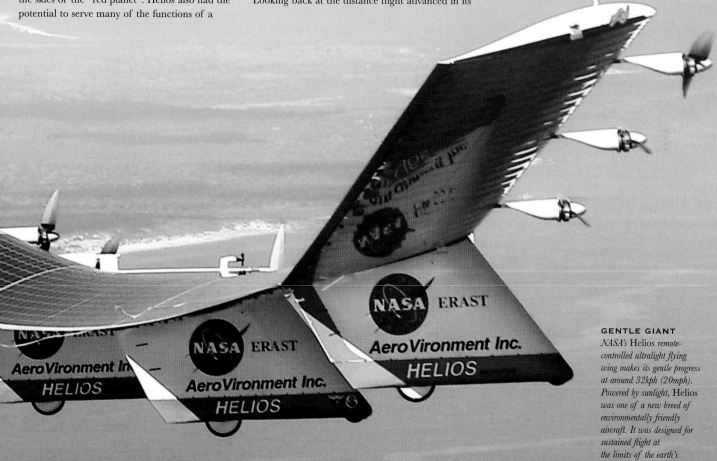

GENTLE GIANT
NASA's Helios *remote-controlled ultralight flying wing makes its gentle progress at around 32kph (20mph). Powered by sunlight, Helios was one of a new breed of environmentally friendly aircraft. It was designed for sustained flight at the limits of the earth's atmosphere.*

The last commercial Concorde flight took place on 24 October 2003. Supersonic passenger flight had proved a technological success story but a commercial failure, demonstrating that the future of aviation did not necessarily lie in ever increasing speed or range.

Continuing innovation

If you were writing a general account of life in the 20th century, aircraft would only figure marginally for the first three decades. Some reputable single-volume histories of World War I barely mention aviation at all. Until the late 1930s, aircraft were a craze that generated heroes, but really had little effect on the lives of any but a small minority of people. It was World War II that truly brought aircraft centre stage, transforming the practice of warfare. Commercial aviation took until the jet age to begin to effect a dramatic change in leisure and business. Even in the United States, in the early 1960s half the population had still never flown. But by the 1990s over a billion passengers were flying worldwide every year.

It was an open question at the start of the third millennium whether flight still had revolutionary possibilities, or whether it had become, like tanks in warfare or railways in passenger transport, an established feature of the landscape that would endure (with improvements) but undergo no further dramatic expansion or

transformation. Alongside constant experimentation with fresh approaches to the technological and commercial aspects of aviation there was an inherent conservatism that had its own rationale. Both civil and military aviation were mature businesses that at times saw little logic in exposing themselves to the expense and risk of major innovation.

Supersonic passenger transports had once seemed the way of the future. It appeared obvious that air travel would simply become faster. But as the sorry commercial history of Concorde and its even less successful supersonic rivals showed, the price for speed could become greater than people were prepared to pay – whether in terms of the cost of a seat or the cost in noise pollution and other environmental damage. Before its withdrawal from service in 2003, Concorde had found a market niche, but it was a small and relatively insignificant one that no new airliner was rushing to fill. With 21st century technology, design of an aircraft to carry about 300 passengers at over twice the speed of sound was perfectly feasible, yet no one expected to see such an airliner brought into service before 2030, if it ever happened at all.

The fate of the Boeing Sonic Cruiser was another case in point. This revolutionary airliner was presented as the future of passenger aviation

in March 2001. It was intended to carry up to 300 passengers at a cruising speed of Mach 0.98 – just below the speed of sound, thus avoiding the "sonic boom". With a range of up to 16,000km (10,000 miles), it would knock hours off long-haul flights. But negative reaction from world airlines soon forced Boeing to abandon their new fast aircraft. Airline bosses were far keener on cutting costs than they were on increasing speed. More seats, lower maintenance costs, and reduced fuel consumption were the kind of unglamorous achievements that would really sell an aircraft.

Money matters

In military aviation, the quest for ever-improved performance came up against cost-conscious politicians who were increasingly inclined to query the need for ever-more-expensive aircraft. These could easily seem like toys for the boys to play with. In the United States, the technological lead over any currently conceivable enemies was the most potent deterrent to investment in expensive high-tech military aviation. It did not pass unnoticed that a successful air attack on America was achieved by a bunch of young Muslims whose technology was limited to knives, box-cutters, and the ability to pilot aircraft.

The controversy that surrounded the Lockheed Martin F-22 Raptor, the sophisticated American fighter that came into service in December 2005, turned precisely on such questions of cost and function. The F-22 was designed as a replacement for the F-15, to guarantee the United States air superiority against any other fighter force. It marked a clear advance over its predecessors and contemporaries in its stealth features and its ability to cruise at supersonic speed, without use of afterburners – all other fighters could only "go supersonic" in short bursts because their afterburners used up fuel so quickly.

But the F-22 was also very expensive indeed. Development and production of the aircraft may have cost $70 billion. If, as appeared possible, less than 200 were to be built, the unit cost would stand at around $350 million. All such figures are

WHITE WHALE
The extraordinary Airbus Beluga is the world's largest transport aircraft by volume. It is basically the bottom half of an A300 widebody airliner with a bulbous cargo hold mounted on top. The Beluga was designed to carry sections of Airbus airliners between factories in different countries.

speculative, but even if in practice the Raptor proved to cost less than half that amount, its value-for-money was bound to remain controversial. It had originally been conceived in the 1980s, in the context of the Cold War confrontation with the Soviet Union. With the ending of the Cold War, it was increasingly difficult to see what enemy such an advanced aircraft would be called upon to fight. Arguably, the Russians still had the ability and the will to press on at the cutting edge of technology, depending on money from export markets in China and India to finance development. But the cheaper Joint Strike Fighter, planned to be mass-produced as NATO's future standard fighter aircraft, seemed fully adequate to cope with the air superiority task as well as other fighter roles. Its deliberate compromise between cost and technology looked more likely to prove successful in the long run. The lesson to be learned was that future developments in aviation could not be safely predicted on the basis of what was technologically feasible. At the time of maximum faith in technological

SLEEK SONIC CRUISER

In 2001 Boeing believed they had a winner in the Sonic Cruiser, an airliner that would carry passengers at approaching the speed of sound. Its twin-engine "canard" design – with the tailplanes mounted in front of the main wing – was predicted to revolutionize air travel. However, the Sonic Cruiser never made it off the drawing board because Boeing's customers did not see high speed as a trump card in the airline business.

progress – an era that effectively ended in the 1970s – it was assumed that anything that could be done, would be done. By the 21st century, a more utilitarian logic was in the ascendant. If a development in aviation could perform a required function for a reasonable price, it would happen. Whether funded by governments or by business, new aircraft projects that did not meet these criteria were destined to run into dead ends or become stranded in backwaters.

Space exploration might be the only area where the technological imperative could still predominate. It remained the open frontier where, at least in theory, boundless possibilities existed for new achievement and funding might materialize for almost any project that sufficiently caught the public imagination.

TAXI TO SPACE

This artist's interpretation depicts a reusable "Space Taxi" riding on the back of its Evolved Expandable Launch Vehicle. Conceived jointly by Orbital Sciences and Northrop Grumman, this design was in the end not the route chosen by NASA for a space shuttle replacement.

Commercial aviation

Predictions about the future of passenger air travel fluctuate with circumstances. In 2000 the aircraft industry was predicting that four billion passengers a year would be carried by 2020 – almost triple the current level of air travel. But then came the terrorist attacks of 11 September 2001 (see page 410) and a sharp economic downturn, which together produced a shocking fall in passenger numbers. Suddenly a permanent end to growth in air traffic did not seem out of the question. By 2006, however, growth in passenger traffic was back to around 5 per cent a year worldwide and airlines were ordering a record number of new aircraft.

Perhaps the most certain guarantee of future growth in passenger air travel was provided by the untapped potential of populous Asian countries, especially China, to become a major new source of air passenger business. In 2000 China had around 527 airliners carrying some 67 million passengers a year. By 2005 passenger numbers had increased to 138 million travelling on board 863 aircraft. The Chinese authorities responded in 2006 by announcing plans to expand further the civil aviation fleet by 100 airliners a year for the following five years. The aim was to double the country's air traffic, making China second only to the United States in the scale of its passenger operations.

TRANSPORTED IN SECTIONS
A convoy carries sections of an A380, made in Britain, France, Germany, and Spain, from a barge terminal on the Garonne river to the Airbus assembly plant at Toulouse, France. Crawling along rural backroads, the massive convoy takes three nights to complete its journey.

ENGINE TESTING
The Airbus A380's massive Rolls-Royce engines underwent rigorous testing to ensure that they met targets for fuel consumption, emissions, and noise, as also safety. One requirement for air safety certification was that the engine would still function if a large goose flew into the fan.

Airbus versus Boeing

If air travel continues its spectacular expansion, what aircraft will future passengers be boarding? In the first 30 years of the jet airline era, Boeing had established a seemingly unshakeable worldwide supremacy in this sector, but the spectacular rise of the European consortium Airbus Industrie shattered Boeing's grip on the market. In 2004 Airbus knocked the mighty Boeing off the top of the international airliner sales and orders league for the first time.

Designed for the market

The boldest Airbus project to date has been the A380 superjumbo, which had its maiden flight in April 2005. As the largest airliner ever built, capable of carrying at least 550 passengers, it aimed to replicate the revolutionary impact the Boeing 747 had had on air travel three decades earlier. The project showed how far Europe had advanced since the days of the Concorde project – not in technological know-how, which was never lacking, but in commercial realism. The A380 was crafted to fulfil the business objectives of its potential airline customers, offering a projected 20 per cent cut in operating cost per seat compared with the Boeing 747-400. Careful thought was given to minimizing the new demands these enormous aircraft would make on airports, virtually ensuring that the superjumbo would be welcome at all major air transportation hubs. The design of the aircraft also reflected the need to meet environmental concerns and to avoid excessive noise pollution. As a result, 150

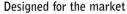

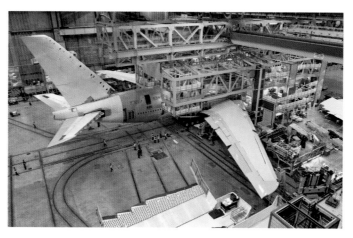

firm orders for the Airbus were on the books even before the aircraft's maiden flight.

Boeing fought back strongly, if somewhat belatedly, with a longer-range version of the existing 777 and, above all, with the 787 Dreamliner, set to enter service in 2008. Despite its kitsch name, the Dreamliner constituted an intelligently pitched answer to the Airbus A380. As a mid-size airliner, it would be able to fly direct into the smaller airports that the A380 could not use, obviating the need for transfer flights. Making extensive use of lightweight materials and aluminium alloys, it promised ultra-efficient fuel use and maintenance, quiet engines, and a range of 15,700 km (9,775 miles).

Global manufacture

The head-to-head between Airbus and Boeing obviously had aspects of an old-style contest between national air industries. Boeing's vociferous complaints about alleged unfair support for Airbus from European governments were countered by allegations of hidden US subsidies to Boeing. Yet both the Airbus A380 and the Boeing 787 demonstrated the impossibility of the pursuit of narrow nationalism in the 21st century global economy. International cooperation was, of course, basic to Airbus Industrie, which prided itself on being a joint European enterprise, but the 787 was also far from being a specifically American piece of work. The Japanese companies Fuji and Kawasaki were major contributors to the aircraft, manufacturing much of the fuselage and wings, and Korean, British, and French companies were among those involved in the project.

Environmental challenge

Ultimately the greatest threat to the future expansion of global air travel probably lies in the rising sensitivity to the environmental damage

AIRBUS ASSEMBLY
The A380 superjumbo comes together on the Airbus assembly line at Toulouse, surrounded by a complex structure of gantries. The enormous assembly hall is one of the largest buildings in the world, covering the area of 20 football pitches. The hall's construction required four times as much steel as the Eiffel Tower.

caused by aircraft. At the start of the 21st century, it seemed as though those governments committed to combating global warming were bound eventually to listen to their scientific advisers telling them that airliners were a large part of the problem of excessive emission of "greenhouse gases". Airlines and aircraft manufacturers were naturally hostile to the introduction of "green" taxes as a means of increasing the cost of air travel and thereby detering people from making unnecessary flights. On the other hand, they were keen

on the concept of fuel efficiency, which would translate not only into lower pollution levels but also into lower operating costs and therefore greater profitability. Both the Boeing Dreamliner and Airbus A380 had fuel efficiency as a key selling point. However, environmental campaigners remained unimpressed. If fuel efficiency were to be used to make flying still cheaper, then it would only lead to more flights rather than a cleaner atmosphere. These issues continue to pose a stimulating challenge to those involved in the development of future patterns of passenger air travel and commercial aircraft technology.

ICE AND SNOW
Ability to operate safely in cold weather conditions is an essential requirement for any modern airliner. Here the A380 is put to the test in extreme snow and ice. The exhaustive flight test process required around 2,500 hours of flying by four A380s carried out over a period of more than a year.

LANDING ON WATER
An Airbus A380 makes high-speed passes through a water-filled trough on the runway to check that spray from the wheels does not impair engine operation.

Airbus A380

THE AIRBUS A380 WAS CONCEIVED in the early 1990s to compete with, and if possible replace, the Boeing 747, and in 1994 work began in earnest on what was then called the A3XX. The main contractors for the A380 are in France, Germany, Britain, and Spain. A custom-built ship carries fuselage sections from Hamburg and Brittany, wings from north Wales, and tailcones from Spain to the French port of Bordeaux, from where they are transported by barge and road to the assembly plant in Toulouse.

The A380's shape is subtly moulded to minimize drag from its ovoid fuselage. The structure makes extensive use of composite materials, such as thermoplastics and GLARE (aluminium and glass fibre). Its four huge engines, the most powerful ever used on an airliner, are surprisingly quiet. When carrying 550 passengers, Airbus claims the superjumbo will be eco-friendly and cheap to run, since it will use only 2.9 litres (⅗ gallon) of fuel per passenger per 100km (60 miles). With 22 wheels to spread its enormous weight the A380 is compatible with most existing runways at major hubs, but its massive wingspan will necessitate widening of taxiways. Airports will also need to upgrade facilities for passenger deboarding and baggage handling. While the additional cabin space available on the superjumbo has the potential to improve passenger comfort, cabin configuration will be decided by the airlines.

Tall vertical tailplane

Split rudder for yaw control

NOSE SECTION
The lower forward part of the A380 houses the nose landing gear in flight. It is unpressurized, and is separated from the pressurized cabin and flight deck by a bulkhead. There is a strengthened bird-impact shield below the cockpit windows.

Specifications A380

ENGINE	4 x 311,000kg (70,000lb) Rolls-Royce Trent 970 or Engine Alliance GP7270
WINGSPAN	79.8m (261ft 10in)
TOTAL LENGTH	72.9m (239ft 2in)
WEIGHT (EMPTY)	277,000kg (610,680lb)
CREW 2 **PASSENGERS**	555 (in typical three-class arrangement)
CRUISING SPEED	927kph (575mph) (Mach 0.85)

MASSIVE FUSELAGE
The Airbus A380 is the world's largest airliner. Its massive fuselage incorporates three decks – upper and main decks for passengers and a lower cargo deck. With 49 per cent more floor space than the Boeing 747, the A380 could theoretically carry up to 853 passengers.

Main deck windows

Jupp-Reeselet winglet

Belly fairing

F-WWOW

COCKPIT
The two-man flightdeck is deliberately conservative, maintaining many common features with earlier Airbus models, including the Honeywell flight management system. As pilot Captain Jacques Rosay said after the superjumbo's maiden flight: "Any Airbus pilot will feel immediately at ease with this aircraft."

Pay point

Display cabinet

Overhead control panel

Primary flight display

On-board information terminal

Fly-by-wire side stick

Pilot's seat

Navigation display

Multifunction display

Power levers

Cursor control device (CCD)

ON-BOARD SHOP
The aft main cabin area has here been mocked up as an in-flight shop. Individual airlines will in fact determine the interior layout themselves.

Cabin lighting that simulates sunset and sunrise

Maximum of four seats in a row

LUXURY TRAVEL
With a floor area of 310 square metres (3,350 square ft), the A380 can be configured to provide comfortable seating for about 550 passengers.

ENGINES AND WHEELS
The Trent 970 has the largest fan diameter of any Rolls-Royce engine. It underwent many subtle modifications to meet targets for noise reduction. The A380 has 22 wheels: two six-wheel bogies make up the body landing gear and two four-wheel bogies the wing landing gear, plus two wheels at the nose.

GLARE upper fuselage

Swept titanium fan blades

Wing landing gear

Antenna

Upper deck passenger door

Flightdeck door

Sculpted wing-root fairing

Nose landing gear

Military aviation

By the early 21st century, air power was seen as the key to power projection and security, suiting citizens in technologically advanced democracies accustomed to peace and reluctant to countenance the level of casualties that a ground war usually involves. The 2003 invasion of Iraq by American and British forces and the subsequent occupation of the country, as well as the worldwide threat of terrorist attacks, became the focus for thinking about future war plans, although planners ran the well-known risk of preparing for the last war rather than the next. Air operations in the 2003 Operation Iraqi Freedom showed how technologies had been refined, extended, and advanced to a far higher pitch of efficiency since the 1991 Gulf War.

"Risk-free" air attack

Precision guided bombs and missiles had constituted less than 10 per cent of munitions used during the Gulf War; in 2003, around two-thirds of airborne munitions were "smart". The use of JDAMs (Joint Direct Attack Munitions) was crucial. A JDAM was a bomb with a guidance package linked to a Global Positioning System (GPS) or Inertial Navigation System (INS). The map coordinates of a target on the ground were fed into the bomb's control system. Once it was released from the aircraft, the JDAM would then steer itself to the designated coordinates without any further guidance. JDAMs were found to be accurate to within around 10m (33ft) and could be used in bad weather conditions that would create problems for laser- or TV-guided munitions.

Other key factors in the 2003 invasion included the increasing use of UAVs (Unmanned Aerial Vehicles), more than 100 of which were deployed. They mostly carried out reconnaissance missions, but Predators on occasion acted as UCAVs (Unmanned Combat Air Vehicles), executing precision air strikes with Hellfire missiles. The introduction of "time-sensitive targeting" – the direction of strike aircraft lurking over the battle area on to a target only minutes after it was identified – was considered a success. In one incident a B-1 bomber struck a house where Saddam Hussein and his sons were said to be meeting within 12 minutes of the target being identified by US intelligence.

Although a number of well-known problems resurfaced during the Iraq invasion, including civilian casualties caused by occasional inaccuracy or mistaken target identification and losses to "friendly fire", the effectiveness of the deployment of air power could not be disputed. Iraqi ground forces simply could not operate without air cover. Command

IRAQ CAMPAIGN MEDAL
Even high-tech campaigns such as Operation Iraqi Freedom involve risk to ground troops and air crews. Among the recipients of this medal were air crew who had flown sorties over Iraq on 30 consecutive days.

SHOCK AND AWE
During the Iraq invasion, media attention was riveted by the spectacle of strikes by aircraft and cruise missiles on Baghdad. Although precisely targeted against official buildings, these air strikes were linked to the fashionable military principle of "shock and awe" – the use of a spectacular display of power to undermine the enemy's will to fight.

and communications could not be maintained in the face of precision air strikes. With its massive costs and its immense organizational and technological demands, air power remained the clearest determinant of power differences between states in the early 21st century. Iraq might have given a superpower a fight on the ground, but in the air it could not compete.

Limited usefulness

The subsequent occupation of Iraq graphically illustrated the limitations of are power. Operating against an enemy indistinguishable from the rest of the local population under circumstances in

TANK DESTRUCTION
Iraqi armoured forces had no defence against pinpoint air strikes once they had been identified as a target by America's "eyes in the sky". The A-10 Warthog was the Americans' primary tank-busting aircraft, but any aeroplane armed with JDAMS or other guided munitions could be directed to strike Saddam's tanks.

2001 were seen as the United States' prime enemies. The Pentagon's 2006 Defense Review Report envisaged a future in which America's "new and elusive foes" would need to be "found, fixed, and finished" anywhere around the globe. An expanded force of UAVs would be a crucial part of a combined intelligence effort to identify targets. Destroying these targets would require a new bomber aircraft capable of hitting any spot on Earth. This newcomer would join existing B-1s, B-2s, and a slimmed-down force of B-52s as a precision strike force that might well attack targets in countries with which America was not at war and might operate in support of non-American ground forces – following the model of the 2001 invasion of Afghanistan.

However, the insistence of the United States on the deployment of the F-22 as primarily an air superiority fighter showed that no one had forgotten the possibility of genuine air combat. The Americans felt the need to be able to take on the Sukhoi fighters of China or even the new Eurofighters of their current European allies, should history take an unexpected turn.

LONG-LIVED WARTHOG
The A-10 Thunderbolt, better known as the Warthog, pummelled Iraqi armour in 2003 as it had in 1991. First flown in 1972, the Warthog is likely to remain in service until 2028 – a striking example of the longevity of some current military aircraft.

DESERT PATROL
US helicopters fly over the Iraqi desert during counter-insurgency operations. The challenge of bringing air power to bear against an elusive enemy was one the United States faced not only in post-invasion Iraq, but also on a wider global scale.

which civilian casualties were a sensitive issue, aircraft were useful but not decisive and, importantly, could not prevent the need for face-to-face fighting in urban areas or stop terrorists from operating. Iraq confirmed the point once made by air historian John Buckley, that "Air power was an excellent weapon of major war, but of limited use in low-intensity conflicts restricted by political considerations."

Future wars

The US Defense Department had the difficult task of conceiving a credible plan for the effective use of America's power and precision in the air against the "dispersed, global terrorist networks" and "rogue states" that after

The Space Shuttle Columbia *disintegrates high in the atmosphere on 1 February 2003, killing the seven crew on board. The disaster was not only a human tragedy but also a severe setback for NASA's ongoing manned space programme. The Shuttle fleet was grounded for the next two and a half years.*

Stalled in space

By the start of the 21st century, manned space flight had not progressed as rapidly as might have been expected, given the achievements of the first decade of space travel. Thirty years after the last man stood on the surface of the Moon there was no sign of a resumption of such long-distance manned space journeys, despite the efforts of enthusiasts such as former astronaut Buzz Aldrin to drum up support for missions to Mars and beyond. What is more, the comparatively simple goal of a safe, fully reusable "space liner" to fly to space stations and back had remained elusive. The second Shuttle disaster, when *Columbia* broke up on re-entry on 1 February 2003, was a massive setback for NASA. It once more saw the space agency accused of sloppiness in its safety procedures and placed a question mark against continuation of Shuttle flights. This in turn put the whole future of the ambitious International Space Station project at risk.

Fresh vision

At the root of the problems threatening to turn manned space flight into a historical dead end was the failure to identify a clear purpose that would guarantee consistent political and financial support. In the 1960s manned missions had had a clear goal for both the Americans and the Soviet Union – to be the winner in a race. But once the competition came to an end in the 1970s, the decisive incentive was lost.

The sense of drift in the American manned space programme was brought to an end on 14 January 2004, when President George W. Bush announced that he was giving NASA "a new focus and vision for future exploration". The president stated: "We will build new ships to carry man forward into the Universe, to gain a new foothold on the Moon and to prepare for new journeys to the worlds beyond our own." Initial reactions suggested that around half the American people were sceptical about the new vision, but NASA reacted with enthusiasm.

Returning the Shuttle to service and completing the ISS remained important concerns, but the overriding goal was now the return to the Moon. Over the following two years, NASA developed its concept of new-style Moon missions. Conscious of cost limitations, they banked on as much re-use of existing technology as was feasible. The result was a plan for a Crew Exploration Vehicle that combined elements of the Shuttle system and the old Apollo technology. The space capsule would be similar to an Apollo capsule, but larger to accommodate four crew members instead of three. To cut costs, it would be expected to serve for ten missions. The capsule would be launched atop a Shuttle rocket booster, with a Shuttle main engine to power its second stage into orbit. For a Moon shot, a second launch by a heavy-lift rocket would follow, carrying a lunar lander and a "departure

This badge bears the names of the seven astronauts who took part in the Shuttle return-to-flight mission in July 2005: Eileen Collins, Jim Kelly, Soichi Noguchi, Steve Robinson, Wendi Lawrence, Charlie Camarda, and Andy Thomas

Space Shuttle Discovery *launches from NASA's Kennedy Space Center on 21 July 2005 on the first mission after the loss of* Columbia. *Unfortunately, the mission revealed that the problem with debris from the external fuel tank striking the orbiter during the launch – the cause of the* Columbia *disaster – had not been solved.*

stage" into orbit. The crew capsule would join up with these bulkier elements and the departure stage would then fire the craft toward the Moon. Once in lunar orbit, the crew would climb into the lunar lander to descend to the surface, leaving the orbiting capsule unmanned. A section of the lunar lander would carry them back to the capsule once they had completed their stay on the Moon. After reentering the Earth's atmosphere, parachutes would slow the capsule's descent.

And so to Mars

NASA is aiming to return to the Moon in around 2018. At some time after that date a permanent moonbase is to be set up, possibly exploiting resources such as frozen water that the apparently barren satellite might have to offer.

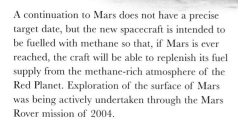

MARS ROVER
The exploration of the surface of Mars by robot vehicles known as "Mars Rovers" has greatly increased our knowledge of the planet. The ability to land such vehicles on Mars indicates that a landing by a manned spacecraft is perfectly feasible.

SHIELD AND CRATER ON MARS
Mars Rover Opportunity, which landed on the planet in January 2004, transmitted this image. It shows the Rover's shattered heat shield, jettisoned during descent, and the crater that the shield made on impact with the red dirt of the planet's surface. The featureless landscape suggests the cheerless prospect facing any long-term human settlers.

Chinese astronaut sets foot on the Moon, the United States will undoubtedly rediscover the motivation to back manned space flight in earnest.

Space goes private

At the start of the third millennium it was still hard to see space travel to the Moon or Mars ever affecting most people's lives, except in science-fiction scenarios of escape from a devastated Earth after some natural or man-made cataclysm. In the first 44 years of state-sponsored space travel, fewer than 450 people had gone outside the Earth's atmosphere. The possibility of large numbers of people experiencing space travel can only become capable of realization when manned space flight begins to open up to private entrepreneurship.

A continuation to Mars does not have a precise target date, but the new spacecraft is intended to be fuelled with methane so that, if Mars is ever reached, the craft will be able to replenish its fuel supply from the methane-rich atmosphere of the Red Planet. Exploration of the surface of Mars was being actively undertaken through the Mars Rover mission of 2004.

Chinese competitors

Given the past record of space exploration, there was still room for scepticism as to whether the will to fulfil this "new vision" can be sustained through decades ahead. What was needed to boost the impulse for state-sponsored manned space exploration was a space race. With the demise of the Soviet Union, who would provide the competition against which the Americans would race to reach Mars? The only possible candidates were the Chinese. In October 2003 China became the third country to put a man in space when 38-year-old fighter pilot Yang Liwei orbited the globe 14 times aboard Shenzhou 5 – an evolved version of the Russian Soyuz spacecraft. Of course, the Chinese were far behind NASA in technological development, but

they had the nationalist motivation to devote major resources to space exploration. The Chinese National Space Administration (CNSA) had as part of its publicly stated mission to "become a world leader in manned space exploration". China in the early 21st century, with its vast resources generated by a rapidly expanding economy at the disposal of a powerful and combative state system, had the potential to surprise everyone with spectacular space ventures in the future. The day that the first

MARS ORBITER
The Mars Reconnaissance Orbiter, which began circling Mars in March 2006, had among its objectives the identification of landing sites for missions to the planet's surface. It was expected to stay in orbit for more than four years.

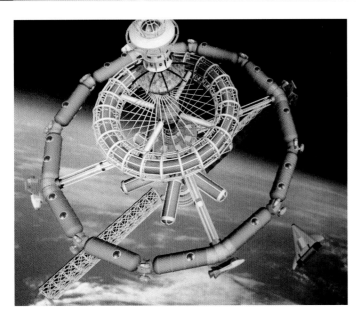

SPACE HOTEL
In the 1990s the Space Island Group produced a plan for a wheel-shaped orbital space hotel that would be constructed out of used Space Shuttle fuel tanks. If suborbital space trips for paying passengers become commonplace, there will inevitably be demand for longer stays in space.

Space tourism

The fact that the exploration of space developed in the context of the Cold War competition between the United States and the Soviet Union ensured that manned space flight was promoted by governments rather than by private entrepreneurs, and that its primary goal was national prestige. Although business became an important source of finance for satellite launches, the privatization of manned space travel did not begin until Dennis Tito made his historic flight as the first space tourist in 2001 (see page 372). Tito was able to travel to the ISS because Russia's post-communist economic debacle had left its space programme chronically short of funds. Milking rich businessmen of millions of dollars in return for a trip into space offered a simple and viable way for the Russians to raise cash.

The model of future space tourism suggested by Tito's trip was complex and expensive. It involved months spent training the paying passengers as cosmonauts and a journey to an orbiting space station. If space tourism took this direction, then the space station would presumably evolve into a space hotel where the customers would spend an interestingly unusual vacation with spectacular views. They would necessarily be ultra-rich individuals, with time and a fortune to spare. The limited scale of the operation made it likely that tourism would continue merely to "piggy-back" upon government programmes.

The Ansari X-prize

This vision of the future of space tourism was overturned by the successful development of the world's first privately financed space vehicle. The leap to private space travel was stimulated by a device that had proved spectacularly successful in the early days of flight: the offer of a significant money prize. Just like Blériot's flight across the English Channel in 1909 and Lindbergh's transatlantic flight in 1927, the first space flight by a privately financed reusable vehicle was to be rewarded in cash. In 1996 the X Foundation offered $10 million for anyone who could reach suborbital space – defined as above 100km (60 miles) altitude – twice in the same craft within a fortnight. A deadline was set of 1 January 2005.

Twenty-six hopeful teams decided to enter for the prize. Argentina, Britain, Canada, Israel, Romania, and Russia were represented, as well as the United States. In the end it was the American designer Burt Rutan who struck the right blend of innovation and feasibility. Focusing on the key issue of safety, Rutan identified the two points in space flight that carried maximum risk: ground launch and re-entry to the atmosphere. To avoid the first problem, he planned to have his spaceship carried to high altitude under the belly of an aircraft. To cope with re-entry, he equipped the wings with folding booms that would slow the craft through a high-drag "shuttlecock" effect as it plunged towards the Earth, avoiding the extremely high temperatures experienced by traditional spacecraft entering the atmosphere. There would also be no need to strike the atmosphere at a precisely calculated angle.

SpaceShipOne

Rutan's airborne launch vehicle, the White Knight, made its first flight in summer 2002. First tests with the three-seater spacecraft, SpaceShipOne (SS1), began the following year. It made its first foray into space on 21 June 2004, with pilot Mike Melvill at the controls. The rocket motor, fuelled by an unusual mix of nitrous oxide and rubber, ignited after the spaceship was released from the White Knight into gliding flight. The spaceship then accelerated upward,

reaching around Mach 2.9 before the motor burned out just over a minute later. The spaceship's momentum carried it onward, gradually decelerating beyond the atmosphere, before gravity took over and pulled it back towards the Earth. The "shuttlecock" effect ensured a safe re-entry, followed by a leisurely glide down to its landing site in the Mojave Desert. The goal of designing a reusable private space vehicle had been achieved. Successfully completing repeat flights on 29 September and 4 October 2004, Rutan duly claimed the prize.

Rush into space

SpaceShipOne was only an experimental space vehicle, but its success led to immediate plans for commercial exploitation. British entrepreneur Sir

SPACESHIPONE
The first privately financed reusable space vehicle, Burt Rutan's SpaceShipOne glides down to land in the Mojave Desert. SS1 alters its configuration for different stages of its flight.

Richard Branson made a deal with Rutan to set up Virgin Galactic as the first practical space tourism company. The plan was to offer suborbital flights in SpaceShipOne derivatives for around $200,000. This would not exactly democratize space travel, but it compared favourably with the reputed $20 million paid by Tito for his trip. Little preparation would be required beyond medical checks and the passengers would not need to wear spacesuits. Admittedly, the experience of weightlessness at the edge of space would last only a few minutes, but no one believed the flights would be short of customers. Virgin Galactic was offering the greatest fairground ride in the world.

With plenty of other private space ventures in the field, it is clear that competition to attract paying customers is going to be stiff. In 2006 more conventional projects such as the Altaris Spaceship and the Canadian Arrow were racing to beat Virgin Galactic into commercial service by exploiting tried and tested rocket technology, basically replicating the way that Alan Shepard was launched into suborbital flight in 1962. Meanwhile Rutan was already working on developing the SS1 concept into a craft that could fly passengers to a space station and back. The drive toward a rapid expansion of private space travel seems unstoppable. NASA's development of manned space flight had been hesitant mostly because of the difficulty in explaining its purpose to the tax-paying public. Entrepreneurs do not need to justify putting resources into space travel. If people are prepared to pay what it costs, that is enough.

CANADIAN ARROW
The Canadian Arrow, an X-Prize entrant, is a conventional two-stage ground-launched rocket. Passengers would splash down in the ocean at the end of their trip.

BURT RUTAN

Born in Oregon in 1943 and raised in California, Burt Rutan is one America's most original aircraft designers. Rutan was a flight-test project engineer for the US Air Force before setting up his own business in the 1970s. He first made his name as a designer of homebuilts, especially the innovatory canard-configuration Vari-eze. In 1982 he set up his Scaled Composites company in the Mojave Desert and, four years later, earned worldwide renown with Voyager, the

first aircraft to fly non-stop around the globe without refuelling. Rutan had more than 20 experimental designs to his credit before embarking on the SpaceShipOne project with which he won the Ansari X-Prize in 2004. He also found time to create GlobalFlyer, in which Steve Fossett made record-breaking flights in 2005–06 (see page 438).

RUTAN IN 2005
Burt Rutan speaking at the Smithsonian's National Air and Space Museum.

WHITE KNIGHT LAUNCHER
The White Knight launcher carries SpaceShipOne under its fuselage. With two afterburning jet engines and a wingspan of 25m (82ft), it lifts SS1 to an altitude of 15,250m (50,000ft).

What next?

US president George W. Bush pitched high objectives in 2004 when he stated that "Humans are headed into the cosmos." Certainly, a return to the Moon seems likely and it is even feasible that humans could set foot on Mars sometime this century. But beyond that, informed speculation stumbles.

Some visionaries have long predicted a mass migration of humans to other worlds. In 1926 Konstantin Tsiolkovsky, the Russian prophet of space travel, had a vision of the colonization of planets on a massive scale, with "the population of the solar system... one hundred thousand million times the current population of the Earth." American rocket pioneer Robert Goddard believed that, when the Sun died, human life could continue through a "last migration" of interstellar "arks" in search of another inhabitable planet. More recently, physicist Stephen Hawking predicted the need for such migration much sooner, asserting that "The human race will not survive the next thousand years unless we spread into space."

Against such visions stand what is known about the impossibly long time required for even the fastest conceivable journey to any other star and the implacable hostility to human life of the environments found everywhere in our Solar System except on planet Earth. One of the most consistent experiences of those who have spent long periods in space has been nostalgia for the simple smells, sights, and sounds of nature. As Russian cosmonaut Vladimir Lyakhov said with telling simplicity, Earth is the place "where we will always strive to return. It is our home."

Remote warfare

Some scenarios for future aerial warfare sound as surreal as interstellar colonization. The notion of a fleet of remote-controlled pilotless aircraft patrolling the planet, ready to assassinate a suspected enemy of the United States anywhere in the world at the press of a button, sounds like science fiction. But it forms a central part of the Pentagon's latest war plans. UAVs such as Global Hawk already looked set to replace high-altitude reconnaissance aircraft like the U-2 and SR-71,

EUROFIGHTER
The first production Eurofighter Typhoon was delivered in 2003. It achieves outstanding agility through an inherently unstable design and lightweight construction. Like the American F-22, it can cruise at supersonic speed without afterburning.

partly because they can carry out missions lasting over 30 hours, beyond the endurance of an aircrew. Increasingly, though, combat is also likely to become the preserve of pilotless aircraft. By the first decade of the 21st century, the Predator UAV had already been used to carry out missile attacks on ground targets and several UCAVs designed specifically for this role were under development in the United States and Europe. The chief competitors for the US Air Force and Navy J-UCAS (Joint Unmanned Combat Air Systems) contract were the Boeing X-45 and the Northrop Grumman X-47. Dassault and Sukhoi

SOLO AROUND THE WORLD

In the early 21st century, millionaire adventurer Steve Fossett identified an unscaled peak in the annals of great flying exploits: no one had ever flown a jet solo, non-stop, and unrefuelled around the world. The aircraft for this exploit, GlobalFlyer, was built by Burt Rutan's company Scaled Composites with the financial backing of Sir Richard Branson's airline Virgin Atlantic. Made of composite materials such as carbon fibre and epoxy, GlobalFlyer was extremely light. Four-fifths of its takeoff weight was fuel. Fossett took off from Salina, Kansas, on 28 February 2005. Despite losing some fuel early in the flight, he was able to continue with the aid of favourable tail winds. Flying at high altitude, he completed the non-stop flight of 36,817km (22,863 miles) in just over 67 hours, landing back at Salina on the evening of 3 March. The following February, in the same aircraft, he made the world's longest ever non-stop flight at 42,469km (26,273 miles), landing at Bournemouth in southern England with 200lb of fuel remaining.

LONE TRAVELLER
Sixty years old at the time of his record-breaking solo round-the-world flight, Steve Fossett has also circumnavigated the globe as a solo balloonist and yachtsman.

GLOBALFLYER
Resembling Rutan's Voyager design (see pp.412–413), Global Flyer's twin tail-booms were large fuel tanks. The single turbofan engine was mounted on the central fuselage, behind the cockpit.

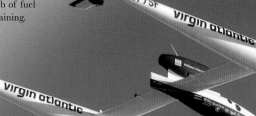

Unmanned remote-controlled helicopters were in operation with American forces in Iraq by 2005. UAV helicopters could perform a variety of functions, including communications, reconnaissance, and supply.

were other manufacturers developing UCAVs. It seemed likely nonetheless that piloted aircraft would continue to perform a large part of military duties. Indeed, one of the prime functions of UCAVs would be to suppress ground-air defences, allowing conventional aircraft to penetrate enemy airspace unscathed.

Gradual progress

Developments in piloted military aircraft will probably not be startling in the foreseeable future. The crop of fighters currently under development or just entering service, including the F-22, the Eurofighter Typhoon, the Dassault Rafale, and the Joint Strike Fighter, are likely to remain state-of-the-art for several decades, and the B-52 bomber, in service since the mid 1950s, seems set to remain in service indefinitely. Similarly, in commercial aviation, no one is likely to rush to build an airliner bigger than the Airbus A380 or to push passenger flight beyond the sound barrier again. Progress will be in reducing fuel and maintenance costs, and in cutting emissions and reducing noise, a response to unrelenting criticism from environmentalists. Unmanned civil aircraft may be a future possibility – the military Global Hawk UAV has been cleared to fly in US civilian air corridors – but passengers were likely to prefer the comforting presence of a pilot.

Human imagination is sometimes more limited than the human technological inventiveness. For example, flight pioneers of the Wright brothers' era were firmly convinced that aircraft would always be small; nothing could have amazed them more than the A380. What was achieved in the first century of flight was literally beyond imagination in the early 1900s. Along the way, a certain elated optimism associated with the early dream of flight has undoubtedly been sullied. When American aviation pioneer Octave Chanute wrote of the future of flight in 1894, he asked his readers to hope "that the advent of a successful flying machine… will bring nothing but good into the world; that it shall abridge distance, make all parts of the globe accessible, bring men into closer relation with each other, advance civilization, and hasten the promised era in which there shall be nothing but peace and goodwill among men." Part of this flight has done, but not the most optimistic part. Yet human flight remains a magical achievement in which imaginative aspiration and technology are

GLOBAL HAWK
A superstar among American UAVs, Global Hawk is a high-altitude reconnaissance aircraft. The basic model is 13.4m (44ft) long, with a wingspan of 35.4m (116ft). Its maximum speed is over 600kph (350mph).

welded, just as aesthetics and functionality have reached perfect accord in the greatest aircraft designs. Inevitably aircraft became a tool of business and war, but at the heart of aviation still lies the unquenched aspiration – as famously expressed by poet-pilot John Magee – to slip "the surly bonds of earth" and tread "the high, untrespassed sanctity of space".

GLOSSARY

AAM (air-to-air missile)
A radar- or heat-guided missile fired at an enemy aircraft or missile, from another aircraft.

Aerodynamics
The physics of the movement of objects through air or gas.

Aerofoil
See AEROFOIL panel, right.

Aeronautics
The science of aerial travel.

Aeronaut
The pilot of any lighter-than-air aircraft, especially a balloon.

Afterburner
A device that injects additional fuel into a specially designed turbojet jetpipe to provide extra thrust. Also called reheat.

Aileron
A control surface on the wing's far trailing edge – operated by moving the control column sideways – which governs roll.

Airframe
An aircraft's body excluding engines.

Altimeter
An instrument used to measure an aircraft's altitude.

Angle of attack
The angle between the centreline of an aircraft's wing and the oncoming airflow. See also AEROFOIL panel, right.

Anti-gravity suit
An anti-gravity or anti-G suit counteracts the effects of very fast flying by squeezing the legs and forcing blood to the brain so that a pilot does not lose consciousness.

Artificial horizon
Primary cockpit flight instrument that indicates the aircraft's attitude in relation to the horizon. Also called a Gyro Horizon.

Astronaut
A person engaged in space travel, derived from Greek words

meaning "star sailor". Cosmonaut is the Russian equivalent.

Attitude
The tilt of an aircraft in relation to the oncoming airflow or to the ground.

Autogyro
An aircraft equipped with a rotating wing, or rotor, to sustain itself in the air, and a propeller to move forwards. De la Cierva's machine was called an Autogiro.

Automatic pilot
Airborne electronic system that automatically stabilizes an aircraft about its three axes, restores it to its original flightpath after a disturbance, and can be preset to make the aircraft follow a particular flight path.

AWACS
AWACS (Airborne Warning and Control System) refers to a modified Boeing airliner with a rotating radar dome (the E-3 Sentry), which provides all-weather

surveillance, command, and communications for allied air defence forces.

Bank
To travel with one side higher than the other when turning. In an aircraft, the pilot must combine roll and yaw to achieve this manoeuvre. See also BASIC CONTROLS panel, right.

Black box
A box – usually bright yellow or orange – containing electronic equipment, which records all the flight details. In the event of a crash, it is recovered from the aircraft's wreckage to ascertain the accident's cause.

Cabane strut
Wing strut attached to the fuselage.

Cantilever wing
A wing that does not require external struts or bracing wires.

Chaff
Radar-reflective particulate matter

released by aircraft to confuse radar detection systems.

Cockpit
The space in the fuselage occupied by a pilot (and/or crew). Usually refers to smaller aircraft.

Cold War
Term given to the post-WWII period of tension between the two superpowers – the USA and the USSR.

Control column
Floor-mounted stick used by pilot to control roll (by moving it sideways) and pitch (by moving it forwards or backwards). Also called a control stick or joystick.

Control surfaces
The moving parts of an aircraft that change the airflow around its trailing edges – e.g. ailerons, elevators, rudders – causing it to roll, pitch, or yaw.

Cowling
Engine covering, often using hinged or removable panels.

Cross-braced
Held rigid by wires attached to the extremities of the structural frame and crossing at its centre.

Dihedral
The angling of the wing or tailplane upwards from root to tip when seen from head on, to maintain lateral stability.

Dirigible
An airship that is capable of being guided by an engine.

Drag
The force (air resistance) that resists the motion of an aircraft.

Drift indicator
An instrument indicating an aircraft's angle of "drift" – the difference between an aircraft's projected flight path and its actual heading, as affected by winds.

Drone
A remotely piloted aircraft, used in a variety of military roles, including reconnaissance. Also known as an Unmanned Aerial Vehicle (UAV) or Remote Piloted Vehicle (RPV).

Duralumin
Wrought alloy of aluminium with small percentages of copper, magnesium, and manganese.

Ejection seat
A special seat on military aircraft which uses a rocket motor to blast the pilot clear in an emergency and parachute him or her to safety.

ANATOMY OF AN AEROPLANE

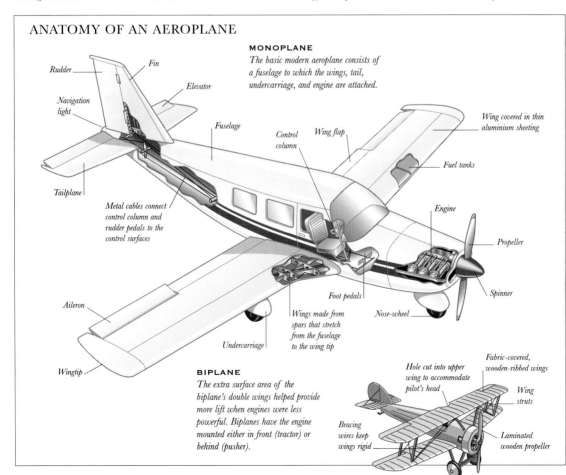

MONOPLANE
The basic modern aeroplane consists of a fuselage to which the wings, tail, undercarriage, and engine are attached.

Rudder

Fin

Elevator

Navigation light

Fuselage

Control column

Wing flap

Wing covered in thin aluminium sheeting

Fuel tanks

Tailplane

Metal cables connect control column and rudder pedals to the control surfaces

Engine

Propeller

Spinner

Aileron

Foot pedals

Nose-wheel

Wings made from spars that stretch from the fuselage to the wing tip

Undercarriage

Wingtip

BIPLANE
The extra surface area of the biplane's double wings helped provide more lift when engines were less powerful. Biplanes have the engine mounted either in front (tractor) or behind (pusher).

Hole cut into upper wing to accommodate pilot's head

Fabric-covered, wooden-ribbed wings

Wing struts

Bracing wires keep wings rigid

Laminated wooden propeller

AEROFOIL

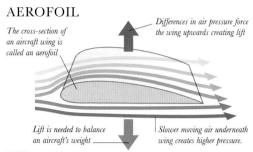

The cross-section of an aircraft wing is called an aerofoil

Differences in air pressure force the wing upwards creating lift

Lift is needed to balance an aircraft's weight

Slower moving air underneath wing creates higher pressure.

LIFT

As the curved wing moves through the air, the air passing over the wing moves faster than the air passing beneath. Fast-moving air has a lower pressure, so slower, high-pressure air beneath the wing forces it upwards.

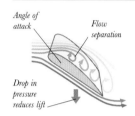

Angle of attack

Flow separation

Drop in pressure reduces lift

STALLING

Stalling (when an aircraft drops suddenly) occurs when the critical angle of attack between the wing's centreline and the airflow direction is exceeded. This causes a breakdown in airflow, leading to a drop in pressure and reducing lift.

Elevator
A control surface on the tailplane's trailing edge, which governs pitch.

Elevon
Wing control surface – mostly found on delta-wing aircraft – that combines the functions of elevator and aileron.

Envelope
The part of an airship that contains the lifting gas.

Fairing
A covering used to streamline any projections in the aircraft structure.

Fin
The vertical section of the tail or wingtip, used to increase the stability of the aircraft. Also known as a vertical stabilizer

Flight-deck
The compartment housing the crew and flight controls in a large aircraft or spacecraft.

Flight simulator
A training device that creates a simulated flying environment and experience within a virtual cockpit.

Fly-by-wire
An electronic flight control system used instead of mechanical controls.

Flying boat
An aircraft with a watertight fuselage, resembling the shape of a boat's hull, allowing it to take off and land in water.

Flying wing
An all-wing aircraft without the customary fuselage, and a tail designed to be aerodynamically efficient.

Fuselage
The main body of an aircraft, to which the wings and tail are attached.

G-force
The force experienced by the crew of a craft undergoing rapid acceleration or deceleration. One "g" is equal to the force normally exerted by Earth's gravity.

Glider
A light aircraft without an engine.

Gondola
The passenger compartment that hangs from a balloon or airship.

Hangar
An enclosed structure with a large floor area used to house aircraft.

Hang-glider
A fabric sailwing attached to a light, triangular frame, beneath which the pilot hangs horizontally,

Head-up display (HUD)
A unit that projects information, such as combat status and aircraft performance data, onto a transparent screen in the pilot's line of sight, lessening the need to look down into the cockpit.

Helicopter
See HELICOPTER *panel, p.431.*

Horsepower
A unit of power devised by the 19th-century British engineer James Watt, which describes the energy required to raise 550lb (250kg) one foot (30cm) in one second.

HOTAS
HOTAS (Hands On Throttle and Stick) describes a system whereby the switches are positioned on the throttle and the control stick so that the pilot does not have to remove his or her hands from them.

Hypersonic
A flight speed equal to or greater than Mach 5 (i.e. five times the speed of sound).

Hypoxia
An oxygen deficiency in the body tissues, which can be brought on by flying at high altitudes (where there is less oxygen) in an unpressurized environment.

INS
INS (Inertial Navigation System) is a cockpit device that provides positional and navigational information without the need for data from external references.

Interrupter gear
A device that allowed a fixed machine gun to fire through a propeller, by halting the gun's fire whenever a blade was in line with the barrel of the gun.

Jet-stream
A high-altitude, high-speed wind flow system.

J-STARS
Joint-STARS (Surveillance Target Attack Radar System) is a battle-management surveillance system carried in a modified Boeing airliner (E-8), which detects, tracks, and targets enemy activity on the ground, relaying the information to US Air Force and US Army command posts.

Landing gear
see **Undercarriage.**

Leading edge
The front, rounded edge of an aerofoil such as a wing or rotor.

Lift
The force exerted on a moving aerofoil that causes a wing to rise. *See also* AEROFOIL *panel, above.*

Longeron
Any main load-bearing fuselage structural member with a fore and aft axis.

Look-down Shoot-down
The capability of an aircraft flying at high altitude to use advanced radar systems to detect a fast, low-flying enemy aircraft against a jumble of radar reflections from the ground.

Mach number
The ratio of an aircraft's speed to the speed of sound. Mach 2, for example, is twice the speed of sound.

MFD (Multi-Function Display)
A cockpit instrument able to display two or more types of information, such as navigational and fuel consumption data.

Monoplane
Aeroplane with one set of wings.

Microlight
A small, powered aircraft, often in the form of a hang-glider.

Monocoque
A structure, such as a fuselage, that has no internal bracing and derives most of its strength from its skin.

NACA
NACA (National Advisory Committee on Aeronautics) was founded during WWI and renamed NASA in 1958.

NACA Cowling
A ring-shaped cowling of a certain design, used to cover radial engines and reduce drag.

Nacelle
A streamlined, protective enclosure housing a part of the aircraft likely to induce drag, such as engines, crew, or armament.

NASA
Founded in 1958, NASA (National Aeronautics and Space Administration) is the US government agency that conducts all non-military spaceflight and aeronautical research.

Ornithopter
An aircraft with oscillating wings.

Parachute
A fabric canopy that allows a person or package to descend safely to earth.

Pay-load
The load carried by an aircraft. including passengers and cargo, from which revenue is obtained.

Pilot
The person with primary responsibility for flying an aircraft.

Piston engine
See PISTON ENGINES *panel, p.430.*

BASIC CONTROLS

PITCH
To pitch (climb or dive), the pilot pushes or pulls the control column, raising or lowering the elevator flaps on the tailplane. This causes movement about the aircraft's lateral axis.

Tail elevator controls pitch

ROLL
To roll, the pilot moves the control column to the left or right, which raises the ailerons on one wing and lowers them on the other. This causes movement about the aircraft's longitudinal axis.

Wing ailerons control roll

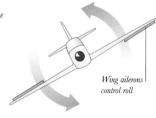

YAW
To yaw left or right, the pilot's feet swivel the rudder bar, turning the upright rudder on the aeroplane's fin. This causes movement about the aircraft's normal axis.

Tail rudder controls yaw

Pitch
See BASIC CONTROLS *panel, p.429*.

Pitot tube
A small, open-ended tube, usually located on the leading edge of the wing, which is used to measure airspeed and is linked to a pressure-measuring device in the cockpit.

Powerplant
The permanent assembly (including propellers and engine) which is responsible for aircraft propulsion.

Pressurized
Describes a cockpit or cabin that maintains a constant pressure, regardless of the aircraft's altitude.

PISTON ENGINES

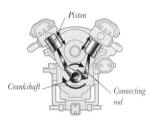

"Master" con-rod
Piston
Cylinder
Crankshaft

RADIAL ENGINE
A radial engine has an odd number of cylinders arranged in a circle around the crankshaft. The "master" con-rod is attached to the crankshaft while the other con-rods are "slaved" off the master rod. The propeller is attached to the rotating crankshaft.

ROTARY ENGINE
Similar to the radial engine, except that the crankshaft remains stationary while the cylinders and propeller rotate around it, cooling themselves in the process.

Propeller attached to cylinders
"Master" con-rod
"Slave" con-rod
Crankshaft remains stationary

INLINE ENGINE
An inline engine has the cylinders in a line – with pistons pumping up and down inside – one behind the other in the crankcase.

Propeller
Piston
Crankcase
Crankshaft

V ENGINE
Two inline banks of cylinders are arranged at an angle to each other. Each pair of pistons pumps diagonally on the same crankshaft, spinning it round.

Piston
Crankshaft
Connecting rod

HORIZONTALLY OPPOSED ENGINE
Also known as the boxer engine, it has two banks of inline cylinders arranged opposite each other. This type is often used on trainers and small aircraft.

Crankshaft
Piston

Propeller
A device consisting of aerofoil-shaped blades that spin around a central hub. As the engine turns the blades, they create thrust by "biting" into the air and forcing it to move back.

Pusher
A configuration in which the propeller is situated at the back of the engine or fuselage and pushes the aircraft through the air.

Radar
Radar (radio direction and ranging) is an instrument used to locate an object, to navigate, and in warning and detection systems. Radar works by transmitting radio waves and noting the time it takes for the reflected waves to return.

Radar signature
The waveform of a radar echo, which can be used to identify an object.

Ramp doors
The doors inside the intake ramps of a jet aircraft's engine. These regulate the flow of air to the engine.

Radial engine
See PISTON ENGINES *panel, left*.

Radome
A protective radar dome – which allows radio waves in and out – covering radar equipment.

Roll
See BASIC CONTROLS *panel, p.429*.

Rotary engine
See PISTON ENGINES *panel, left*.

Rotocraft
See HELICOPTER *panel, right*.

Rotors
Aerofoils, used on a helicopter, that rotate at high speed to create thrust and lift.

Ruddervator
A movable structure that combines the functions of elevator and rudder.

Rudder
A control surface on the fin's trailing edge which governs yaw.

SAM (surface-to-air missile)
A missile fired at an enemy aircraft or missile, launched from the ground or from a ship.

Satellite
A spacecraft or other body orbiting the Earth.

Seaplane
An aircraft that can land on or take off from water.

Skid
Sledge-runners used as part of an aircraft's undercarriage.

JET ENGINES

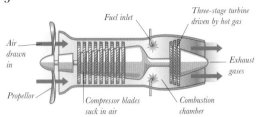

Air drawn in
Fuel inlet
Three-stage turbine driven by hot gas
Exhaust gases
Propellor
Compressor blades suck in air
Combustion chamber

TURBO-PROP
Fast-spinning compressor blades draw air into the combustion chamber, where it is heated by burning fuel. As expanding gases escape through the exhaust, they spin a turbine which powers the propeller.

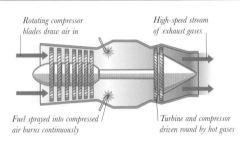

Rotating compressor blades draw air in
High-speed stream of exhaust gases
Fuel sprayed into compressed air burns continuously
Turbine and compressor driven round by hot gases

TURBOJET
The simplest jets – "turbojets" – work by pushing a stream of hot exhaust gases out the back. This hits the air behind so fast that the reaction thrusts the plane forward like a deflating balloon.

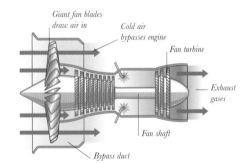

Giant fan blades draw air in
Cold air bypasses engine
Fan turbine
Exhaust gases
Fan shaft
Bypass duct

TURBOFAN
The turbofan is quieter and more fuel-efficient than the pure jet or turbo-prop at airliner speeds (high subsonic) because of the combination of cold and hot thrust. The bypass air helps to keep the engine cool.

Smart weapon
A precision-guided munition that is directed to its target using laser guidance or, more recently, satellite-linked GPS (global positioning system) guidance.

Sonic boom
The thunder-like "clap" heard when an aircraft flies faster than the speed of sound. This is caused by sudden changes in air pressure as the craft pushes air molecules out of its path.

Spar
The main structural element that run the span of a wing. The ribs and other secondary structural parts are attached to the spars.

Splitter plate
A plate located between a jet engine's air intake and the fuselage, which evens out the air flow to the engine.

SST (Supersonic transport)
Commercial airliners that can carry passengers at supersonic speeds.

Stall
See AEROFOIL *panel, p.429*.

HELICOPTER

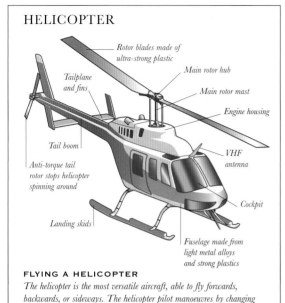

Rotor blades made of ultra-strong plastic

Main rotor hub

Main rotor mast

Engine housing

Tailplane and fins

VHF antenna

Tail boom

Cockpit

Anti-torque tail rotor stops helicopter spinning around

Landing skids

Fuselage made from light metal alloys and strong plastics

FLYING A HELICOPTER

The helicopter is the most versatile aircraft, able to fly forwards, backwards, or sideways. The helicopter pilot manoeuvres by changing the blades' pitch (angle), while the tail rotor stops it spinning round.

Stealth technology
A combination of geometric shaping and material technologies aimed at producing an aircraft that has low-observability, primarily in relation to enemy radar detection.

STOL
STOL (short take-off and landing) refers to a system by which aircraft take off and land over a short distance.

Streamline
To shape an object, such as an aerofoil or fuselage, so that it creates less drag and moves smoothly through the air.

Subsonic
Slower than the speed of sound.

Supercharger
A type of compressor in a piston engine used to boost power by increasing the amount of air fed to the cylinders.

Supersonic
Faster than the speed of sound.

Swept wing
A wing that it is angled towards the rear of the aircraft, in order to reduce drag.

Swing-wing
Also known as Variable-sweep. See WING SHAPES panel, below.

Tachometer
The gauge that displays an aircraft's engine speed.

Tailplane
A fixed, horizontal structure attached to the tail or the rear part of the fuselage. Also know as a horizontal stabilizer.

Throttle
The lever that regulates the amount of fuel reaching an engine, which in turn affects the amount of thrust generated.

Thrust
The force, generated by an engine, that pushes an aircraft through the air.

Thrust vectoring
Controlling a jet aircraft's movement by altering the angle of propulsive thrust, via a jet engine's rotating nozzles.

Torque
The twisting force created by a turning component, (such as a propeller) powered by an engine.

Tractor
A configuration in which the propeller is situated in front of the engine or fuselage and pulls the aircraft through the air.

Trailing edge
The rear edge of an aerofoil, usually thin and sharp.

Triplane
An aircraft with three sets of wings, arranged one on top of the other.

Turbofan
See JET ENGINES panel, left.

Turbo-prop
See JET ENGINES panel, left.

Turbulence
Irregular air flow, which can cause an aircraft to fly unpredictably.

Undercarriage
The unit beneath an aircraft – consisting of wheels (or skis or floats), shock absorbers, and support struts – that supports the aircraft on the ground.

Variable-geometry
Also known as Variable-sweep. See WING SHAPES panel, below.

Variable-sweep
See WING SHAPES panel, below.

VTOL
VTOL (vertical take-off and landing) is a system that allows aircraft to take off and land vertically, eliminating the need for a runway altogether.

Widebody
A type of airliner with a passenger cabin sufficiently wide for it to be divided into three sets of multiple-seat columns.

Wing flap
The movable part of the wing's near trailing edge, which is used to increase lift at slower air speeds and slow the aircraft on landing.

Winglet
A small, upright structure on the wingtip, used to lessen tip vortices and therefore reduce drag.

Wing-warping
A system invented by the Wright brothers to provide lateral control to their biplane aircraft, by twisting its wingtips (via wires in the wings). It was later replaced by movable, hinged ailerons.

Wire-braced
The term given to any part of an aircraft, such as a wing or fuselage, that uses tensioned wire bracing to maintain rigidity.

Yaw
See BASIC CONTROLS panel, p.429.

Zeppelin
Any of the airships built by the Zeppelin company, founded by Count von Zeppelin (1838–1917).

ROCKET

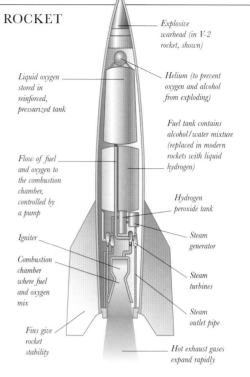

Explosive warhead (in V-2 rocket, shown)

Liquid oxygen stored in reinforced, pressurized tank

Helium (to prevent oxygen and alcohol from exploding)

Fuel tank contains alcohol/water mixture (replaced in modern rockets with liquid hydrogen)

Flow of fuel and oxygen to the combustion chamber, controlled by a pump

Hydrogen peroxide tank

Igniter

Steam generator

Combustion chamber where fuel and oxygen mix

Steam turbines

Steam outlet pipe

Fins give rocket stability

Hot exhaust gases expand rapidly

ROCKET ENGINE

The rocket is the simplest and most powerful kind of engine – a reaction engine. Fuel (solid or liquid) is burned in an open-ended combustion chamber and the resulting heated, high-pressure gases escape at high speed out of the chamber, providing thrust.

WING SHAPES

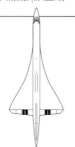

STRAIGHT
Short, straight wings produce good lift and low drag at medium speed. Propellers or jet engines provide the power that produces the lift.

SWEPT
Swept-back wings minimize drag at high speed. However, lift is also reduced, requiring high take-off and landing speeds.

VARIABLE-SWEEP
Hinged wings can be slanted backwards during flight to reduce drag. The wings are at right-angles for take-off and landing to increase lift.

DELTA
Supersonic aircraft often have delta wings to help retain control of the aircraft within the shock wave that forms around it at such high speeds.

INDEX

Page numbers in **bold** refer to catalogue, profile, or feature entries; those in *italics* refer to other illustrated entries.

ACKNOWLEDGEMENTS

Dorling Kindersley would like to thank the following for help with photography: British Airways; Cindy Dawson at the Imperial War Museum Duxford; the Imperial War Museum London; Michel Thourin and Philippe Gras at the Musée de l'Air et de l'Espace; Ray Thomas at Aces High Ltd.; Sarah Hanna at Old Flying Machine Company; JU-AIR, Air Force Centre Dübendorf; Lincolnshire Aviation Heritage Centre. For editorial help: Sean O'Connor, John Mapps, Peter Frances, and Georgina Garner; Jane Parker, for the index; Liz Bowers (co-ordinator) and David Parry (curator) at the Imperial War Museum, London; Chris Savill for additional picture research; Ellen Nanney and Robyn Bissette at the Smithsonian Institution, and the Archives department at the National Air and Space Museum for their patience, dedication, and hard work.

QUOTATIONS

Every effort has been made to identify and contact copyright holders. The publishers will be
glad to rectify, in future editions, any omissions or corrections brought to their notice.

On page 108, from Beryl Markham's *West With the Night* used by permission of Virago Press on behalf of the author; on page 152, from Graham Coster's *Corsairville* used by permission of Penguin Books on behalf of the author; on pages 191 and 213, from Roald Dahl's *Going Solo* used by permission of David Higham Associates on behalf of the author; on page 402, from Julian Barnes's *Staring at the Sun* reprinted by permission of PFD on behalf of the author.

PICTURE CREDITS

Key: l = left, r = right, c = centre, t = top, b = bottom.

Abbreviations:
AB – Austin Brown / Aviation Picture Library; **DK** – DK Picture Library; **DFRC:** Dryden Flight Research Center Photo Collection; **HG** – Hulton Getty Archive; **IWM** – Imperial War Museum; **JSC** – Johnson Space Center; **LOC** – Library of Congress, Washington, D.C.; **MEPL** – Mary Evans Picture Library; **MSFC** – Marshall Space Flight Center; **MUS** – Musée de L'Air et de L'Espace/le Bourget; **NASM** – National Air and Space Museum, Smithsonian Institution; **SS** – Science & Society Picture Library.

1 DK: Martin Cameron. **2/3** MUS (MC 11631), courtesy **NASM** (90-15729). **4/5 Aviation Images:** John Dibbs. **5 HG:** Archive Photos ca; **Image Bank / Getty Images:** Andy Caulfield bc; **IWM:** (C5422) c; (Q27633) tcb; **NASA:** JSC (S69-39962) bca; **NASM:** (SI 71-2699) cb; (SI A-4384-C) tl. **6/7 HG:** Topical Press. **8/9 NASM:** (SI A-4387-C-Q). **NASM:** (85-10844) c; (SI 86-12762) cr; (SI 87-17029) cl. **10 DK:** Peter Chadwick bl; **NASM:** Photozincograph copied from Christopher Hatton Turner, Astra Castra, Chapman and Hall (London) 1865, frontispiece (NASM 9A00238) tl. **10/11 NASM:** (SI 87-17029). **11 HG:** tc; **SS:** Science Museum br. **12 AB:** Science Museum bc, bcr; **DK:** Mike Dunning tr. John J Ide Collection, courtesy **NASM** (SI 96-16071) tl. **13 MEPL:** tc; **NASM:** (SI 85-3941) cr; (SI A-32051-A) br. **14 AB:** Musée de L'Air clb; **Philip Jarret:** t; **SS:** Science Museum bc. **15 AB:** Science Museum br; **NASM:** (SI 97-15315) t. **16 HG:** tl. **16/17 NASM:** (SI A-30908-A). **17** Sherman Fairchild Collection of Aeronautical Photographs, **NASM** (SI 2001-1004) br. **18 NASM:** (SI A-18792) bl; (SI A-18800) t; (SI A-18841) bcl; (SI A-18862) bc. **19 NASM:** (SI 93-4212) br; (SI A-38877) tr. **20 DK:** bl; **National Air and Space Museum, Smithsonian Institution:** (SI 85-6235) tl; (SI A-43268) tc. **20/21 NASM:** (SI 85-10844). **22 HG:** Library of Congress bl; **NASM:** (SI 85-11473) tr. **23 NASM:** (SI 78-159) bc; (SI 86-9864) tc; (SI 97-16784) br; (SI A-43056-A) tr. **24 NASM:** (SI 93-15834) clb; (SI A-41898-E) br. **25 AB:** t; **NASM:** (SI A-41898-D) bc; (SI A-41898-E) br. **26 LOC:** (LC-W861-22) tcb; **NASM:** (SI A-4814-H) tl. **26/27 NASM:** (SI A-26767-B) b. **27 LOC:** (LC-USZ62) tcb; (LC-W861-23) cl; **NASM:** (SI 9A00038) br; Photo by Eric F Long, copyright 1997 **NASM** (SI 97-16653) tc. **28 NASM:** (SI 7A43009) cl; (SI 88-6579) br; (SI A-38388-B) c. **28/29 DK:** Martin Cameron b. **29 DK:** Martin Cameron c. **30 AB:** Bibliothèque Nationale, Paris tl; The John Stroud Collection cb; **NASM:** (SI 95-8347) bcr. **31 HG:** Illustrated London News t; **Popperfoto:** br. **32 Aspect Picture Library:** Derek Bayes bc. **32 National Air and Space Museum, Smithsonian Institution:** (NASM 00158570) bl. **32/33 NASM:** (SI 88-16089) t. **33 NASM:** (SI 93-7192) br. **34 NASM:** (SI A-3173) tc; Library of Congress, courtesy **NASM** (SI A-42869-A) c. **34/35 NASM:** (SI A-42689-D).**35 NASM:** cr; (SI A-43148-B) tc; (SI A-8678-A) cra. **36 NASM:** (SI 88-8142) tr; (SI A-44339) br; **SS:** Science Museum c, bl. **37 AB:** The John Stroud Collection bl, bc; **Corbis:** Bettmann br; **HG:** tr; **SS:** Science Museum tl, c. **38 DK:** Martin Cameron bl; **NASM:** (SI 80-12292) cra; **TRH Pictures:** Science Museum tl. **38/39 NASM:** (SI 86-12762). **40 Offentliche Kunstsammlung, Basel:** *Homage à Blériot*, 1914 by DELAUNAY, Robert. (Inv. Nr. G 1962.6) Photo: Martin Buhler c; **HG:** Topical Press Agency bl; **NASM:** (SI 80-12296) tl. **41 HG:** Daily Mirror b; **SS:** Science Museum tr. **42 AB:** The John Stroud Collection tl, tr. **42/43 DK:** Dave King b. **43 DK:** Dave King tc, cb; **SS:** Science Museum tl. **44 NASM:** (SI 78-11939) tl; (SI 95-2260) tcb; **SS:** Science Museum bc. **45 NASM:** (SI 87-10389). **46 AB:** The John Stroud Collection c; **SS:** Science Museum bl. **46/47 AB:** The John Stroud Collection t.**47 NASM:** (SI 77-9038) cb; (SI 89-21352) tr; (SI A-3491) br. **48 NASM:** (SI 72-11124) tl; (SI 80-4128) /49 b. **50 NASM:** (SI 80-14849) bl; (SI A-47151) tl. **51 NASM:** (SI 72-10099) tc; (SI 73-4023) br. **52 NASM:** (NASM 00163377) bl; (SI A-33652-E) bc. **52/53 AB:** The John Stroud Collection t. **53 DK:** Peter Chadwick cr. **54** Roger B Whitman Early Aviators Photograph Collection, courtesy **NASM** (NASM 2B30296) tl; **54/55 TRH Pictures:** Musée de L'Air /55 b. **55** Hammer Collection, courtesy **NASM** (SI 91-17327) tl; Roger B Whitman Early Aviation Photograph Collection, courtesy **NASM** (SI 00180433) cr. **56 NASM:** (NASM 00105150) tl; **Quadrant Picture Library:** The Flight Collection b. **57 NASM:** (SI 90-3800-A) cr; **Peter Newark's Pictures:** bc; **TRH Pictures:** tr. **58 Advertising Archives**. **59 NASM:** (NASM 00144974) tc; Poster Collection, **NASM** (SI 98-20413) br. **60 DK:** Musee de l'Air et de l'Espace / Gary Ombler cr; **MUS:** (11723) tl. **60/61 DK:** Musee de l'Air et de l'Espace / Gary Ombler b. **61 DK:** Musee de l'Air et de l'Espace / Gary Ombler ca, cb; **MUS:** (11694) tc; (11698) tl. **62** United Technologies Corp., courtesy **NASM** (SI 83-16525). **63 NASM:** (SI 83-16519) tr; (SI 90-2793) cra; (SI 90-3716) br; James Dietz, Breakthrough Over Kiev (1989) **NASM** (SI 90-2792) tl. **64 DK:** Martin Cameron c, br; **HG:** cr; **NASM:** (SI A-5186-E) cl; L'Aerophile Collection, Library of Congress, courtesy **NASM** (SI 2001-11633) tr; **Quadrant Picture Library:** The Flight Collection bl. **65 AB:** cl, bl; **Corbis:** National Aviation Museum tl; **HG:** c; **NASM:** (NASM 00158541) tr; (SI 86-9865) br; L'Aerophile Collection, Library of Congress, courtesy **NASM** (SI 2001-11569) cr. **66 IWM:** (Q27633). **67 Illustrated London News Picture Library:** Postcard. *Dueling in Cloudland* by CH Davis © The Sphere, Valentine's Series (ref.4500) c; **NASM:** (NASM 00090721) cr; Riverside Keystone Mast Collection, University of California, courtesy **NASM** (NASM 9A00062) cl. **68 AB:** bl; **NASM:** (SI 80-3101) c; Poster Collection, **NASM** (NASM 98-20430) tl. **69** Riverside Keystone Mast Collection, courtesy **NASM** (NASM 9A00062). **70 DK:** IWM / Gary Ombler ca; **IWM:** (HU 64358) tl; (Q12288) bl; World War I Photography Collection (Driggs), courtesy **NASM** (NASM 9A-00225) tc. **71 AB:** c, cr; IWM, Duxford bl; Musee de l'Air cl; The John Stroud Collection br; **Corbis:** Bettmann tr. **72 IWM:** (Q23779) bc; (Q33847) t. **73 DK:** IWMs / Gary Ombler cr; **IWM:** (Q65882) bc; **LOC:** (Z29597) bcr; Robert Soubiran Collection, courtesy **NASM** (NASM Videodisc No 2B-25828) tr. **74 DK:** Gary Ombler bl; Mike Dunning c. **75 IWM:** (Q69447) b; **Philip Jarret:** RAF Museum, Hendon tr. **76 IWM:** (Q10331) tc. **76/77 DK:** Gary Ombler b. **77 DK:** Dave King cra; Gary Ombler crb; Martin Cameron tl; **IWM:** (Q12018) tr. **78 IWM:** (Q58391) tr. **78/79 IWM:** (E(AUS) 2875) b. **79 DK:** IWM tr; **Peter Newark's Pictures:** c. **80 IWM:** (Q114172) c; **Philip Jarret:** bl; **Peter Newark's Pictures:** tl. **81 Illustrated London News Picture Library:** Postcard. *Dueling in Cloudland* by CH Davis © The Sphere, Valentine's Series (ref.4500). **82 DK:** IWM / Gary Ombler cr; **IWM:** (Q63129) c; (Q63147) tc; Smithsonian Institution Libraries, courtesy **NASM** (SI 2001-7669) tcb. **83 IWM:** (Q69593) tr; IWM, courtesy **NASM** (NASM 9A00057) b. **84** Robert Soubiran Collection **NASM** (SI A-48746-A) tl. **84/85 IWM:** (Q12050). **85 DK:** IWM / Gary Ombler tr. **86 DK:** RAF Museum, Hendon b; **NASM:** (NASM 00167741) tc. **87 NASM:** (SI 96-16296) cr; Phillips Petroleum Company, courtesy **NASM** (SI 98-20424) b. **88 DK:** Gary Ombler cl, b; **IWM:** (Q63153) tl. **89 DK:** Gary Ombler cla; **IWM:** (Q66377) tr. **90 IWM:** (Q10918) cr; **NASM:** (SI 76-13317) cl; (SI 98-15058) b; SI 86-5656) tc. **91 IWM:** (Q63127) tl; **NASM:** (SI A-53691) br. **92 Corbis:** Museum of Flight c; **Hugh Cowin:** cl, cr; Shell Companies Foundation Inc., courtesy **NASM** (1A16565) tr; **Quadrant Picture Library:** The Flight Collection bl, br. **93 AB:** tl, tr, br; **Hugh Cowin:** cl. **NASM:** (SI 85-11071) cr; Photo by Mark Avino **NASM** (SI 86-12092) bc; **Quadrant Picture Library:** The Flight Collection c, bl. **94 NASM:** (SI 88-11874) c; Photo by Eric F Long, copyright 1994 **NASM** (SI 94-2230) bl; **Peter Newark's Pictures:** tl. **95 NASM:** (NASM 00090721). **96 DK:** IWM / Gary Ombler tl; Douglas H Robinson, University of Texas, Dallas, courtesy **NASM** (NASM 9A00051) c; IWM, courtesy **NASM** (NASM 9A00050) bl. **97 DK:** IWM / Gary Ombler cr, cb; **IWM:** (H (am) 2185) cl. **NASM:** (NASM 00105510) b. **98 NASM:** (NASM 9A00063). **99 Deutsches Museum München:** cl; (NASM Videodisc No. 2A-02002) tl; Krainik Lighter-than-Air Collection **NASM** (NASM Videodisc No. 7A47185) br. **100 IWM:** (Q12033). **101 Philip Jarret:** c; **NASM:** (SI 81-1136) tr. **102 IWM:** (HU 68240) bc; (Q20637) tr; **NASM:** (SI 90-8270) clb. **103 Philip Jarret:** tr. **104 AB:** cr; **CMS:** MAP cl; **Hugh Cowin:** c, bl, br; **NASM:** (SI 83-16519) tr. **105 CMS:** MAP tr, cr; **Hugh Cowin:** tl, bl, br. **106 HG:** Archive Photos. **107 NASM:** (NASM 00145060) crr; (SI 92-15635) cr; (SI 96-15030) cl; (SI A-4463) cll. **108** (SI 85-14508) tc; Photo by Dane Penland, **NASM** (SI 80-2092) bl; Poster Collection, **NASM** (SI 98-20396) tl. **109 NASM:** (SI A-4463). **110 NASM:** (SI 86-4482) r; (SI 93-7758) tl; Ted Wilbur, courtesy **NASM** (SI 92-6534) bl. **111 NASM:** (SI 88-17783) cr; (SI 94-12) tr; (SI 97-15205) br. **112 HG:** bl; Topical Press Agency c. **NASM:** (SI 83-390) tl; (SI 92-16996) br. **113 NASM:** (NASM 7B01562) br; (SI 76-15516) cr; Photo by Mark Avino, **NASM** (SI 97-15335) t. **114 AB:** bl. Poster Collection, **NASM** (SI 98-20132) bc; **Topham Picturepoint:** tr; Associated Press, Paris tc. **115 NASM:** (SI 89-147) bkgnd; (SI 95-8335) crb; (SI A-1954) tl; (SI A-32862-A) bc. **116** Nathaniel L Dewell Collection, courtesy **NASM** (SI 92-16282) c. **116/117 NASM:** (NASM 00138183) b. **117 NASM:** (SI 80-3912) br; Photo by Richard Kuyli, **NASM** (SI 85-2016) tr. **118 NASM:** (SI A-42065-A) t; Photo by Dane Penland, **NASM** (SI 79-763) bl; Photo by Eric F Long, copyright 2001 **NASM** (SI2001-136) cb. **119 NASM:** (82-11095) tr; (SI 78-1771) bl; (SI A-14405) cl; (SI A-4819-A) br. **120 AB:** c, bc; **NASM:** (NASM 00154982) tr; (SI 86-4482) cr; **SS:** Science Museum br. **121 AB:** bl; The John Stroud Collection tr; **Hugh Cowin:** cr; Photo by Dane Penland **NASM** (SI 79-763) c; Photo by Dane Penland, copyright 1980 **NASM** (SI 80-2082) cl; Photo by Eric F Long **NASM** (SI 79-831) tl; **Quadrant Picture Library:** The Flight Collection br. **122 DK:** Dave Rudkin tr; Unisys Corporation, courtesy **NASM** (SI 81-879) bl. **123 NASM:** (SI 75-5419) b. **124** WideWorld Photos, Los Angeles, courtesy **NASM** (NASM 00170182) c. **124/125 LOC:** Underwood & Underwood (LC-USZ62-105052) /125. **126 AB:** The John Stroud Collection tr; The John Stroud Collection / Air Canada tc; **HG:** Keystone tl. **126/127 DK:** Science Museum / Mike Dunning b. **127 DK:** Steve Gorton tc; **TRH Pictures:** cl. **128 AB:** cr; **Hugh Cowin:** cl, bl, br; **NASM:** (SI 75-14898) tr; (SI A-43352) c. **129 Hugh Cowin:** tl, cl, bl; **NASM:** SI (93-15838) br; **SS:** Science Museum tc. **130 AB:** c; **NASM:** (SI 73-4032) bl. **131 NASM:** (NASM 2A34118) tr; Lufthansa Archiv, courtesy **NASM** (SI 80-14394) b. **132** Boeing Air Transport System, courtesy **NASM** (SI 90-5378) c; Poster Collection, **NASM** (SI 98-20283) tl; **TRH Pictures:** MacDonnell Douglas bl. **133 NASM:** (SI 96-15030). **134** Lufthansa, courtesy **NASM** (NASM 2A32385) b; Poster Collection, **NASM** (SI 98-20181) tr. **135 HG:** Topical Press Agency tl; **NASM:** (SI 76-2425-A) cr; Poster Collection, **NASM** (SI 98-20199) br. **136 NASM:** (SI 77-5844) bl; **Courtesy of United Airlines:** tl. **136/137** (NASM 00058404) t.**137 HG:** Topical Press Agency / E Bacon tr; **TRH Pictures:** br. **138 Corbis:** Bettmann clb, bl. **139 Corbis:** Bettmann tl. Courtesy TWA, **NASM** (NASM 2A38566) b. **140 NASM:** (NASM 2A33759) tc; (SI 78-1484) c; (SI 91-14177) tl. **140/141 DK:** Gary Ombler b. **141 DK:** Gary Ombler tc; **NASM:** (SI 92-16279) trb. **142** United Air Lines via **NASM** (SI 2000-9721) bl. **142 TRH Pictures:** Lufthansa c. **143 HG:** London Express. **144 NASM:** (SI 83-9838). **145 AB:** The John Stroud Collection tr; **HG:** Archive Photos bl; **NASM:** (TMS-A19710687022) cr. **146 NASM:** (SI 77-5801) clb; (SI 82-6659) tc; **SS:** Science Museum bc. **147 NASM:** (SI 97-15046) t; (SI A-11037-L) br. **148 Hugh Cowin:** tr. **148 DK:** Gary Ombler c**; Robert Hunt Library:** tl. **148/149 DK:** Gary Ombler b. **149 Corbis:** Museum of Flight br; **DK:** Gary Ombler tc; **Robert Hunt Library:** c; **NASM:** (SI 76-3137) cl. **150 AB:** cr; The John Stroud Collection bl; **NASM:** (NASM 2A34132) tr; **TRH Pictures:** cl, br. **151 AB:** The John Stroud Collection tl, tr; **Hugh Cowin:** cr; **NASM:** (SI 89-1) cl; **Quadrant Picture Library:** The Flight Collection br; **SS:** Science Museum bl. **152 IWM:** (DXP 92-46-31) bl; **NASM:** (SI 85-19406) c; Poster Collection, **NASM** (SI 98-20245) tl. **153 NASM:** (SI 92-15635). **154 NASM:** (NASM 00105620) c; (SI 88-6826) tr. **155** 'International Newsreel' photo (ref. 36469E), courtesy **NASM** (SI 74-1039) tr; Poster Collection, **NASM** (SI 98-20504) br; SI A-945-A) bl. **156 HG:** Lambert / Archive Photos bl; **NASM:** (SI 77-15140) c. **157 NASM:** (SI 73-8701) tl; (SI 73-8702) c; Photo by Arthur Coford Jr., via US Air Force, courtesy **NASM** (SI 98-15068) tlb. **158 NASM:** (SI 89-9947) cr; Acme Newspapers Inc, courtesy **NASM** (SI 88-17651) t; Poster Collection, **NASM** (SI 98-20488) cl; Poster Collection, **NASM** (SI 98-20490) c. **159 NASM:** (NASM 9A00217) b; (SI 79-10994) tr; (SI 90-14359) tl; TIME-Warner Corporation, courtesy **NASM** (NASM 2B27470) cr. **160 HG:** Fox Photos ca; **NASM:** (SI 80-17077) bl; Poster Collection, **NASA** (SI 98-20242) tcl; **160/161 TRH Pictures:** Archives and Special Collections, University of Miami Library, Coral Gables, Florida b. **161 NASM:** (NASM 2A33098) br; (SI 80-9016) tr. **162 NASM:** (NASM 1A14330) tl; Boeing Archives, courtesy **NASM** (SI 75-12105) b. **163 HG:** Topical Press Agency br. **163** Poster Collection, **NASM** (SI 98-20248) tc. **164 AB:** br; **Copyright The Boeing Company:** cl; **Hugh Cowin:** bl, bc; **HG:** Topical Press Agency tr. **165 Corbis:** Underwood & Underwood tr; **Hugh Cowin:** tl, cl, cr; **Quadrant Picture Library:** AJ Jackson Collection br; **Copyright Igor I Sikorsky Historical Archives:** bl, bc. **166 Foto Saporetti:** c; **Peter Newark's Pictures:** tl; **Courtesy of U.S. Air Force Museum:** bl. **167** Boeing Airplane Company, courtesy **NASM** (SI 00145060). **168 NASM:** (NASM 00169975) b; (NASM 2A45822) tc; (SI A-31261-A) tl. **169 NASM:** (SI 78-7047) tr; (SI 87-15135) br; Museo Aeronautico Caproni, Milan, Italy, courtesy **NASM** (NASM 2B01247) c; Rudy Arnold Photo Collection (RA-95) **NASM** crb; **TRH Pictures:** tl. **170 HG:** US Air Force tr; **NASM:** (SI 86-12543) c. **171 HG:** General Photographic Agency cr; **NASM:** (SI A-43567-A) c. **172 NASM:** (SI 95-2267) tl. **172/173 NASM:** (NASM 9A00085). **173 Philip Jarret:** tr; **NASM:** (SI 83-15030) br. **174 AB:** The John Stroud Collection br; **Hugh Cowin:** cl, c, cr, bc; **HG:** Topical Press Agency tr. **175 AB:** tc; **Corbis:** bc; Bettmann bl; **Hugh Cowin:** tl, tr, cr, br; **Quadrant Picture Library:** The Flight Collection cl. **176 MEPL:** tc. **176/177** WideWorld Photos Inc., courtesy **NASM** (NASM 00146018) /177 b. **177 Hugh Cowin:** tc. **Peter Newark's Pictures:** cr. **178 AB:** The John Stroud Collection tc; **Corbis:** Bettmann c; **HG:** bl. **178/179 DK:** Gary Ombler b. **179 DK:** Gary Ombler tl, tr, c. **180 NASM:** (NASM 1B08865) bl; (NASM 7A27834) c. **181 Hulton Getty Archive**.**182 MEPL:** tr; **Robert Hunt Library:** bl. **183 AKG London:** bl; **Peter Newark's Pictures:** tr, cr. **184 Peter Newark's Pictures**. **185 DK:** Dave King tr; **Ronald Grant Archive:** br; Warren M Bodie, via **NASM** (SI 84-10658) c. **186/187 IWM:** (C5422). **187 HG:** cr; **NASM:** (NASM 7A27846) cll; (SI 85-7275) cl; US Air Force, courtesy **NASM** (SI A-45879-F) crr. **188 IWM:** (LDP 304) tl; **NASM:** (NASM 7A37803) bl. **188/189 NASM:** (NASM 7A27846). **189 Topham Picturepoint:** tc. **190 MEPL:** Unsere Wehrmacht. **191 AB:** The John Stroud Collection tr; **MEPL:** Signal, January 1941 bl. **192 DK:** Gary Ombler c; **Peter Newark's Pictures:** tc. **192/193 DK:** Gary Ombler b. **193 DK:** Gary Ombler tl, c; **NASM:** (SI 74-2974) tc; **TRH Pictures:** tr. **194 IWM:** (HU 63656). **195 IWM:** (CBI 35851) tr; (COL 196) br. **196 Corbis:** Yevgeny Khaldei bl; **IWM:** (NYF 72692) tc. **197 DK:** Airbourne Forces Museum, Aldershot / Dave Rudkin br; **HG:** tr; **IWM:** (HH541) bl. **198 DK:** Musee de l'Air et de l'Espace / Gary Ombler c, 199 b. **198 MUS:** (14391) tc. **198/199 DK:** Musee de l'Air et de l'Espace / Gary Ombler b. **199 AB:** tr; **DK:** Musee de l'Air et de l'Espace / Gary Ombler tc; **MUS:** (42942) br. **200 HG:** Archive Photos / Robert F Sargent bl; **IWM:** (C4571) bc; **Royal Air Force Museum, Hendon:** (P100408) c. **201 IWM:** (TR 1411) tr; Musee de l'Air et de 'Espace, courtesy **NASM** (SI 76-2767) br. **202 AB:** tr, br; **Hugh Cowin:** cl, bl; **Quadrant Picture Library:** The Flight Collection cr. **203 AB:**

ACKNOWLEDGEMENTS

tl, br; **CMS:** tr; **Hugh Cowin:** cr; US Navy, courtesy **NASM** (NASM 00057582) bl; **Royal Air Force Museum, Hendon:** Charles E Brown tl. **204 Hugh Cowin:** cr, br; **NASM:** (SI 1B-14679) c; (SI 71-338) bl; William Green, courtesy **NASM** (SI 91-7990) t. **205 Hugh Cowin:** tl; Republic Aviation Corp. photo, courtesy **NASM** (NASM 7A37803) bl; US Air Force, courtesy **NASM** (SI 85-17696) br; US Navy, courtesy **NASM** (NASM 7A35628) c; **Royal Air Force Museum, Hendon:** tr. **206 HG:** Fox Photos bl; **IWM:** (COL 30) cr; **Public Record Office:** Roy Nockolds tl. **207 NASM:** (SI 85-7275). **208 HG:** br; **IWM:** (CH13680) tl; **Courtesy of The Museum of World War II:** tc. **209 Corbis:** Bettmann br; **HG:** tl. **210 AB:** tl. **210/211 DK:** Gary Ombler b. **211 DK:** Gary Ombler tr, c; US Air Force, courtesy **NASM** (SI 71-338) bl; **HG:** Topical Press Agency c; **Popperfoto:** tcb. **212 Camera Press:** IWM bl. **213 AB:** The John Stroud Collection. **214 Dungarvan Museum, Ireland:** cl; **HG:** Picture Post / Haywood Magee b; **NASM:** (NASM 2B34210) cl. **215 IWM / Gary Ombler** tl, bl; **HG:** br. **216 DK:** IWM / Andy Crawford cr; IWM, courtesy **NASM** (SI 75-15871) cl; **Courtesy of The Museum of World War II:** br. **217 DK:** IWM / Andy Crawford cr; IWM, courtesy **NASM** (SI 75-15871) cl; **Courtesy of The Museum of World War II:** tr. **218** Photo by Eric F Long, copyright 2000 **NASM** (SI 2000-9387) bl; **Peter Newark's Pictures:** tl. **218/219 HG:** tr. **219** US Navy, courtesy **NASM** (NASM 9A00154) tl. **220 Corbis:** Hulton-Deutsch Collection tc, cl; **HG:** b. **221 AB:** tl; **Hugh Cowin:** c, cr; **IWM:** (TR 26) bl; US Navy, courtesy **NASM** (SI A-4127) cl; **Quadrant Picture Library:** The Flight Collection cr; **Aeroplane** t; **Royal Air Force Museum, Hendon:** br. **222 Corbis:** Bettmann tl; **HG:** b; **Courtesy of The Museum of World War II:** cl. **223 Hugh Cowin:** tr; **IWM:** (HU 63027) br; Photo by Dane Penland, **NASM** (SI 80-2093) cr; RAF Museum, Hendon & Charles E Brown, courtesy **NASM** (SI 83-4524) tl. **224 Corbis:** Bettmann b; **Courtesy of The Museum of World War II:** br. **225 HG:** MPI bc; **NASM:** (NASM 9A00199) cr. **226 HG:** c. **226** US Navy, courtesy **NASM** (SI 90-4357) tc. **226/227** US Navy, courtesy **NASM** (NASM 2B34012). **227** RG Smith, Douglas SBD-3 copyright **NASM**, gift of the MPB Corporation (SI 95-8196) cr. **228 Corbis:** cl; Hulton-Deutsch Collection tl; **DK:** Gary Ombler c. **228/229 DK:** Gary Ombler b. **229 DK:** Gary Ombler tc, cl. **230 NASM:** (SI 85-7301). **231** US Navy, courtesy **NASM** (SI 91-18760) tl; US Navy, courtesy **NASM** (NASM 9A00174) c. **232 Corbis:** Bettmann l; **HG:** MPI tc; **Courtesy of The Museum of World War II:** br. **233 Corbis:** Bettmann tr; US Navy, courtesy **NASM** (NASM 9A00165) cr. **234 HG:** Topical Press Agency / JA Hampton cr; **Robert Hunt Library:** tr; **NASM:** (NASM 1A29944) br. **235 AB:** cr. **NASM:** l; Photo by Rudy Arnold, courtesy **NASM** (NASM 7A25896) br; **TRH Pictures:** cl. **236 AB:** bl; **DK:** IWM / Gary Ombler tl; **HG:** Keystone c. **237** US Air Force, courtesy **NASM** (SI A-15879-F). **238** DK: IWM / Gary Ombler bl. **239 HG:** IWM: (CH 10801) cb; (CL 3399) br. **240 DK:** Gary Ombler cl; **TRH Pictures:** tl. **240/241 DK:** Gary Ombler b. **241 DK:** Gary Ombler tr; **HG:** br; Picture Post tl, tc. **242 IWM:** tl; **Courtesy of Norman Groom:** bl. **242/243 IWM:** (CH11641) c. **243 IWM:** (C 3371) br. **244 NASM:** (NASM 1B21835) t; (SI 74-3069) bl. **245 DK:** cl; **IWM:** (W 9687) bkgnd; (HU 69915) bl; (IWMFLM 2338) bcl; (IWMFLM 2339) bc; (IWMFLM 2340) bcr; (IWMFLM 2342) br; (TR 1127) tr; **245 Peter Newark's Pictures:** c. **246 DK:** IWM / Gary Ombler cl. **246/247 HG:** c; Boeing Airplane Company, courtesy **HG** (SI 86-13281) tr. **248 DK:** Gary Ombler c; **Topham Picturepoint:** tl. **248/249 DK:** Gary Ombler b. **249 DK:** Gary Ombler cl; US Air Force, courtesy **NASM** (SI 79-17239) cr; US Air Force, courtesy **NASM** (SI 2001-2047) tr. **249 Popperfoto:** tc. **250 NASM:** (NASM 7A04843). **251 DK:** IWM / Andy Crawford cr; Boeing Airplane Company,

courtesy **NASM** (NASM 7A04716) br; Wright / McCook Field Still Photograph Collection (WF-223205) **NASM** bl. **252** US Air Force, courtesy **NASM** (SI 97-17480) bl; US Air Force, courtesy **NASM** (SI 99-15473) c. **253 NASM:** (SI 77-6797) bl; US Air Force, courtesy **NASM** (SI 98-15407) t. **254 Hugh Cowin:** cl, cr; **DK:** Gary Ombler cr; **HG:** Photo by Dale Hrabak, copyright 1979 **NASM** (SI 79-4623) br; Warren M Brodie, via **NASM** (SI 84-10658) bl; **SS:** Science Museum bc. **255 HG:** Keystone Features / Fred Ramage tl. **NASM:** (SI 87-9656) br. **256 DK:** IWM / Andy Crawford cl; US Air Force, courtesy **NASM** (SI A-38635) tr. **257 IWM:** (S-8018). **258 AB:** John Dibbs cl; **Aviation Images:** John Dibbs cr; **IWM:** (COL 185) br; **TRH** (SI 168) tr; **TRH Pictures:** bl. **259 The All Transport Media Picture Library:** Charles E Brown br; **AB:** tl; **Aviation Images:** John Dibbs cl; **CMS:** MAP bl; **Corbis:** cr; **DIZ Munchen SV-Bilderdienst:** Presse-Hoffman tr. **260/261 NASM:** (SI 71-2699). **261 Corbis:** Bettmann cr; George Hall crr; **NASM:** (SI 87-6747) cll; Boeing Airplane Company, courtesy **NASM** (SI 72-7758) cl. **262 Aviation Images:** Mark Wagner tl; **DK:** Gary Ombler bl. **262/263 NASM:** (SI 87-6747. **264 NASM:** (SI 94-7714) bl; US Navy, courtesy **NASM** (SI 99-40502) tc. **265 Corbis:** Hulton-Deutsch Collection b; **NASM:** Rudy Arnold Photo Collection, courtesy **NASM** (SI 2001-5328) tl. **266 Corbis:** Bettmann tl; **NASM:** (SI 79-14868). **266/267 NASM:** (SI 86-1008). **267** Photo by Mark Avino and Eric F Long, copyright 2001 **NASM** (SI 2001-1353) cr; US Air Force, courtesy **NASM** (SI 97-17485) c. **268** Photo by Eric F Long, copyright 1998 **NASM** (SI 98-16048) b; Photo by Mark Avino and Eric F Long, copyright 2001 **NASM** cr; Photo by Mark Avino and Eric F Long, copyright 2001 **NASM** (SI 2001-134) cl; US Air Force, courtesy **NASM** (SI 81-12615) tl. **269 Corbis:** Bettmann br, t. **270 NASM:** (SI 97-15068) bc; The Polestar Group, HelioCollect, courtesy **NASM** (NASM 00032959) t. **271 DK:** Steve Gorton bc, br; **HG:** Keystone / Douglas Miller tl; Courtesy of Martin-Baker Aircraft Company Limited: cr. **272 Hugh Cowin:** cr, bl; **SS:** Science Museum br. **273 AB:** tl; **Hugh Cowin:** tc, c, cr, br; **Jakob Dahlgaard Kristensen:** Airliners.net b; **NASM:** (C-20208) cl; Photo by Carolyn Russo, copyright 1996 **NASM** 9A00350 bl. **274 Boeing Airplane Co.**, courtesy **NASM** (SI 98-15162) tl; US Air Force, courtesy **NASM** (SI 92-1117) bl. **274/275** Boeing Airplane Company, courtesy **NASM** (SI 72-7758). **276/277 Corbis:** Bettmann b. **276 HG:** Keystone cla. **277 HG:** Keystone tc, br. **278** Boeing Airplane Company, courtesy **NASM** (SI 80-5653). **279 Corbis:** Bettmann tc; **HG:** Archive Photos br. **280 AB:** tc. **280/281 DK:** Gary Ombler b. **281 AB:** br; **Corbis:** Museum of Flight tl; **DK:** Gary Ombler tr, cl. **282 DK:** Mike Dunning br; **HG:** Hauser, Paris cr. **283 AB:** RAF Museum, Hendon br; **Hugh Cowin:** cr; **HG:** Topical Press Agency / Brooke cl; **Copyright Igor I Sikorsky Historical Archives:** bc. **284 Corbis:** Lake County Museum bcl; US Air Force, courtesy **NASM** (NASM 9A00365) t. **285 Corbis:** Bettmann tr; **Tony Rogers:** Airliners.net br. **286 DK:** Gary Ombler c; **HG:** US Airforce / Archive Photos tl. **286/287 DK:** Gary Ombler b **287 Corbis:** George Hall tl, tr, tcb; **DK:** Gary Ombler cl. **288 Corbis:** Bettmann bl; **Military Picture Library:** cr; **NASM:** (NASM 7B14999) bc. **289 HG:** Keystone c; **Courtesy of Lockheed Martin Aeronautics Company, Palmdale:** t; **NASM:** (NASM 9A00030) cr; Photo by Eric F Long, copyright 1999 **NASM** (SI 99-15007) cl. **290 Courtesy of Lockheed Martin Aeronautics Company, Palmdale:** tr; Tony Landis cr; **NASA:** (EC92-1284) clb; (EC94-42883-4) tl. **290/291 DK:** Gary Ombler b. **291 DK:** Gary Ombler cl; **Courtesy of Lockheed Martin Aeronautics Company, Palmdale:** tr; Photo by Eric F Long, copyright 2001 **NASM** (SI 2001-650) cl. **292 AB:** tr; **Hugh**

Cowin: c, br; **NASA:** (E-1044) bl. **293 AB:** b; **Hugh Cowin:** tr, cl, cr; Boeing Historical Archives, courtesy **NASM** (SI 90-16864) tl. **294 AB:** br; **Aviation Images:** br; **Corbis:** George Hall tr, c; **Hugh Cowin:** cl; **Royal Air Force Museum, Hendon:** cr. **295 AB:** cl, bl; **Hugh Cowin:** cr, br; **NASM:** (SI A-4113-G) tl; US Air Force, courtesy **NASM** (NASM 1B20632) tr. **296 Corbis:** Bettmann tl; **NASM:** (NASM 9A00278) bl. **296/297 Corbis:** Bettmann. **298 HG:** Photo by Lt. Col. SF Watson (US Army), **NASM** (NASM 9A00343) cr. **299 HG:** Archive Photos / Patrick Christain b; Photo by Eric F Long, copyright 2002 **NASM** (SI-2002-2216) tc. **300 TRH Pictures:** Bell tc. **300/301 DK:** Gary Ombler b. **301 DK:** Gary Ombler tr, cl; **TRH Pictures:** Bell tl. **302 HG:** Central Press bl; **NASM:** (NASM 7B12043) tl; Courtesy of Fairchild, via **NASM** (NASM 7B12049) tr. **303 AB:** tr, c, br; **Corbis:** George Hall c; **Copyright Igor I Sikorsky Historical Archives:** b, bc; **Skip Robinson:** Airliners.net cl. **304 Agence France Presse:** Vuk Brankovic cr; **Aviation Images:** Mark Wagner cr; **TRH Pictures:** bl. **304/305 DK:** Gary Ombler b. **305 DK:** Gary Ombler tr; **TRH Pictures:** E Nevill r. **306 Military Picture Library:** David Hunter tc, ca; William F Bennett cl. **307 Corbis:** George Hall tr; **Ronald Grant Archive:** Courtesy of H.J. Saunder Insignia Inc.: tc; **Courtesy of US Navy:** tl. **308 Corbis:** Tim Page bl; **HG:** U.S. Air Force / Archive Photos t. **309 Corbis:** br; Bettmann br; **NASM:** (NASM 7A05825) trb. **310 AB:** tr, c; **Copyright The Boeing Company:** cr; **NASA:** (EC62-144) br; **NASM:** (NASM 9A00499) bc; **Courtesy of Northrop Grumman:** br; **Quadrant Picture Library:** The Flight Collection / Graham Laughton tl. **311 AB:** bl; **Corbis:** b; **HG:** Keystone br; **NASM:** (SI 89-5947) cl. **312 AB:** British Aerospace. **312/313 DK:** b. **313 The All Transport Media Picture Library:** r; **Corbis:** Military Picture Library / Robin Adshead tr; **DK:** c; **Military Picture Library:** Robert Kravitz tcb. **314/315 Military Picture Library:** Robert Kravitz. **316 Aviation Images:** Mark Wagner tl. **316 Corbis:** Military Picture Library / Robin Adshead bl. **316/317 Corbis:** George Hall. **318 Corbis:** Montgomery; Don s., USN (Ret.) bc; **Military Picture Library:** John Skliros c. **319 Military Picture Library:** Martin McKenzie and Mike Green tr; Photo by Eric F Long, copyright 2001 **NASM** (SI 2001-1822) l. **320 Copyright The Boeing Company:** tc; **TRH Pictures:** U.S. Air Force / Department of Defense tcb. **320/321 DK:** Gary Ombler b. **321 AB:** AeroGraphics Inc tl; **DK:** Gary Ombler cr; **TRH Pictures:** McDonnell Douglas tr, tl. **322 AB:** bl; **Hugh Cowin:** cl, br; **NASA:** (EC94-42883-4) tr; **Courtesy of Northrop Grumman:** cr. **323 AB:** cr; **Corbis:** George Hall cl; Lockheed Martin, courtesy **NASM** (SI 93-13704) bl. **324 Corbis:** AeroGraphics Inc tr. **324/325 DK:** Lockheed Martin / Gary Ombler b. **325 Aviation Images:** Mark Wagner br; **DK:** Lockheed Martin / Gary Ombler tl, c; **HG:** MPI tr. **326 Agence France Presse:** tcb; **Aviation Images:** John Dibbs cl; **Copyright The Boeing Company:** tc. **327 Associated Press AP:** Laurent Rebours c; **Corbis:** b; **Courtesy of U.S. Air Force:** tl. **328 Hugh Cowin:** bl; **Courtesy of U.S. Air Force:** tl. **329 Courtesy of Lockheed Martin Aeronautics Company, Palmdale.** 330 **Aviation Images:** Mark Wagner cr; **Corbis:** AeroGraphics Inc br; **Hugh Cowin:** c; **NASA:** (EC92-10061-10) br; (ECN-2092) cr. **331 AB:** bl; bc; **Aviation Images:** Mark Wagner tr, cl, br; **Corbis:** tc; Military Picture Library cr; **NASA:** (EC96-43830-12) tl. **332/333 NASA:** JSC (S69-39962). **333 Corbis:** Baldwin H Ward & Kathryn C Ward tr; **HG:** MPI cr. **334 MEPL:** tl; **NASA:** Stennis (00-068D-01) bl; **Novosti** (London): cr. **334/335 Corbis:** Baldwin H Ward & Kathryn C Ward. **336 NASM:** (SI 79-13155) cl; Copyright 1940 National Geographic Collection, courtesy **NASM** (SI A-840) tl. **336/337 NASA:** Kennedy Space Center (66P-0631. **337 NASA:** MSFC (MSFC-6407244) br. **338**

Corbis: Bettmann br; **NASA:** MSFC (MSFC-5800669) l. **339 Corbis:** Bettmann bl; **Hulton Getty Archive:** Blank Archives / Archive Photos bla; **Novosti** (London): tl, tr, br. **340 NASA:** Dryden Flight Research Center (ECN-225) t; Langley Research Center (L-1989-00361) bl. **341 Corbis:** Bettmann tr, tl; **NASA:** JSC (S65-39907) c. **342 Corbis:** Bettmann cl; Roger Ressmeyer bl; **Novosti** (London): tr. **343 Corbis:** Bettmann br; Rykoff Collection crb; **HG:** Keystone cr. **344 NASA:** JSC (S61-01928) cl; JSC (S88-31378) cbl; Kennedy Space Center (71P-0263) br; Photo by Eric F. Long copyright 2001 **NASM** (SI 2001-1831) cr. **345 NASA:** JSC (S62-00304) tr; Kennedy Space Center (62PC-0011) l; **NASM:** (99035 NOVA) c. **346/347 NASA:** JSC (S65-30432). **347 HG:** Central Press cl; **NASA:** JSC (S65-63221) bc; Kennedy Space Center (67PC-0016) tr; MSFC (MSFC-9903826) bl. **349 NASA:** MSFC (MSFC-6754387) c, (MSFC-6761894) bc **350 NASA:** JSC (S69-39957-39963) t; JSC (S69-42583) c. **352/353 NASA:** Langley Research Center (EL-2001-00427). **353 NASA:** JSC (S69-31740) bl; Kennedy Space Center (69PC-0435) c; Photo by Eric F Long, copyright 2001 **NASM** (SI 2001-2974) tc. **354 NASA:** JSC (AS11-37-5445) br; JSC (AS11-44-6642) t. **355 NASA:** JSC (69-HC-916) br; JSC (S69-21365) trb; JSC (S69-21698) tc. **356 NASA:** JSC (AS13-59-8500) bl; JSC (S69-60662) tca; JSC (S69-62224) tc; JSC (S70-35139) cb. **357 NASA:** JSC (S72-55422 & S72-55421) bl; MSFC (MSFC-7022489) t. **358 DK:** Space and Rocket Center, Alabama / Bob Gunthany b; **NASA:** Hubble Space Telescope Center (PR95-44D) b; JSC (S79-31775) ca. **358/359 Corbis:** Museum of Flight. **360 NASA:** JSC (S74-17843) clb; JSC (S75-20361) cla; **Novosti** (London): tr. **360/361 DK:** t. **361 NASA:** JSC (S73-20236) cr; JSC (S73-23952) cl; JSC (S74-17843) br; JSC (SL2-02-157) c. **362 NASA:** Kennedy Space Center (96PC-0997). **363 NASA:** JSC (EC01-0129-17) bl; JSC (S85-46260) tc; JSC (S86-38989) tr; **NASM:** (SI 5/27) cl. **364 DK:** NASA br; **NASA / Finley Holiday Films bl. 364 NASA:** Kennedy Space Center (KSC-02PD-0043) tl. **364/365 NASA:** (ISS01E-6128) c **365 NASA:** (STS100-S-020) br; (STS105-343-010) tr. Photo by Eric F Long and Mark Avino, copyright 2001 **NASM** (SI 2001-2277) cr. **366 NASA:** JSC (S98-04614) br; JSC (STS061-98-050) bc. **367 NASA:** JSC (STS061-98-050). **368 NASA:** JSC (S94-36965) bcl; JSC (S87-45890) tc; JSC (STS071-S-072) bl; **Novosti** (London): tl. **368/369 NASA:** MSFC (MSFC-9613399). **370 NASA:** MSFC (MSFC-0101424) bc. **370/371 NASA:** MSFC (MSFC-9503941). **371 NASA:** (STS104-E-5121) bc; MSFC (MSFC-9706217) tr. **372 Courtesy of Lockheed Martin Aeronautics Company, Palmdale:** b; **Popperfoto:** Reuters / RTV c. **373 NASA:** (Mars Exploration) br; MSFC (MSFC-9906385) bra; (ED99-45243-01) tr. **374/375 Image Bank / Getty Images:** Andy Caulfield cr. **375 Corbis:** Bryn Colton / Assignment Photographers; Dewitt Jones cl; James A Sugar cr; **NASA:** (ED01-0209-6) cr. **376 Advertising Archives:** tl; **TRH Pictures:** bl. **376/377 Corbis:** Dewitt Jones. **378 Corbis:** Alain le Garsmeur bl; Bettmann bc; **HG:** c. **379 AB:** The John Stroud Collection tr; **Courtesy of Lufthansa AG:** br; **TRH Pictures:** cb. **380 AB:** Stephen Piercey Collection tr; **Copyright The Boeing Company:** cr; **Hugh Cowin:** br. **381 AB:** tl, br; The John Stroud Collection tr, cr; **Hugh Cowin:** cl; **Courtesy of Lockheed Martin Aeronautics Company, Palmdale:** b. **382 HG:** Topical Press / J A Hampton tl. **383 AB:** Pan American tcb; **Corbis:** Bettmann b. **384 Agence France Presse:** Archives cl; **AB:** The John Stroud Collection tl. **385 DK:** Gary Ombler tr; **384/385 DK:** Gary Ombler b. **385 DK:** Gary Ombler tr, cl; **TRH Pictures:** James H Stevens tl. **386**

Corbis: Carl & Ann Purcell bl; **Courtesy of Southwest Airlines, Texas:** tc. **387 Associated Press AP:** tr; **Corbis:** Yann Arthus-Bertrand b. **388 AB:** cl, br; **Copyright The Boeing Company:** c, cr, bl; **TRH Pictures:** tr. **389 AB:** tr, cl, bl, br; **Copyright The Boeing Company:** tl; **Hugh Cowin:** br. **390 Agence France Presse:** Martyn Hayhow cr; **Popperfoto:** Reuters / Jean-Paul Pelissier t; **Royal Aeronautical Society:** br. **390/391 DK:** Gary Ombler b. **391 DK:** Gary Ombler tl, tr, cr; **Rex Features:** cl. **392 Hulton Getty Archive:** Central Press bl. **392/393 Corbis:** Bettmann t. **393 Corbis:** Bettmann b; **Popperfoto:** Reuters / Andras Kisgely cr. **394 AB:** tl; **Corbis:** Bettmann b; **Quadrant Picture Library:** The Flight Collection tr. **394/395 DK:** Gary Ombler b. **395 AB:** Boeing tr; **DK:** Gary Ombler d; **Richard Leeney c. 396** George Hall c. **396 Quadrant Picture Library:** The Flight Collection / Tony Hobbs tcb. **397 HG:** Keystone / Alan Band br; **TRH Pictures:** tr. **398 Airbus Industrie:** bl; **DK:** Dave Rudkin cl. **399 Popperfoto:** Reuters tr, c, trb. **400 AB:** c; **Copyright The Boeing Company:** cr, bl, br; **Hugh Cowin:** br; c; U.S. Air Force courtesy **NASM** (NASM 9A00007) tr. **401 AB:** tl, cl, cr, bc, br; **Copyright The Boeing Company:** tl, tc; **NASA:** (EC95-43247-4) bl. **402 Courtesy of McDonnell Douglas:** bl; **Science Photo Library:** Jerry Mason tl. **402/403 Corbis:** Bryn Colton / Assignment Photographers. **404 Katz/Frank Spooner Pictures:** tl; **Popperfoto:** Reuters cr. **404/405 FPG International / Getty Images:** John Neubauer. **406 Agence France Presse:** Matt Campbell b; **DK:** Peter Chadwick cr. **408 Copyright TRH Pictures:** United Press International b. **409 Corbis:** Bryn Colton / Assignment Photographers; David Rubinger br. **410 Agence France Presse:** Frederic J Brown clb; **Popperfoto:** Reuters / Joe Skipper bc. **411 Agence France Presse:** Seth McCallister. **412 Corbis:** Bill Ross d; **DK:** Gary Kevin bl. **412/413 Corbis:** James A Sugar tc. **413 Corbis:** James A Sugar tc. **414 NASM:** (SI 2001-10543) bl; **414/415 TRH Pictures. 415 Corbis:** Charles O'Rear cr; **TRH Pictures:** Mike Smith; The Daedalus Project tl. **416 Agence France Presse:** Abdelhak Senna br; Breitling Orbiter II bc; Copyright Breitling SA, courtesy **NASM** (NASM 9A-00019) b. **417 Courtesy of AeroVironment Inc.. 418 Corbis:** Morton Beebe c; Museum of Flight clb, bl. **419 Popperfoto:** Reuters / Nikolai Ignatiev cr; **Ray Buckland:** bc. **420 AB:** John Wendover cl, b; **Courtesy of Bombardier Learjet:** tr; **TRH Pictures:** Learjet t. **421 Courtesy of Bombardier Learjet:** tl; **Corbis:** br; **NASM:** (NASM 2B16292) cr. **422 AB:** cl, c, cr, bc, br; **Hugh Cowin:** bl; **TRH Pictures:** Learjet tr. **423 AB:** tl, tr, cl, c, cr, bl, br. **424 Corbis:** Forrest J Ackerman Collection tl; **Courtesy of The Joint Strike Fighter Program:** bl. **424/425 NASA:** (ED01-0209-6). **426 Agence France Presse:** Wolfgang Kumm bl; **Empics/AP:** Jeff Christensen tl **427 Copyright The Boeing Company:** tr; **Getty Images News Service:** NASA bl. **428 aviation-images.com** tr. **428/429 Airbus. 430/431 Airbus. 432 Empics/AP/Jerome Delay bl; Empics/AP/Andrew Milligan br; 432–433 aviation-images.com. 433 Empics/AP/Murad Sezer t; Empics/AP/Anja Niedringhaus b. 434 Empics/AP/Dr Scott Lieberman t; NASA cr; NASA:** KCS tr; **435 NASA:** JPL t; **NASA:** JPL c; **NASA:** JPL b. **436 Empics/AP/The Space Island Group tl; X-Prize Foundation b. 437 Canadian Arrow tr; Empics/AP/Jim Cambell tl; Empics/AP/Haraz N. Ghanbari br. 438 Eurofighter GmbH tl; Virgin Atlantic Global Flyer cl, cr, b. 439 Northrop Grumman t, cr

Endpapers: DK: Gary Ombler

Jacket: Front **HG:** Archive Photos; Back **The Image Bank/Getty Images:** Luis Castaneda Inc.; Spine: **Skyscan.**